Lecture Notes in Com

Founding Editors

Gerhard Goos
Juris Hartmanis

Editorial Board Members

Elisa Bertino, *Purdue University, West Lafayette, IN, USA*
Wen Gao, *Peking University, Beijing, China*
Bernhard Steffen ⓘ, *TU Dortmund University, Dortmund, Germany*
Moti Yung ⓘ, *Columbia University, New York, NY, USA*

The series Lecture Notes in Computer Science (LNCS), including its subseries Lecture Notes in Artificial Intelligence (LNAI) and Lecture Notes in Bioinformatics (LNBI), has established itself as a medium for the publication of new developments in computer science and information technology research, teaching, and education.

LNCS enjoys close cooperation with the computer science R & D community, the series counts many renowned academics among its volume editors and paper authors, and collaborates with prestigious societies. Its mission is to serve this international community by providing an invaluable service, mainly focused on the publication of conference and workshop proceedings and postproceedings. LNCS commenced publication in 1973.

Xuanang Xu · Zhiming Cui · Islem Rekik ·
Xi Ouyang · Kaicong Sun
Editors

Machine Learning in Medical Imaging

15th International Workshop, MLMI 2024
Held in Conjunction with MICCAI 2024
Marrakesh, Morocco, October 6, 2024
Proceedings, Part I

Editors
Xuanang Xu
Rensselaer Polytechnic Institute
Troy, NY, USA

Zhiming Cui
ShanghaiTech University
Shanghai, China

Islem Rekik
Imperial College London
London, UK

Xi Ouyang
Shanghai United Imaging Intelligence Co., Ltd.
Shanghai, China

Kaicong Sun
ShanghaiTech University
Shanghai, China

ISSN 0302-9743　　　　　　　　ISSN 1611-3349　(electronic)
Lecture Notes in Computer Science
ISBN 978-3-031-73283-6　　　　ISBN 978-3-031-73284-3　(eBook)
https://doi.org/10.1007/978-3-031-73284-3

© The Editor(s) (if applicable) and The Author(s), under exclusive license to Springer Nature Switzerland AG 2025

This work is subject to copyright. All rights are solely and exclusively licensed by the Publisher, whether the whole or part of the material is concerned, specifically the rights of translation, reprinting, reuse of illustrations, recitation, broadcasting, reproduction on microfilms or in any other physical way, and transmission or information storage and retrieval, electronic adaptation, computer software, or by similar or dissimilar methodology now known or hereafter developed.

The use of general descriptive names, registered names, trademarks, service marks, etc. in this publication does not imply, even in the absence of a specific statement, that such names are exempt from the relevant protective laws and regulations and therefore free for general use.

The publisher, the authors and the editors are safe to assume that the advice and information in this book are believed to be true and accurate at the date of publication. Neither the publisher nor the authors or the editors give a warranty, expressed or implied, with respect to the material contained herein or for any errors or omissions that may have been made. The publisher remains neutral with regard to jurisdictional claims in published maps and institutional affiliations.

This Springer imprint is published by the registered company Springer Nature Switzerland AG
The registered company address is: Gewerbestrasse 11, 6330 Cham, Switzerland

If disposing of this product, please recycle the paper.

Preface

The 15th International Workshop on Machine Learning in Medical Imaging (MLMI 2024) was held in Marrakesh, Morocco, on October 6, 2024, in conjunction with the 27th International Conference on Medical Image Computing and Computer Assisted Intervention (MICCAI 2024).

As artificial intelligence (AI) and machine learning (ML) continue to significantly influence both academia and industry, MLMI 2024 aimed to facilitate new cutting-edge techniques and their applications in the medical imaging field, including but not limited to medical image reconstruction, medical image registration, medical image segmentation, computer-aided detection and diagnosis, image fusion, image-guided intervention, image retrieval, etc. MLMI 2024 focused on major trends and challenges in this area and facilitated translating medical imaging research into clinical practice. Topics of interest included deep learning, generative adversarial learning, ensemble learning, transfer learning, multi-task learning, manifold learning, and reinforcement learning, along with their applications to medical image analysis, computer-aided diagnosis, multi-modality fusion, image reconstruction, image retrieval, cellular image analysis, molecular imaging, digital pathology, etc.

The MLMI workshop attracted original, high-quality submissions on innovative research work in medical imaging using AI and ML. MLMI 2024 received a large number of submissions (100 in total). All the submissions underwent a rigorous double-blind peer-review process, with each paper being reviewed by at least two members of the Program Committee, composed of 71 experts in the field. Based on the reviewing scores and critiques, 63 papers were accepted for presentation at the workshop and chosen to be included in two Springer LNCS volumes, which resulted in an acceptance rate of 63%. It was a tough decision and many high-quality papers had to be rejected due to the page limitation.

We are grateful to all Program Committee members for reviewing the submissions and giving constructive comments. We also thank all the authors for making the workshop very fruitful and successful.

October 2024

Xuanang Xu
Islem Rekik
Zhiming Cui
Xi Ouyang
Kaicong Sun

Organization

Workshop Organizers

Xuanang Xu	Rensselaer Polytechnic Institute, USA
Islem Rekik	Imperial College London, UK
Zhiming Cui	ShanghaiTech University, China
Xi Ouyang	Shanghai United Imaging Intelligence Co., Ltd., China
Kaicong Sun	ShanghaiTech University, China

Steering Committee

Dinggang Shen	ShanghaiTech University, China
	Shanghai United Imaging Intelligence Co., Ltd., China
Pingkun Yan	Rensselaer Polytechnic Institute, USA
Kenji Suzuki	Tokyo Institute of Technology, Japan
Fei Wang	Visa Research, USA

Program Committee

Anna Banaszak	Technical University of Munich, Germany
Bin Wang	Northwestern University, USA
Caiwen Jiang	ShanghaiTech University, China
Caner Özer	Istanbul Technical University, Turkey
Chongyue Zhao	University of Pittsburgh, USA
Chuang Niu	Rensselaer Polytechnic Institute, USA
Cong Cong	Macquarie University, Australia
Dingyi Hu	Beihang University, China
Fan Li	Shanghai Jiao Tong University, China
Gang Li	University of North Carolina at Chapel Hill, USA
Han Wu	ShanghaiTech University, China
Haoshen Wang	ShanghaiTech University, China
Hengtao Guo	Rensselaer Polytechnic Institute, USA
Janne J. Nappi	Massachusetts General Hospital, USA
Jiale Cheng	University of North Carolina at Chapel Hill, USA

Jiameng Liu	ShanghaiTech University, China
Jiayu Huo	King's College London, UK
Jun Shi	Hefei University of Technology, China
Junghwan Lee	Georgia Institute of Technology, USA
Jungwook Lee	Rensselaer Polytechnic Institute, USA
Kai Xuan	Nanjing University of Information Science and Technology, China
Kun Wu	Beihang University, China
Lin Teng	ShanghaiTech University, China
Linkai Peng	Northwestern University, USA
Lintao Zhang	University of North Carolina at Chapel Hill, USA
Linwei Wang	Rochester Institute of Technology, USA
Masoud Mokhtari	University of British Columbia, Canada
Mingquan Lin	Weill Cornell Medicine, USA
Minhui Tan	Southern Medical University, China
Nathan Lampen	Rensselaer Polytechnic Institute, USA
Pu Huang	Nanjing Audit University, China
Reza Azad	RWTH Aachen University, Germany
Ruilong Dan	ShanghaiTech University, China
Ruoyu Wang	University of Warwick, UK
Samuel Joutard	ImFusion, Germany
Sayed Mohammad Mostafavi Isfahani	Lunit, South Korea
Shaoteng Zhang	Northwestern Polytechnical University, USA
Sheng Wang	Shanghai Jiao Tong University, China
Wanyu Bian	Neuro42, Inc., USA
Wenzheng Tao	University of Utah, USA
Xi Fang	Rensselaer Polytechnic Institute, USA
Xi Ouyang	Shanghai United Imaging Intelligence Co., Ltd., China
Xiangyu Zhao	Shanghai Jiao Tong University, China
Xiao Zhang	Northwest University, USA
Xin Yang	Shenzhen University, China
Xingyue Wang	ShanghaiTech University, China
Xinrui Song	Rensselaer Polytechnic Institute, USA
Xuanang Xu	Rensselaer Polytechnic Institute, USA
Xukun Zhang	Fudan University, China
Yang Liu	King's College London, UK
Yanyun Jiang	Shandong Normal University, China
Yi Zhang	Sichuan University, China
Yilan Zhang	King Abdullah University of Science and Technology, Saudi Arabia

Yongsong Huang	Tohoku University, Japan
Yu Guo	Tianjin University, China
Yuanwang Zhang	ShanghaiTech University, China
Yue Sun	University of North Carolina at Chapel Hill, USA
Yulong Dou	ShanghaiTech University, China
Yuning Gu	United Imaging Intelligence, China
Yunxiang Li	UT Southwestern Medical Center, USA
Yushan Zheng	Beihang University, China
Yuxiao Liu	ShanghaiTech University, China
Yuxuan Liang	Rensselaer Polytechnic Institute, USA
Yuyan Ge	Xi'an Jiaotong University, China
Zaixin Ou	ShanghaiTech University, China
Ze Jin	Tokyo Institute of Technology, Japan
Zefan Yang	Rensselaer Polytechnic Institute, USA
Zhentao Liu	ShanghaiTech University, China
Zheyuan Zhang	Northwestern University, USA
Zixin Tang	ShanghaiTech University, China
Zixu Zhuang	Shanghai Jiao Tong University, China

Contents – Part I

A Novel Momentum-Based Deep Learning Techniques for Medical Image
Classification and Segmentation 1
    *Koushik Biswas, Ridam Pal, Shaswat Patel, Debesh Jha,
Meghana Karri, Amit Reza, Gorkem Durak, Alpay Medetalibeyoglu,
Matthew Antalek, Yury Velichko, Daniela Ladner, Amir Borhani,
and Ulas Bagci*

Generalizable Lymph Node Metastasis Prediction in Pancreatic Cancer 12
    Jiaqi Qu, Xunbin Wei, and Xiaohua Qian

IRUM: An Image Representation and Unified Learning Method for Breast
Cancer Diagnosis from Multi-View Ultrasound Images 22
    *Haoyuan Chen, Yonghao Li, Jiadong Zhang, Qi Xu, Meiyu Li,
Zhenhui Li, Xuejun Qian, and Dinggang Shen*

Classification, Regression and Segmentation Directly from K-Space
in Cardiac MRI 31
    *Ruochen Li, Jiazhen Pan, Youxiang Zhu, Juncheng Ni,
and Daniel Rueckert*

DDSB: An Unsupervised and Training-Free Method for Phase Detection
in Echocardiography 42
    *Zhenyu Bu, Yang Liu, Jiayu Huo, Jingjing Peng, Kaini Wang,
Guangquan Zhou, Rachel Sparks, Prokar Dasgupta,
Alejandro Granados, and Sebastien Ourselin*

Mitral Regurgitation Recogniton Based on Unsupervised
Out-of-Distribution Detection with Residual Diffusion Amplification 52
    *Zhe Liu, Xiliang Zhu, Tong Han, Yuhao Huang, Jian Wang, Lian Liu,
Fang Wang, Dong Ni, Zhongshan Gou, and Xin Yang*

Deep Reinforcement Learning with Multiple Centerline-Guidance
for Localization of Left Atrial Appendage Orifice from CT Images 63
    Jongum Yoon, Sunghee Jung, and Byunghwan Jeon

Lung-CADex: Fully Automatic Zero-Shot Detection and Classification
of Lung Nodules in Thoracic CT Images 73
    *Furqan Shaukat, Syed Muhammad Anwar, Abhijeet Parida,
Van Khanh Lam, Marius George Linguraru, and Mubarak Shah*

CIResDiff: A Clinically-Informed Residual Diffusion Model for Predicting
Idiopathic Pulmonary Fibrosis Progression 83
  Caiwen Jiang, Xiaodan Xing, Zaixin Ou, Mianxin Liu, Walsh Simon,
  Guang Yang, and Dinggang Shen

Vision Transformer Model for Automated End-to-End Radiographic
Assessment of Joint Damage in Psoriatic Arthritis 94
  Darshana Govind, Zijun Gao, Chaitanya Parmar, Kenneth Broos,
  Nicholas Fountoulakis, Lenore Noonan, Shinobu Yamamoto,
  Natalia Zemlianskaia, Craig S. Meyer, Emily Scherer, Michael Deman,
  Pablo Damasceno, Philip S. Murphy, Terence Rooney, Elizabeth Hsia,
  Anna Beutler, Robert Janiczek, Stephen S. F. Yip, and Kristopher Standish

CorticalEvolve: Age-Conditioned Ordinary Differential Equation Model
for Cortical Surface Reconstruction 104
  Wenxuan Wu, Tong Xiong, Dongzi Shi, Ruowen Qu, Xiangmin Xu,
  Xiaofen Xing, and Xin Zhang

CSR-dMRI: Continuous Super-Resolution of Diffusion MRI
with Anatomical Structure-Assisted Implicit Neural Representation
Learning 114
  Ruoyou Wu, Jian Cheng, Cheng Li, Juan Zou, Jing Yang, Wenxin Fan,
  Yong Liang, and Shanshan Wang

Atherosclerotic Plaque Stability Prediction from Longitudinal Ultrasound
Images 124
  Jan Kybic, David Pakizer, Jiří Kozel, Patricie Michalčová,
  František Charvát, and David Školoudík

Leveraging IHC Staining to Prompt HER2 Status Prediction
from HE-Stained Histopathology Whole Slide Images 133
  Yuping Wang, Dongdong Sun, Jun Shi, Wei Wang, Zhiguo Jiang,
  Haibo Wu, and Yushan Zheng

VIMs: Virtual Immunohistochemistry Multiplex Staining via Text-to-Stain
Diffusion Trained on Uniplex Stains 143
  Shikha Dubey, Yosep Chong, Beatrice Knudsen, and Shireen Y. Elhabian

Structural-Connectivity-Guided Functional Connectivity Representation
for Multi-modal Brain Disease Classification 156
  Zhaoxiang Wu, Biao Jie, Wen Li, Wentao Jiang, Yang Yang,
  and Tongchun Du

Clinical Brain MRI Super-Resolution with 2D Slice-Wise Diffusion Model 166
  Runqi Wang, Zehong Cao, Yichu He, Jiameng Liu, Feng Shi,
  and Dinggang Shen

Low-to-High Frequency Progressive K-Space Learning for MRI
Reconstruction 177
  Xiaohan Xing, Liang Qiu, Lequan Yu, Lingting Zhu, Lei Xing,
  and Lianli Liu

LSST: Learned Single-Shot Trajectory and Reconstruction Network
for MR Imaging 187
  Hemant Kumar Aggarwal, Sudhanya Chatterjee, Dattesh Shanbhag,
  Uday Patil, and K.V.S. Hari

7T-Like T1-Weighted and TOF MRI Synthesis from 3T MRI
with Multi-contrast Complementary Deep Learning 197
  Zheng Zhang, Zechen Zhou, Lei Xiang, Kelei He, Zhiqing Zhu,
  Xingang Wang, Zhiming Zeng, Hongqin Liang, and Chen Liu

A Probabilistic Hadamard U-Net for MRI Bias Field Correction 208
  Xin Zhu, Hongyi Pan, Batuhan Gundogdu, Debesh Jha, Yury Velichko,
  Adam B. Murphy, Ashley Ross, Baris Turkbey, Ahmet Enis Cetin,
  and Ulas Bagci

Structure-Preserving Diffusion Model for Unpaired Medical Image
Translation 218
  Haoshen Wang, Xiaodong Wang, and Zhiming Cui

Simultaneous Image Quality Improvement and Artefacts Correction
in Accelerated MRI 228
  Georgia Kanli, Daniele Perlo, Selma Boudissa, Radovan Jiřík,
  and Olivier Keunen

Full-TrSUN: A Full-Resolution Transformer UNet for High Quality PET
Image Synthesis 238
  Boyuan Tan, Yuxin Xue, Lei Bi, and Jinman Kim

TS-SR3: Time-Strided Denoising Diffusion Probabilistic Model for MR
Super-Resolution 248
  Zejun Wu, Samuel W. Remedios, Blake E. Dewey, Aaron Carass,
  and Jerry L. Prince

PDM: A Plug-and-Play Perturbed Multi-path Diffusion Module
for Simultaneous Medical Image Segmentation Improvement
and Uncertainty Estimation 259
  Bo Zhou, Tianqi Chen, Jun Hou, Yinchi Zhou, Huidong Xie, Chi Liu,
  and James S. Duncan

DyNo: Dynamic Normalization based Test-Time Adaptation for 2D
Medical Image Segmentation 269
  Yihang Fu, Ziyang Chen, Yiwen Ye, and Yong Xia

Accurate Delineation of Cerebrovascular Structures from TOF-MRA
with Connectivity-Reinforced Deep Learning 280
  Shoujun Yu, Cheng Li, Yousuf Babiker M. Osman, Shanshan Wang,
  and Hairong Zheng

Learning Instance-Discriminative Pixel Embeddings Using Pixel Triplets 290
  Long Chen and Dorit Merhof

Geo-UNet: A Geometrically Constrained Neural Framework
for Clinical-Grade Lumen Segmentation in Intravascular Ultrasound 300
  Yiming Chen, Niharika S. D'Souza, Akshith Mandepally,
  Patrick Henninger, Satyananda Kashyap, Neerav Karani, Neel Dey,
  Marcos Zachary, Raed Rizq, Paul Chouinard, Polina Golland,
  and Tanveer F. Syeda-Mahmood

Domain Influence in MRI Medical Image Segmentation: Spatial Versus
k-Space Inputs 310
  Erik Gösche, Reza Eghbali, Florian Knoll, and Andreas M. Rauschecker

Enhanced Small Liver Lesion Detection and Segmentation Using
a Size-Focused Multi-model Approach in CT Scans 320
  Abdullah F. Al-Battal, Van Ha Tang, Steven Q. H. Truong,
  Truong Q. Nguyen, and Cheolhong An

Generation and Segmentation of Simulated Total-Body PET Images 331
  Arnau Farré-Melero, Pablo Aguiar-Fernández, and Aida Niñerola-Baizán

Integrating Convolutional Neural Network and Transformer for Lumen
Prediction Along the Aorta Sections 340
  Yichen Yang, Pengbo Jiang, Xiran Cai, Zhong Xue, and Dinggang Shen

CSSD: Cross-Supervision and Self-denoising for Hybrid-Supervised
Hepatic Vessel Segmentation 350
  Qiuting Hu, Li Lin, Pujin Cheng, and Xiaoying Tang

Calibrated Diverse Ensemble Entropy Minimization for Robust Test-Time
Adaptation in Prostate Cancer Detection 361
    Mahdi Gilany, Mohamed Harmanani, Paul Wilson,
    Minh Nguyen Nhat To, Amoon Jamzad, Fahimeh Fooladgar,
    Brian Wodlinger, Purang Abolmaesumi, and Parvin Mousavi

SpineStyle: Conceptualizing Style Transfer for Image-Guided Spine
Surgery on Radiographs 372
    R. Neeraja, S. Devadharshiniinst, N. Venkateswaran,
    Vivek Maik, Aparna Purayath, Manojkumar Lakshmanan,
    and Mohanasankar Sivaprakasam

SGSR: Structure-Guided Multi-contrast MRI Super-Resolution
via Spatio-Frequency Co-Query Attention 382
    Shaoming Zheng, Yinsong Wang, Siyi Du, and Chen Qin

Knowledge Distillation Based Dual-Branch Network for Whole Slide
Image Analysis 392
    Weiheng Fu, Meilan Xu, Jie Wu, Xiaoshuang Shi, Kang Li,
    and Xiaofeng Zhu

DHSampling: Diversity-Based Hyperedge Sampling in GNN Learning
with Application to Medical Imaging Classification 402
    Jiameng Liu, Furkan Pala, Islem Rekik, and Dinggang Shen

Author Index 413

Contents – Part II

Robust Box Prompt Based SAM for Medical Image Segmentation 1
   Yuhao Huang, Xin Yang, Han Zhou, Yan Cao, Haoran Dou, Fajin Dong, and Dong Ni

Multi-task Learning Approach for Intracranial Hemorrhage Prognosis 12
   Miriam Cobo, Amaia Pérez del Barrio, Pablo Menéndez Fernández-Miranda, Pablo Sanz Bellón, Lara Lloret Iglesias, and Wilson Silva

Mitigating False Predictions in Unreasonable Body Regions 22
   Constantin Ulrich, Catherine Knobloch, Julius C. Holzschuh, Tassilo Wald, Maximilian R. Rokuss, Maximilian Zenk, Maximilian Fischer, Michael Baumgartner, Fabian Isensee, and Klaus H. Maier-Hein

`UniFed`: A Universal Federation of a Mixture of Highly Heterogeneous
Medical Image Classification Tasks 32
   Atefe Hassani and Islem Rekik

Tackling Domain Generalization for Out-of-Distribution Endoscopic
Imaging 43
   Mansoor Ali Teevno, Gilberto Ochoa-Ruiz, and Sharib Ali

Benchmarking Dependence Measures to Prevent Shortcut Learning
in Medical Imaging 53
   Sarah Müller, Louisa Fay, Lisa M. Koch, Sergios Gatidis, Thomas Küstner, and Philipp Berens

Selective Classifier Based Search Space Shrinking for Radiographs
Retrieval 63
   Teo Manojlović, Ivo Ipšić, and Ivan Štajduhar

Pseudo-rendering for Resolution and Topology-Invariant Cortical
Parcellation 74
   Pablo Blasco Fernandez, Karthik Gopinath, John Williams-Ramirez, Rogeny Herisse, Lucas J. Deden-Binder, Dina Zemlyanker, Theressa Connors, Liana Kozanno, Derek Oakley, Bradley Hyman, Sean I. Young, and Juan Eugenio Iglesias

Partially Supervised Unpaired Multi-modal Learning for Label-Efficient
Medical Image Segmentation 85
  Lei Zhu, Yanyu Xu, Huazhu Fu, Xinxing Xu, Rick Siow Mong Goh,
  and Yong Liu

VIS-MAE: An Efficient Self-supervised Learning Approach on Medical
Image Segmentation and Classification 95
  Zelong Liu, Andrew Tieu, Nikhil Patel, George Soultanidis,
  Louisa Deyer, Ying Wang, Sean Huver, Alexander Zhou, Yunhao Mei,
  Zahi A. Fayad, Timothy Deyer, and Xueyan Mei

Transformer-Based Parameter Fitting of Models Derived
from Bloch-McConnell Equations for CEST MRI Analysis 108
  Christof Duhme, Chris Lippe, Verena Hoerr, and Xiaoyi Jiang

Probabilistic 3D Correspondence Prediction from Sparse Unsegmented
Images 117
  Krithika Iyer and Shireen Y. Elhabian

StoDIP: Efficient 3D MRF Image Reconstruction with Deep Image Priors
and Stochastic Iterations 128
  Perla Mayo, Matteo Cencini, Carolin M. Pirkl, Marion I. Menzel,
  Michela Tosetti, Bjoern H. Menze, and Mohammad Golbabaee

Detection of Emerging Infectious Diseases in Lung CT Based on Spatial
Anomaly Patterns 138
  Branko Mitic, Philipp Seeböck, Jennifer Straub, Helmut Prosch,
  and Georg Langs

Data Alchemy: Mitigating Cross-site Model Variability Through Test
Time Data Calibration 148
  Abhijeet Parida, Antonia Alomar, Zhifan Jiang, Pooneh Roshanitabrizi,
  Austin Tapp, María J. Ledesma-Carbayo, Ziyue Xu,
  Syed Muhammed Anwar, Marius George Linguraru, and Holger R. Roth

Noise-Robust Conformal Prediction for Medical Image Classification 159
  Coby Penso and Jacob Goldberger

Identifying Critical Tokens for Accurate Predictions in Transformer-Based
Medical Imaging Models 169
  Solha Kang, Joris Vankerschaver, and Utku Ozbulak

Resource-Efficient Medical Image Analysis with Self-adapting
Forward-Forward Networks 180
  Johanna P. Müller and Bernhard Kainz

SDF-Net: A Hybrid Detection Network for Mediastinal Lymph Node
Detection on Contrast CT Images 191
    Jiuli Xiong, Lanzhuju Mei, Jiameng Liu, Dinggang Shen, Zhong Xue,
    and Xiaohuan Cao

Arges: Spatio-Temporal Transformer for Ulcerative Colitis Severity
Assessment in Endoscopy Videos 201
    Krishna Chaitanya, Pablo F. Damasceno, Shreyas Fadnavis,
    Pooya Mobadersany, Chaitanya Parmar, Emily Scherer,
    Natalia Zemlianskaia, Lindsey Surace, Louis R. Ghanem,
    Oana Gabriela Cula, Tommaso Mansi, and Kristopher Standish

Characterizing the Histology Spatial Intersections Between
Tumor-Infiltrating Lymphocytes and Tumors for Survival Prediction
of Cancers Via Graph Contrastive Learning 212
    Yangyang Shi, Qi Zhu, Yingli Zuo, Peng Wan, Daoqiang Zhang,
    and Wei Shao

Identifying Nonalcoholic Fatty Liver Disease and Advanced Liver Fibrosis
from MRI in UK Biobank 222
    Rami Al-Belmpeisi, Kristine Aavild Sørensen,
    Josefine Vilsbøll Sundgaard, Puria Nabilou, Monica Jane Emerson,
    Peter Hjørringgaard Larsen, Lise Lotte Gluud,
    Thomas Lund Andersen, and Anders Bjorholm Dahl

Explainable and Controllable Motion Curve Guided Cardiac Ultrasound
Video Generation 232
    Junxuan Yu, Rusi Chen, Yongsong Zhou, Yanlin Chen, Yaofei Duan,
    Yuhao Huang, Han Zhou, Tao Tan, Xin Yang, and Dong Ni

Author Index 243

A Novel Momentum-Based Deep Learning Techniques for Medical Image Classification and Segmentation

Koushik Biswas[1]($\boxtimes$), Ridam Pal[2], Shaswat Patel[3], Debesh Jha[1], Meghana Karri[1], Amit Reza[4], Gorkem Durak[1], Alpay Medetalibeyoglu[1], Matthew Antalek[1], Yury Velichko[1], Daniela Ladner[1], Amir Borhani[1], and Ulas Bagci[1]

[1] Machine and Hybrid Intelligence Lab, Northwestern University, Chicago, USA
{koushik.biswas,debesh.jha,meghana.karri,medetalibeyoglu.alpay,y-velichko,
dladner,amir.borhani,ulas.bagci}@northwestern.edu, matthew.antalek@nm.org
[2] IIIT Delhi, New Delhi, India
ridamp@iiitd.ac.in
[3] NSIT Delhi, New Delhi, India
[4] Space Research Institute (IWF) of Austrian Academy of Sciences, Graz, Austria

Abstract. Accurately segmenting different organs from medical images is a critical prerequisite for computer-assisted diagnosis and intervention planning. This study proposes a deep learning-based approach for segmenting various organs from CT and MRI scans and classifying diseases. Our study introduces a novel technique integrating momentum within residual blocks for enhanced training dynamics in medical image analysis. We applied our method in two distinct tasks: segmenting liver, lung, & colon data and classifying abdominal pelvic CT and MRI scans. The proposed approach has shown promising results, outperforming state of-the-art methods on publicly available benchmarking datasets. For instance, in the lung cancer segmentation dataset, our approach yielded significant enhancements over the TransNetR model, including a 5.72% increase in dice score, a 5.04% improvement in mean Intersection over Union (mIoU), an 8.02% improvement in recall, and a 4.42% improvement in precision. Hence, incorporating momentum led to state-of-the art performance in both segmentation and classification tasks, representing a significant advancement in the field of medical imaging. The code is available at https://github.com/koushik313/momentum.

Keywords: Liver segmentation · Lung Cancer Segmentation · Polyp Segmentation · Medical Image Classification

---

K. Biswas, R. pal, S. patel, D.Jha, and M. Karri—Contributed equally to this work.

Supplementary Information The online version contains supplementary material available at https://doi.org/10.1007/978-3-031-73284-3_1.

1 Introduction

In modern medicine, medical imaging plays an important role in bridging visual data and clinical insights. Computer vision plays a key role in improving the interpretation of complex medical images, including CT, X-rays, and MRIs. This transformational field plays a pivotal role in automating abnormality detection, classifying anatomical structures, and quantifying disease features. Real-time surgical guidance, minimally invasive procedures, and ongoing research into interpretability and deployment further emphasize its significance. Healthcare professionals gain the ability to make informed decisions based on a deeper understanding of patients' conditions, even at the early stages. This transformative capability encourages collaboration among computer scientists, clinicians, and researchers, driving innovations with great promise for healthcare outcomes.

Deep learning has emerged as a significant tool in clinical support for medical imaging, revolutionizing disease detection, segmentation, and classification [27]. By automating feature extraction, deep learning models, particularly convolutional neural networks (CNNs), have demonstrated the ability to learn hierarchical features from raw medical images. Techniques such as segmentation and classification enable these models to extract critical visual cues from various imaging modalities, including CT, MRI, and endoscopy. These advancements assist clinicians in disease staging, surgical planning, and assessing treatment responses. Furthermore, deep learning is good at analyzing patterns in extensive datasets, which can aid in early disease detection.

However, medical images are more complex than standard images, which makes analyzing them thoroughly quite challenging. Researchers have been working hard to solve these challenges, especially in areas like early diagnosis and quantitative imaging, where mistakes can be really risky. Colorectal cancer is one of the most common cancers worldwide. Detecting polyps early is crucial, as some types can develop into cancer if not addressed at an early stage. However, sometimes, it's tough for doctors to differentiate polyps from normal tissue visually. Fortunately, deep learning (DL) models have become a powerful tool for identifying and categorizing abnormalities from medical images.

The introduction of residual connection and self-attention mechanisms has further enhanced deep learning and computer vision domain, allowing them to focus on crucial clinical regions within an image and leading to more robust architectures [5,21]. These innovative methods have greatly enhanced medical imaging, leading to more efficient diagnoses and streamlined workflows, thus reducing the strain on healthcare professionals and clinical resources over the past decade.

Our study introduces a method that utilizes the power of the **momentum** term within the design of the residual block. Incorporating momentum within the residual blocks offers effective network training to enhance the learning algorithm. Our experiments demonstrated that this enhancement led to faster convergence and improved stability, which had the potential to achieve superior performance. The efficacy of our proposed method is supported by extensive evaluations across various tasks, including lung, liver, and polyp segmentation, as well as the classification of abdominal pelvic CT and MRI scans (on RadImageNet Data). Based on these extensive experiments, it is clear that including a momentum-based method in residual block outperforms the current state-of-the-art methods.

2 Related Works and Motivation

Accurate segmentation and classification of medical images are crucial for computer-aided diagnosis and treatment planning but remain challenging due to factors like low contrast, noise, and patient variability. Convolutional neural networks (CNNs) have shown promise in these tasks, automating diagnosis and aiding medical decision-making [18,26]. With advancements in deep learning, skip-connection was crucial in addressing the degradation problem arising from vanishing gradients. Introducing skip connections in architectures like ResNet has addressed the vanishing gradient problem, improving the training of deep networks [11]. The use of residual connections has become widespread across a variety of image classification models, including CapsuleNet [23], PreactResNet [12], MobileNet V2 [25], and ShuffleNet V2 [28], among others. Drozdzal et al. [8] demonstrated that incorporating long and short skip connections in Fully Convolutional Networks (FCNs) enhances biomedical image segmentation without additional post-processing. U-Net's architecture captured high-level features and helped in reconstructing the segmentation map [22]. The drawbacks have been further improved by ResUNet, combining the strengths of Residual Networks and U-Net for better performance [29]. These advancements highlight the critical role of deep learning in enhancing medical image analysis. In addition, this connection has been extensively utilized in many segmentation models, such as TransNetR [16], ResUNet++ [14], and PVTformer [15], among others. Due to its versatility and effectiveness, it has become a crucial component in modern deep-learning models, especially in computer vision and medical imaging applications.

Building on previous research, we have integrated the momentum term into ResNet blocks for both segmentation and classification tasks across various models. For segmentation, we have integrated into architectures like UResNet and ResUnet++, while for image classification, we used models such as Mobile Net V2 and ShuffleNet V2. Our study aims to enhance training dynamics, improve generalization, boost performance metrics, and increase the efficiency and scalability of deep neural networks. The addition of momentum can improve convergence, and potentially lead to state-of-the-art results in segmentation and classification tasks, offering a novel contribution to the field of medical imaging with an impact on future neural network architectures and training strategies. Our contributions to this paper are as follows:

1. We have integrated the momentum term in the resnet block in various neural network architectures for both segmentation and classification tasks.
2. Our extensive experiments on large datasets demonstrate that our proposed momentum-based architecture significantly enhances the ability of previous

models to identify complex data patterns, leading to more accurate predictions on unseen data or test datasets.

3 Method

Deep convolutional neural networks have demonstrated exceptional performance in various computer vision tasks and have become state-of-the-art in image classification problems. AlexNet [18], VGG [26], ResNet [11], Vision Transformer [7] are some popular architectures for image classification problem. However, as the number of layers increases, the problem of vanishing gradients becomes increasingly prevalent, which leads to a drop in training accuracy beyond a certain depth. We present our approach in the next subsection.

3.1 Residual Network

In 2015, a group of researchers from Microsoft Research introduced ResNet [11]. This architecture is designed to overcome the problem of degradation that deep neural networks often face. As the number of layers increases, the accuracy of the network can either saturate or degrade. ResNet addresses this issue by introducing residual connections, also known as skip connections. These connections allow the network to bypass one or more layers during training and inference. Residual connections work by adding the output of a layer to the output of a few layers ahead, creating a shortcut path for gradient flow. This enables the training of much deeper networks. ResNets have been widely adopted in various computer vision tasks such as image classification, object detection, and semantic segmentation. They have achieved state-of-the-art performance on several benchmark datasets.

3.2 Proposed Momentum-Based Approach for Medical Images

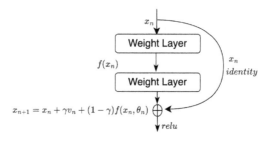

Fig. 1. An integration of the momentum term in the ResNet Block

The momentum ResNet [24] relies on the integration of the momentum within the Residual Block for image classification, as shown in Fig. 1. Our methodology is designed to optimize network performance in medical image classification and semantic segmentation tasks. The feed-forward residual block at layer n is defined as follows:

$$x_{n+1} = x_n + f(x_n, \theta_n) \quad (1)$$

The velocity equation is defined as follows:

$$v_{n+1} = \gamma v_n + (1-\gamma)f(x_n, \theta_n) \quad (2)$$

The momentum equation with the residual block is defined as

$$x_{n+1} = x_n + v_{n+1} \quad (3)$$

In particular, if we consider $\gamma = 0$, we got the classical ResNet architecture, and for $\gamma = 1$, we got the RevNet [9] architecture.

3.3 Reversible Property

In the realm of deep learning, a neural network is considered reversible if all of its activations can be recalculated when performing a backward pass. By contrast, a network that is not reversible requires saving activations from the forward pass, leading to increased memory usage. Reversible or invertible networks have the unique advantage of being able to perform backpropagation without storing the outputs of activation function, thus significantly reducing the memory footprint of models that employ this approach [4,9,10,24]. Momentum-residual block is invertible. We can invert this equation as follows:

$$x_n = x_{n+1} - v_{n+1} \quad (4)$$

$$v_n = \frac{1}{\gamma}(v_{n+1} - (1-\gamma)f(x_n, \theta_n)). \quad (5)$$

4 Experiments and Results

We have reported results on medical image segmentation problems on different organs like lungs (medical decathlon data), liver (medical decathlon data), and polyps (Kvasir-SEG dataset). For classification, we have considered RadImageNet data. Please see the **supplementary material** for more experimental results.

4.1 Medical Image Segmentation

We evaluated the efficacy of our architecture by considering two tasks: segmentation and classification. We consider the decathlon Segmentation Benchmark [1-3] and the Kvasir-Seg [13] datasets for the segmentation task. The Decathlon is a comprehensive collection of medical image segmentation datasets covering various anatomies, modalities, and sources, including the brain, heart, liver, hippocampus, prostate, lung, pancreas, hepatic vessel, spleen, and colon. For our experiments, we consider the Liver [2,3] and the Lung [1,2] data. The Kvasir-SEG dataset consists of 1000 images, of which 880 were used for training and the remaining for testing. Please see the **supplementary material** for more experimental results.

Table 1. Comparison of different segmentation models and our proposed momentum-based approach on Lung cancer segmentation benchmark dataset.

Model	mDSC	mIoU	Rec.	Prec.	F2	HD
ResUNet++ [14]	32.78	25.88	35.11	**84.29**	33.88	2.66
Momentum-ResUNet++ (**Ours**)	**43.39**	**32.32**	**52.75**	82.09	**46.57**	**2.25**
ResUNet [29]	42.02	31.62	43.07	74.57	41.97	2.36
Momentum-ResUNet (**Ours**)	**43.55**	**32.35**	**44.78**	**75.45**	**42.67**	**2.34**
TransNetR [16]	45.82	35.07	44.30	75.41	44.45	2.36
Momentum-TransNetR (**Ours**)	**51.54**	**40.11**	**52.32**	**79.83**	**50.88**	**2.23**
PVTFormer [15]	26.92	18.50	27.90	49.43	26.47	3.54
Momentum-PVTFormer (**Ours**)	**29.47**	**20.70**	**30.05**	**55.85**	**28.88**	3.52

To avoid bias, we divided the Liver data into independent training (70 patients), validation (30 patients), and test (30 patients) sets. The volumetric CT scans were processed slice-by-slice to fit into regular computer hardware (GPU). Prior to segmentation, we extracted healthy liver masks for unbiased results.

The liver, Lung, and Kvasir data segmentation experiments were conducted using the PyTorch framework [20]. We consider a batch size of 16 and a learning rate of $1e^{-4}$ for the segmentation tasks. We trained the network for 500 epochs with an early stopping patience of 50 to fine-tune the network parameters. To enhance the network performance further, we used a hybrid loss function combining binary cross-entropy and dice loss and an Adam optimizer for updating the parameters. The data was divided into three sets: 80% for training, 10% for validation, and 10% for testing. We resized the image to 256 × 256 pixels in-plane resolution to balance the training time and model complexity. All the segmentation experiments were conducted on the A100 GPU server.

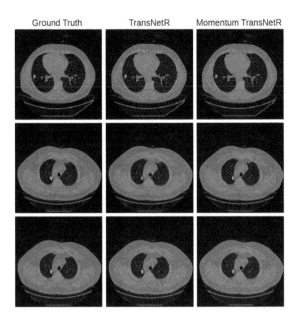

Fig. 2. Qualitative results of models trained Lung cancer dataset on the TransNetR model. It can be observed that the Momentum-based method produces a more accurate segmentation map in all the cases.

4.2 Medical Image Classification

For the image classification task, we consider the RadImageNet [19] dataset, which is a medical imaging database that is publicly available and designed to improve transfer learning capabilities in medical imaging applications. It is one of the largest medical imaging classification datasets currently available and is intended for use by professionals in the field of healthcare. We conducted experiments on CT and MRI abdominal/pelvis using the entire dataset. The dataset comprises 28 disease classes, each with an average class size of 4994 and a total of 139,825 slices. The dataset is specifically designed to have slices per disease, although the overall scans are in 3D volumes. We conducted our experiments on the dataset of MRI images of the abdomen and pelvis. The dataset consisted of 26 different classes of diseases, with an average class size of 3513 slices and a total of 91,348 slices. Although there is some overlap between this dataset and the CT dataset, the MRI dataset has some unique disease classes, such as enlarged organs and liver disease, which are not found in the CT dataset. On the other hand, the CT dataset has a specific class for entire abdominal organs. Please see the **supplementary material** for more experimental results.

We consider the Tensorflow-Keras [6] framework to run the experiments, with MobileNet V2 [25] and ShuffleNet V2 [28] serving as baseline models classification networks. The networks are trained with a batch size of 32, an initial learning rate set at 0.00001, Adam [17] optimizer, and a weight decay rate of

Table 2. Comparison of different segmentation models and our proposed momentum-based approach on liver cancer segmentation benchmark (LiTS) dataset.

Model	mDSC	mIoU	Rec.	Prec.	F2	HD
ResUNet++ [14]	73.82	70.63	**76.10**	91.13	72.21	1.23
Momentum-ResUNet++ (**Ours**)	**74.53**	**71.88**	74.98	**95.46**	**72.83**	**1.22**
ResUnet [29]	74.47	71.25	76.58	92.32	72.97	1.14
Momentum-ResUnet (**Ours**)	**76.22**	**72.79**	**77.45**	**92.76**	**74.10**	**1.08**
TransNetR [16]	**78.74**	75.18	**78.16**	**95.59**	76.86	**1.10**
Momentum-TransNetR (**Ours**)	78.50	**75.20**	78.00	95.40	**77.86**	1.14
PVTFormer [15]	77.11	73.66	77.59	**94.35**	75.36	1.14
Momentum-PVTFormer (**Ours**)	**79.26**	**75.67**	**80.74**	93.87	**77.87**	**1.06**

$1e^{-4}$. The data is partitioned into three sets, with 80% used for training, 10% for validation, and 10% for testing. The results obtained from CT scan image data are presented in Table 3, while those from the MRI image data are presented in the **supplimentary material**. All the classification experiments are conducted on an NVIDIA RTX 3090 GPU system.

Table 3. Baseline models and the proposed method and their impact on RadImageNet 28 classes Abdominal/Pelvis CT Scans.

Method	Accuracy	MCC
MobileNet V2 [25]	58.68	35.21
Momentum-MobileNet V2 (**Ours**)	**60.79**	**37.89**
ShuffleNet V2 [28]	62.70	43.20
Momentum-ShuffleNet V2 (**Ours**)	**64.91**	**45.59**

4.3 Performance Evaluation

We have examined the momentum-based approach in various situations. We carried out experiments to explore the model's ability to learn on the test set of Liver, Lung, and Kvasir-SEG datasets for image segmentation and RadImageNet data for CT and MRI image classification. The outcomes pertaining to the observed dataset on the segmentation task have been presented in Table 1, and Table 2, and classification results are presented in Table 3. Our proposed methodology has yielded the most favourable results in terms of mIoU, dice score, precision, recall, F2, and HD score when compared to other models. For example, in the lung cancer segmentation benchmark dataset, we got a 5.72% improvement in dice score, a 5.04% improvement in mIoU, an 8.02% improvement in recall, a 4.42% improvement in precision, a 6.43% improvement in F2 score compared to the TransNetR model. In the medical image classification task, we consider the RadImageNet dataset. With our proposed method, in Abdominal/pelvic CT scan data, we got a 2.11% improvement on MobileNet V2 and a 2.21% improvement on the ShuffleNet V2 model.

Figure 2 shows the outcomes of lung cancer segmentation on the TransNetR model and the proposed method. It is noticeable that the proposed method has a higher segmentation accuracy in comparison to the state-of-the-art baselines.

5 Conclusion

This study presents a novel momentum-based segmentation and classification approach that effectively segments liver, lung, and polyps using the momentum equation and residual block. The findings from various publicly available data demonstrate the efficacy of the proposed classification and segmentation approach. Based on a comprehensive comparison of our momentum algorithm on other datasets, our approach has consistently shown superior performance over our competitors. The quantitative and qualitative analysis results indicate that the momentum-based approach is more generalizable to most datasets, making it a suitable tool for clinical settings. Therefore, the proposed momentum-based approach provides a strong benchmark for developing algorithms that can assist clinicians in early lung cancer, liver cancer, polyp detection, and medical image classification.

Acknowledgements. This project is supported by NIH funding: R01-CA246704, R01-CA240639, U01-DK127384-02S1, and U01-CA268808.

Disclosure of Interests. The authors declare no competing interests.

References

1. The cancer imaging archive. https://www.cancerimagingarchive.net/ (2024). ISSN: 2474-4638
2. Antonelli, M., et al.: The medical segmentation decathlon. Nat. Commun. **13**(1), 4128 (2022)
3. Bilic, P., et al.: The liver tumor segmentation benchmark (lits). Med. Image Anal. **84**, 102680 (2023)
4. Chang, B., Meng, L., Haber, E., Ruthotto, L., Begert, D., Holtham, E.: Reversible architectures for arbitrarily deep residual neural networks (2017)
5. Chen, X., et al.: Recent advances and clinical applications of deep learning in medical image analysis. Med. Image Anal. **79**, 102444 (2022)
6. Chollet, F., et al.: Keras. https://keras.io (2015)
7. Dosovitskiy, A., et al.: An image is worth 16x16 words: Transformers for image recognition at scale. arXiv preprint arXiv:2010.11929 (2020)
8. Drozdzal, M., Vorontsov, E., Chartrand, G., Kadoury, S., Pal, C.: The importance of skip connections in biomedical image segmentation. In: International workshop on deep learning in medical image analysis, international workshop on large-scale annotation of biomedical data and expert label synthesis, pp. 179–187 (2016)

9. Gomez, A.N., Ren, M., Urtasun, R., Grosse, R.B.: The reversible residual network: Backpropagation without storing activations. Adv. Neural Inf. Process. Syst. **30** (2017)
10. Gugglberger, J., Peer, D., Rodríguez-Sánchez, A.: Momentum capsule networks (2022)
11. He, K., Zhang, X., Ren, S., Sun, J.: Deep residual learning for image recognition. In: Proceedings of the IEEE Conference on Computer Vision and Pattern Recognition, pp. 770–778 (2016)
12. He, K., Zhang, X., Ren, S., Sun, J.: Identity mappings in deep residual networks (2016)
13. Jha, D., et al.: Kvasir-seg: a segmented polyp dataset. In: Proceedings of the 26th International Conference, on MultiMedia Modelling, pp. 451–462 (2020)
14. Jha, D., et al.: H.D.: Resunet++: an advanced architecture for medical image segmentation. In: 2019 IEEE international symposium on multimedia (ISM) (2019)
15. Jha, D., et al.: CT liver segmentation via PVT-based encoding and refined decoding (2024)
16. Jha, D., Tomar, N.K., Sharma, V., Bagci, U.: TransNetR: transformer-based residual network for polyp segmentation with multi-center out-of-distribution testing. In: Medical Imaging with Deep Learning, pp. 1372–1384 (2023)
17. Kingma, D.P., Ba, J.: Adam: A method for stochastic optimization (2017)
18. Krizhevsky, A., Sutskever, I., Hinton, G.E.: Imagenet classification with deep convolutional neural networks. Adv. Neural Inf. Process. Syst. **25** (2012)
19. Mei, X., et al.: RadImageNet: an open radiologic deep learning research dataset for effective transfer learning. Radiol. Artif. Intell. **4**(5), e210315 (2022)
20. Paszke, A., et al.: Pytorch: an imperative style, high-performance deep learning library. Adv. Neural Inf. Process. Syst. **32** (2019)
21. Rao, A., Park, J., Woo, S., Lee, J.Y., Aalami, O.: Studying the effects of self-attention for medical image analysis. In: Proceedings of the IEEE/CVF International Conference on Computer Vision, pp. 3416–3425 (2021)
22. Ronneberger, O., Fischer, P., Brox, T.: U-net: Convolutional networks for biomedical image segmentation. In: Medical image computing and computer-assisted intervention–MICCAI 2015: 18th international conference, Munich, Germany, October 5-9, 2015, proceedings, part III 18, pp. 234–241 (2015)
23. Sabour, S., Frosst, N., Hinton, G.E.: Dynamic routing between capsules. Adv. Neural Inf. Process. Syst. **30** (2017)
24. Sander, M.E., Ablin, P., Blondel, M., Peyré, G.: Momentum residual neural networks. In: International Conference on Machine Learning, pp. 9276–9287 (2021)
25. Sandler, M., Howard, A., Zhu, M., Zhmoginov, A., Chen, L.C.: Mobilenetv2: inverted residuals and linear bottlenecks. In: Proceedings of the IEEE conference on computer vision and pattern recognition, pp. 4510–4520 (2019)
26. Simonyan, K., Zisserman, A.: Very deep convolutional networks for large-scale image recognition. arXiv preprint arXiv:1409.1556 (2014)
27. Zhang, H., Qie, Y.: Applying deep learning to medical imaging: a review. Appl. Sci. **13**(18), 10521 (2023)

28. Ma, N., Zhang, X., Zheng, H.T., Sun, J.: ShuffleNet V2: practical guidelines for efficient CNN architecture design. In: Computer Vision and Pattern Recognition (2018). https://doi.org/10.48550/arXiv.1807.11164
29. Zhang, Z., Liu, Q., Wang, Y.: Road extraction by deep residual u-net. In: IEEE Geoscience and Remote Sensing Letters **15**(5), 749–753 (2018)

Generalizable Lymph Node Metastasis Prediction in Pancreatic Cancer

Jiaqi Qu[1], Xunbin Wei[1,2,3,4,5(✉)], and Xiaohua Qian[1(✉)]

[1] School of Biomedical Engineering, Shanghai Jiao Tong University, Shanghai, China
xwei@bjmu.edu.cn, xiaohua.qian@sjtu.edu.cn
[2] Peking University Cancer Hospital & Institute, Beijing, China
[3] Biomedical Engineering Department, Peking University, Beijing, China
[4] Institute of Medical Technology, Peking University Health Science Center, Beijing, China
[5] International Cancer Institute, Peking University, Beijing, China

Abstract. Pancreatic cancer is an aggressive malignancy with an extremely poor prognosis, where lymph node metastasis (LNM), as a critical prognostic factor, is closely associated with patient recurrence and overall survival. In clinical practice, preoperative assessment of lymph node status is critical for treatment decisions. However, diverse acquisition conditions, scarce data resources, and limited utilization of 3D information pose challenges for imaging-based preoperative assessment algorithms. In this work, we propose a generalizable automated method for LNM prediction in pancreatic cancer. Specifically, to address the variability in sampling conditions, a sampling simulation generation model is proposed. This model simulates the acquisition process of enhanced CT to generate latent modalities, thereby improving the model adaptability to diverse data. Then, a spiral transformation is introduced to project the 3D CT images onto a 2D plane, which fully leverages the 3D information and enhances the representation of tumor features. Finally, class-guided contrastive learning is developed, which clusters together the features of each class, thereby enhancing the discriminative of the learned fine-grained features. Extensive experiments demonstrated that this method achieved remarkable prediction performance, while its stability and generalizability were validated through independent tests. Therefore, this method provides a valuable potential clinical tool for predicting LNM in pancreatic cancer.

Keywords: Lymph Node Metastasis · Pancreatic Cancer · Deep Learning · Generalizable

1 Introduction

Pancreatic cancer is a highly aggressive malignant tumor with a five-year survival rate of about 10%, often accompanied by metastasis [1]. Among these, lymph node metastasis (LNM) is a common form of metastasis which holds significant value for treatment decision and prognosis assessment. Currently, the LNM diagnosis relies on histopathologic examination after surgical lymphadenectomy [2]. However, for patients with LNM, preoperative chemotherapy can effectively improve their overall survival [3]. Therefore,

LNM prediction should be preferably scheduled preoperatively. Nowadays, CT is recommended as the preferred imaging modality for preoperative cancer staging of pancreatic cancer [4]. Thus, an objective and reliable automated method is critically required for LNM prediction of pancreatic cancer patients before surgery.

The risk of LNM is fundamentally driven by primary tumors. Therefore, imaging analysis of the primary tumor can reveal subtle distinctions associated with the risk of LNM in patients [5]. Recently, imaging-based preoperative LNM prediction methods have been emerging, which can be divided into two categories: 1) radiomics-based methods and 2) deep learning models. The first category typically involves two steps: extracting radiomics and constructing a classifier, which has been utilized in various cancers, including prostate cancer [6], colorectal cancer [7], thyroid cancer [8], and others. For example, Ke et al. [3] extracted radiomics from the tumor region of contrast-enhanced CT, used the least absolute shrinkage and selection operator for feature selection, and constructed a multivariate logistic regression model for LNM prediction. Wang et al. [9] collected data from gastric cancer patients and employed the intraclass correlation coefficient for preliminary selection on radiomic features. Then, they constructed a random forest model to predict LNM. Compared to radiomics-based methods, deep learning models have attracted increasing interest due to their ability to automatically learn clinically relevant features. A study on LNM in breast cancer [10] indicates that, within the same dataset, the performance of deep learning models surpasses that of radiomics-based models. Qiao et al. [11] proposed a multimodal LNM prediction model that enhances prediction performance by deeply mining tumor information across different modalities. Liu et al. [12] introduced a multi-task residual cross-attention network that leverages segmentation tasks to improve the extraction of LNM-related features.

The above approaches provide valuable insights; however, there are still challenges in LNM prediction based on deep learning: 1) **Variable imaging acquisition conditions.** This lead to appearance differences, degrading the generalization performance on unseen datasets and limiting the clinical applicability. 2) **Poor utilization of 3D information.** Due to the large parameter size of 3D neural networks and the limited information in 2D slices, it is difficult to make full use of 3D medical image information. This results in the model's inability to extract sufficient features from tumor images. 3) **Difficult-to-obtain discriminative features.** LNM prediction is a fine-grained prediction task involving classification of cancer subtypes, making it difficult to extract sufficiently discriminative features from the limited data.

Therefore, we propose a deep learning model combined with sampling simulation generation model and spiral transformation for generalizable LNM prediction in pancreatic cancer. Specifically, 1) to cope with varied sampling conditions, a sampling simulation generation model is proposed to generate latent arterial and venous phase of CT images, adapting the model to complex and diverse unseen data, thus enhancing the generalizability of the model on unseen data. 2) A spiral transformation method which can effectively retain information is introduced to transform 3D lesions into 2D views, make full use of 3D spatial information, and improve the extraction of tumor internal heterogeneity. 3) A class-guided contrastive learning is proposed, where features from the same class are encouraged to cluster while features from different classes are pushed far apart, effectively utilizing limited prior knowledge to enhance the discriminative fine-grained features.

In summary, this paper proposed a generalizable and automated model for predicting LNM in pancreatic cancer, with significant clinical application potential. Specifically, the technical contributions of this paper are as follows:

1. A sampling simulation generation model is proposed to generate latent data by simulating varied sampling conditions, enriching data diversity and improving the generalization performance on unseen data.
2. A class-guided contrastive learning is designed to obtain high-quality fine-grained features and enhance the discriminativeness of the learned features.

2 Method

Figure 1 illustrates the proposed generalizable method for LNM prediction in pancreatic cancer. Firstly, sampling simulation generation model is employed to generate latent modalities (Sect. 2.1). Then, the multimodal 3D CT data is transformed into 2D planes through spiral transformation (Sect. 2.2). Finally, a class-guided contrastive learning is introduced to enhance the extraction of discriminative fine-grained features (Sect. 2.3).

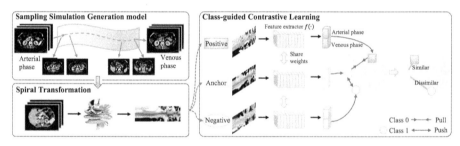

Fig. 1. The proposed model for the generalizable LNM prediction in pancreatic cancer.

2.1 Sampling Simulation Generation Model

Due to the variability in sampling environments and operators, contrast-enhanced CT often exhibits imaging biases. Contrast-enhanced CT follows a temporal sequence of first capturing the arterial phase and then the venous phase. Hence, we leverage this to construct a generation model, which simulates the sampling process and generates latent modalities.

Taking the direction from arterial phase to venous phase as an example. The arterial phase-enhanced CT is considered as the source modality $x^s \sim P_s$, venous phase as the target modality $x^t \sim P_t$, and an intermediate variable $w \in [0,1]$ is introduced as a correlation coefficient between the source and target to obtain the intermediate modality x^w. When $w = 0$, x^w refers to arterial phase; when $w = 1$, it refers to venous phase. By adjusting the w, a various intermediate modality between the arterial and venous phases are generated [13, 14].

Specifically, this generation model consists of two generators, $G_{ST}(x^s, w)$ and $G_{TS}(x^t, w)$, and two discriminators, D_S and D_T. Here, G_{ST} represents the transformation from the source modality to the target modality, while G_{TS} is the reverse. D_S is used to discriminate between x^w and the source modality x^s, and D_T serves the reverse. Thus, the adversarial loss for x^s and x^w is as follows:

$$L_{adv}(G_{ST}, D_S) = \mathbb{E}_{x^s:P_s}[\log(D_S(x^s))] + \mathbb{E}_{x^s:P_s}[\log(1 - D_S(G_{ST}(x^s, w)))] \quad (1)$$

x^w and x^t are as similar. Combining the two losses:

$$L_{adv} = (1 - w)L_{adv}(G_{ST}, D_S) + wL_{adv}(G_{ST}, D_T) \quad (2)$$

To ensure the semantic content is well-preserved, the image cycle consistency loss which similar to CycleGAN [15] is employed, where data x^s transformed after model G_{ST} and G_{TS} should be consistent with source data x^s: $x^s \to G_{ST} \to G_{TS} \approx x^s$, and the loss is as follows:

$$L_{cyc} = \mathbb{E}_{x^s \sim P_s}[\|G_{TS}(G_{ST}(x^s, w), w) - x^s\|_1] \quad (3)$$

We integrate them to obtain the final loss function as follows:

$$L = L_{adv} + \lambda L_{cyc} \quad (4)$$

where λ is the hyperparameter to balance the two losses. By changing the w, we can use the sampling simulation generation model to obtain potential arterial and venous phase sampling data.

2.2 Spiral Transformation

To fully exploit the 3D CT information and extract tumor internal heterogeneity features, inspired by [16, 17] we introduce spiral transformation. This transformation samples the target region based on a spiral line in 3D space and projects the 3D image onto a 2D plane, as showed in Fig. 2. Specifically, we take the approximate center of the tumor in the original 3D image as the origin O for the spiral transformation and use R as the maximum radius. The spiral line A is determined by the azimuth angle Ψ, elevation angle Θ, and the distance to the origin O.

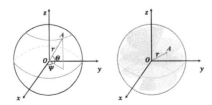

Fig. 2. Coordinate system of spiral transformation.

During the sampling process, the arc length d between adjacent points is defined by the distance between two points on the equator. Assuming $2N$ sampling points on the equator of a circle with a radius r, then the distance $d = 2\pi r/2N = \pi r/N$. Therefore, the number of sampling points on the horizontal plane relative to the angle Θ is given by $\pi r|cos\Theta|/d = 2\pi r|cos\Theta|/(\pi r/N) = 2N|cos\Theta|$. If Θ is divided into N angles, the sampling points on the spiral line:

$$m = \int_0^N 2N \cos\frac{k\pi}{2N} dk = 2N \int_0^N \frac{2N}{\pi} \cos\frac{k\pi}{2N} d\frac{k\pi}{2N} = \frac{4N^2}{\pi} \int_0^{\pi/2} \cos dx = \frac{4N^2}{\pi} \quad (5)$$

Since the number of sampled points for each r is the same, and $r \in [-R, R]$, the size of the transformed 2D view is $2R \times m$. As the 2D-transformed images depend on the direction of the coordinate system and the transformation parameters, an infinite number of images can be generated.

In conclusion, the spiral transformation sequentially unfolds the image from 3D to 2D, which effectively preserves the 3D spatial information, enhancing the extraction of fine-grained features.

2.3 Class-Guided Contrastive Learning

Although the sampling simulation generation model and spiral transformation can enhance the represented features of tumors, the automatic prediction algorithm still faces challenges. Since LNM prediction is a fine-grained tumor subtype classification, the model struggles to extract discriminative features. Therefore, in this section, we propose a class-guided contrastive learning to reduce the distances between samples of the same class and increases the distances between samples of different classes in the embedding space, enhancing the discriminative of learned features.

Let (x_i, y_i) be the i-th sample in the training set $D = \{(x_i, y_i)\}_{i=1}^N$. We employ a feature extractor $f(\cdot)$ to obtain the features $z_i \in \mathbb{R}^d$ from x_i, where d represents the feature dimension. During the training process, samples are randomly selected to construct the triplet $T = (x_a, x_p, x_n)$, including an anchor sample x_a, a positive sample x_p, and a negative sample x_n, where $y_a = y_p \neq y_n$. The features corresponding to the i-th triplet are denoted as $T_z = (z_i^a, z_i^p, z_i^n)$. Ideally, positive pairs have high similarity, and negative pairs have low similarity. Therefore, we construct a class-guided contrastive loss (L_{cont}) as follows:

$$L_{cont} = \frac{1}{N} \sum_{i=1}^N \left[\|z_i^a - z_i^p\|^2 - \|z_i^a - z_i^n\|^2 + \alpha \right]_+ \quad (6)$$

where N is the batch size, and $[\cdot]_+ = \max(0, \cdot)$. α is a margin constraint that requires the distance between negative pairs $\|z_i^a - z_i^n\|^2$ larger than the distance between positive pairs $\|z_i^a - z_i^p\|^2$.

Since the LNM prediction process is trained with strongly supervised, cross-entropy is used as the prediction loss (L_{main}). The overall loss is composed of prediction term (L_{main}) and class-guided contrastive learning term (L_{cont}):

$$Loss = L_{main} + \lambda L_{cont} \quad (7)$$

where λ is the hyperparameter to balance each part loss.

3 Experiments and Results

3.1 Datasets

In this study, we collected two CT datasets: 1) Dataset I: 300 pancreatic cancer patients from Shanghai Jiao Tong University School of Medicine (157 patients have a positive status of LNM and 143 patients in negative status). 2) Dataset II: 104 pancreatic cancer patients from the First Affiliated Hospital at Sun Yat-sen University (57 patients have a positive status of LNM and 47 patients in negative status). All these collected data were contrast-enhanced CT scans in arterial and venous phases. CT radiodensity values were truncated to the range of [−100, 240] Hounsfield Units (HU) based on empirical knowledge.

3.2 Implementation Details

All models were implemented using NVIDIA TITAN X (Pascal) GPUs in the PyTorch library. The framework for sampling simulation generation model followed the settings of DLOW [13]. ResNet-18 with parameters initialized by pre-trained results from ImageNet was used as the backbone for feature extraction. The constraint term coefficient (λ) was 0.00001, and a batch size was set to 64. SGD with a base learning rate of 0.0001 is used to optimize network. Accuracy (Acc), the area under the receiver operating characteristic curve (AUC), sensitivity (Sen), specificity (Spe), precision (Prec), and F1 score (F1) are used as performance metrics.

3.3 Ablation Study

Table 1. Ablation study results based on five-fold CV.

Methods	5fold-cv on Dataset I				5fold-cv on Dataset II			
	Acc	AUC	Sen	Spe	Acc	AUC	Sen	Spe
ST	0.677	0.735	0.705	0.649	0.618	0.630	0.646	0.568
ST + SSGM	0.689	0.744	0.616	0.732	0.646	0.646	0.714	0.537
ST + CGCL	0.693	0.754	0.722	0.666	0.655	0.650	0.766	0.509
Ours (ST + SSGM + CGCL)	0.711	0.769	0.730	0.699	0.687	0.680	0.756	0.568

ST: Spiral transformation.
SSGM: Sampling simulation generation model.
CGCL: Class-guided contrastive learning.

To verify the LNM prediction performance of the proposed method and the effectiveness of each component, we conducted five-fold cross-validation (CV) on two datasets and performed ablation experiments, as shown in Table 1. The reference baseline (ST) was implemented on the scheme which takes 2D views after spiral transformation as input and employs ResNet18 as the backbone. The ablated components included the sampling simulation generation model (SSGM) and class-guided contrastive learning (CGCL). As shown in the last row of Table 1, the collaborative strategy achieved an AUC of 0.769 on Dataset I and an AUC of 0.680 on Dataset II. Furthermore, each component contributed to the improvement of evaluation metrics. Particularly, on Dataset II, the introduction of SSCM and CGCL resulted in accuracy improvements of 2.8% and 3.7%, respectively, along with AUC improvements of 1.6% and 2.0%. When all strategies were employed, the prediction accuracy increased from 0.618 to 0.687, representing a growth of 6.9%, and the AUC also increased by 5.0%. These results strongly demonstrate the effectiveness of the proposed strategies in enhancing the LNM prediction performance of the model.

3.4 Generalization Performance on External Dataset

To assess the generalizability of the proposed method, we trained the model on Dataset I and conducted independent tests on Dataset II, as shown in Table 2. Ablation settings were consistent with Sect. 3.3. Compared to the reference baseline, the introduction of SSGM led to an increase about 2.0% in accuracy, while CGCL contributed to an enhancement of nearly 4.0%, and their joint scheme led to a 5.7% increase in accuracy.

Table 2. Results of the generalization experiments (trained on Dataset l, tested on Dataset ll)

Methods	Independent test on Dataset II			
	Acc	AUC	Sen	Spe
ST	0.606	0.629	0.603	0.608
ST + SSGM	0.625	0.643	0.638	0.609
ST + CGCL	0.644	0.629	0.689	0.587
Ours (ST + SSGM + CGCL)	0.663	0.643	0.689	0.630

ST: Spiral transformation
SSGM: Sampling simulation generation model
CGCL: Class-guided contrastive learning

Moreover, we visualized the normalized results using a radar chart (Fig. 3), where different colors represent distinct model configurations. The collaborative strategy, depicted in red on the outermost perimeter, intuitively illustrates a significant improvement in multiple metrics compared to the baseline.

3.5 Comparison with State-of-the-Art Methods

We compared our approach with four robust and widely applicable methods (Radiomics [18], 3D ResNet [19], 3D attenMIL [20], and transformer [21]). Given the involvement of two phases, we also compared it with several multimodal methods (EmbraceNet [22], Two-Stream CNN [23], MM_Dynamic [24]). Specifically, 1) five-fold CV on Dataset

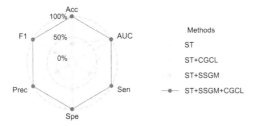

Fig. 3. Generalization experiment results displayed in radar charts.

I to assess the LNM prediction performance, and 2) generalization experiment which trained the models on Dataset I and independently tested them on Dataset II to evaluate the generalization performance. For a fair comparison, all methods focused on the tumor region and were trained on both arterial and venous phase CT scans.

These 3D imaging-based methods [18–21] did not reach an LNM prediction accuracy of 0.65. However, by employing a spiral transformation (ST) to map the tumor from 3D to a 2D plane, an accuracy of 0.677 and an AUC of 0.735 were achieved, demonstrating the effectiveness of the spatial transformation. Compared to multimodal methods [22–24], the proposed model demonstrates higher prediction accuracy and AUC. In contrast to all these comparative methods [18–24], the proposed model exhibits a more balanced Sen and Spe in the generalization experiment, confirming the superior stability and generalization capability of the proposed approach (Table 3).

Table 3. Comparison with the existing state-of-the-art methods

Methods	5fold-cv on Dataset I				Independent test on Dataset II			
	Acc	AUC	Sen	Spe	Acc	AUC	Sen	Spe
Radiomic [18]	0.620	0.625	0.530	0.720	0.596	0.624	0.603	0.587
3D ResNet [19]	0.635	0.692	0.693	0.570	0.563	0.540	0.842	0.217
3D attenMIL [20]	0.574	0.581	0.788	0.339	0.553	0.523	0.632	0.457
Transformer [21]	0.613	0.686	0.554	0.677	0.573	0.531	0.807	0.283
EmbraceNet [22]	0.603	0.641	0.663	0.539	0.577	0.555	0.741	0.370
Two-Stream CNN [23]	0.567	0.600	0.573	0.560	0.558	0.558	0.552	0.565
MM_Dynamic [24]	0.617	0.629	0.623	0.612	0.567	0.563	0.603	0.522
ST	0.677	0.735	0.705	0.649	0.606	0.629	0.603	0.608
Ours (ST + SSGM + CGCL)	0.711	0.769	0.730	0.699	0.663	0.643	0.689	0.630

4 Conclusion

In this study, we proposed a generalizable disease assessment method for the LNM prediction in pancreatic cancer patients. The novelty of our model can be summarized as follows: 1) a sampling simulation generation model is constructed to simulate the CT sampling process and enrich data diversity. 2) a class-guided contrastive learning is proposed to enhance feature specificity of fine-grained classes by distance constraints in feature space. Extensive experiments in cross-validation and independent tests validate the prediction and generalization performance of the proposed method. Overall, our approach provides an effective solution to the fine-grained generalization problem in oncology within the artificial intelligence community.

Acknowledgments. This work was supported by grants from the National Key Research and Development Program of China (Grant No. SQ2021YFF0500363), the Special Fund for Research on National Major Research Instruments of China (Grant No. 62027824), Natural Science Foundation of Shanghai (No. 22ZR1432100), National Natural Science Foundation of China (No. 62171273).

References

1. Mizrahi, J.D., Surana, R., Valle, J.W., Shroff, R.T.: Pancreatic cancer. Lancet **395**, 2008–2020 (2020)
2. Lee, Y.-W., Huang, C.-S., Shih, C.-C., Chang, R.-F.: Axillary lymph node metastasis status prediction of early-stage breast cancer using convolutional neural networks. Comput. Biol. Med. **130**, 104206 (2021)
3. Li, K., et al.: Contrast-enhanced CT radiomics for predicting lymph node metastasis in pancreatic ductal adenocarcinoma: a pilot study. Cancer Imaging **20**, 1–10 (2020)
4. Gao, J., Han, F., Jin, Y., Wang, X., Zhang, J.: A radiomics nomogram for the preoperative prediction of lymph node metastasis in pancreatic ductal adenocarcinoma. Front. Oncol. **10**, 1654 (2020)
5. Jin, C., et al.: Deep learning analysis of the primary tumour and the prediction of lymph node metastases in gastric cancer. Br. J. Surg. **108**, 542–549 (2021)
6. Peeken, J.C., et al.: A CT-based radiomics model to detect prostate cancer lymph node metastases in PSMA radioguided surgery patients. Eur. J. Nucl. Med. Mol. Imaging **47**, 2968–2977 (2020)
7. Liu, H., et al.: Preoperative prediction of lymph node metastasis in colorectal cancer with deep learning. BME frontiers (2022)
8. Yu, J., et al.: Lymph node metastasis prediction of papillary thyroid carcinoma based on transfer learning radiomics. Nat. Commun. **11**, 4807 (2020)
9. Wang, Y., et al.: CT radiomics nomogram for the preoperative prediction of lymph node metastasis in gastric cancer. Eur. Radiol. **30**, 976–986 (2020)
10. Sun, Q. et al.: Deep learning vs. radiomics for predicting axillary lymph node metastasis of breast cancer using ultrasound images: don't forget the peritumoral region. Front. Oncol. **10**, 53 (2020)
11. Qiao, M., et al.: Breast tumor classification based on MRI-US images by disentangling modality features. IEEE J. Biomed. Health Inform. **26**, 3059–3067 (2022)

12. Liu, S., Fang, M., Dong, D., Shang, W., Tian, J.: Multi-task residual cross-attention network for tumor segmentation and lymph node metastasis prediction in cervical cancer. In: 2023 IEEE 20th International Symposium on Biomedical Imaging (ISBI), pp. 1–5. IEEE (2023)
13. Gong, R., Li, W., Chen, Y., Gool, L.V.: Dlow: domain flow for adaptation and generalization. In: Proceedings of the IEEE/CVF Conference on Computer Vision and Pattern Recognition, pp. 2477–2486 (2019)
14. Qu, J., Xiao, X., Wei, X., Qian, X.: A causality-inspired generalized model for automated pancreatic cancer diagnosis. Med. Image Anal. **94**, 103154 (2024)
15. Zhu, J.Y., Park, T., Isola, P., Efros, A.A.: Unpaired image-to-image translation using cycle-consistent adversarial networks. In: Proceedings of the IEEE International Conference on Computer Vision, pp. 2223–2232 (2017)
16. Wang, J., Engelmann, R., Li, Q.: Segmentation of pulmonary nodules in three-dimensional CT images by use of a spiral-scanning technique. Med. Phys. **34**, 4678–4689 (2007)
17. Chen, X., Lin, X., Shen, Q., Qian, X.: Combined spiral transformation and model-driven multi-modal deep learning scheme for automatic prediction of TP53 mutation in pancreatic cancer. IEEE Trans. Med. Imaging **40**, 735–747 (2020)
18. Gao, J., et al.: Differentiating TP53 mutation status in pancreatic ductal adenocarcinoma using multiparametric MRI-derived radiomics. Front. Oncol. **11**, 632130 (2021)
19. Ebrahimi, A., Luo, S., Chiong, R.: Introducing transfer learning to 3D ResNet-18 for Alzheimer's disease detection on MRI images. In: 2020 35th International Conference on Image and Vision Computing New Zealand (IVCNZ), pp. 1–6. IEEE (2020)
20. Ilse, M., Tomczak, J., Welling, M.: Attention-based deep multiple instance learning. In: International Conference on Machine Learning, pp. 2127–2136. PMLR (2018)
21. Dosovitskiy, A., et al.: An image is worth 16x16 words: Transformers for image recognition at scale. arXiv preprint arXiv:2010.11929 (2020)
22. Choi, J.-H., Lee, J.-S.: EmbraceNet: a robust deep learning architecture for multimodal classification. Inf. Fusion **51**, 259–270 (2019)
23. Wang, W., et al.: Two-stream CNN with loose pair training for multi-modal AMD categorization. In: Medical Image Computing and Computer Assisted Intervention–MICCAI 2019: 22nd International Conference, Shenzhen, China, October 13–17, 2019, Proceedings, Part I 22, pp. 156–164. Springer (2019)
24. Han, Z., Yang, F., Huang, J., Zhang, C., Yao, J.: Multimodal dynamics: dynamical fusion for trustworthy multimodal classification. In: Proceedings of the IEEE/CVF Conference on Computer Vision and Pattern Recognition, pp. 20707–20717 (2022)

IRUM: An Image Representation and Unified Learning Method for Breast Cancer Diagnosis from Multi-View Ultrasound Images

Haoyuan Chen[1], Yonghao Li[1], Jiadong Zhang[1,2], Qi Xu[3], Meiyu Li[1], Zhenhui Li[4], Xuejun Qian[1], and Dinggang Shen[1,5,6(✉)]

[1] School of Biomedical Engineering and State Key Laboratory of Advanced Medical Materials and Devices, ShanghaiTech University, Shanghai 201210, China
dgshen@shanghaitech.edu.cn
[2] Department of Biomedical Engineering, City University of Hong Kong, Hong Kong 999077, China
[3] Department of Ultrasound, Wenzhou People's Hospital, Wenzhou 325000, China
[4] Department of Radiology, Yunnan Cancer Hospital, Kunming 650118, China
[5] Shanghai United Imaging Intelligence Co., Ltd., Shanghai 200230, China
[6] Shanghai Clinical Research and Trial Center, Shanghai 201210, China

Abstract. Multi-view breast ultrasound imaging has been routinely performed in clinical settings to ensure comprehensive disease evaluation. Recently, artificial intelligence (AI) has been developed to interpret medical images; however, most of the current AI models are restricted to single-view images, resulting in weak representation of breast 3D tissues. Here, we develop an Image Representation and Unified learning Method (IRUM) on a dataset comprising 3800 ultrasound images from 1900 patients with an accuracy of 86.8%. Owing to the design of four distinct learning modules, the proposed IRUM is *not only* able to predict breast cancer risk using multi-view inputs, *but also* compatible with single-view input (a commonly encountered situation in clinical practice). We demonstrate that the IRUM achieves superior performance to conventional single-view and multi-view approaches to a certain degree.

Keywords: Multi-view diagnosis · Image representation · Consistency learning · Ultrasound · Breast cancer

1 Introduction

Breast cancer is a leading cause of cancer-related death among women globally [24]. Identifying breast cancer at an early stage could considerably improve

---

Supplementary Information The online version contains supplementary material available at https://doi.org/10.1007/978-3-031-73284-3_3.

patients' outcomes and has therefore attracted sustained attention [15]. Owing to unique advantages of low cost, real-time and lack of ionizing radiation, ultrasound has long been the most widely utilized imaging technique in diagnosing breast diseases, especially for women with dense breast tissue [5,11]. Despite standardized terminology and management recommendations established by the American College of Radiology, interpreting breast ultrasound remains a challenging task due to high false-positive rates and intra- or inter-reader variability in clinical decision-making [1].

Artificial intelligence (AI) system has been proposed to assist radiologists in the interpretation of ultrasound images over a decade ago, while has been recently revolutionized by deep learning technique [20]. Notably, studies have demonstrated that deep learning improves the efficiency and robustness of ultrasound image analysis with comparable sensitivity and specificity to those of radiologists [16,19]. To offer an advanced approach with comprehensive diagnostic capability, multi-view (i.e., using transverse and longitudinal views, which are routinely performed in clinical workflows for disease evaluation) AI models have recently been proposed instead of conventional single-view AI models. For instance, Qian et al. [17] first leveraged multi-view breast ultrasound images to enhance the performance of a clinically applicable deep learning system via a late fusion strategy. Huang et al. [8] introduced a multi-view feature personalized distribution method for thyroid ultrasound image diagnosis. Despite impressive results, the inherent feature correlations among different views and their ability to address missing views in clinical practice remain unresolved.

In this study, we propose a novel Image Representation and Unified learning Method (IRUM), which consists of four individual learning modules, including a Feature Translation Module (FTM), a lesion localization module (LLM), and a multi-view feature fusion process via a Unified Consistency Learning Module (UCLM) and an Attention Weighting Module (AWM). We demonstrate that our IRUM achieves a promising result on a large-scale breast ultrasound dataset (Fig. 1).

2 Method

An overview of the proposed IRUM framework with four distinct modules is shown in Fig. 1. To be specific, IRUM first extracts multi-view features (transverse view I_t and longitudinal view I_l) via a weighted-shared encoder $E(\cdot)$, followed by Global Average Pooling (GAP) layers. Next, a UCLM module is employed to extract the unified features f_u from f_t and f_l. Finally, f_t, f_l, and f_u are simultaneously fed into the multilayer perceptron (MLP) classification head. The weights of those features are automatically selected via an attention weighting module $A(\cdot)$.

Given that lesion regions should receive more attention than surrounding tissue, we additionally implement an LLM module to guide the focus of the model by performing lesion segmentation. The optimization of our IRUM framework is based on the integration of three loss functions, defined as L_{rec}, L_{ce}, and L_{seg}

in Eq. 1, respectively. The detailed description can be found in the following subsections.

$$\mathcal{L}_{total} = \alpha \mathcal{L}_{rec} + \beta \mathcal{L}_{ce} + \gamma \mathcal{L}_{seg}. \quad (1)$$

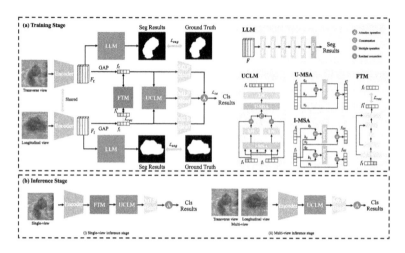

Fig. 1. An overview of the proposed IRUM. (a) Training stage, where multi-view images serve as inputs to the encoders with shared weights. Initially, FTM is trained to enable inference with single-view input. Subsequently, unified features are utilized for classification. Additionally, IRUM addresses lesion segmentation using LLM. (b) Inference with single- or multi-view inputs. During single-view inference, a single-view input image is processed by FTM to impute features of the missing view before feature aggregation. 'Cls' denotes the classification task, while 'Seg' denotes the segmentation task.

2.1 Feature Translation Module

Considering that multi-view data is not commonly preserved in clinical practice, we thus conditionally utilize an FTM module to address potential missing view issue during inference stage. In other words, FTM is designed to leverage pre-established correlation between paired multi-view images to generate features of the missing views. To achieve this goal, FTM is designed as an auto-encoder with a two-layer skip-connection structure. During the training stage, the weights of FTM are learned by evaluating the similarity between f'_l (transformed by f_t) and f_l. The reconstruction loss $\mathcal{L}_{rec}$ is computed using mean square error (MSE) as expressed in Eq. 2. N represents the feature dimension of both $f'li$ and fli.

$$\mathcal{L}_{rec} == \frac{1}{N}\sum_{i=1}^{N}(f'_{li} - f_{li})^2. \quad (2)$$

During the single-view inference stage, missing view features are imputed from the single-view input image with FTM.

2.2 Lesion Localization Module

The main objective of LLM module is to guide the classification process by performing lesion segmentation. LLM comprises five transposed convolutional layers and a 1×1 convolutional layer. Specifically, F_t and F_l, extracted by $E(\cdot)$ from I_t and I_l, respectively, serve as inputs to LLM. The segmentation probability maps S_t and S_l are aligned with their ground truths G_t and G_l, which are used to calculate the segmentation loss $\mathcal{L}_{seg}$ using the Dice loss. The pertinent formula is shown in Eq. 3. N denotes the number of pixels of the segmentation and ground truth maps.

$$\mathcal{L}_{seg} = 1 - (\frac{\sum_i^N S_{ti} G_{ti}}{\sum_i^N S_{ti} + \sum_i^N G_{ti}} + \frac{\sum_i^N S_{li} G_{li}}{\sum_i^N S_{li} + \sum_i^N G_{li}}). \tag{3}$$

2.3 Multi-View Feature Fusion

Unified Consistency Learning Module. UCLM is designed to address the challenge of inherent feature correlations across multi-view images. Initially, f_t and f_l are derived from F_t and F_l using GAP. Subsequently, Interaction Multi-head Self-Attention (I-MSA) is employed to separate f_t and f_l into q_t, k_t, v_t, and q_l, k_l, v_l, ensuring consistent interaction. The interacting features f_{t2l} and f_{l2t} undergo stabilization of the forward input distribution through Layer Normalization (LN). Finally, f_{t2l} and f_{l2t} are concatenated to form f_c as illustrated in Eq. 4.

$$f_c = C(\text{LN}(s(q_l k_t) \cdot v_t), \text{LN}(s(q_t k_l) \cdot v_l)), \tag{4}$$

where $s(\cdot)$ denotes the softmax function. After the interaction, the concatenated feature f_c is disentangled into q_c, where k_c and v_c are further fused using Unified Multi-head Self-Attention (U-MSA). The resulting distribution is stabilized through LN, yielding the unified feature f_u, as expressed in Eq. 5.

$$f_u = \text{LN}(s(q_c k_c) \cdot v_c). \tag{5}$$

Attention Weighting Module. The extracted features f_t, f_l, and unified feature f_u are classified by MLP to yield their corresponding classification confidences p_t, p_l, and p_u, respectively. Recognizing that each view may contribute differently to the final diagnostic decision, we incorporate an AWM to effectively weigh significance associated with each view. Specifically, an MLP is employed to transform f_t, f_l, and f_u into three distinct weights. Then, AWM applies these weights to the diagnostic outcomes obtained from each view, ensuring a comprehensive assessment. The classification confidence $\hat{p}$ is computed through weighted summation. The cross-entropy loss $\mathcal{L}_{ce}$ is computed by comparing $\hat{p}$ with the ground-truth p in the training stage, as illustrated in Eq. 6.

$$\mathcal{L}_{ce} = -[p \cdot \log(\hat{p}) + (1-p) \cdot \log(1-\hat{p})]. \tag{6}$$

3 Experiments

3.1 Dataset and Implementation Details

We collected multi-view ultrasound B-mode images from 1900 (1064 benign and 836 malignant) lesions from Yunnan Cancer Hospital. The dataset is randomly divided into a development set and an independent test set at a ratio of 80%: 20%. The development set is further divided into a training set and a validation set using five-fold cross-validation. A total of four evaluation metrics, including accuracy (ACC), sensitivity (SEN), specificity (SPE), and F1-score (F1), are implemented to assess the classification performance of the proposed IRUM. As for lesion segmentation, Dice Similarity Coefficient (DSC) and mean Intersection over Union (mIoU) are employed.

The experiments are conducted on an Ubuntu 18.0 operating system with an NVIDIA Tesla A100 80G HPC cluster. All experiments are carried out using Python 3.8 and PyTorch 1.13.0. During training, a configuration of 100 epochs, a batch size of 16, and a learning rate of 0.0001 are employed. The AdamW optimizer is utilized for optimization, and the learning rate is reduced by a factor of 10 every 10 epochs. The hyper-parameters for the loss function are set as follows: α, β, and γ are 0.1, 1 and 1.5, respectively.

3.2 Breast Cancer Diagnosis with Multi-View Inputs

To substantiate the outstanding diagnostic efficacy of IRUM for breast cancer leveraging multi-view input, we execute the comparative experiments with other classification methods. Specifically, ConvNeXt [23] exhibits the most favorable outcomes among other single-view classification models [6,7,12–14] as the baseline. Subsequently, ConvNeXt is selected as the backbone for IRUM, and a comparative analysis is conducted with multi-modal (AW3M [9], DRLP [10]), multi-view (MMDLM [17], PDT [8], DLGNet [22]), and multi-task (MIB-Net [21], PANet [4]) models employing the same backbone.

The results of the image classification task for multi-view breast cancer diagnosis are depicted in Table 1. Compared with the single-view approaches (traditional classification [6,7,12–14,23] and multi-task [4,21] models), IRUM improves ACC, F1, and SPE by 4.8%, 4.0%, and 1.7% through multi-view information, respectively. Compared with the multi-view approaches, IRUM considers the effects of feature interactions on diagnosis, which leads to the a better diagnostic accuracy (Table 1).

3.3 Breast Cancer Diagnosis with Single-View Only

To validate the effectiveness of FTM in acquiring additional prior knowledge and its ability to handle single-view image input, we conduct experiments using IRUM with FTM for transverse and longitudinal view images separately, and calculate the average performance of the two views. As shown in Table 2, the ACC and F1 improve by 3.2% and 4.1% in comparison to single-view input,

Table 1. The performance of comparative experiments in classification tasks. [In (%)]

Method	ACC	SEN	SPE	F1
Baseline	81.0±2.5	73.6±7.7	88.5±3.0	79.3±4.0
AW3M [9]	83.8±1.7	80.0±1.1	87.7±2.9	83.4±1.9
DRLP [10]	83.1±1.5	79.3±1.4	87.1±2.2	82.4±1.4
MMDLM [17]	84.2±1.8	79.4±2.6	88.9±1.9	83.6±1.9
PDT [8]	83.9±1.4	79.2±1.4	88.6±2.1	83.1±1.4
DLGNet [22]	82.6±2.5	77.7±2.6	87.5±3.3	81.7±2.7
MIB-Net [21]	84.6±1.0	82.3±4.1	86.9±3.5	84.2±1.2
PANet [4]	82.0±1.3	**84.1±1.7**	80.0±1.8	82.4±1.2
IRUM	**86.8±1.1**	83.5±1.0	**90.2±1.7**	**86.4±1.1**

respectively. This suggests that IRUM with FTM is capable of learning prior knowledge from multi-view perspectives, thereby enhancing the overall diagnostic capability of the model (Table 2).

Table 2. The performance of single-view experiments in classification tasks. [In (%)]

Method	ACC	SEN	SPE	F1
Baseline	81.0±2.5	73.6±7.7	88.5±3.0	79.3±4.0
IRUM (w FTM)	**84.2±1.3**	**79.5±1.8**	**89.0±1.1**	**83.4±1.4**

3.4 Ablation Studies

To assess the impact of the three modules (AWM, UCLM, and LLM) in IRUM on multi-view breast cancer diagnosis, an ablation experiment is conducted, and the results are presented in Table 3. Initially, when using a single module, AWM exhibits the most favorable performance, whereas UCLM shows relatively poorer results. Subsequently, upon combining the two modules, the performance of AWM and LLM slightly improves with LLM. The optimal performance is eventually achieved when all three modules are utilized concurrently in IRUM.

To verify that IRUM using LLM can also have good performance in segmentation tasks, we present the results of the image segmentation task compared with other segmentation methods in Table 4. Due to the scarcity of tasks for multi-view segmentation, IRUM is compared with single-view methods (traditional segmentation [2,3,18,25], and multi-task [4,21] models). Compared to these methods, IRUM demonstrates superior capability in identifying lesion regions and edges through multi-view interaction, resulting in improvements of 1.0% and 0.8% in DSC and mIoU, respectively (Table 4).

Table 3. The performance of ablation experiments. AWM represents the attention weighting mechanism, LLM means lesion localization module, and UCLM is the unified consistency learning module (Table 3). [In (%)]

Method			Classification				Segmentation	
AWM	LLM	UCLM	ACC	SEN	SPE	F1	DSC	mIoU
✓			85.0±0.7	80.6±5.5	89.4±4.8	84.3±1.2	-	-
	✓		-	-	-	-	77.6±4.2	65.5±5.4
		✓	84.6±1.3	81.5±2.2	87.5±5.2	85.7±1.0	-	-
	✓	✓	86.0±1.1	84.4±4.2	87.5±5.2	85.7±1.0	81.1±3.8	70.2±5.1
✓		✓	85.4±1.3	82.7±5.7	88.1±3.6	85.9±1.9	-	-
✓	✓		86.0±1.2	**85.2±3.7**	86.8±2.8	85.9±1.5	79.1±3.4	67.5±4.6
✓	✓	✓	**86.8±1.1**	83.4±1.0	**90.2±1.7**	**86.4±0.1**	**82.1±2.3**	**71.2±3.2**

Table 4. The performance of comparative experiments in segmentation task. [In (%)]

Method	DSC	mIoU
Baseline	76.9±0.3	66.0±0.3
MIB-Net [21]	80.8±2.8	69.9±3.9
PANet [4]	81.1±1.4	70.4±1.8
IRUM	**82.1±2.3**	**71.2±3.2**

4 Conclusion

In this study, we propose IRUM to assess breast cancer risk by utilizing complementary information from multi-view ultrasound images. With the proposed unified consistency learning module, we achieve better both inter-view and intra-view feature aggregation, which significantly improves overall diagnostic performance. Additionally, the feature translation module addresses the challenge of missing views, making the proposed method suitable for clinical practice. Moreover, we employ a localization module for lesion segmentation to guide the focus of the network. Experimental results demonstrate that IRUM outperforms other methods in breast cancer diagnosis.

Acknowledgments. This work was supported in part by National Natural Science Foundation of China (grant numbers 62131015, 62250710165, U23A20295), the STI 2030-Major Projects (No. 2022ZD0209000), Shanghai Municipal Central Guided Local Science and Technology Development Fund (grant number YDZX20233100001001), and The Key R&D Program of Guangdong Province, China (grant numbers 2023B0303040001, 2021B0101420006).

Disclosure of Interests.. The authors have no competing interests to declare that are relevant to the content of this article.

References

1. Berg, W.A., et al.: Prospective multicenter diagnostic performance of technologist-performed screening breast ultrasound after tomosynthesis in women with dense breasts (the DBTUST). J. Clin. Oncol. **41**(13), 2403–2415 (2023)
2. Cao, H., et al.: Swin-unet: Unet-like pure transformer for medical image segmentation. In: European Conference on Computer Vision, pp. 205–218. Springer (2022)
3. Chen, J., et al.: Transunet: Transformers make strong encoders for medical image segmentation. arXiv preprint arXiv:2102.04306 (2021)
4. Fan, Z., et al.: Joint localization and classification of breast masses on ultrasound images using an auxiliary attention-based framework. Med. Image Anal. **90**, 102960 (2023)
5. Farhadi, A., Ho, G.H., Sawyer, D.P., Bourdeau, R.W., Shapiro, M.G.: Ultrasound imaging of gene expression in mammalian cells. Science **365**(6460), 1469–1475 (2019)
6. He, K., Zhang, X., Ren, S., Sun, J.: Deep residual learning for image recognition. In: Proceedings of CVPR, pp. 770–778 (2016)
7. Huang, G., Liu, Z., Van Der Maaten, L., Weinberger, K.Q.: Densely connected convolutional networks. In: Proceedings of the CVPR, pp. 4700–4708 (2017)
8. Huang, H., et al.: Personalized diagnostic tool for thyroid cancer classification using multi-view ultrasound. In: Proceedings of MICCAI, pp. 665–674. Springer (2022)
9. Huang, R., et al.: Aw3m: an auto-weighting and recovery framework for breast cancer diagnosis using multi-modal ultrasound. Med. Image Anal. **72**, 102137 (2021)
10. Huang, Y., et al.: Deep learning radiopathomics based on preoperative us images and biopsy whole slide images can distinguish between luminal and non-luminal tumors in early-stage breast cancers. EBioMedicine **94** (2023)
11. Liao, J., et al.: Artificial intelligence-assisted ultrasound image analysis to discriminate early breast cancer in Chinese population: a retrospective, multicentre, cohort study. EClinicalMedicine **60** (2023)
12. Liu, X., Peng, H., Zheng, N., Yang, Y., Hu, H., Yuan, Y.: Efficientvit: memory efficient vision transformer with cascaded group attention. In: Proceedings of the CVPR (2023)
13. Liu, Z., et al.: Swin transformer: hierarchical vision transformer using shifted windows. In: Proceedings of the CVPR, pp. 10012–10022 (2021)
14. Liu, Z., et al.: Fastvit: a fast hybrid vision transformer using structural reparameterization. In: Proceedings of the ICCV (2023)
15. Messas, E., et al.: Treatment of severe symptomatic aortic valve stenosis using non-invasive ultrasound therapy: a cohort study. Lancet **402**, 2317–2325 (2023)
16. Mo, Y., et al.: Hover-trans: anatomy-aware hover-transformer for roi-free breast cancer diagnosis in ultrasound images. IEEE Trans. Med. Imaging **42**(6), 1696–1706 (2023)
17. Qian, X., et al.: Prospective assessment of breast cancer risk from multimodal multiview ultrasound images via clinically applicable deep learning. Nat. Biomed. Eng. **5**(6), 522–532 (2021)
18. Ronneberger, O., Fischer, P., Brox, T.: U-net: convolutional networks for biomedical image segmentation. In: Proceedings of the MICCAI, pp. 234–241. Springer (2015)
19. Shareef, B., Xian, M., Vakanski, A., Wang, H.: Breast ultrasound tumor classification using a hybrid multitask CNN-transformer network. In: Proceedings of MICCAI, pp. 344–353. Springer (2023)

20. Wang, G., et al.: Development of metaverse for intelligent healthcare. Nat. Mach. Intell. **4**(11), 922–929 (2022)
21. Wang, J., et al.: Information bottleneck-based interpretable multitask network for breast cancer classification and segmentation. Med. Image Anal. **83**, 102687 (2023)
22. Wang, K.N., et al.: DLGNET: a dual-branch lesion-aware network with the supervised Gaussian mixture model for colon lesions classification in colonoscopy images. Med. Image Anal. **87**, 102832 (2023)
23. Woo, S., et al.: Convnext v2: co-designing and scaling convnets with masked autoencoders. In: Proceedings of the CVPR, pp. 16133–16142 (2023)
24. Zhang, J., Wu, J., Zhou, X.S., Shi, F., Shen, D.: Recent advancements in artificial intelligence for breast cancer: image augmentation, segmentation, diagnosis, and prognosis approaches. In: Seminars in Cancer Biology. Elsevier (2023)
25. Zhou, Z., Rahman Siddiquee, M.M., Tajbakhsh, N., Liang, J.: Unet++: a nested u-net architecture for medical image segmentation. In: Proceedings of LMIA, pp. 3–11. Springer (2018)

Classification, Regression and Segmentation Directly from K-Space in Cardiac MRI

Ruochen Li[1(✉)], Jiazhen Pan[1], Youxiang Zhu[2], Juncheng Ni[1], and Daniel Rueckert[1,3]

[1] Technical University of Munich, Munich, Germany
ruochen.li@tum.de
[2] University of Massachusetts Boston, Boston, MA, USA
[3] BioMedIA, Imperial College London, South Kensington, UK

Abstract. Cardiac Magnetic Resonance Imaging (CMR) is the gold standard for diagnosing cardiovascular diseases. Clinical diagnoses predominantly rely on magnitude-only Digital Imaging and Communications in Medicine (DICOM) images, omitting crucial phase information that might provide additional diagnostic benefits. In contrast, k-space is complex-valued and encompasses both magnitude and phase information, while humans cannot directly perceive. In this work, we propose KMAE, a Transformer-based model specifically designed to process k-space data directly, eliminating conventional intermediary conversion steps to the image domain. KMAE can handle critical cardiac disease classification, relevant phenotype regression, and cardiac morphology segmentation tasks. We utilize this model to investigate the potential of k-space-based diagnosis in cardiac MRI. Notably, this model achieves competitive classification and regression performance compared to image-domain methods e.g. Masked Autoencoders (MAEs) and delivers satisfactory segmentation performance with a myocardium dice score of 0.884. Last but not least, our model exhibits robust performance with consistent results even when the k-space is 8× undersampled. We encourage the MR community to explore the untapped potential of k-space and pursue end-to-end, automated diagnosis with reduced human intervention. Codes are available at https://github.com/ruochenli99/KMAE_cardiac.

1 Introduction

Cardiac Magnetic Resonance Imaging (CMR) serves as the gold standard for diagnosing and treating cardiovascular diseases, offering a comprehensive view of

---

R. Li and J. Pan—Equal contribution.

Supplementary Information The online version contains supplementary material available at https://doi.org/10.1007/978-3-031-73284-3_4.

the heart's morphology and function. This non-invasive method enables detailed assessments of myocardial viability, ventricular function, and vascular anatomy. While Digital Imaging and Communications in Medicine (DICOM) protocol images are the prevalent format for storage and visualization, they consist solely of magnitude data derived from the real and imaginary components of the original complex data. Crucially, the phase information omitted in DICOM images holds potential value for tasks such as image reconstruction, segmentation, and the evaluation of flow dynamics and tissue movement [9, 25, 28]. Meanwhile, accelerated MR scans are preferred in clinics to reduce scan time and enhance patient comfort, leading to undersampled k-space(frequency domain representation of MR signal), which results in corrupted/blurred CMR images and deteriorates the follow-up downstream tasks [7, 10, 19, 21, 22].

Recently, methods that perform downstream tasks, e.g., motion estimation [15, 16] and segmentation [25, 26, 30] directly from k-space data gained attention. K-space, being complex-valued, encapsulates phase information and remains an intact and reliable data source with no corruptions, despite some acquisition lines that could be missing in undersampling. However, humans may struggle to perceive k-space data since they are not visually understandable to humans. Conversely, deep learning models excel in processing these data, as their computational frameworks readily handle complex values. Given the complexity and rich content of k-space data, selecting appropriate methods to effectively process and utilize this data is crucial for optimizing the diagnostic capabilities of cardiac MRI and for a comprehensive assessment of cardiovascular health.

Transformers are highly proficient in capturing long-range dependencies [17] and handling complex data structures, making them well-suited for modeling the temporal dynamics and global information present in k-space data [20]. Pan et al. proposed the Transformer-based K-GIN model [23], showing outstanding performance in MRI reconstruction solely using k-space data, highlighting the strong capabilities of its encoders in feature extraction and representation learning. We argue that this learned representation is not limited to the reconstruction tasks, but can be leveraged to more diverse tasks such as classification and segmentation. In this work, we propose KMAE, a versatile model that takes (undersampled) k-space data as inputs and can handle various downstream tasks, including disease classification, relevant phenotype regression, and cardiac segmentation. It leverages the pre-trained K-GIN encoders to attain rich representation and applies different decoders to carry out diverse downstream tasks. This adaptation facilitates efficient and accurate diagnostics and analyses based on k-space data. The contributions of this study can be summarised as follows:

1. We propose KMAE, a Transformer-based method for processing cardiac MR k-space data. KMAE can perform multiple downstream tasks, including disease classification, phenotype regression, and cardiac segmentation. To the best of our knowledge, we are the first to conduct disease classification directly from k-space data.

2. Unlike Convolutional Neural Networks (CNNs), which use local convolutional windows, we demonstrate that Transformers, which capture long-range dependencies, are more effective and robust for k-space data.
3. KMAE achieves competitive classification and regression performance compared to image-domain methods such as Masked Autoencoders (MAEs). It also provides satisfactory segmentation results with a myocardium dice score of 0.884, matching the quality of image-domain segmentation. Our model exhibits robust performance, maintaining consistent results even with 8× undersampled k-space data.

2 Related Work

K-space Interpolation: Previous methods typically leverage auto-calibration signals (ACS) in the k-space center to carry out k-space interpolation [8,18]. RAKI [2,14] improved the ACS-based methods by implementing CNNs. However, these approaches did not fully exploit the global dependencies present in k-space. Recently, a Transformer-based k-space interpolation method considering k-space global dependencies for dynamic CMR reconstruction was introduced by Pan et al. [23], achieving superior performance compared to baselines.

Downstream Tasks directly from K-space: Schlemper et al. [26] proposed CNN-based models with an end-to-end synthesis network and a latent feature interpolation network, predicting cardiac segmentation maps directly from undersampled dynamic MRI data. Kuestner et al. introduced LAP-Net [16], which can estimate the cardiac motion from the k-space of Cardiac MR. Moritz Rempe et al. proposed k-strip model [25], a complex-valued CNN-based algorithm for skull stripping in MRI, skipping operations in the image domain. Nevertheless, these methods are built upon CNNs and may not be able to fully exploit the global dependencies in k-space. Concurrently, Zhang et al. [30] proposed to use Transformers to directly derive segmentation from undersampled k-space data.

Masked Image Modelling: Vision Transformers(ViTs) [5] adapted Transformers from natural language processing to computer vision. Unlike CNNs that rely on local convolutions, the global self attention mechanism of ViTs allows for the modeling of long-range dependencies within images [13]; Masked Autoencoders (MAEs) [11], was introduced based on ViTs to extract the representation using masked image modeling in a self-supervised manner. Its versatility is further demonstrated in cardiac MR imaging analysis [29]. Recently, K-GIN [23] was introduced based on MAEs to learn k-space representation and conduct cardiac MR reconstruction, presenting robust and superior performance. We argue that its learned representation is not limited to the reconstruction tasks, but also the other tasks such as classifications and segmentation.

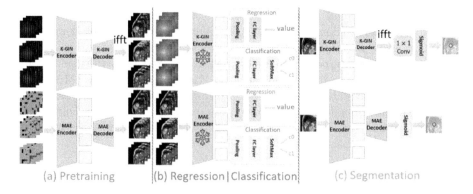

Fig. 1. An overview of KMAE and MAEs with downstream tasks. The upper section depicts KMAE, a modification of the K-GIN model. The lower section illustrates the modification of MAEs. (a) The pre-training of KMAE and MAEs for MRI reconstruction. KMAE processes under-sampled k-space data, while MAEs handle in the image domain with masked-out patches. (b) In downstream task fine-tuning, we freeze encoders of KMAE and MAEs while their decoders are modified for regression and classification. (c) We adapt decoders of KMAE and MAEs for segmentation tasks, with the upper section highlighting our newly proposed k-space segmentation method.

3 Methods

3.1 Pre-Training

Models: **K-GIN** [23], designed for MRI reconstruction, processes undersampled k-space data (e.g., Cartesian undersampling) and performs k-space interpolation to predict fully sampled k-space data, and converts it to MRI via an inverse fast Fourier transform. **MAEs** [11] features an asymmetric design. The model inputs images with most patches masked, exposing only a few. The encoder processes these visible patches and passes them to a smaller decoder, whose primary task is reconstructing the original image pixels, resulting in high-quality MRI images.
Pre-Training Process: We utilize k-space interpolation / image reconstruction tasks to pre-train KMAE / MAEs, as illustrated in Fig. 1(a). During the pre-training phase, our KMAE model rigorously adheres to the foundational principles and procedures established by the K-GIN architecture.
Evaluation Metrics Meaning: After pre-training, both KMAE and MAEs achieve high Peak Signal-to-Noise Ratio (PSNR) values, showcasing the excellent quality of their reconstructed images. This performance illustrates their encoders' effectiveness in extracting meaningful representations from raw data, facilitating subsequent tasks. Moreover, this pre-training approach significantly reduces training times, enabling faster adaptation to various downstream tasks.

3.2 Regression and Classification

After pre-training, we freeze the model's encoder and discard the reconstruction decoder. The trained encoder is then used to extract valuable feature representations for downstream tasks such as regression and classification, as shown in Fig. 1(b). Both KMAE and MAEs adopt a consistent architectural framework for downstream tasks.

For **regression** tasks, we employ a pooling layer to transform the extracted features into a feature vector. This vector is then fed into a fully connected (FC) layer to predict the regression value for each subject. Regarding **classification** tasks, we incorporate a final layer equipped with a SoftMax function, which computes the probability of each class to categorize the subjects.

3.3 Segmentation

In regression and classification tasks, outputs are numerical values and probabilities, so the decoder used for image reconstruction in pre-training is removed as it is unnecessary. However, image segmentation tasks remain closely linked to reconstruction tasks as they both require pixel-level prediction. Therefore, we adopted the same reconstruction decoder to accomplish CMR segmentation.

Regarding **MAEs**, it processes in the image domain and we can adapt it to segmentation tasks by simply replacing the final layer with a Sigmoid function. On the other hand, **KMAE** handles in the frequency domain, therefore we first Fourier transfer the reconstructed k-space to MR images. These images are then processed through a 1x1 convolution layer, followed by a Sigmoid function to effectively segment the myocardium. Both structures are described in Fig. 1(c).

4 Data and Experiments

4.1 Dataset

Dataset. We used short-axis cardiac MR images provided by UKBioBank [24] and corresponding clinical information, which provide a cross-sectional view of the left and right ventricles of the heart. We applied center-cropped CMR images with a matrix size of 128×128 across 25 temporal cardiac phases (we used every two temporal frames). Since the original CMR from UKBioBank are magnitude-only images, we created synthetic k-space data for each 2D+time scan by applying additional Gaussian B0 variations in real-time to remove the conjugate symmetry of k-space [26], thus simulating fully sampled single-coil acquisitions. We stacked 11 slices along the long axis. Additionally, we applied VISTA Cartesian undersampling masks [1] to generate the accelerated k-space and the corresponding MRI.

Filter Data and Label Strategy. The UK Biobank CMR dataset initially comprises 47,097 subjects. We identified three distinct subsets for our study: **Healthy Subgroup**, consisting of 2,660 individuals without risk factors such as obesity, myocardial infarction, acute myocardial infarction, insulin-dependent diabetes mellitus, or physician-diagnosed vascular or heart conditions. This subgroup only includes individuals rated as "Excellent" or "Good" in overall health who also reported never having smoked tobacco [27]. **Cardiopathy Subgroup** [4] comprises 1,340 subjects with diagnosed heart conditions, including heart attacks, myocardial infarction, and angina. **Left Ventricular Dysfunction Subgroup** [6] includes 937 subjects with a Left Ventricular Ejection Fraction (LVEF) below 50%. For **regression**, we selected 2,000 subjects from the Healthy Subgroup to calculate cardiac age based on birth year and scan date [12]. We used LVEF and LVEDV(Left Ventricular End-Diastolic Volume) labels from 1,000 healthy subjects sourced from [3]. For **classification**, we compared 937 subjects from the Left Ventricular Dysfunction Subgroup to an equal number from the Healthy Subgroup. Similarly, 1,340 subjects from the Cardiopathy Subgroup were matched with an equivalent number from the Healthy Subgroup.

Table 1. Comparison of different models using two types of inputs: using original k-space data (first and third rows) and using undersampled k-space data (second and fourth rows). The evaluation metrics include Mean Absolute Error for regression and accuracy for classification. R=4 denotes the acceleration rate for undersampling k-space. '†' means the encoder is not frozen. The best results are marked in bold.

	Regression			Classification	
	Age ↓	LVEF ↓	LVEDV ↓	LV Dys ↑	Cardiopathy ↑
ResNet	6.031	5.887	27.188	63.31%	72.54%
ResNet(R=4)	6.559	5.443	22.012	65.09%	72.95%
KMAE	5.840	4.547	23.917	68.64%	75.41%
KMAE(R=4)	5.690	4.568	22.591	69.82%	75.00%
KMAE(R=4,†)	**4.439**	**4.128**	**16.452**	**76.33%**	**77.46%**

4.2 Implementing Details

Pre-training. We trained MAEs and K-GIN on data from the Healthy Subgroup, which consisted of MRI datasets with 5 slices and 25 temporal frames. Both models were tasked with image reconstruction, achieving PSNR of 38.846 for MAEs and 38.755 for K-GIN. The MAEs used a patch size of 2, while all other hyperparameters remained consistent with the original MAE specifications. Similarly, K-GIN adhered to its original configurations. Details of the implementation are disclosed in our code repository.

Training Strategy. We employed an NVIDIA A40 GPU to train our framework, configuring the setup with a single batch and a learning rate scheduler, peaking at 0.0001. Our Transformer architecture utilized 8 layers, 8 heads, and an embedding dimension of 512, while the ResNet model was trained without pre-trained weights from cardiac MRI data. For **classification** and **regression** tasks, we processed 5 MRI slices per subject, each containing 25 frames, and averaged the results from each slice to compute final regression scores or classification probabilities. The KMAE and MAEs' encoder were frozen, with only training on subsequent layers, as shown in Fig. 1(b). Moreover, we performed a comprehensive performance comparison by training the full KMAE pipeline without freezing any components. For **segmentation** tasks, we used a single MRI slice with 25 frames per subject to accurately segment myocardial regions with no encoder freezing, as illustrated in Fig. 1(c).

ResNet Baseline. Our k-space data includes 2D spatial and temporal dimensions (2D+t). So, we adapted ResNet50 by modifying the channel dimensions of its 2D convolutional layers to match the number of cardiac SAX slices.

Table 2. Comparison of one model using three types of inputs: using original MRI image (first row), original k-space data (second row), and using undersampled k-space data (third and fourth rows). The metrics evaluated include Mean Absolute Error for regression and accuracy for classification. The best results are marked in bold.

	Regression			Classification	
	Age ↓	LVEF ↓	LVEDV ↓	LV Dys ↑	Cardiopathy ↑
MAEs	**5.553**	**4.511**	23.545	**78.36%**	**76.45%**
KMAE	5.840	4.547	23.917	68.64%	75.41%
KMAE(R=4)	5.690	4.568	**22.591**	69.82%	75.00%
KMAE(R=8)	5.669	4.610	22.694	69.23%	74.59%

Table 3. Comparison of one model using three types of inputs for segmentation task: using original MRI image(first column), original k-space data (second column), and using undersampled k-space data (third and fourth columns).

	MAEs	KMAE	KMAE(R=4)	KMAE(R=8)
DICE	0.941	0.884	0.873	0.870

Metrics. In accelerated CMR, where CMR imaging employs acceleration techniques, higher acceleration factors (R=4 or R=8) lead to increased undersampling of k-space data. For **regression** tasks, we used Huber loss to train and evaluated performance by Mean Absolute Error (MAE). Lower MAE values indicate better regression performance. For **classification** tasks, performance was assessed using cross-entropy loss and accuracy, with higher accuracy indicating better classification results. For **segmentation** task, we employed binary cross-entropy loss during training and gauged effectiveness with the Dice score, with higher scores indicating enhanced segmentation performance.

5 Results and Discussion

In table 1, KMAE generally exhibits lower MAE values for regression tasks, indicating more accurate predictions for variables such as age, LVEF, and LVEDV.

Even with undersampling k-space data, KMAE tends to outperform ResNet. KMAE consistently achieves higher accuracy for classification tasks than ResNet, regardless of undersampling or freezing layers.

Table 2 demonstrates that the Transformer model performs comparably well with undersampled k-space data as input, even when compared to original MRI images. This adaptability is evident in the first two rows of the table. Even when undersampled k-space data is used as input (KMAE at R=4 and KMAE at R=8), the model still achieves competitive performance, with only slight variations compared to the implementation on the full sampled k-space data.

Table 3 shows that MAEs achieves the highest Dice coefficient. Meanwhile, KMAE and its undersampled version also exhibit reasonably high Dice coefficients, demonstrating that they are capable of producing accurate segmentation results, albeit slightly lower than the MAEs model.

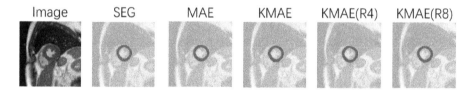

Fig. 2. Comparison of segmentation methods for delineating myocardial regions

Figure 2 shows that MAEs, utilizing CMR images as the input, delivers optimal segmentation performance by precisely delineating the myocardium within the heart. Furthermore, KMAE employing undersampled k-space inputs also exhibits impressive segmentation capabilities.

Our experiments show that the Transformer architecture is highly effective for processing k-space data, particularly when undersampled. The self-attention mechanism in Transformers efficiently handles global information in k-space, in which each k-space measurement point contributes to forming every pixel in the image domain. This capability enables the model to assimilate information from diverse positions, effectively capturing global correlations. Specifically in cardiac applications, Transformers treats k-space frames as time sequences, integrating temporal information by segmenting frames into patches, thus capturing cardiac dynamics over time. Notably, predictions derived from undersampled k-space outperform those from fully-sampled k-space, as the latter often includes irrelevant details to classification / segmentation. At the same time, the former focuses more on low-frequency components, which is the most critical information contributing to the downstream tasks.

Our findings suggest that predictions based on MAEs for fully sampled images represent the **upper bound** of performance comparison. Remarkably, even with undersampled k-space data, the results are comparable to those from fully sampled images. This underscores the robustness of KMAE and the potential for direct application in k-space-based diagnostics, cardiac assessment, and other CMR applications.

Outlook. The current work only verified the feasibility of the k-space analysis method with single-coil-acquired CMR data. Future work will extend it to multi-coil CMR scans, thereby allowing for the incorporation of more redundant information and further improvement of the estimation accuracy compared to image-domain-based methods.

6 Conclusion

In this study, we introduce KMAE model, designed to utilize k-space data for tasks such as disease classification, phenotype regression, and cardiac segmentation. Our findings reveal that Transformer-based architectures effectively process k-space data, achieving comparable classification and regression performance to image-domain models and successfully emulating image-domain segmentation techniques. Moreover, KMAE maintains consistent performance with undersampled k-space data, underscoring its robustness and potential for accelerated MRI applications. This research also highlights the considerable promise of employing k-space data in cardiac MRI and confirms the suitability of Transformer architectures for such applications. Future research should extend to multi-coil CMR scans and explore further downstream tasks to validate and expand these findings in clinical settings.

Acknowledgements and Disclosure of Interests. This research has been conducted using the UK Biobank Resource under Application Number 87802. This work is funded by the European Research Council (ERC) project Deep4MI (884622). The authors have no competing interests to declare that are relevant to the content of this article.

References

1. Ahmad, R., Xue, H., Giri, S., Ding, Y., Craft, J., Simonetti, O.P.: Variable density incoherent spatiotemporal acquisition (VISTA) for highly accelerated cardiac MRI. Magn. Reson. Med. **74**(5), 1266–1278 (2015)
2. Akçakaya, M., Moeller, S., Weingärtner, S., Uğurbil, K.: Scan-specific robust artificial-neural-networks for k-space interpolation (RAKI) reconstruction: database-free deep learning for fast imaging. Magn. Reson. Med. **81**(1), 439–453 (2019). https://doi.org/10.1002/mrm.27420
3. Bai, W., Suzuki, H., Huang, J., et al.: A population-based phenome-wide association study of cardiac and aortic structure and function. Nat. Med. **26**(10), 1654–1662 (2020)
4. Clough, J.R., et al.: Global and local interpretability for cardiac MRI classification. In: International Conference on Medical Image Computing and Computer-Assisted Intervention, pp. 656–664. Springer International Publishing (2019)
5. Dosovitskiy, A., et al.: An image is worth 16x16 words: Transformers for image recognition at scale. arXiv preprint arXiv:2010.11929 (2020)

6. Elghazaly, H., McCracken, C., Szabo, L., et al.: Characterizing the hypertensive cardiovascular phenotype in the UK biobank. Eur. Heart J. Cardiovasc. Imaging (2023). https://doi.org/10.1093/ehjci/jead123
7. Gong, S., Lu, W., Xie, J., Zhang, X., Zhang, S., Dou, Q.: Robust cardiac mri segmentation with data-centric models to improve performance via intensive pre-training and augmentation. In: International Workshop on Statistical Atlases and Computational Models of the Heart, pp. 494–504. Springer (2022)
8. Griswold, M.A., et al.: Generalized autocalibrating partially parallel acquisitions (GRAPPA). Magn. Reson. Med. **47**(6), 1202–10 (2002)
9. Haji-Valizadeh, H., Guo, R., Kucukseymen, S., et al.: Comparison of complex k-space data and magnitude-only for training of deep learning-based artifact suppression for real-time cine MRI. Front. Phys. **9**, 684184 (2021)
10. Hammernik, K., Pan, J., Rueckert, D., Küstner, T.: Motion-guided physics-based learning for cardiac MRI reconstruction. In: 2021 55th Asilomar Conference on Signals, Systems, and Computers, pp. 900–907. IEEE (2021)
11. He, K., et al.: Masked autoencoders are scalable vision learners. In: Proceedings of the IEEE/CVF Conference on Computer Vision and Pattern Recognition, pp. 16000–16009 (2022)
12. Inácio, M.H.A., et al.: Cardiac age prediction using graph neural networks. medRxiv (2023). 10.1101/2023.04.19.23287590
13. Islam, S., et al.: A comprehensive survey on applications of transformers for deep learning tasks. Expert Syst. Appl. 122666 (2023)
14. Kim, T., Garg, P., Haldar, J.: Loraki: Autocalibrated recurrent neural networks for autoregressive mri reconstruction in k-space. arXiv preprint arXiv:1904.09390 (2019)
15. Küstner, T., et al.: Self-supervised motion-corrected image reconstruction network for 4d magnetic resonance imaging of the body trunk. APSIPA Trans. Sign. Inf. Process. **11**(1) (2022)
16. Küstner, T., et al.: LAPNET: non-rigid registration derived in k-space for magnetic resonance imaging. IEEE Trans. Med. Imaging **40**(12), 3686–3697 (2021)
17. Li, F., Zhou, L., Wang, Y., et al.: Modeling long-range dependencies for weakly supervised disease classification and localization on chest X-ray. Quant. Imaging Med. Surg. **12**(6), 3364 (2022)
18. Lustig, M., Pauly, J.M.: SPIRiT: iterative self-consistent parallel imaging reconstruction from arbitrary k-space. Magn. Reson. Med. **64**(2), 457–471 (2010). https://doi.org/10.1002/mrm.22428
19. Lyu, J., et al.: The state-of-the-art in cardiac mri reconstruction: Results of the cmrxrecon challenge in miccai 2023. arXiv preprint arXiv:2404.01082 (2024)
20. Oh, C., Kim, D., Chung, J.Y., et al.: A k-space-to-image reconstruction network for MRI using recurrent neural network. Med. Phys. **48**(1), 193–203 (2021)
21. Pan, J., Huang, W., Rueckert, D., Küstner, T., Hammernik, K.: Reconstruction-driven motion estimation for motion-compensated MR CINE imaging. IEEE Trans. Med. Imaging (2024)
22. Pan, J., Rueckert, D., Küstner, T., Hammernik, K.: Learning-based and unrolled motion-compensated reconstruction for cardiac MR CINE imaging. In: International Conference on Medical Image Computing and Computer-Assisted Intervention, pp. 686–696 (2022)
23. Pan, J., et al.: Global k-space interpolation for dynamic MRI reconstruction using masked image modeling. In: International Conference on Medical Image Computing and Computer-Assisted Intervention, pp. 228–238. Springer (2023)

24. Petersen, S.E., et al.: UK Biobank's cardiovascular magnetic resonance protocol. JCMR pp. 1–7 (2015)
25. Rempe, M., Mentzel, F., Pomykala, K.L., et al.: k-strip: a novel segmentation algorithm in k-space for the application of skull stripping. Comput. Methods Programs Biomed. **243**, 107912 (2024)
26. Schlemper, J., et al.: Cardiac MR segmentation from undersampled k-space using deep latent representation learning. In: Medical Image Computing and Computer Assisted Intervention–MICCAI 2018: 21st International Conference, Granada, Spain, September 16-20, 2018, Proceedings, Part I. pp. 259–267. Springer International Publishing (2018).
27. Shah, M., et al.: Environmental and genetic predictors of human cardiovascular ageing. Nat. Commun. **14**(1), 4941 (2023)
28. Sriram, A., et al.: End-to-end variational networks for accelerated MRI reconstruction. In: Medical Image Computing and Computer Assisted Intervention–MICCAI 2020: 23rd International Conference, Lima, Peru, October 4–8, 2020, Proceedings, Part II 23. pp. 64–73. Springer (2020)
29. Zhang, Y., Chen, C., Shit, S., Starck, S., Rueckert, D., Pan, J.: Whole heart 3d+t representation learning through sparse 2d cardiac mr images. arXiv preprint arXiv:2406.00329 (2024)
30. Zhang, Y., Stolt-Ansó, N., Pan, J., Huang, W., Hammernik, K., Rueckert, D.: Direct cardiac segmentation from undersampled k-space using transformers. arXiv preprint arXiv:2406.00192 (2024)

DDSB: An Unsupervised and Training-Free Method for Phase Detection in Echocardiography

Zhenyu Bu[2], Yang Liu[1(✉)], Jiayu Huo[1], Jingjing Peng[1], Kaini Wang[2], Guangquan Zhou[2], Rachel Sparks[1], Prokar Dasgupta[1], Alejandro Granados[1], and Sebastien Ourselin[1]

[1] King's College London, London, UK
{yang.9.liu,jiayu.huo,jingjing.peng,rachel.sparks,
prokar.dasgupta,alejandro.granados,sebastien.ourselin}@kcl.ac.uk
[2] Southeast University, Nanjing, China
{zybu,230218244,guangquan.zhou}@seu.edu.cn

Abstract. Accurate identification of End-Diastolic (ED) and End-Systolic (ES) frames is key for cardiac function assessment through echocardiography. However, traditional methods face several limitations: they require extensive amounts of data, extensive annotations by medical experts, significant training resources, and often lack robustness. Addressing these challenges, we proposed an unsupervised and training-free method, our novel approach leverages unsupervised segmentation to enhance fault tolerance against segmentation inaccuracies. By identifying anchor points and analyzing directional deformation, we effectively reduce dependence on the accuracy of initial segmentation images and enhance fault tolerance, all while improving robustness. Tested on Echo-dynamic and CAMUS datasets, our method achieves comparable accuracy to learning-based models without their associated drawbacks. Our code is open-sourced and available at https://github.com/MRUIL/MaxMin-Room-Detection.

Keywords: Frame detection · Unsupervised · Training-free

1 Introduction

Cardiovascular diseases represent a major global health issue, underscoring the need for early detection [15]. As a cost-effective and real-time diagnostic technique, echocardiography plays a vital role in the diagnosis of these conditions [16]. Accurately identifying the End-Diastolic (ED) and End-Systolic (ES) phases on echocardiograms, as shown in Fig. 1, is crucial for computing critical clinical metrics such as ejection fraction and global longitudinal strain. These metrics are critical in assessing cardiac functionality [7]. Identifying ED and ES

Z. Bu and Y. Liu—Equal contribution.

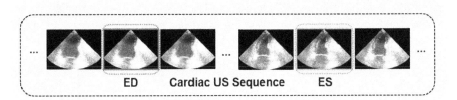

Fig. 1. Example of cardiac ultrasound sequence with ED in Red and ES in Green. (Color figure online)

phases poses significant challenges, primarily due to the inherent variability in heart shapes, sizes and movement patterns among individuals. Moreover, the quality of echocardiographic images can be affected by factors such as imaging conditions and patient anatomy, further complicating the detection process.

Early techniques for identifying ED and ES frames in echocardiograms primarily relied on manual selection or simplistic automated criteria, often failing to accurately capture the heart's complex dynamics [14]. The innovative use of the QRS complex onset and the T wave's end as markers for ED and ES did not account for regional motion irregularities, proving to be impractical in emergency scenarios requiring rapid diagnosis [14]. Nadjia et al. [9] attempted to quantify the similarity between the ED and ES frames using the correlation coefficient, an approach that still necessitated manual selection of the ED frame. Meanwhile, Barcaro et al. [2], along with Darvishi et al. [4] and Abboud et al. [1], explored automated segmentation techniques to delineate the left ventricle, identifying the ED and ES frames by the largest and smallest ventricular cross sections, respectively. However, these methods heavily depended on the accuracy of segmentation results, and have a very poor tolerance for segmentation errors.

With the rapid advancement of deep learning technologies, methods for detecting ED and ES frames in echocardiograms have significantly evolved. These approaches generally fall into two main categories: classification-based [18] and regression-based [5]. In classification models, ED, ES, and other frames are categorized into distinct labels, yet this often leads to class imbalance due to the singular occurrence of ED and ES frames within a series of intermediate frames. To counter this imbalance, some researchers have pivoted towards regression tasks, using interpolation to assign unique values to each frame, a strategy gaining popularity for its effectiveness in the field. Further innovation has been seen in the application of Recurrent Neural Networks (RNNs) [6], traditionally used in natural language processing to understand character sequence correlations. Kong et al. [11] introduced TempReg-Net, integrating Convolutional Neural Networks (CNNs) [10] with RNNs to pinpoint specific frames within MRI sequences, demonstrating a novel approach to leveraging deep learning for temporal and spatial feature extraction. Dezaki et al. [5] demonstrated how combining traditional CNN architectures with RNNs for temporal analysis, particularly employing DenseNet [8] and GRU [3] models, could yield optimal results. This methodology, however, entails extensive trial and error to identify the most effective CNN and RNN combinations. Wang et al. [20] proposed a dual-branch feature extraction model, defining the task of identifying ED/ES frames as a curve regression problem, thus moving away from direct index regression. Meanwhile, Li et al. [13] explored a semi-supervised approach for ED/ES detection, requiring only a portion of labeled data, thereby reducing the dependency on extensively annotated datasets. Singh et al. [19] advanced the methodology by combining CNN with Bidirectional Long Short-Term Memory (BLSTM) networks [21], offering an enhanced solution over previous models. These deep learning strategies require extensive annotated datasets. Acquiring such data and securing annotations from medical professionals present substantial challenges due to the lack of resources and the intensive workload required.

To overcome the above drawbacks, we proposed an unsupervised and training-free method, called **DDSB** (Directional Distance to Segmentation Boundary), for phase detection. Our contributions are three-fold:

- We have proposed a novel unsupervised and training-free approach for phase detection, which can recognize the ED/ES phase in the cardiac cine without the need for annotated datasets for training and avoid the GPU resource wastage caused by the training process.
- We employed a distance-based strategy to formulate proposed modifications from diverse perspectives, utilizing initial segmentation outcomes and serving as the frame representation. This approach is designed to fight the inherent limitations of coarse segmentation results, thereby strengthening the robustness of our model.
- We demonstrate the effectiveness and advantages of our method on two public datasets (Echo-dynamic [17] and CAMUS [12]), achieving comparable performance compared to other deep learning based approaches.

2 Method

In this work, we delve into the challenge of unsupervised detection of echocardiography phases, with a keen focus on identifying the End-Diastolic (ED) and End-Systolic (ES) stages. Our objective is to devise an algorithm f that can reliably approximate the moments of ED and ES, denoted as (t_{ed}, t_{es}), from a given sequence of T-frame echocardiographic video, $X_T = {\boldsymbol{x}_j}_{j=1}^{T}$. Here, $\boldsymbol{x}_j$ signifies the j-th frame within the video stream. The proposed methodology, referred to as DDSB (Directional Distance to Segmentation Boundary), encapsulates three integral components, as shown in Fig. 2: an unsupervised segmentation optimizer, an anchor points picker, and a temporal expansion-contraction discriminator. DDSB stands out as an innovative, unsupervised, and training-free approach in the realm of cardiac phase detection.

2.1 Unsupervised Cavity Segmentation

For the effective identification of ED and ES frames in echocardiography videos, the primary step involves the precise segmentation of interested regions, specifically, the heart chambers. We achieve this using an unsupervised adaptive thresh-

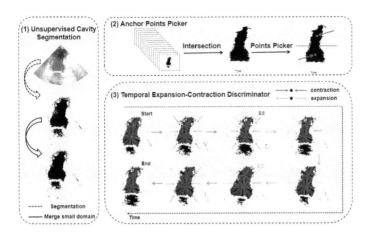

Fig. 2. The overview of our DDSB for unsupervised and training-free ED/ES detection.

old segmentation algorithm[1]. The result of this process is an initial segmentation sequence, $\mathcal{S}_i$, where the ventricular cavities are marked as '0' (indicating the absence of cardiac tissue), and the myocardial along with other areas are marked as '1' (indicating their presence).

To address the challenge posed by noise within the ventricular cavities, which can significantly disrupt further processing, we apply a threshold, s, to filter based on the area of the connected non-cavity regions. The process for obtaining the filtered segmentation, $\mathcal{S}_f$, is succinctly formulated as follows:

$$\mathcal{S}_f(p) = \mathcal{S}_i(p) \cdot \mathbf{1}_{\{\mathcal{A}(p) \geq s\}},$$

where $p = (x, y)$ denotes the pixel coordinates, $\mathcal{A}(p)$ quantifies the area of the connected non-cavity domain at pixel p, and $\mathbf{1}_{\{\cdot\}}$ equals 1 if '·' is true, otherwise it is 0. This strategy ensures that $\mathcal{S}_f$ achieves a refined segmentation by proficiently minimizing noise impacts.

2.2 Anchor Points Picker

For effective analysis of cardiac chamber dynamics, especially to distinguish the dilation and contraction phases, identifying anchor point(s) P_a within the heart chamber is crucial. These points act as references for observing boundary movements in relation to P_a. Initially, we select these points based on their persistent presence within the cardiac cavity across the video sequence. Hence, the potential

---

[1] 'cv2.adaptiveThreshold()' function as the OpenCV package, which adjusts thresholds based on local intensity variations to accurately segment cardiac structures.

positions, M_a, are identified as:

$$M_a = \left\{ p \mid \sum_{i=1}^{T} \mathcal{S}_f(i,p) \leq \text{Percentile}\left(\sum_{i=1}^{T} \mathcal{S}_f(i,\cdot), 1\%\right)\right\},$$

where Percentile$(\cdot, 1\%)$ calculates the 1% percentile over the set of summed values. To ensure that P_a is located within the cardiac cavity and not in other cavities, we further refine our selection. Among the four largest connected domains by area, we opt for the one nearest to the image's top center $p_{top} = (0, \frac{W}{2})$, where W is the frame width, as the final anchor point candidate. This decision takes advantage of the proximity of the center of the domain to the top center of the image, ensuring an optimal representation of the cardiac cavity:

$$P_a = \underset{\mathcal{C} \subseteq \mathcal{L}_4(M_a)}{\mathrm{argmin}}\ d(\text{Center}(\mathcal{C}), p_{top}),$$

with $\mathcal{L}_4(M_a)$ representing the four largest connected domains within M_a, and $d(\cdot, \cdot)$ denoting the distance between two points.

The ventricular center point serves as an optimal reference due to its stability and unobstructed perspective on various boundary locations, minimizing the likelihood of occlusion. To fully capture the deformation, the incorporation of additional anchor points is beneficial. Given the predominantly longitudinal configuration of the heart, we propose a strategy to partition P_a into t_a vertical segments. From each segment, we select a central point as an anchor, thereby acquiring t_a anchor points. We divide P_a into t_a regions vertically, denoted by $\{R_1, R_2, \ldots, R_{t_a}\}$. For each region R_i, we determine the central anchor point C_i as:

$$C_i = \text{Center}(R_i), \quad \text{for } i = 1, 2, \ldots, t_a,$$

where Center(R_i) computes the centroid of region R_i. Consequently, the set of anchor points $\{C_1, C_2, \ldots, C_{t_a}\}$ provides a detailed representation of cardiac deformation, enhancing the analysis of ventricular dynamics.

2.3 Temporal Expansion-Contraction Discriminator

To elucidate heart deformation, we introduce the *change description element*, $\delta(\theta, C_i, j)$, to quantify boundary distance changes between consecutive frames (x_j, x_{j+1}) along a specific direction θ, using the reference point C_i. Positive values of δ signify deformation, indicating either expansion or contraction of the heart. Given the potential challenges associated with the fidelity of unsupervised segmentation, we enhance the method's robustness by analyzing k directions, equally spaced, from each anchor point. Consequently, for each pair of adjacent frames, we compile:

$$L_i = \left\{ \delta\left(\frac{2\pi k_0}{k}, C_i, j\right) \mid k_0 \in \{1, \ldots, k\}, C_i \in \{C_1, \ldots, C_{t_a}\}\right\}.$$

Detecting a significant change in δ between two frames that exceeds a predefined threshold α is considered an anomaly. The expansion rate between frames is calculated by balancing the sums of positive and negative δ values, normalized by the total count of valid δ:

$$E_j = \frac{\sum_{\alpha > \delta(\cdot,\cdot,j) > 0} - \sum_{-\alpha < \delta(\cdot,\cdot,j) < 0}}{\sum_{|\delta(\cdot,\cdot,j)| < \alpha} + 1e-6}.$$

A negative E_j implies contraction, offering an immediate assessment of deformation dynamics between the frames.

To accurately capture the heart's relative size at a specific frame j, we define $A_j = \sum_{i=1}^{j} E_i$, where E_i represents the expansion rate between consecutive frames. This summation reflects the heart's cyclical pattern of expansion and contraction. Identifying the ED/ES phases is achieved by pinpointing the indices (i_0, j_0) that maximize the absolute difference in cumulative expansion rates, as given by:

$$(i_0, j_0) = \underset{i<j}{\operatorname{argmax}} |2A_i - 2A_j + A_T|,$$

where A_T is the total cumulative expansion rate at the final frame. The criterion $2A_{i_0} - 2A_{j_0} + A_T > 0$ indicates an initial phase of contraction followed by expansion, denoting (i_0, j_0) as (t_{ed}, t_{es}). Conversely, a negative value suggests an initial expansion followed by contraction, leading to $(t_{ed}, t_{es}) = (j_0, i_0)$. This methodology adeptly delineates the ED and ES phases by examining the heart deformation pattern across the sequence.

3 Dataset and Metrics

Dataset. In our research, we used the Echo-dynamic and CAMUS datasets for validation. However, we needed to adjust them for our experiments. For CAMUS, where ED and ES phases are at the start and end of the cycle, we mixed up the sequence to prevent models from just focusing on these endpoints. We did this by randomly selecting two points in the cycle, flipping the frames between them to form a varied sequence. In the Echo-dynamic dataset, where only one ED/ES pair is labeled, we removed sequences with too short intervals between these phases for consistency. Then, we chose sequences that included the labeled ED/ES pair and added variability by cutting the sequence at two random points. Regarding sequence length, the original CAMUS dataset had 500 samples, each 10 to 32 frames long. After our modifications, the average sequence length increased to about 36 frames, making the data more suitable for our analysis.

Metrics. In our evaluation, we measure accuracy using the mean absolute error (MAE) in frames, denoted by μ. This is calculated as the average absolute difference between the predicted frame index ($\tilde{t}$) and the true frame index (t) across all N samples in the dataset, specifically for the ED or ES frames.

Table 1. Comparative analysis of experimental results: A perspective on the necessity of training and the application of supervised learning approaches

Datasets (Training Set Num.)	Dataset Scale	Methods	Supervised	Trained	$\mu_{ED}\downarrow$	$\mu_{ES}\downarrow$
CAMUS (450)	100%	Kong et al. [11]	✓	✓	1.59	2.31
		Dezaki et al. [5]	✓	✓	1.44	1.99
		Singh et al. [19]	✓	✓	1.77	2.59
		Li et al. [13]	Semi	✓	2.23	2.73
	50%	Dezaki et al. [5]	✓	✓	2.97	3.48
	0%	Size-based	✗	✗	6.09	3.68
		DDSB (Ours)	✗	✗	2.27	1.29
Echo-dynamic (5970)	100%	Kong et al. [11]	✓	✓	1.35	2.76
		Dezaki et al. [5]	✓	✓	1.32	2.04
		Singh et al. [19]	✓	✓	1.98	2.56
		Li et al. [13]	Semi	✓	2.10	1.70
	50%	Dezaki et al. [5]	✓	✓	4.12	3.79
	0%	Size-based	✗	✗	12.54	9.24
		DDSB (Ours)	✗	✗	3.84	4.62

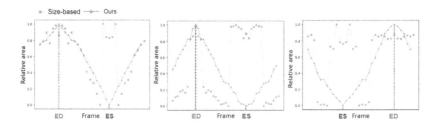

Fig. 3. Comparison of the size-based method with our DDSB.

4 Results and Discussion

4.1 Comparisons with State-of-the-Art Methods

In-Dataset Evaluation. We compared methods for detecting ED and ES in echocardiograms on the CAMUS and Echo-dynamic datasets. We rebuilt each deep learning model from the literature for fair comparison and set a baseline using a size-based approach, the size of the connected area with our anchor point, as shown in Table 1. On CAMUS, our unsupervised method not only competes well but also beats the top supervised model by 0.7 in μ_{ES}. Even with half the training data, it still surpasses Dezaki et al.'s model, reducing ED and ES errors by 0.7 and 2.19 frames, respectively. Compared to the size-based baseline, our method significantly cuts down errors by 3.82 frames for ED and 2.39 for ES. On the Echo-dynamic dataset, our method is slightly less effective than deep learning models trained with 5970 samples. However, with a smaller dataset of

Table 2. Cross-dataset evaluation results. We evaluate performance by testing a model trained on one dataset on a different dataset.

Metrics	CAMUS → Echo-dynamic		Echo-dynamic → CAMUS	
	μ_{ED}	μ_{ES}	μ_{ED}	μ_{ES}
Dezaki et al. [5]	4.24	**3.99**	3.13	4.21
DDSB (Ours)	**3.84**	4.62	**2.27**	**1.29**

2985 samples, our performance matches the best models. This highlights our method's value in scenarios with limited labeled data.

We compared our method to the size-based one through visuals in Fig. 3. Our method deals better with errors in the mask and shows fewer, smoother changes, proving its reliability. Overall, it is more precise and stable than the size-based method. Also, unlike deep learning methods that only offer a final result, our method can show the heart's changes dynamically, making it more flexible for different uses.

Cross-Dataset Evaluation. We tested our method's generalization against Dezaki et al.'s by training on one dataset and testing on another. Our method showed better accuracy in this cross-dataset setting, while it was slightly less effective in in-dataset evaluations, highlighting our strong generalization compared to deep learning-based methods in this task (Table 2).

4.2 Ablation Study

We conducted ablation studies on the CAMUS dataset to understand how certain hyper-parameters affect our method.

Effects of k Directions. We calculated k directional distances from each anchor point to the segmentation boundary. Our findings, shown in Table 3, indicate that our method's performance is stable across different k values, highlighting its robustness.

Effects of Change Threshold α. We defined any distance change between adjacent frames exceeding α as invalid. This approach led to an improvement

Table 3. Effects of k directions.

k	$\mu_{ED} \downarrow$	$\mu_{ES} \downarrow$
72	2.36	1.46
180	2.63	1.39
360	2.75	1.62

Table 4. Effects of change threshold α.

α	$\mu_{ED} \downarrow$	$\mu_{ES} \downarrow$
None	2.77	1.46
5	**2.27**	1.29
10	2.36	1.46
15	2.57	**1.28**

of approximately 0.5 for μ_{ED} and 0.17 for μ_{ES}, as documented in Table 4 Our method's effectiveness is not heavily dependent on the exact value of α, showcasing its flexibility.

5 Conclusion

We introduced DDSB, an innovative unsupervised and training-free method for phase detection. Our approach yields results that are on par with the latest supervised deep learning methods but without the need for training resources, such as computational power, data labeling, or extensive datasets. This advantage becomes particularly clear when data availability is limited. Moreover, DDSB allows for dynamic visualization of results, unlike deep learning methods that offer a single result, broadening its application. DDSB does not always outperform supervised learning with certain datasets. Yet, our framework has the potential to integrate with deep learning techniques, offering new perspectives for improving them, which is also in our plan.

References

1. Abboud, A.A., et al.: Automatic detection of the end-diastolic and end-systolic from 4D echocardiographic images. J. Comput. Sci. **11**(1), 230 (2015)
2. Barcaro, U., Moroni, D., Salvetti, O.: Automatic computation of left ventricle ejection fraction from dynamic ultrasound images. Pattern Recognit Image Anal. **18**, 351–358 (2008)
3. Chung, J., Gulcehre, C., Cho, K., Bengio, Y.: Empirical evaluation of gated recurrent neural networks on sequence modeling. arXiv preprint arXiv:1412.3555 (2014)
4. Darvishi, S., Behnam, H., Pouladian, M., Samiei, N.: Measuring left ventricular volumes in two-dimensional echocardiography image sequence using level-set method for automatic detection of end-diastole and end-systole frames. Res. Cardiovascular Med. **2**(1), 39–45 (2013)
5. Dezaki, F.T., et al.: Cardiac phase detection in echocardiograms with densely gated recurrent neural networks and global extrema loss. IEEE Trans. Med. Imaging **38**(8), 1821–1832 (2018)
6. Elman, J.L.: Finding structure in time. Cogn. Sci. **14**(2), 179–211 (1990)
7. Halliday, B.P., Senior, R., Pennell, D.J.: Assessing left ventricular systolic function: from ejection fraction to strain analysis. Eur. Heart J. **42**(7), 789–797 (2021)
8. Huang, G., Liu, Z., Van Der Maaten, L., Weinberger, K.Q.: Densely connected convolutional networks. In: Proceedings of the IEEE Conference on Computer Vision and Pattern Recognition, pp. 4700–4708 (2017)
9. Kachenoura, N., Delouche, A., Herment, A., Frouin, F., Diebold, B.: Automatic detection of end systole within a sequence of left ventricular echocardiographic images using autocorrelation and mitral valve motion detection. In: 2007 29th Annual International Conference of the IEEE Engineering in Medicine and Biology Society, pp. 4504–4507. IEEE (2007)
10. Kim, Y.: Convolutional neural networks for sentence classification. arXiv preprint arXiv:1408.5882 (2014)

11. Kong, B., Zhan, Y., Shin, M., Denny, T., Zhang, S.: Recognizing end-diastole and end-systole frames via deep temporal regression network. In: Ourselin, S., Joskowicz, L., Sabuncu, M.R., Unal, G., Wells, W. (eds.) MICCAI 2016. LNCS, vol. 9902, pp. 264–272. Springer, Cham (2016). https://doi.org/10.1007/978-3-319-46726-9_31
12. Leclerc, S., et al.: Deep learning for segmentation using an open large-scale dataset in 2D echocardiography. IEEE Trans. Med. Imaging **38**(9), 2198–2210 (2019)
13. Li, Y., Li, H., Wu, F., Luo, J.: Semi-supervised learning improves the performance of cardiac event detection in echocardiography. Ultrasonics p. 107058 (2023)
14. Mada, R.O., Lysyansky, P., Daraban, A.M., Duchenne, J., Voigt, J.U.: How to define end-diastole and end-systole? impact of timing on strain measurements. JACC: Cardiovascular Imaging **8**(2), 148–157 (2015)
15. Nabel, E.G.: Cardiovascular disease. N. Engl. J. Med. **349**(1), 60–72 (2003)
16. Otto, C.M.: Textbook of clinical echocardiography. Elsevier Health Sciences (2013)
17. Ouyang, D., et al.: Video-based AI for beat-to-beat assessment of cardiac function. Nature **580**(7802), 252–256 (2020)
18. Pu, B., Zhu, N., Li, K., Li, S.: Fetal cardiac cycle detection in multi-resource echocardiograms using hybrid classification framework. Futur. Gener. Comput. Syst. **115**, 825–836 (2021)
19. Singh, G., Darji, A.D., Sarvaiya, J.N., Patnaik, S.: Preprocessing and frame level classification framework for cardiac phase detection in 2D echocardiography (2023)
20. Wang, Z., Shi, J., Hao, X., Wen, K., Jin, X., An, H.: Simultaneous right ventricle end-diastolic and end-systolic frame identification and landmark detection on echocardiography. In: 2021 43rd Annual International Conference of the IEEE Engineering in Medicine & Biology Society (EMBC), pp. 3916–3919. IEEE (2021)
21. Zhang, S., Zheng, D., Hu, X., Yang, M.: Bidirectional long short-term memory networks for relation classification. In: Proceedings of the 29th Pacific Asia Conference on Language, Information and Computation, pp. 73–78 (2015)

Mitral Regurgitation Recogniton Based on Unsupervised Out-of-Distribution Detection with Residual Diffusion Amplification

Zhe Liu[1,2], Xiliang Zhu[1,2], Tong Han[1,2], Yuhao Huang[1,2], Jian Wang[3], Lian Liu[1,2,4], Fang Wang[5], Dong Ni[1,2], Zhongshan Gou[5(✉)], and Xin Yang[1,2(✉)]

[1] National-Regional Key Technology Engineering Laboratory for Medical Ultrasound, School of Biomedical Engineering, Medical School, Shenzhen University, Shenzhen, China
[2] Medical Ultrasound Image Computing (MUSIC) Lab, Shenzhen University, Shenzhen, China
xinyang@szu.edu.cn
[3] Key Laboratory for Bio-Electromagnetic Environment and Advanced Medical Theranostics, School of Biomedical Engineering and Informatics, Nanjing Medical University, Nanjing, China
[4] Shenzhen RayShape Medical Technology Co., Ltd, Shenzhen, China
[5] Center for Cardiovascular Disease, The Affiliated Suzhou Hospital of Nanjing Medical University, Suzhou, China
gzhongshan1986@163.com

Abstract. Mitral regurgitation (MR) is a serious heart valve disease. Early and accurate diagnosis of MR via ultrasound video is critical for timely clinical decision-making and surgical intervention. However, manual MR diagnosis heavily relies on the operator's experience, which may cause misdiagnosis and inter-observer variability. Since MR data is limited and has large intra-class variability, we propose an unsupervised out-of-distribution (OOD) detection method to identify MR rather than building a deep classifier. To our knowledge, we are the first to explore OOD in MR ultrasound videos. Our method consists of a feature extractor, a feature reconstruction model, and a residual accumulation amplification algorithm. The feature extractor obtains features from the video clips and feeds them into the feature reconstruction model to restore the original features. The residual accumulation amplification algorithm then iteratively performs noise feature reconstruction, amplifying the reconstructed error of OOD features. This algorithm is straightforward yet efficient and can seamlessly integrate as a plug-and-play component in reconstruction-based OOD detection methods. We validated the proposed method on a large ultrasound dataset containing 893 non-MR and 267 MR videos. Experimental results show that our OOD detection method can effectively identify MR samples.

Z. Liu, X. Zhu and T. Han contribute equally to this work.

1 Introduction

Mitral regurgitation (MR) is a significant heart valve disorder that becomes more common with advancing age. It may cause heart size changes, decreased cardiac function, and even life-threatening consequences [5]. Early detection and accurate assessment of MR enables timely intervention, holding vital clinical significance. Color Doppler echocardiography is the primary tool to diagnose MR, enabling visualization of blood flow direction and velocity within the heart that is not visible in standard B-mode ultrasound [22]. However, manual assessment is subjective and easily affected by experts' experience, potentially leading to diagnostic errors and observer inconsistency. Figure 1 shows the challenges in MR recognition, where similar features appear in both negative and positive cases.

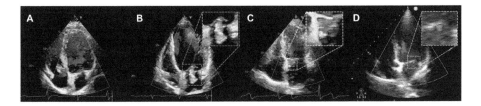

Fig. 1. The challenges encountered in OOD detection for our task. **A)** Negative sample, **B)** Positive sample, **C)** Negative sample with artifacts similar to regurgitation, **D)** Positive samples of reflux bundles resembling artifacts.

Deep learning techniques can potentially address the above problems, e.g., by training a classifier to recognize MR. However, the limited availability of MR data and its significant intra-class variability compared to normal data can constrain the performance of binary classifier [12]. For example, the imbalanced data distribution caused by the lack of MR data can severely affect the performance of supervised methods [8,13,21]. Recently, there have been significant advancements in out-of-distribution (OOD) detection, opening up new possibilities for MR recognition. The goal of OOD detection is to identify samples that are inconsistent with the distribution of in-distribution (ID) data (i.e., training data) [20]. It is noted that in MR recognition, numerous normal samples with similar features are regarded as ID data, and minimal but diverse MR cases are considered OOD data. Although promising, we found that studies applying OOD detection to MR recognition have not yet been reported.

Applying existing OOD research to MR recognition without modification may face several challenges. Unsupervised OOD detection methods for videos showed great potential, and they mainly included two main streams: frame prediction and generation techniques. Among frame prediction methods, those using optical flow maps [1,4] and long sequences of observations [2,18] were deemed inadequate for MR ultrasound videos due to noise, artifacts, and the cardiac cycle's motion pattern. Besides, generation-based approaches may encounter information bottlenecks caused by discrepancies between potential input dimensions and the dimensions reconstructed by the model [6,15]. The diffusion model solely employed the mean and standard deviation for denoising the reconstructed image, thereby mitigating the bottleneck issue mentioned previously to ensure the reconstruction of echocardiographic videos [6,13,19].

In this study, we propose a diffusion-based unsupervised OOD detection method for recognizing MR from echocardiography videos. Specifically, we first employ pre-trained models as feature extractors to extract features from video clips. Second, we train a diffusion model for feature-level reconstruction. Last, we design a residual accumulation amplification algorithm during the testing phase. This algorithm iteratively performs noise feature selection and reconstruction, amplifying the reconstructed error of OOD features. We validate the proposed method on a large four-chamber cardiac (4CC) ultrasound video dataset containing 893 non-MR and 267 MR videos, and the experimental results show that the proposed method is effective. We believe we are the first to explore OOD detection in MR recognition with echocardiography videos.

2 Methodology

As shown in Fig. 2, our method consists of a feature extractor, a feature reconstruction model, and a residual accumulation amplification algorithm. The feature extractor catches features from video clips and inputs them into the reconstruction model to rebuild the features. Then, during testing, under the proposed residual accumulation amplification algorithm, the OOD data can gradually be more distinguishable from the ID data.

2.1 Representative Feature Extractor

The feature extractor aims to capture essential knowledge from video clips. Let $V \in R^{k \times 3 \times h \times w}$ denote a video clip consisting of k consecutive RGB images with a resolution of $h \times w$ pixels, the feature $v \in R^l$ is obtained by feeding V into the feature extractor $f(\cdot)$. The architecture of the feature extractor is flexible and can be convolutional neural networks, long short-term memory, visual transformers, and their hybrids. Here, we tried several models, including 3D-ResNet18 [7], 3D-ResNet101 [7], Video MAE [16], and X-CLIP [14]. These models were pretrained on Kinetics-400 [3], a large-scale dataset containing more than 160,000 videos with 400 categories. We froze the models' weights without any fine-tuning.

2.2 Feature Reconstruction Model

The diffusion model gains the capability to generate diverse samples by disrupting training samples with noise and learning the reverse process. For a given video clip V, features are obtained by $v = f(V)$. The process of feature disruption can be formulated as a progressive addition of Gaussian noise (standard deviation: σ)

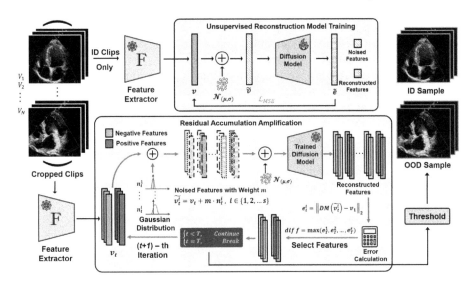

Fig. 2. Overview of our proposed method.

to the input data v_T obtained by sampling from a data distribution $p_{data}(v)$ with standard deviation σ_{data}. After adding noise, the data distribution could be represented as $p(v;\sigma)$ and $p(v;\sigma_{min}) \approx p_{data}(v)$ ideally. If $\sigma_{max} \gg \sigma_{data}$, the $p(v;\sigma)$ becomes isotropic Gaussian and allows to sample a data $v_0 \sim \mathcal{N}(0,\sigma_{max}I)$. The reversed process of diffusion model can be formulated as v_0 being gradually denoised with noise levels $\sigma_0 = \sigma_{max} > \sigma_1 > \ldots > \sigma_T = 0$ into new samples, until it becomes a data with data distribution of $p_{data}(v)$.

In detail, we follow the diffusion model in [10], treating $s_\theta(\tilde{v},\sigma)$ as a denoising function that directly estimates denoised samples with target function $\mathcal{L}$:

$$\mathcal{L} = E_{v \sim p_{data}} E_{\epsilon \sim \mathcal{N}(0,\sigma I)} \|s_\theta(v+\epsilon;\sigma) - v\|_2. \quad (1)$$

Besides, in [10], researchers untangled the design choices of previous diffusion models and offered a framework where each component (Sampling, Network and preconditioning, Training) can be adjusted independently. Therefore, in our study, the denoising function $s_\theta(\cdot)$ is formulated as follows:

$$s_\theta(v;\sigma) = c_{skip}(\sigma)v + c_{out}(\sigma) * F_\theta(c_{in}(\sigma)v; c_{noise}(\sigma)), \quad (2)$$

where $F_\theta(\cdot)$ is a multi-layer perceptron (MLP) with an encoder-decoder structure. $c_{skip}(\cdot)$ modulates the skip connection. $c_{out}(\cdot)$ and $c_{in}(\cdot)$ scale magnitudes of the noise variance. $c_{noise}(\cdot)$ scales θ to a suitable value for the input of $F_\theta(\cdot)$.

The reverse process of disrupting samples in the feature reconstruction of video clips can be formulated as:

$$\tilde{v}_t = \tilde{v}_{t-1} + \frac{\varepsilon}{2} s_\theta(\tilde{v}_{t-1},\sigma) + \sqrt{\varepsilon} z_t, \quad (3)$$

where $\varepsilon > 0$ is a predefined step size and $z_t \sim \mathcal{N}(0, I)$ is random term. With certain constraints, reversing multiple times at very small step sizes results in the final generated sample obeying the distribution $p_{data}(v)$.

The diffusion process is governed by multiple hyperparameters, with our main focus being the exploration of the effect of noise distribution on model performance. For fair comparisons, we follow [17] to determine other settings, including model structure, learning rate scheduler, etc. The diffusion model is central to the OOD detection model, used to recover the samples after being corrupted with Gaussian noise during the testing phase. The OOD information is derived by evaluating the discrepancy between samples before and after restoration.

2.3 Residual Accumulation Amplification

Inspired by the genetic accumulation of mutated genes with inheritance in genetics [9], we designed a residual accumulation amplification algorithm used in the testing stage. The core of the algorithm involves selecting MR recognition-related features and amplifying the reconstruction error via multiple iterations of noise feature reconstruction to enhance the distinction between OOD and ID data. In each iteration, we first randomly sample s Gaussian noises from the normal distribution $\mathcal{N}(\mu, \sigma)$, and then add them to the anchor feature to randomly destroy features that represent different information. The formula is as follows:

$$\widetilde{v}_t^i = v_t + m \cdot n_t^i, \quad i \in \{1, 2, ...s\}, \tag{4}$$

where n^i denotes the sampled Gaussian noise and m denotes the weight of noise. v and $\widetilde{v}^i$ represent the anchor and noise features, respectively. t is the current iteration number. In the first iteration, the anchor feature (v_1) is extracted from the video clips using the feature extractor. Those noise features are fed into the reconstruction model, and the reconstructed error is calculated as follows:

$$e_t^i = \|DM(\widetilde{v}_t^i) - v_1\|_2, \quad i \in \{1, 2, ...s\}, \tag{5}$$

where $DM(\cdot)$ denotes the reconstruction model and e_t^i denotes the reconstructed error in the t-th iteration. $\|\cdot\|_2$ represents the L_2 norm.

As shown in Fig. 3 (D), OOD data is more sensitive to reconstruction, and the difference between the reconstruction error of ID and OOD data will increase after multiple reconstructions. Therefore, we continue to add Gaussian noise to the current noise features for the next reconstruction. The noise feature corresponding to the maximum error is selected as the anchor feature in the next iteration. We believe that the larger error arises from the destruction of features unrelated to MR recognition. As these irrelevant features are eliminated after multiple iterative selections, the remaining features that are more related to MR recognition can enhance the detection efficacy:

$$v_{t+1} = \arg\max_{\widetilde{v}_t}(e_t). \tag{6}$$

The reconstruction error continues to be amplified until the stopping condition is reached, that is, $t = T$. The maximum error in the last iteration is regarded as the difficulty of feature reconstruction: $diff = max(e_T^1, e_T^2, ...e_T^s)$. The sample is classified according to a data-driven threshold $(thre)$, defined as $thre = \mu_{diff} + 0.001\sigma_{diff}$, where μ_{diff} and σ_{diff} are the mean and standard deviation of the difficulty of feature reconstruction in the validation set.

3 Experimental Results

3.1 Dataset and Implementations

The 4CC ultrasound video dataset was collected, comprising 893 non-MR and 267 MR videos. Regions of interest were extracted from frames originally sized at 1016×708 and resized to 224×224 for further processing. Multiple non-overlapping video clips were sampled from each video, and each clip consists of 16 consecutive frames, resulting in 3480 clips. Those clips cropped from MR videos are designated as positive samples (i.e., OOD samples), while others are taken as negative samples (i.e., ID samples). The dataset was divided into a training set with 1506 ID clips, a validation set with 621 clips (372 ID/249 OOD), and a testing set with 1353 clips (801 ID/552 OOD).

We adopted log-normal sampling (mean: −0.05, var: 1.5) for noise generation. Training epoch is set to 100 with *batch size* = 64, using the AdamW optimizer with *learning rate* = 0.0002. We used five metrics for evaluation: F1-score, Recall, Precision, Specificity, and Accuracy. All experiments were conducted on the *PyTorch* 2.0.1 framework with NVIDIA Tesla A40 GPUs.

3.2 Performance Comparisons

We compared our proposed method with a basic supervised classification method (3D-ResNet50) and other reconstruction-based unsupervised OOD detection methods under the same settings. In 3D-ResNet50, we gradually increased the number of OOD clips N in the training set. As shown in Table 1, the proposed method surpasses previous unsupervised methods and even surpasses 3D-ResNet50 with limited OOD data.

3.3 Comparison of Feature Extractor

We compared the MR recognition performance under different feature extractors. Thus, we used X-CLIP as the backbone in the following experiments for it achieved the best results in Table 2. The X-CLIP extractor with Transformer modules pretrained on a vast video dataset has a stronger capability to capture rich semantic information. This may be because text-video model fusion allows a more comprehensive understanding of complex features by capturing diverse aspects of the data. The performance gap between Convolution-based models and Transformer-based models can be attributed to the superior ability of Transformer with self-attention modules in learning long-range dependence features compared to convolutional modules. Particularly in video tasks with temporal information, Transformers exhibit advantages due to their enhanced capability to capture temporal dependencies and patterns.

Table 1. Comparison of our method with 3D-ResNet50 and unsupervised methods.

Supervised Methods	F1-score	Recall	Precision	Specificity	Accuracy
3D-ResNet50 (N = 240) [7]	0.0246	0.0128	0.3333	**0.9845**	0.6167
3D-ResNet50 (N = 360) [7]	0.7338	0.8213	0.6632	0.7461	0.7746
3D-ResNet50 (N = 480) [7]	**0.7549**	**0.8255**	0.6953	0.7798	**0.7971**
Unsupervised Methods	F1-score	Recall	Precision	Specificity	Accuracy
Diffusion + Star [17]	0.4921	0.5545	0.4423	0.5822	0.5719
Diffusion + Dyn [17]	0.4964	0.5446	0.4561	0.6118	0.5867
HF^{2}VAD [11]	0.4800	0.5347	0.4355	0.5858	0.5667
Ours	**0.5545**	**0.5883**	**0.5244**	**0.6342**	**0.6156**

Table 2. Performance comparison of OOD models with different feature extractors.

Model	F1-score	Recall	Precision	Specificity	Accuracy
3D-ResNet18 [7]	0.3193	0.255	0.4268	0.7653	0.5578
3D-ResNes101 [7]	0.3218	0.2459	0.4655	0.8065	0.5785
VideoMAE [16]	0.4124	0.4226	0.4028	0.5705	0.5104
X-CLIP [14]	**0.5116**	**0.5228**	**0.5009**	**0.6429**	**0.5941**

3.4 Diffusion Model Analysis

We tend to find the optimal T and $\mathcal{N}(\mu, \sigma)$ to preserve sufficient structural information of the video clip while disrupting potential anomaly information. Then, higher reconstruction errors can be obtained to judge the associated video frames as abnormal. For Gaussian distributions $\mathcal{N}(\mu, \sigma)$, μ ranges from -0.05 to -4 and σ ranges from 0 to 2. Besides, the start points T of the reverse process are set to 1, 5 and 10. For each value of T, 25 different Gaussian noise samples can be produced, as shown in Fig. 3 (A-C). It can be observed that $\mu = -0.05$, $\sigma = 1.5$ and $T = 5$ obtain the best score. Overall, increasing the T value while fixing μ

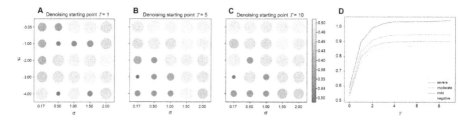

Fig. 3. (A-C) The effect of noise and the starting point of the reverse process. The larger the circle area, the greater the difference between the features. (D) Reconstruction difference among different subclasses of MR samples with feature amplification.

and σ can improve the OOD detection results. σ is the primary determinant of the model. Specifically, the performance gap for changing the value of σ is 3% in the F1-score, with all other hyperparameters remaining constant.

Table 3. Performance Comparison with different parameters. Bold: the final model adopted in our work. The feature extractor is a pretrained X-CLIP.

T	s	m	F1-score	Recall	Precision	Specificity	Accuracy
1	1	0	0.5116	0.5228	0.5009	0.6429	0.5941
2	1	0	0.5081	0.5118	0.5045	0.6554	0.5970
3	1	0	0.5036	0.5064	0.5009	0.6542	0.5941
4	1	0	0.5041	0.5046	0.5036	0.6592	0.5963
5	1	0	0.5246	0.5337	0.5158	0.6567	0.6067
10	1	0	0.5132	0.5118	0.5147	0.6650	0.6022
5	3	1	0.5426	0.5683	0.5191	0.6392	0.6104
5	3	1.5	0.5530	0.5847	0.5245	0.6367	0.6156
5	3	1.6	0.5547	0.5865	0.5261	0.6380	0.6170
5	3	1.7	0.5547	0.5865	0.5261	0.6380	0.6170
5	3	1.8	**0.5545**	**0.5883**	**0.5244**	**0.6342**	**0.6156**
5	3	1.9	0.5515	0.5847	0.5220	0.6330	0.6133
5	3	2	0.5526	0.5883	0.5210	0.6292	0.6126

Table 4. OOD performance of the reconstruction model with different modules.

Model	F1-score	Recall	Precision	Specificity	Accuracy
Baseline	0.5116	0.5228	0.5009	0.6429	0.5941
Baseline+RA	0.5246	0.5337	0.5158	**0.6567**	0.6067
Baseline+RA+NS	**0.5545**	**0.5883**	**0.5244**	0 6342	**0.6156**

3.5 Residual Accumulation Amplification Analysis

Residual accumulation amplification consists of residual accumulation (RA) and noise sampling (NS). This section analyzes the impact of both components on MR recognition performance. For RA, the reconstruction is iterated T times (i.e., 1, 2, 3, 4, 5, and 10). Table 3 shows that when T is set to 5, the diffusion model achieves optimal detection results on all metrics, and $T = 10$ causes a slight performance drop. For NS, we use the control variable method to assess how different parameters affect diffusion model performance. First, we set the noise sampling number $s = 3$ as a trade-off for computational efficiency. Then we change the weight of noise m to analyze its impact. Results in Table 3 show that consistent optimal results across all metrics can not be obtained with a single set of parameters. But in general, the number of iterations $T = 5$ can get good results easier. Finally, we choose the optimal hyperparameters ($m = 1.8$, $s = 3$, $T = 5$) to construct the OOD detection model. The impact of each module on model performance is shown in Table 4. RA improves the model performance by amplifing the difference between ID and OOD data reconstruction errors, which is consistent with Fig. 3 (D). However, its impact is limited by irrelevant features in the extracted information. NS addresses this issue effectively, significantly boosting performance by retaining MR recognition-relevant features through iterative selections.

4 Conclusion

We propose an innovative model tailored for recognizing MR echocardiographic videos using an unsupervised OOD detection approach. Our model exclusively leverages features extracted from ID samples by a fixed-weight feature extractor to train the diffusion model. Subsequently, by iteratively introducing disturbance, the model selects MR recognition-related features and amplifies the difference between reconstruction errors of OOD and ID data. Comprehensive experimentation conducted on MR datasets substantiates the efficacy of our proposed method. Future endeavors may explore expanding our approach to encompass multi-classification within the OOD samples.

Disclosure of Interests. The authors have no competing interests to declare that are relevant to the content of this article.

Acknowledgement. This work was supported by the grant from National Natural Science Foundation of China (12326619, 62101343, 62171290), Science and Technology Planning Project of Guangdong Province (2023A0505020002), Shenzhen-Hong Kong Joint Research Program (SGDX20201103095613036) and Suzhou Gusu Health Talent Program (GSWS 2022071, GSWS 2022072).

References

1. Baradaran, M., Bergevin, R.: Object class aware video anomaly detection through image translation. In: 2022 19th Conference on Robots and Vision (CRV), pp. 90–97. IEEE (2022)
2. Baradaran, M., Bergevin, R.: Future video prediction from a single frame for video anomaly detection. In: Bebis, G., et al. (eds.) ISVC 2023. LNCS, vol. 14361, pp. 472–486. Springer, Cham (2023). https://doi.org/10.1007/978-3-031-47969-4_37
3. Carreira, J., Zisserman, A.: Quo vadis, action recognition? A new model and the kinetics dataset. In: proceedings of the IEEE Conference on Computer Vision and Pattern Recognition, pp. 6299–6308 (2017)
4. Duman, E., Erdem, O.A.: Anomaly detection in videos using optical flow and convolutional autoencoder. IEEE Access **7**, 183914–183923 (2019)

5. El Sabbagh, A., Reddy, Y.N., Nishimura, R.A.: Mitral valve regurgitation in the contemporary era: insights into diagnosis, management, and future directions. JACC: Cardiovascular Imaging **11**(4), 628–643 (2018)
6. Graham, M.S., Pinaya, W.H., Tudosiu, P.D., Nachev, P., Ourselin, S., Cardoso, J.: Denoising diffusion models for out-of-distribution detection. In: Proceedings of the IEEE/CVF Conference on Computer Vision and Pattern Recognition, pp. 2947–2956 (2023)
7. Hara, K., Kataoka, H., Satoh, Y.: Learning spatio-temporal features with 3d residual networks for action recognition. In: Proceedings of the IEEE International Conference on Computer Vision Workshops, pp. 3154–3160 (2017)
8. Hendrycks, D., Gimpel, K.: A baseline for detecting misclassified and out-of-distribution examples in neural networks. arXiv preprint arXiv:1610.02136 (2016)
9. Holland, J.H.: Adaptation in Natural and Artificial Systems: An Introductory Analysis with Applications to Biology, Control, and Artificial Intelligence. MIT Press (1992)
10. Karras, T., Aittala, M., Aila, T., Laine, S.: Elucidating the design space of diffusion-based generative models. In: Advances in Neural Information Processing Systems, vol. 35, pp. 26565–26577 (2022)
11. Liu, Z., Nie, Y., Long, C., Zhang, Q., Li, G.: A hybrid video anomaly detection framework via memory-augmented flow reconstruction and flow-guided frame prediction. In: Proceedings of the IEEE/CVF International Conference on Computer Vision, pp. 13588–13597 (2021)
12. Lu, J.C., et al.: Simplified rheumatic heart disease screening criteria for handheld echocardiography. J. Am. Soc. Echocardiogr. **28**(4), 463–469 (2015)
13. Mishra, D., Zhao, H., Saha, P., Papageorghiou, A.T., Noble, J.A.: Dual conditioned diffusion models for out-of-distribution detection: application to fetal ultrasound videos. In: Greenspan, H., et al. (eds.) MICCAI 2023. LNCS, vol. 14220, pp. 216–226. Springer, Cham (2023). https://doi.org/10.1007/978-3-031-43907-0_21
14. Ni, B., et al.: Expanding language-image pretrained models for general video recognition. In: Avidan, S., Brostow, G., Cissé, M., Farinella, G.M., Hassner, T. (eds.) ECCV 2022. LNCS, vol. 13664, pp. 1–18. Springer, Cham (2022). https://doi.org/10.1007/978-3-031-19772-7_1
15. Serrà, J., Álvarez, D., Gómez, V., Slizovskaia, O., Núñez, J.F., Luque, J.: Input complexity and out-of-distribution detection with likelihood-based generative models. arXiv preprint arXiv:1909.11480 (2019)
16. Tong, Z., Song, Y., Wang, J., Wang, L.: Videomae: masked autoencoders are data-efficient learners for self-supervised video pre-training. Adv. Neural. Inf. Process. Syst. **35**, 10078–10093 (2022)
17. Tur, A.O., Dall'Asen, N., Beyan, C., Ricci, E.: Unsupervised video anomaly detection with diffusion models conditioned on compact motion representations. In: Foresti, G.L., Fusiello, A., Hancock, E. (eds.) pp. 49–62. Springer, Cham (2023). https://doi.org/10.1007/978-3-031-43153-1_5
18. Wang, X., et al.: Robust unsupervised video anomaly detection by multipath frame prediction. IEEE Trans. Neural Networks Learn. Syst. **33**(6), 2301–2312 (2021)
19. Wyatt, J., Leach, A., Schmon, S.M., Willcocks, C.G.: Anoddpm: anomaly detection with denoising diffusion probabilistic models using simplex noise. In: Proceedings of the IEEE/CVF Conference on Computer Vision and Pattern Recognition, pp. 650–656 (2022)
20. Yang, J., Zhou, K., Li, Y., Liu, Z.: Generalized out-of-distribution detection: a survey. arXiv preprint arXiv:2110.11334 (2021)

21. Zhang, J., Inkawhich, N., Linderman, R., Chen, Y., Li, H.: Mixture outlier exposure: towards out-of-distribution detection in fine-grained environments. In: Proceedings of the IEEE/CVF Winter Conference on Applications of Computer Vision, pp. 5531–5540 (2023)
22. Zoghbi, W.A., et al.: Recommendations for evaluation of the severity of native valvular regurgitation with two-dimensional and doppler echocardiography. J. Am. Soc. Echocardiogr. **16**(7), 777–802 (2003)

Deep Reinforcement Learning with Multiple Centerline-Guidance for Localization of Left Atrial Appendage Orifice from CT Images

Jongum Yoon[1], Sunghee Jung[2], and Byunghwan Jeon[1]($\boxtimes$)

[1] Division of Computer Engineering, Hankuk University of Foreign Studies, Yongin 17035, South Korea
bhjeon@hufs.ac.kr
[2] Department of Electronics Engineering, Kookmin University, Seoul 02727, Republic of Korea

Abstract. In patients with Atrial Fibrillation (AF), the Left Atrial Appendage (LAA) is a known site for blood clot formation, which can lead to strokes. The LAA closure procedure is effective in reducing the risk of stroke in patients with AF, and its success hinges on the precise placement of the closure device at the LAA orifice. In this paper, we introduce a novel framework for localizing the LAA orifice, based on Deep Reinforcement Learning (DRL) with guidance from multiple LAA centerlines. Our framework initiates with sophisticated segmentation and extraction of LAA centerlines, followed by the integration of these centerlines with the agent's estimated location determined by DRL. The proposed method is evaluated against a recent approach for LAA orifice localization and another existing DRL-based method, demonstrating superior performance with an accuracy of 2.76 mm $\pm$ 2.01. Our method is robust and can be seamlessly integrated into workstations for diagnosis and procedural planning.

Keywords: Left Atrial Appendage · Orifice Localization · Reinforcement Learning · Segmentation · Multiple Centerlines

1 Introduction

The left atrial appendage (LAA) exhibits a complex structure [1], and in patients with atrial fibrillation (AF), vortices within the LAA can lead to blood clot formation. These clots can travel through the bloodstream and potentially cause a stroke if they block the carotid artery [2,3].

LAA closure is a procedure in which an occlusion device is inserted into the left atrial appendage through a catheter to prevent stroke. Prior to the procedure, the size of the device is determined based on 3D CT images, enabling the creation of an accurate diagnosis and treatment plan, with particular emphasis on the placement of the device. As the device is typically positioned near the orifice of the LAA, accurately localizing the orifice and determining the vertical orientation of the orifice plane is crucial [4]. Clinical experts manually undertake this process using a workstation.

Several approaches have been proposed to detect and measure LAA landmarks from CT images. In early studies, most classical methods required user interaction due to the LAA's complex morphology [5,6]. Additionally, classical machine learning methods have been proposed for the automatic segmentation of the LAA [7,8]. Zheng et al. [8] introduced marginal space learning to estimate bounding boxes for multiple heart anatomies, followed by the application of region growing to segment the LAA within the localized bounding box. Cheng et al. [9] defined the ROI of the LAA based on four manual seeds and implemented a 2D-based convolutional neural network for the LAA region. Subsequently, a 3D-based conditional random field was applied to refine local details. More recently, You et al. [10] proposed a sophisticated segmentation method that focuses on the uncertain boundary of the LAA, utilizing a semantic difference module.

Leventić et al. [11] developed a semi-automatic method to segment the LAA using adaptive thresholding and sphere fitting, subsequently identifying the LAA centerline. The orifice's neck was estimated using a gradient-based weighting function. However, the centerline's accuracy is contingent on the precision of LAA segmentation. Moreover, for non-vascular types of LAA, such as cauliflower or cactus shapes, the LAA may not be adequately represented by a single centerline. Consequently, localizing the optimal LAA orifice location based solely on a single centerline may have inherent limitations.

Deep reinforcement learning (DRL)-based methods have recently been introduced to rapidly localize specific anatomies in 3D images, leveraging a learned policy [12–16]. DRL agents generally analyze 3D boxed local contexts, leading to effective convergence when targeting smaller anatomies with distinct features, such as coronary ostia [15]. However, for larger anatomies like the LA, optimal action selection becomes challenging due to the homogeneous intensity of surrounding information. Additionally, setting termination conditions within the DRL framework poses difficulties in real-world scenarios [13]. Given that agents typically circulate near target locations, accuracy is significantly influenced by the defined terminal state. The primary challenging issues that our work tries to solve are summarized as follows:

(1) A single CNN model may not sufficiently segment the LAA due to its complex morphology.
(2) Certain types of LAA cannot be represented by a single centerline, including cauliflower and cactus types.
(3) It is challenging to define termination conditions for DRL agents that wander around targets.

In this paper, we propose a fully automatic system to localize the LAA orifice, leveraging deep reinforcement learning (DRL) with guidance from multiple LAA centerlines. Figure 1 illustrates the workflow diagram of our proposed method. Initially, our system conducts LA segmentation, including the LAA, using a 2D U-Net, followed by a sophisticated refinement process with a 3D

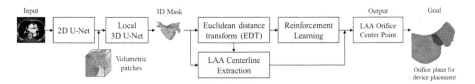

Fig. 1. A workflow of the proposed deep reinforcement learning approach with multiple centerline-guidence for LAA orifice center point detection.

U-Net to enhance the local details of the LAA. Subsequently, a Euclidean distance transform (EDT) map within the LA/LAA segment is generated. Based on this EDT map, several potential centerlines from the LAA tip to the LA center are extracted utilizing a sphere fitting method inspired by Leventić et al. [11]. The DRL framework then sequentially guides the agent towards the LAA orifice, integrating the EDT map with the extracted LAA centerlines to ensure robust termination at the LAA orifice. Compared to other recent methods, our proposed network demonstrates superior performance in the LAA orifice localization challenge.

2 Methods

LA/LAA Segmentation Using Multi-Context U-Net for Local Detail Refinement. For segmentation of left atrium (LA) and left atrial appendage (LAA), we first apply a U-Net [17] that takes 2-dimensional (2D) input to obtain initial segmentation mask M_{2D}. Subsequently, to address the relatively complex morphology of the LAA region, a local 3D U-Net [18] model, which leverages 3D volumetric data, is separately employed to refine the segmentation by correcting missing areas. For a 3D U-Net, volumetric patches of size 32^3 are extracted from the points $P = \{p_i | p_i = p_c + \mathbf{n}\}$ that are randomly chosen around the center of the LAA p_c, which is determined through Bayesian analysis of geometric relationships as described in Jeon et al. [19]. The points are not only randomly chosen but also fall within the prediction mask generated by the 2D U-Net as $p_i \in M_{2D}$. The predicted regions by 3D U-Net can be overlapped. The predicted softmax values are accumulated at the overlap regions and divided by a count map to find average softmax values. The binary mask M can be obtained based on the average softmax values. Additionally, the segmentation results M are further refined by eliminating false positive areas through connected component analysis (CCA).

Euclidean Distance Transform. The Euclidean distance transform (EDT) map is calculated from the obtained mask M by the method in the previous Sect. 2. In our framework, the EDT map is not only utilized to extract the centerline of LAA, but also serves as guidance for the agents to observe the local context. The EDT computes for each pixel in the mask the minimum Euclidean distance to the nearest foreground pixel. Given a binary mask M where foreground and background pixels are denoted by 1 and 0, respectively,

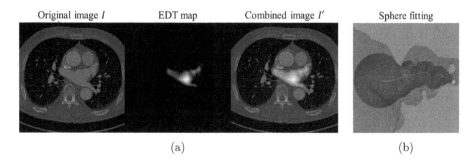

(a) (b)

Fig. 2. (a) shows the EDT mapping used for centerline extraction and agent guidance. (b) displays a visual representation of extracted centerline using the sphere fitting method.

the distance $D(x, y)$ for each background pixel at position (x, y) can be computed as follows:

$$D(x,y) = \min_{(x_f, y_f) \in F} \sqrt{(x - x_f)^2 + (y - y_f)^2}$$

where F represents the set of all foreground pixel positions, and (x_f, y_f) denotes the position of a foreground pixel. The original CT image I are combined with EDT map by $I' = \{I \cup \text{EDT map}\}$, as presented in Fig. 2a.

LAA Centerline Extraction. The LAA morphology is diverse and complex and is not actually a type of vascular structure [1]. In particular, in the case of LAA classified as cauliflower or cactus type, it is difficult to uniquely define a tip location of the LAA, so it is also ambiguous to define an unique centerline that is from an LAA tip to the center of LA. However, the neck area where the LAA and LA are connected has an anatomical structure that gradually narrows, therefore, there might be multiple paths that converge from multiple seed points to a center point of the LAA orifice. From the assumption above, we achieve multiple seed points of LAA from $(t) = p_c + \frac{v}{||v||}t$ where $v = p_c - q_c$, and p_c and q_c are a center points of LAA and LA, respectively. (t) casts a ray by increasing the parameter t to find a point s_c where it is located around the LAA tip but is inside of foreground based on M. Then, multiple seed points S are simply achieved by adding random noise to $S = \{s_i | s_i = s_c + \mathbf{n}\}$. s_i is uniformly spaced using Poisson disk sampling method and is also inside of foreground based on M.

Mainly two steps are repeated to find a centerline. From each seed point s_i, first, the radius size of a sphere model is increased to fit the sphere to the foreground region. Then, the furthest location is found among the locations with the maximum value of the EDT map within the fitted sphere. the mentioned steps are repeated while the radius size of spheres is lesser than a specific value. After the sphere fitting process is completed, all the center points of the spheres along the process are interpolated using B-spline curves $\gamma_i(t)$, a visualization of which is provided in Fig. 2b.

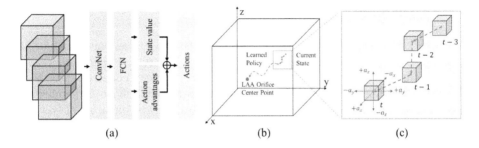

Fig. 3. (a) dueling DQN, Q-approximator in our framework, (b) environment with an example of a trajectory from a random location to a target location, (c) state s_t with three histories and directional actions are illustrated.

Multi-centerline Guided DRL for LAA Orifice Detection. We designed Markov decision process (MDP) for localization of LAA orifice and employed a DRL framework inspired by a dueling deep Q-learning method [13] to learn the agent to directly search to LAA orifice from a center of LA (Fig. 3(b)).

In a 3D CT images, state $s_t \in S$ is defined as a 3D boxed region of interest (RoI) centered around the agent location p_t and last three previous states in the sequence of the search trajectory (Fig. 3(c)). Considering some previous states together can stabilize the entire trajectory. Agents can observe combined image I' (Fig. 2a) as 3D contextual information based on these 3D boxed RoI as shown in Fig. 3.

The action set A consists of 3D directional actions and a stationary action as $A = \{\pm a_x, \pm a_y, \pm a_z, 0\}$ as shown in Fig. 3(c). Each action is represented as a 3D unit displacement vector $\bm{a}$ and the agent can change its state as $p_{t+1} = p_t + \rho\bm{a}$ where ρ is a step size ($\rho = 3$).

The reward function is employed for agents to have larger values when agents approaches to targets rather than a previous state as $R_{t+1} = ||\bm{p}_t - \bm{p}_{GT}||_2^2 - ||\bm{p}_{t+1} - \bm{p}_{GT}||_2^2$. This function can provide the distance based feedback to agents for every steps.

The dueling deep Q-network [20] is employed as a Q-approximator that separately modeled into the action advantage and state value as shown in Fig. 3(a). Assuming the agent is situated in a spacious anatomical area like LA, characterized by homogeneous image distribution, the agent's focus is solely on the features related to the direction towards the goal, considering the state value. The anatomical formations near the agent may act as either guidance or barriers. It's presumed that action advantages allows the agent to pay attention to the surrounding anatomical structures and select actions that facilitate the most efficient, shorter path to the target object. Each important point $\hat{p}_i$ along

$\gamma_i(t)$ is computed based on the estimated location by the agent with minimum distance. The average of $\hat{p}_i$ is our final estimated center point of LAA orifice.

3 Experiments and Results

Dataset. The evaluation of our proposed methodology was conducted using a dataset comprising 30 CT images, each from a different patient undergoing LAA closure procedures. This study, using retrospectively collected data from Severance Hospital, South Korea, was exempt from ethical review as it relied on anonymized pre-existing data for research purposes. For the purposes of training and validation, we allocated 20 CT images to the training set and designated the remaining 10 scans for testing. The images in this dataset feature a resolution range of 512×512 pixels, with varying slice thicknesses between 166 and 640 slices. The original voxel spacings across CT images varied within the ranges of 0.25 to 0.5 mm in the axial plane and 0.3 to 0.5 mm in both the coronal and sagittal planes. All the images are converted to isotropic based on the minimum spacing size of the axis.

Implementation Details. For the 2D U-Net model [17], the input dimensions were set to 512×512 pixels, and the model was trained over 100 epochs with a batch size of 8 and a learning rate of 1e−4. For the 3D U-Net [18], the input was adjusted to $32 \times 32 \times 32$ voxels, and it was trained for 300 epochs with a batch size of 48 and the same learning rate. Both models employed binary cross-entropy as the loss function and utilized the Adam optimizer for optimization.

In the reinforcement learning framework, the batch size was set to 48, and the replay memory capacity was 13,000. The target network update rate was 0.9, with a learning rate of 0.0005. The mean squared error (MSE) was used as the loss function, with the epsilon value for the exploration strategy decaying from 0.9 to 0.1. The initial memory was populated with 10,000 states, and the training extended over 10,000 epochs.

The computational experiments were facilitated by a NVIDIA Quadro RTX 8000 GPU. The software environment was TensorFlow 2.7.0, supported by CUDA 11.7.

Refinement of Segmentation. Segmentation performance was assessed against a Ground-Truth dataset manually annotated by medical experts. The evaluation metric employed for this comparison was the dice similarity coefficient (DSC), which quantifies the overlap between the predicted and true binary segmentation masks. By leveraging a multi-context framework that integrates a 2D U-Net with a local 3D U-Net, this method achieves an aggregate DSC of 90.22% across 10 test datasets, illustrating a significant improvement in accurately delineating complex anatomical structures. This approach outperforms traditional 2D-based methods, particularly in segmenting intricate regions, where the depth information provided by 3D data is crucial for enhanced precision. Quantitative details are presented in Table 1a, while Fig. 4a visually demonstrates the segmentation quality against expert-defined Ground-Truth.

Table 1. Performance of the proposed method on 10 test scans. (a) The DSC for the segmentation results. (b) The detection error of the LAA orifice center point in comparison with other methods.

	(a) DSC (%)		(b) Detection Error (mm)			
	2D U-Net	3D Refinement	Leventić et al. [11]	RL [13]	RL+1C	RL+5C
D1	87.30	94.88	6.25	4.07	3.37	3.87
D2	86.31	90.72	9.81	3.65	3.75	2.28
D3	85.89	87.24	11.06	3.35	2.43	1.77
D4	90.51	92.85	12.31	3.56	3.31	3.28
D5	90.71	92.33	0.86	2.20	1.40	1.40
D6	81.12	91.34	3.68	3.94	4.01	3.87
D7	90.90	91.28	5.27	9.80	13.49	7.56
D8	89.87	92.09	5.33	1.35	2.20	0.60
D9	72.87	83.36	2.19	3.86	2.19	1.23
D10	81.76	86.13	2.10	4.19	1.71	1.83
Avg	85.72 ± 5.74	90.22 ± 3.52	5.88 ± 3.97	3.99 ± 2.22	3.78 ± 3.52	2.76 ± 2.01

D: Data, RL: Reinforcement Learning, 1C: 1 Centerline, 5C: 5 Centerlines.

LAA Orifice Detection. Detection errors for the LAA orifice's center point were evaluated using the Euclidean distance between locations proposed by our method and those suggested by experts. Our approach, guided by 5 centerlines within a reinforcement learning framework, demonstrated an average distance error of 2.76 mm, marking a significant accuracy enhancement. In comparison, the rule-based method by Leventic reported a higher average error of 5.88 mm. It means that the generation of over-segmented masks, resulting from setting a low threshold in CT images and relying on EDT image-derived values for rule application, becomes particularly problematic in complex morphological structures.

A purely reinforcement learning-based method resulted in an average error of 3.99 mm. The nature of reinforcement learning, which requires distinct features for accurate target identification, poses challenges in large, feature-sparse areas, leading to approximate vicinity-based target point definitions and the associated errors. Nevertheless, the integration of 1 or 5 centerlines, as detailed in Table 1b, markedly enhances accuracy by diminishing the average distance error. As described in Sect. 2, the centerline extraction commences at the LAA's tip and extends towards the neck area. The use of a single centerline, particularly in cases with cauliflower or cactus-like morphologies of the LAA, complicates the definition of a singular seed point for initiating the centerline. This complexity resulted in an average error of 3.78 mm, with deviations significantly influenced by the initial seed points, notably in anatomically intricate LAAs. These deviations highlight the limitations inherent in the single centerline approach, where the initial positioning critically affects accuracy, especially in complex anatomical contexts. The adoption of the 5-centerline strategy advocated in this study mitigated the average error to 2.76 mm, evidencing its efficacy in achieving precise anatomical segmentation. Cases illustrating these complexities are presented in Fig. 4b.

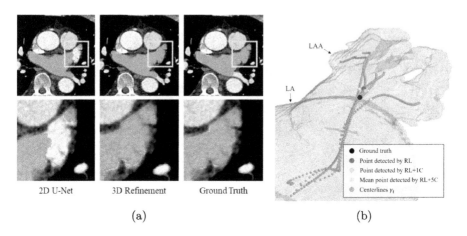

Fig. 4. Qualitative samples of the proposed method's results. (a) Results with only 2D U-Net, and results after applying 3D refinement for enhanced local detail. The bottom row shows an enlarged LAA view of the yellow bounded boxes. (b) Detected center points of the LAA orifice in a cauliflower-shaped LAA. Points detected using only RL compared with the mean point detected based on the centerline extracted from 5 initial seed points as proposed. (Color figure online)

4 Conclusion

This study introduces a novel DRL-based system augmented by multiple LAA centerline guidance for the precise localization of the LAA orifice. Our methodology, which synergizes advanced segmentation techniques through 2D U-Net with 3D U-Net refinement and sophisticated centerline extraction from the EDT map, demonstrates superior accuracy in LAA orifice detection, achieving a notable DSC of 90.22% and reducing the average distance error to 2.76 mm. These enhancements address key challenges in LAA segmentation and orifice detection, especially in anatomically complex cases, showcasing the proposed system's robustness and its potential for seamless integration into diagnostic and procedural planning workstations.

Acknowledgment. This study was supported by the Hankuk University of Foreign Studies Research Fund (2024).

References

1. Słodowska, K., et al.: Morphology of the left atrial appendage: introduction of a new simplified shape-based classification system. Heart Lung Circul. **30**(7), 1014–1022 (2021)
2. Di Biase, L., et al.: Stroke risk in patients with atrial fibrillation undergoing electrical isolation of the left atrial appendage. J. Am. Coll. Cardiol. **74**(8), 1019–1028 (2019)
3. Lakkireddy, D., et al.: Left atrial appendage ligation and ablation for persistent atrial fibrillation: the LAALA-AF registry. JACC Clin. Electrophysiol. **1**(3), 153–160 (2015)
4. Cabrera, J.A., Saremi, F., Sánchez-Quintana, D.: Left atrial appendage: anatomy and imaging landmarks pertinent to percutaneous transcatheter occlusion. Heart **100**(20), 1636–1650 (2014)
5. Otton, J.M., et al.: Left atrial appendage closure guided by personalized 3d-printed cardiac reconstruction. JACC Cardiovasc. Interv. **8**(7), 1004–1006 (2015)
6. Song, H., et al.: Morphologic assessment of the left atrial appendage in patients with atrial fibrillation by gray values-inverted volume-rendered imaging of three-dimensional transesophageal echocardiography: a comparative study with computed tomography. J. Am. Soc. Echocardiogr. **29**(11), 1100–1108 (2016)
7. Zheng, Y., Barbu, A., Georgescu, B., Scheuering, M., Comaniciu, D.: Four-chamber heart modeling and automatic segmentation for 3-d cardiac ct volumes using marginal space learning and steerable features. IEEE Trans. Med. Imaging **27**(11), 1668–1681 (2008)
8. Zheng, Y., Yang, D., John, M., Comaniciu, D.: Multi-part modeling and segmentation of left atrium in c-arm CT for image-guided ablation of atrial fibrillation. IEEE Trans. Med. Imaging **33**(2), 318–331 (2013)
9. Jin, C., et al.: Left atrial appendage segmentation using fully convolutional neural networks and modified three-dimensional conditional random fields. IEEE J. Biomed. Health Inform. **22**(6), 1906–1916 (2018)
10. You, X., et al.: Semantic difference guidance for the uncertain boundary segmentation of CT left atrial appendage. In: Greenspan, H., et al. (eds.) MICCAI 2023. LNCS, vol. 14226, pp. 121–131. Springer, Cham (2023). https://doi.org/10.1007/978-3-031-43990-2_12
11. Leventić, H., et al.: Left atrial appendage segmentation from 3D CCTA images for Occluder placement procedure. Comput. Biol. Med. **104**, 163–174 (2019)
12. Al, W.A., Yun, I.D., Chun, E.J.: Centerline depth world for left atrial appendage orifice localization using reinforcement learning. Comput. Med. Imaging Graph. **106**, 102201 (2023)
13. Alansary, A., et al.: Evaluating reinforcement learning agents for anatomical landmark detection. Med. Image Anal. **53**, 156–164 (2019)
14. Ghesu, F.-C., et al.: Multi-scale deep reinforcement learning for real-time 3D-landmark detection in CT scans. IEEE Trans. Pattern Anal. Mach. Intell. **41**(1), 176–189 (2017)
15. Jang, Y., Jeon, B.: Deep reinforcement learning with explicit spatio-sequential encoding network for coronary ostia identification in CT images. Sensors **21**(18), 6187 (2021)

16. Zhou, S.K., Le, H.N., Luu, K., Nguyen, H.V., Ayache, N.: Deep reinforcement learning in medical imaging: a literature review. Medical image analysis **73**, 102193 (2021)
17. Ronneberger, O., Fischer, P., Brox, T.: U-Net: convolutional networks for biomedical image segmentation. In: Navab, N., Hornegger, J., Wells, W.M., Frangi, A.F. (eds.) MICCAI 2015, Part III. LNCS, vol. 9351, pp. 234–241. Springer, Cham (2015). https://doi.org/10.1007/978-3-319-24574-4_28
18. Çiçek, Ö., Abdulkadir, A., Lienkamp, S.S., Brox, T., Ronneberger, O.: 3D U-Net: learning dense volumetric segmentation from sparse annotation. In: Ourselin, S., Joskowicz, L., Sabuncu, M.R., Unal, G., Wells, W. (eds.) MICCAI 2016. LNCS, vol. 9901, pp. 424–432. Springer, Cham (2016). https://doi.org/10.1007/978-3-319-46723-8_49
19. Jeon, B., Jang, Y., Shim, H., Chang, H.-J.: Identification of coronary arteries in CT images by Bayesian analysis of geometric relations among anatomical landmarks. Pattern Recogn. **96**, 106958 (2019)
20. Wang, Z., Schaul, T., Hessel, M., Hasselt, H., Lanctot, M., Freitas, N.: Dueling network architectures for deep reinforcement learning. In: International Conference on Machine Learning, pp. 1995–2003. PMLR (2016)

Lung-CADex: Fully Automatic Zero-Shot Detection and Classification of Lung Nodules in Thoracic CT Images

Furqan Shaukat[1,2](✉), Syed Muhammad Anwar[3,4], Abhijeet Parida[3], Van Khanh Lam[3], Marius George Linguraru[3,4], and Mubarak Shah[2]

[1] Faculty of Electronics and Electrical Engineering, University of Engineering and Technology, Taxila 47080, Pakistan
furqan.shoukat@uettaxila.edu.pk
[2] Center for Research in Computer Vision, University of Central Florida, Orlando, USA
[3] Sheikh Zayed Institute for Pediatric Surgical Innovation, Childrens National Hospital, Washington, DC, USA
[4] School of Medicine and Health Sciences, George Washington University, Washington, DC, USA

Abstract. Lung cancer has been one of the major threats to human life for decades. Computer-aided diagnosis can help with early lung nodule detection and facilitate subsequent nodule characterization. Large Visual Language models (VLMs) have been found effective for multiple downstream medical tasks that rely on both imaging and text data. However, lesion level detection and subsequent diagnosis using VLMs have not been explored yet. We propose CADe, for segmenting lung nodules in a zero-shot manner using a variant of the Segment Anything Model called MedSAM. CADe trains on a prompt suite on input computed tomography (CT) scans by using the CLIP text encoder through prefix tuning. We also propose, CADx, a method for the nodule characterization as benign/malignant by making a gallery of radiomic features and aligning image-feature pairs through contrastive learning. Training and validation of CADe and CADx have been done using one of the largest publicly available datasets, called LIDC. To check the generalization ability of the model, it is also evaluated on a challenging dataset, $LUNG_x$. Our experimental results show that the proposed methods achieve a sensitivity of 0.86 compared to 0.76 that of other fully supervised methods. The source code, datasets and pre-processed data can be accessed using the link: https://github.com/Precision-Medical-Imaging-Group/Lung-CADex.

Keywords: Lung Nodule Detection · Computer Aided Diagnosis · Malignancy prediction

---

Supported by Higher Education Commission Pakistan under NRPU Project no: 17019.

© The Author(s), under exclusive license to Springer Nature Switzerland AG 2025
X. Xu et al. (Eds.): MLMI 2024, LNCS 15241, pp. 73–82, 2025.
https://doi.org/10.1007/978-3-031-73284-3_8

1 Introduction

Lung cancer is one of the most commonly occurring cancers worldwide, with around 2.2 million new cases recorded in 2020 [22]. Approximately 225,000 people are diagnosed with lung cancer every year in the United States costing $12 billion [10,21]. A report published by the European Society for Medical Oncology (ESMO) [14], indicates that the highest incidence of lung cancer is found in central/eastern European and Asian populations. The situation in developing countries (such as in Asia) is particularly dire [11]. Studies have shown that the survival rate can be significantly improved by early detection of lung nodules [21]. However, detecting lung cancer at an early stage is challenging due to 1) a lack of symptoms in most patients 2) an extensive amount of data in terms of computed tomography (CT) scans, and 3) the interobserver variability in nodule detection. Computer-aided diagnosis (CAD) can help with early lung nodule detection and its subsequent nodule characterization. Generally, lung nodules have been characterized as the primary symptom of lung cancer, forming within and in the peripheries of the lungs. In radiology, CAD systems assist clinical experts in the analysis of medical images [20], to detect and localize structures of interest in a semi- or fully-automated manner.

With the five-year survival rate for patients diagnosed with lung cancer being the lowest among all other cancers, there is a clear need for the design of automated and robust systems that facilitate its early detection, diagnosis, and treatment. A lot of research has been done on developing a lung nodule detection system using conventional machine learning and deep learning techniques [8,12,17,23,24]. However, the availability of labeled data has been a major bottleneck for the generalization of these methods. Since, manually annotating this extensive amount of data present in CT scans is laborious and requires trained personnel. In addition, the diagnosis part has not been investigated that much, even though only a robust end-to-end system coupled with detection and diagnosis can be useful in real-time clinical scenarios [13].

With the recent advent of large visual language models (VLMs) and their ability to generalize to unseen tasks, there has been an effort within the healthcare community to adapt them to various medical downstream tasks. Specifically, the Segment Anything Model (SAM) [7] has performed exceptionally well on different segmentation tasks on normal real-world images. The domain gap present in these real-world and medical (such as radiology) images and its adaptability have been investigated in different variants of SAM [5,9,15,19]. However, these variants have largely been investigated at the anatomical level, and the lesion-level detection and subsequent characterization have not been investigated yet.

To solve the problem we propose solutions for both computer aided detection (CADe) and computer aided diagnosis (CADx) for lung cancer. These methods effectively segment nodules in a CT scan and can automatically generate the radiomic features associated with the lung nodules to be used to classify if a nodule is benign or malignant. **Our contribution** can be summarized as follows:

1. Development of a fully automatic end-to-end pipeline for lung cancer diagnosis.

2. Zero-shot detection and segmentation of nodules from lung CT images.
3. Design of a textual prompt suite and adaptation of MedSAM via prefix tuning.
4. Subsequent characterization of segmented lung nodules into benign/ malignant via image-feature contrastive learning.

2 Methodology

2.1 Dataset Curation and Pre-Processing

We have used the Lung Image Database Consortium (LIDC) dataset [2] for training and validation purposes. The LIDC data contains 1018 scans, along with nodule annotations by four expert radiologists in a double-blind fashion. We have considered a subset of LIDC called LUNA [18], consisting of 888 scans, which removes the inconsistent cases from the original dataset. The nodule inclusion criteria of LUNA have been followed in subsequent nodules' evaluations which gives a total of 1186 nodules. In addition to the nodule annotations, each radiologist has provided the radiological assessment of the respective nodules. Each nodule has been scored on a scale of 1–5 with respect to different radiological features namely subtlety, internal structure, roundness/sphericity, calcification, margin, lobulation, spiculation, and internal texture. The malignancy rating (1, 2, 3, 4, and 5) of the nodules is also given. We have taken four sets of values (by four radiologists) and averaged them to make a single reading for each nodule patch. For training purposes in the diagnosis (CADx) part, we have segmented the nodule patches from their median slice for their ground truth. We have made a gallery of the radiomic features of each nodule patch, which is being used as an input for the diagnosis model along with the nodule patches.

2.2 Network Architecture

The block diagram of our proposed model is shown in Fig. 1. Our proposed method consists of two stages. The first stage consists of the nodule detection module, whereas the second stage consists of the diagnosis module.

Nodule Detection (CADe): For the detection module, we have used MedSAM [9] in a zero-shot manner via prefix tuning and replaced its visual prompt with a textual prompt. The rationale behind doing this is to enforce the concept of fully automatic analysis for this downstream task of identification and classification of lung nodules. In its current setting, MedSAM provides excellent segmentation results; however, its semi-automatic nature can add the time and effort required for traversing through a complete scan, which can hamper the concept of computer-aided detection in real-time clinical scenarios. For example, if the radiologist or clinician has to provide the bounding box for each slice of CT by first looking into the targeted areas, then its utility for this specific downstream task remains limited. To overcome this challenge, in our proposed strategy we replace the visual prompt with the textual prompt suite.

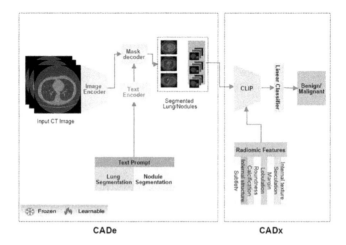

Fig. 1. Flow chart of the proposed method which consists of two stages namely CAD_e which refers to the detection phase and CAD_x which refers to the diagnosis phase.

With our modified architecture, a clinician or radiologist can see the segmented region of interest (i.e., lung/nodule) by giving textual prompts to the model, respectively. Keeping in view the nature of the task, we have trained the text encoder and fine-tuned it with our textual prompts, while the image encoder and mask decoder have been kept frozen. For quick reference, the image encoder (`ViT_Base: 12 layers`) transforms the input image into a high-dimensional image embedding space. The prompt encoder converts the user-provided textual prompt into feature representations using positional encoding. The mask decoder (`Lightweight: 2 Transformer layers`) combines the image embedding and prompt features through cross-attention. We would like to highlight that we have used a 2D MedSAM model here because of its application in the diagnosis pipeline, where slice-by-slice traversing can be more deterministic in the subsequent characterization of nodules. During the training phase, the model is trained with text-image pairs, and during inference, the model can identify and segment the nodules present in respective CT slices.

Nodule Classification (CADx): The second stage of our proposed method consists of the diagnosis module. Leveraging the exceptional alignment power of CLIP [16], we have treated the classification task as a retrieval task. We introduce incremental novelties to the CLIP model with modifications to efficiently perform nodule classification. In particular, the overreaching idea is to form a gallery of radiomic features associated with segmented nodule patches and align the model with text (radiomic features)-image (nodule patch) pairs. Once the model is trained, during inference, the model should output the most similar class from the radiomic feature gallery against the given nodule patch. Finally, this is then fed to a linear classifier for binary classification, i.e., benign or malignant. The rationale behind using this model rather than inputting the radiomic features directly into a linear classifier is to leverage the learned representations of the model, which can significantly improve the classification performance. We have used the Resnet50 [6] as an image encoder. To this end, the radiomic features are fed directly to the projection head, and the nodule patch is input to an image encoder. Both of these feature representations are then fused, a cosine similarity matrix is computed, and the most similar class is given as an output, which is finally fed to a linear classifier for final classification. Figure 2 shows the detailed architecture of the diagnosis part.

Overall, we develop a complete end-to-end pipeline, that includes both detection and diagnosis. One advantage of this strategy is that any false positives during the detection stage would be reduced or eliminated at the diagnosis stage. To ensure this, during the detection phase, we kept the precision high so that the model would not miss any potential nodules, and the false positives were reduced at the diagnosis stage.

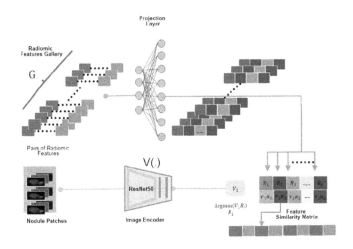

Fig. 2. Detailed architecture of the lung nodule classification method.

2.3 Loss Function

Nodule Segmentation: For the task of nodule segmentation, we have used the same loss function as in MedSAM. For reference, the unweighted sum of cross-entropy loss and dice loss were selected because of their robustness and wide adaptation in medical image segmentation tasks. Specifically, let S and G represent the segmentation result and ground truth, respectively. s_i, g_i represent the predicted segmentation and ground truth of voxel i, respectively. N is the number of voxels in the image I, the binary cross-entropy loss is given as:

$$L_{BCE} = -\frac{1}{N} \sum_{i=1}^{N} [g_i log s_i + (1 - g_i) log(1 - s_i)], \tag{1}$$

and dice loss is defined by

$$L_{Dice} = 1 - \frac{2\sum_{i=1}^{N} g_i s_i}{\sum_{i=1}^{N}(g_i)^2 + \sum_{i=1}^{N}(s_i)^2}, \quad (2)$$

The final loss L is defined by

$$L = L_{BCE} + L_{Dice}, \quad (3)$$

Nodule Classification: For nodule classification as benign or malignant, we used the loss function defined in CLIP [16] as:

$$L_{SCE} = \alpha \cdot L_{CE}(p,q) + \beta \cdot L_{RCE}(p,q), \quad (4)$$

where:

- $L_{CE}(p,q) = -\sum_i p_i \log(q_i)$ denotes the cross-entropy loss,
- $L_{RCE}(p,q) = -\sum_i q_i \log(p_i)$ denotes the reverse cross-entropy loss,
- α and β are the weighting coefficients for the cross-entropy loss and the reverse cross-entropy loss, respectively,
- p_i and q_i represent the predicted probabilities and the true probabilities, respectively.

3   Experimental Results and Discussion

The experimental setup consists of two parts. In the first phase, the detection model was trained to highlight areas within the input CT image that contain nodules, and hence segment the nodule patches using a text prompt. In the second phase, the segmented nodule patches were fed to the diagnosis model for final prediction as benign/malignant.

For the nodule detection model, we prefix-tune the text encoder of the Med-SAM with *"nodules"*, *"nodule"*, *"lung nodule"*, *"LUNG NODULE"*, *"Nodule"*, *"segment nodule"* as prompts to generate nodule segmentation. 70% of the LIDC scans were used for prefix-tuning and the rest were reserved for inference testing. The CTs are converted to slices and normalized with window level 40 and window width 400. The MedSAM model was optimized to minimize Dice and BCE loss using AdamW optimizer with a learning rate $5e^{-5}$ and batch size of 16 for 500 epochs. We train the diagnosis model using contrastive loss between image-radiomic feature pairs. A feature gallery was created using the average radiomic assessments of four expert radiologists for each nodule patch. The median nodule slice from the CT was resized to 96 × 96 to generate an image pair for the radiomic features. We train the Resnet50 image encoder and the radiomic projection layer to optimize for increased feature similarity. The image embedding dimension of 2048 and a radiomic feature embedding dimension of 8 were used. The model was trained to minimize the loss function from Eq. 4 for 500 epochs using a batch size of 8 (Fig. 3).

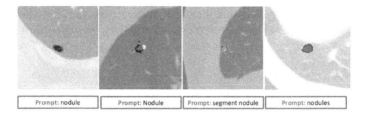

Fig. 3. Few examples of zero-shot segmentation results generated using the segmentation model and corresponding textual prompts. Green represents the segmented regions, and white represents the ground truth. (Color figure online)

During inference, the output of the detection phase, i.e., the segmented nodule patches, was given as input to the diagnosis model, and the output was then fed to a linear classifier for binary classification i.e., benign/malignant test nodule patches. Malignancy label (scaled 1–5) was used to generate a ground truth for the samples by thresholding. Sample scores greater than three were considered malignant. Samples with an average malignancy value of three were left out during the evaluation.

We have done our evaluations on two different datasets 1) 30% holdout samples of LIDC and 2) LUNG$_x$ [1] which contains hard malignancy labels (pathology proven) of 73 nodules. Standard performance metrics, namely area under the curve, sensitivity, accuracy, and specificity, were used for evaluation. We have compared our test results with two notable recent studies [3,4]. However, we would like to note that this is not a one-to-one comparison, keeping in mind the semi-automatic and fully supervised nature of the other two studies. The test results of the two studies reported by [4] along with ours are shown in Table 1 for comparison. It can be seen that our proposed method performs at par with these fully supervised methods. For the LIDC test dataset, our method outperforms other methods in terms of sensitivity and accuracy, even though our test dataset consists of 264 nodules (4 times large) as compared to their results, which were reported on a dataset containing 72 nodules with pathology-given labels. For this, we selected LUNG$_x$ dataset, which contains 60 contrast-enhanced CT scans, with 73 pathology-proven nodules with their hard labels. Keeping in view the notion of zero-shot learning, we do not use the 10 CT scans of this dataset given for calibration which might be a factor in a slight drop of AUC and sensitivity as compared to the LIDC test dataset shown in Table 1. However, the overall results on this dataset show the ability of our method to generalize even to unseen data coming from different image acquisition protocols and with other parameters. Our method outperforms [4] in terms of accuracy and specificity and performs at par with [3] in terms of AUC for LUNG$_X$.

Ablation Studies: Since we can query top-k radiomic feature matches from CADe, we conduct an ablation study for the 'k' radiomic feature required for a good malignancy prediction. The results have been summarized in Table 2. It can be seen that k=5 (our picked model) gives us the best sensitivity compared to others, which reflects that our diagnosis model performs five-shot learning. In addition, **k=9** achieves the best **AUC of 0.81** which outperforms all other baselines, but due to its lower sensitivity, we have not picked that model as the best-performing model.

Table 1. Comparison of malignancy prediction metrics, **N** stands for number of nodules. AUC: area under receiver operating characteristic curve.

Test results on LIDC dataset				
Network	AUC	Accuracy	Sensitivity	Specificity
CIRD [4] (N = 72)	0.73	0.68	0.81	0.57
LungX [3] (N = 72)	0.68	0.68	0.78	0.55
Ours (N=264)	0.69	**0.71**	**0.86**	0.56
Test results on LUNGx (N = 73)				
Network	AUC	Accuracy	Sensitivity	Specificity
LungX [3]	0.670	–	–	–
CIRD [4]	0.733	68.49	80.56	56.76
Ours	0.656	**70.59**	66.67	**73.33**

Table 2. Comparison of number of learning examples used for CLIP contrastive learning. AUC: area under receiver operating characteristic curve, ACC: accuracy.

	Number of learning examples (k)								
	1	2	3	4	5	6	7	8	9
AUC	0.621	0.721	0.765	0.698	**0.698**	0.748	0.763	0.792	0.810
Sensitivity	86.67	73.33	66.67	100.00	**86.77**	80.00	66.67	66.67	66.67
Specificity	37.50	68.62	75.00	50.00	**56.33**	56.33	62.50	62.50	62.50
F1	68.40	70.96	68.95	78.94	**74.25**	70.58	64.50	64.50	64.50
ACC	61.29	70.96	70.96	74.16	**70.96**	67.74	64.51	64.50	64.50

Model limitations: One of the main limitations of this work is the limited annotated data which would have affected the system's overall performance. Another limitation could be the reliance on weak labels for classification phase. With these large foundation models, large, balanced, and well-annotated/strong-labeled data can certainly increase the performance of the system. Another direction for future work can be the integration of Electronic Medical Records into the pipeline for final diagnosis to fully exploit the potential of these multimodal models.

4 Conclusions

In this paper, we present an end-to-end pipeline using both CADe and CADx for lung nodule detection and its subsequent malignancy characterization. We used a variant of the Segment Anything Model called MedSAM in a zero-shot manner for the detection part and a CLIP model for further characterization of nodules into benign and malignant. We replaced the visual prompts of MedSAM with textual prompts and designed a prompt suite for this specific downstream task. After segmenting the nodule patches, we formed a radiomic feature gallery and trained the modified CLIP model with nodule patches and their associated radiomic feature sets. During inference, the model gave the most similar class fed to a linear classifier for the final decision. Our results have shown significant value in the detection of large and diverse data. The proposed tool can be used for early lung cancer screening in an end-to-end manner.

References

1. Armato, S.G., III., et al.: Lungx challenge for computerized lung nodule classification. J. Med. Imaging **3**(4), 044506–044506 (2016)
2. Armato, S.G., III., et al.: The lung image database consortium (LIDC) and image database resource initiative (IDRI): a completed reference database of lung nodules on CT scans. Med. Phys. **38**(2), 915–931 (2011)
3. Causey, J.L., et al.: Highly accurate model for prediction of lung nodule malignancy with CT scans. Sci. Rep. **8**(1), 9286 (2018)
4. Choi, W., Dahiya, N., Nadeem, S.: Cirdataset: a large-scale dataset for clinically-interpretable lung nodule radiomics and malignancy prediction. In: Wang, L., Dou, Q., Fletcher, P.T., Speidel, S., Li, S. (eds.) MICCAI 2022. LNCS, vol. 13435, pp. 13–22. Springer, Cham (2022). https://doi.org/10.1007/978-3-031-16443-9_2
5. Gong, S., et al.: 3dsam-adapter: holistic adaptation of SAM from 2D to 3D for promptable medical image segmentation. arXiv preprint arXiv:2306.13465 (2023)
6. He, K., Zhang, X., Ren, S., Sun, J.: Deep residual learning for image recognition. In: Proceedings of the IEEE Conference on Computer Vision and Pattern Recognition, pp. 770–778 (2016)
7. Kirillov, A., et al.: Segment anything. arXiv preprint arXiv:2304.02643 (2023)
8. Liu, S., et al.: No surprises: training robust lung nodule detection for low-dose CT scans by augmenting with adversarial attacks. IEEE Trans. Med. Imaging **40**(1), 335–345 (2020)
9. Ma, J., He, Y., Li, F., Han, L., You, C., Wang, B.: Segment anything in medical images. Nat. Commun. **15**(1), 654 (2024)
10. Mariotto, A.B., Robin Yabroff, K., Shao, Y., Feuer, E.J., Brown, M.L.: Projections of the cost of cancer care in the united states: 2010–2020. J. Natl Cancer Inst. **103**(2), 117–128 (2011)
11. Moore, M.A., et al.: Cancer epidemiology in mainland south-east Asia-past, present and future. Asian Pac. J. Cancer Prev. **11**(Suppl 2), 67–80 (2010)
12. Niu, C., Wang, G.: Unsupervised contrastive learning based transformer for lung nodule detection. Phys. Med. Bio. **67**(20), 204001 (2022)
13. Ozdemir, O., Russell, R.L., Berlin, A.A.: A 3D probabilistic deep learning system for detection and diagnosis of lung cancer using low-dose CT scans. IEEE Trans. Med. Imaging **39**(5), 1419–1429 (2019)

14. Planchard, D., et al.: Metastatic non-small cell lung cancer: ESMO clinical practice guidelines for diagnosis, treatment and follow-up. Ann. Oncol. **29**, iv192–iv237 (2018)
15. Qiu, Z., Hu, Y., Li, H., Liu, J.: Learnable ophthalmology SAM. arXiv preprint arXiv:2304.13425 (2023)
16. Radford, A., et al.: Learning transferable visual models from natural language supervision. In: International Conference on Machine Learning, pp. 8748–8763. PMLR (2021)
17. Setio, A.A.A., et al.: Pulmonary nodule detection in CT images: false positive reduction using multi-view convolutional networks. IEEE Trans. Med. Imaging **35**(5), 1160–1169 (2016)
18. Setio, A.A.A., et al.: Validation, comparison, and combination of algorithms for automatic detection of pulmonary nodules in computed tomography images: the luna16 challenge. Med. Image Anal. **42**, 1–13 (2017)
19. Shaharabany, T., Dahan, A., Giryes, R., Wolf, L.: Autosam: adapting SAM to medical images by overloading the prompt encoder. arXiv preprint arXiv:2306.06370 (2023)
20. Shaukat, F., Raja, G., Frangi, A.F.: Computer-aided detection of lung nodules: a review. J. Med. Imaging **6**(2), 020901–020901 (2019)
21. Siegel, R.L., Miller, K.D., Jemal, A.: Cancer statistics, 2019. CA: Can. J. Clin. **69**(1), 7–34 (2019)
22. Sung, H., et al.: Global cancer statistics 2020: globocan estimates of incidence and mortality worldwide for 36 cancers in 185 countries. CA Can. J. Clin. **71**(3), 209–249 (2021)
23. Tang, H., Zhang, C., Xie, X.: NoduleNet: decoupled false positive reduction for pulmonary nodule detection and segmentation. In: Shen, D., et al. (eds.) MICCAI 2019, Part VI. LNCS, vol. 11769, pp. 266–274. Springer, Cham (2019). https://doi.org/10.1007/978-3-030-32226-7_30
24. Zhu, W., Liu, C., Fan, W., Xie, X.: Deeplung: deep 3D dual path nets for automated pulmonary nodule detection and classification. In: 2018 IEEE Winter Conference on Applications of Computer Vision (WACV), pp. 673–681. IEEE (2018)

CIResDiff: A Clinically-Informed Residual Diffusion Model for Predicting Idiopathic Pulmonary Fibrosis Progression

Caiwen Jiang[1,2], Xiaodan Xing[2], Zaixin Ou[1], Mianxin Liu[4], Walsh Simon[2], Guang Yang[2,5,7,8], and Dinggang Shen[1,3,6(✉)]

[1] School of Biomedical Engineering and State Key Laboratory of Advanced Medical Materials and Devices, ShanghaiTech University, Shanghai, China
{jiangcw,dgshen}@shanghaitech.edu.cn
[2] Bioengineering Department and Imperial-X, Imperial College London, London, UK
g.yang@imperial.ac.uk
[3] Shanghai Clinical Research and Trial Center, Shanghai 201210, China
[4] Shanghai Artificial Intelligence Laboratory, Shanghai 200232, China
[5] National Heart and Lung Institute, Imperial College London, London, UK
[6] Shanghai United Imaging Intelligence Co., Ltd., Shanghai, China
[7] Cardiovascular Research Centre, Royal Brompton Hospital, London, UK
[8] School of Biomedical Engineering and Imaging Sciences, King's College London, London, UK

Abstract. The progression of Idiopathic Pulmonary Fibrosis (IPF) significantly correlates with higher patient mortality rates. Early detection of IPF progression is critical for initiating timely treatment, which can effectively slow down the advancement of the disease. However, the current clinical criteria define disease progression requiring two CT scans with a one-year interval, presenting a dilemma: *a disease progression is identified only after the disease has already progressed*. To this end, in this paper, we develop a novel diffusion model to accurately predict the progression of IPF by generating patient's follow-up CT scan from the initial CT scan. Specifically, from the clinical prior knowledge, we tailor improvements to the traditional diffusion model and propose a Clinically-Informed Residual Diffusion model, called CIResDiff. The key innovations of CIResDiff include 1) performing the **target region pre-registration** to align the lung regions of two CT scans at different time points for reducing the generation difficulty, 2) adopting the **residual diffusion** instead of traditional diffusion to enable the model focus more on differences (i.e., lesions) between the two CT scans rather than the largely identical anatomical content, and 3) designing the **clinically-informed process** based on CLIP technology to integrate lung function information which is highly relevant to diagnosis into the reverse process for assisting generation. Extensive experiments on clinical data demonstrate that our approach can outperform state-of-the-art methods and effectively predict the progression of IPF.

G. Yang and D. Shen are co-senior last authors.

Keywords: Prediction of pulmonary fibrosis progression · Residual diffusion model · Clinically Informed · CLIP-based text processing

1 Introduction

Idiopathic Pulmonary Fibrosis (IPF) is a severe and irreversible lung disease that scars and thickens lung tissues, leading to respiratory difficulties [14,20]. Timely treatment of IPF can effectively slow down its process and improve patients' quality of life [4,21]. The progression of IPF can either remain stable or exacerbate over time. Consequently, to save healthcare expenses, in some countries (especially those with universal healthcare like the UK), antifibrotic treatment is initiated only if the IPF is confirmed to exacerbate over time. This approach presents a challenge: treatment is delayed until the fibrosis has advanced, losing early intervention opportunities for patients at high risk but not yet showing significant progression. In this context, predicting the progression of IPF in advance is extremely important for enabling timely treatment and reducing healthcare costs.

A feasible approach for predicting the progression of IPF is to generate the follow-up CT scan from the initial CT scan. This method of disease prediction or diagnosis through generation has already achieved success in numerous studies [5,6,10,11,13]. For instance, Han et al. adopt the regularized generative adversarial networks to generate images of future time points for predicting the risk of osteoarthritis [6]. Jiang et al. employ a transformer-based generative adversarial network to generate dual-energy CT images from single-energy CT for diagnosing postoperative cerebral hemorrhage [10]. Moreover, we choose to generate the follow-up CT scan rather than directly predicting disease progression from the initial CT scan considering the following facts: 1) The use of follow-up CT scan is more clinically natural in terms of diagnosis and thus more subjectively convincing. 2) The generation of follow-up CT scan can better exploit in the information entailed initial CT scan, thus more likely to produce a correct diagnosis. 3) The generated follow-up CT scan can provide additional information, such as the location of the lesion area.

Among the existing image generation techniques, the diffusion model has shown large potential and obtained great success [7,19,23]. They accomplish this by converting complex generation tasks into a series of simpler denoising tasks, enabling more stable and detailed generation. Meanwhile, clinical observations reveal the following prior information. **First**, due to patient posture and physiological movement, there is a significant spatial difference between the two CT scans. **Second**, as both scans are from the same individual, the majority of their content (i.e., the anatomical structure) is identical, while the lesion areas we focus on only constitute a minor part. **Lastly**, patients typically undergo lung function tests during their initial CT scan, and the information from these tests is highly relevant to the progression of IPF.

Taking all into consideration, in this paper we propose a Clinically-Informed Residual Diffusion model (CIResDiff) based on the traditional diffusion model to predict the progression of IPF. Specifically, we improve the traditional diffusion model by 1) performing target region pre-registration for the two CT scans to enable voxel-to-voxel generation for reducing the complexity, 2) adopting residual diffusion to make the model's generation process focus more on the differences (i.e., the lesion areas) between the two CT scans for more precise generation of target areas and fast model inference, and 3) designing a clinically-informed process based on CLIP [17] technology to capture the correlation between lung function test information and corresponding CT scans, and utilizing this correlation to guide the generation for producing images with higher diagnostic value.

The main contributions of our work include: i) the first attempt to predict the progression of IPF by generating follow-up CT scans from initial scans, ii) developing a novel diffusion model (CIResDiff) tailored for this task based on clinical prior knowledge, and iii) demonstrating the effectiveness of our CIResDiff on collected datasets by extensive experiments.

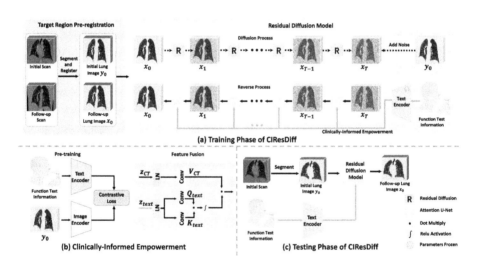

Fig. 1. Overview of proposed CIResDiff. (a) and (c) provide the framework of CIResDiff as well as depict its implementation during both the training and testing phases, and (b) illustrates the details of clinically-informed process.

2 Method

Our method is illustrated in Fig. 1. First, we extract and align the lung region from the two CT scans by the target region pre-registration. Then, in the residual diffusion process, the follow-up lung image x_0 is converted to the noisy initial lung image x_T by incrementally incorporating the differences of two CT scans. Subsequently, the noisy initial lung image x_T undergoes a series of reconstruction steps guided by the clinically-informed process until it's converted back to the follow-up lung image x_0. Note that only the reverse process is involved during the testing phase. In the following, we introduce the details of target region pre-registration, residual diffusion, and CLIP-based text processing module.

2.1 Target Region Pre-registration

Voxel-to-voxel supervised generation is less challenging compared to direct unsupervised generation. Hence, we employ a target area pre-registration to obtain spatially-aligned lung image pairs from the initial and follow-up CT scans for model training.

Specifically, for the input of two CT scans, to eliminate background interference, we first use the TotalSegmentator [22], an open-access tool based on nnU-Net and trained with more than one thousand samples, to segment the left and right lung regions from both CT scans. Then, we perform a dilation operation on the segmented results to preserve lung-surrounding tissues relevant to diagnosis. Subsequently, we apply an affine registration method [1] to align the segmented left and right lung regions individually. Finally, we merge the aligned left and right lung regions to obtain spatially-aligned lung region image pairs for model training.

2.2 Residual Diffusion

In the process of generating follow-up scans from initial scans, the main learning objective for the model is the lesion areas that are crucial for diagnosis but occupy a small proportion. Thus, we integrate residual learning with the diffusion model to enhance the model's ability to learn changes in such lesion areas, proposing a residual diffusion strategy. This strategy enables the diffusion model to exclusively learn the residuals between the initial and follow-up scans.

Unlike the traditional diffusion process, which disrupts the input initial lung image x_0 into pure Gaussian noise, our proposed residual diffusion converts the input initial lung image x_0 into noisy follow-up lung image x_T. Specifically, we first calculate the difference $e_0 = y_0 - x_0$ between the initial lung image y_0 and the corresponding follow-up lung image x_0, and then apply a shifting sequence $\{\eta_t\}_{t=1}^{T}$ to incrementally add this difference e_0 to x_0. In this way, the residual diffusion process can be formulated as follows:

$$q(x_t|x_0, y_0) = \mathcal{N}(x_t; x_0 + \eta_t e_0, k^2 \eta_t I), \quad t = 1, 2, ..., T, \quad (1)$$

where t is the timestep which satisfies $\eta_1 \to 0$, $\eta_T \to 1$, k is a hyper-parameter controlling the noise variance, and I is the identity matrix.

In this context, the reverse process, corresponding to the residual diffusion process, involves recovering the follow-up lung image from the noisy initial lung image rather than from pure noise. Such reverse process can estimate the posterior distribution $p(x_0|y_0)$ via the following formulation:

$$p(x_0|y_0) = \int p(x_T|y_0) \prod_{t=1}^{T} p_\theta(x_{t-1}|x_t, y_0) \mathrm{d}x_{1:T}, \qquad (2)$$
$$p_\theta(x_{t-1}|x_t, y_0) = \mathcal{N}(x_{t-1}; \mu_\theta(x_t, y_0, t), \Sigma_\theta(x_t, y_0, t)),$$
$$p(x_T|y_0) \approx \mathcal{N}(x_T; y_0, k^2 I),$$

where $\Sigma_\theta(x_t, y_0, t)$ is a fixed variance and $\mu_\theta(x_t, y_0, t)$ can be reparameterized as follows:

$$\mu_{\theta(x_t,y_0,t)} = \frac{\eta_{t-1}}{\eta_t} x_t + \frac{\eta_t - \eta_{t-1}}{\eta_t} f_\theta(x_t, y_0, t). \qquad (3)$$

Here, f_θ is a deep neural network with parameter θ, aiming to $\hat{x}_0$, In our implementation, we utilize the attention U-Net [15] as f_θ. The reverse process can be trained through the following reconstruction loss:

$$\mathcal{L}_{rec} = \min_\theta \sum_t \|f_\theta(x_t, y_0, t) - x_0\|_2^2 \qquad (4)$$

2.3 Clinically-Informed Process

During the diagnosis of IPF, lung function tests are typically conducted in addition to two CT scans. The information derived from these tests is highly relevant to the lung anatomical structure. Therefore, we believe that capturing the correlation between function test information and the corresponding CT scan, and using it to guide the generation process, can aid in producing CT scans with higher diagnostic value. To achieve this, we design a clinically-informed process based on Contrastive Language-Image Pretraining (CLIP) [17] technology.

Details of the clinically-informed process are shown in Fig. 1 (b), which includes pre-training and feature fusion stages. In the pre-training stage, we extract textual and image features from function text information and the corresponding initial scan using text and image encoders, respectively. Then, we constrain these features to be aligned using the contrastive loss. In this way, we can obtain the pre-trained text encoder for incorporating function text information into the reverse process. Note that the samples used for pre-training are exclusively from the training set and do not involve any test samples.

In the feature fusion stage, for a particular reconstruction in the reverse process, the pre-trained text encoder first extracts textual features z_{text} from function test information. Then z_{text} is fed into a denoising attention U-Net [15] at each step for calculation of cross-attention, where the query Q and key K are calculated from z_{text} while the value V is still calculated from the output of the previous layer because our final goal is CT estimation. Denoting the output of previous layer as z_{CT}, the CT-guided cross-attention can be formulated as follows:

$$Output = softmax(\frac{Q_{text} K_{text}^T}{\sqrt{d}} + B) \cdot V_{PET}, \qquad (5)$$
$$Q_{text} = Conv_Q(z_{text}), \quad K_{text} = Conv_K(z_{text}), \quad V_{CT} = Conv_V(z_{CT}),$$

Table 1. Quantitative results of ablation analysis, in terms of PSNR and SSIM.

Method	PSNR [dB]↑	SSIM [%]↑
DM	25.84 ± 0.93	98.19 ± 1.24
DM-CIP	26.12 ± 1.07	98.75 ± 1.44
DM-R	26.84 ± 0.92	99.03 ± 1.16
DM-R-CIP	**27.52 ± 0.83**	**99.25 ± 0.92**

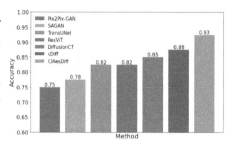

Fig. 2. Diagnostic evaluation.

where d is the number of channels, B is the position bias, and $Conv(\cdot)$ denotes the $1 \times 1 \times 1$ convolution with stride of 1.

3 Experiments

3.1 Dataset and Implementation

Our dataset originates from OSIC[1], an openly accessible global database containing numerous CT scans related to Idiopathic Pulmonary Fibrosis (IPF). We collect a total of 200 samples, each comprising of two CT scans with a 46.3 ± 7.8 weeks interval, along with corresponding lung function test information (including physiological indicators such as vital capacity, peak expiratory flow, etc.) and diagnostic labels. Among these 200 samples, 160 samples are used for training and the remaining 40 samples are used for testing. During the evaluation, we conduct five-fold cross-validation to exclude randomness.

In our implementation, experiments were conducted on the PyTorch platform using two NVIDIA Tesla A100 GPUs and an Adam optimizer with initial learning rate of 0.001. All images are resampled to voxel spacing of $1 \times 1 \times 1$ mm^3 and resolution of $512 \times 512 \times 256$, while their intensity range is normalized to $[0, 1]$ by min-max normalization. For increasing the training samples and reducing the dependence on GPU memory, we extract the overlapped patches of size $96 \times 96 \times 96$ from every whole CT scan. We evaluate the quantitative results by two commonly used quantitative metrics, including Peak Signal to Noise Ratio (PSNR) [9] and Structural Similarity Index (SSIM).

3.2 Ablation Analysis

To verify the effectiveness of our proposed strategies, i.e., *residual diffusion* and *clinically-informed process*, we design another four variant diffusion models (DMs) including: 1) DM: standard DM; 2) DM-CIP: DM with clinically-informed process; 3) DM-R: DM with residual diffusion; 4) DM-R-CIP: DM with residual diffusion and clinically-informed process. All methods use the same experimental settings, and their quantitative results are given in Table 1.

Table 2. Quantitative comparison of our CIResDiff with several state-of-the-art generation methods, in terms of PSNR and SSIM.

Method	PSNR [dB]↑	SSIM [%]↑
Pix2Pix-GAN [8]	22.53 ± 2.13	94.16 ± 1.59
SAGAN [12]	22.93 ± 1.98	94.84 ± 1.44
TransUNet [2]	23.35 ± 1.72	95.49 ± 1.24
ResViT [3]	23.62 ± 1.65	95.96 ± 1.32
DiffusionCT [18]	25.21 ± 0.92	97.82 ± 1.21
cDiff [16]	25.84 ± 0.93	98.19 ± 1.24
CIResDiff	**27.52 ± 0.83**	**99.25 ± 0.92**

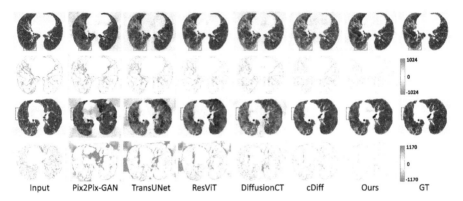

Fig. 3. Visual comparison of follow-up lung images produced by six different methods. From left to right are the input (initial scan), results of five other comparison methods (2nd-6th columns) and our CIResDiff (7th column), and the ground truth (follow-up scan). The corresponding difference maps between the generated results and GT are shown in the 2nd and 4th rows, where darker colors indicate larger differences. Red boxes show the lesion areas for detailed comparison. (Color figure online)

The quantitative results are provided in Table 1, from which, we can find the following observations. (1) DM-CIP with the clinically-informed process achieves better performance than DM. This proves that integrating highly relevant clinical information (i.e., function test information) into the reverse process is beneficial for the generation of follow-up CT scans. (2) DM-R, employing residual diffusion, achieves better performance than DM employing the traditional diffusion process, proving that residual diffusion is more appropriate for learning the mapping between highly similar initial and follow-up CT scans. (3) DM-R-CIP achieves better results than all other variants on both PSNR and SSIM, which shows both of our proposed strategies contribute to the final performance. These three comparisons conjointly verify the effective design of our proposed CIResD-

---
[1] https://www.osicild.org.

iff, where the *residual diffusion* and *clinically-informed process* both benefit our generation task.

3.3 Comparison with State-of-the-Art Methods

We further compare our CIResDiff with six state-of-the-art generation methods, which can be divided into three classes: 1) GAN-based methods, including Pix2Pix-GAN [8] and SAGAN [12]; 2) transformer-based methods, including TransUNet [2] and ResViT [3]; 3) diffusion model-based methods, including DiffusionCT [18] and conditional Diffusion (cDiff) [16]. The quantitative and qualitative results are provided in Table 2 and Fig. 3, respectively.

Quantitative Comparison: The quantitative results are provided in Table 2. From the table, it is evident that diffusion model-based methods generally outperform GAN-based and transformer-based methods, validating our selection of the diffusion model as the baseline. Moreover, among all diffusion model-based methods, our CIResDiff achieves the optimal results, with improvements in PSNR and SSIM over the sub-optimal cDiff by 1.68 dB and 1.06%, respectively. This demonstrates the effectiveness of our targeted improvements to the traditional diffusion model, including residual diffusion and clinically-informed process.

Qualitative Comparison: We provide a visual comparison of follow-up lung images generated by six different methods in Fig. 3. First, compared to other methods, our CIResDiff can generate the overall optimal images, characterized by the least noise, fewest artifacts but clearest structure. Second, in terms of detail, our CIResDiff can also most accurately generate the lesion areas (i.e., areas marked by red boxes) that are crucial for predicting the progression of IPF. Finally, the lightest color in the difference map demonstrates our CIResDiff can generate lung images with the smallest difference from the ground truth. Such key observations demonstrate that our CIResDiff is superior to those state-of-the-art methods.

3.4 Diagnostic Evaluation

Our ultimate goal is to predict the progression of IPF using generated follow-up lung images. Therefore, we design relevant downstream diagnostic tasks to assess the diagnostic value of follow-up lung images generated by different methods. Specifically, we first train a ResNet-based classifier using real image pairs (i.e., real follow-up lung image and real initial lung image) from the training set. During the training process, follow-up and initial lung images are concatenated together as input to predict diagnostic labels. Then, we use the pre-trained classifier to assess the diagnostic value of images generated by different methods. During evaluation, the input is the fake image pairs (i.e., generated follow-up lung image and real initial lung image). The prediction results are provided in Fig. 2.

Figure 2 shows that the follow-up lung images generated by our CIResDiff yield the best results for predicting IPF compared to other methods. Specifically, compared to Pix2Pix-GAN, which obtains the worst results, our CIResDiff shows a significant improvement in accuracy by 18%. This may be due to our method's incorporation of highly relevant lung function information into the generation process, thereby producing images more beneficial for diagnosis. These results further confirm the effectiveness of using generation for IPF prediction and highlight the clinical application potential of our CIResDiff.

4 Conclusion

In this paper, to achieve early prediction of IPF progression, we develop a novel diffusion model named CIResDiff to predict patients's follow-up CT scan from the initial CT scan. To facilitate more precise and effective generation, our CIResDiff employs three strategies: 1) pre-aligning lung regions in both CT scans to reduce the complexity of generation; 2) learning residuals between initial and follow-up CT scans to focus more on lesion areas critical for prediction; 3) integrating clinical lung function test information to help generate results with greater diagnostic value. Extensive experiments demonstrate that our method outperforms state-of-the-art approaches and can generate images with higher diagnostic value.

Acknowledgments. This work was supported in part by National Natural Science Foundation of China (grant numbers U23A20295, 62131015, 62250710165), the STI 2030-Major Projects (No. 2022ZD0209000), Shanghai Municipal Central Guided Local Science and Technology Development Fund (grant number YDZX20233100001001), the China Ministry of Science and Technology (STI2030-Major Projects-2022ZD0213100), The Key R&D Program of Guangdong Province, China (grant numbers 2023B0303040001, 2021B0101420006), the ERC IMI (10100 5122), the H2020 (952172), the MRC (MC/PC/21013), the Royal Society (IECn NS FCn211235), the NVIDIA Academic Hardware Grant Program, the SABER project supported by Boehringer Ingelheim Ltd, NIHR Imperial Biomedical Research Centre (RDA01), Wellcome Leap Dynamic Resilience, UKRI guarantee funding for Horizon Europe MSCA Postdoctoral Fellowships (EP/Z002206/1), and the UKRI Future Leaders Fellowship (MR/V023799/1). This work was completed under the close collaboration between Caiwen Jiang and Xiaodan Xing, and they contributed equally to this work.

Declaration of Competing Interest. The authors declare that they have no known competing financial interests or personal relationships that could have appeared to influence the work reported in this paper.

References

1. Avants, B., Tustison, N., Song, G., et al.: Advanced normalization tools (ANTS). Insight j **2**(365), 1–35 (2009)
2. Chen, J., et al.: TransuNet: transformers make strong encoders for medical image segmentation. arXiv preprint arXiv:2102.04306 (2021)

3. Dalmaz, O., Yurt, M., Çukur, T.: ResViT: residual vision transformers for multimodal medical image synthesis. IEEE Trans. Med. Imaging **41**(10), 2598–2614 (2022)
4. Finnerty, J., Ponnuswamy, A., Dutta, P., Abdelaziz, A., Kamil, H.: Efficacy of antifibrotic drugs, nintedanib and pirfenidone, in treatment of progressive pulmonary fibrosis in both idiopathic pulmonary fibrosis (IPF) and non-IPF: a systematic review and meta-analysis. BMC Pulm. Med. **21**(1), 411 (2021)
5. Frid-Adar, M., Diamant, I., Klang, E., Amitai, M., Goldberger, J., Greenspan, H.: GAN-based synthetic medical image augmentation for increased CNN performance in liver lesion classification. Neurocomputing **321**, 321–331 (2018)
6. Han, T., et al.: Image prediction of disease progression for osteoarthritis by style-based manifold extrapolation. Nat. Mach. Intell. **4**(11), 1029–1039 (2022)
7. Ho, J., Jain, A., Abbeel, P.: Denoising diffusion probabilistic models. Adv. Neural. Inf. Process. Syst. **33**, 6840–6851 (2020)
8. Isola, P., Zhu, J., Zhou, T., Efros, A.: Image-to-image translation with conditional adversarial networks. In: Proceedings of the IEEE conference on Computer Vision and Pattern Recognition, pp. 1125–1134 (2017)
9. Jiang, C., Pan, Y., Cui, Z., Nie, D., Shen, D.: Semi-supervised standard-dose PET image generation via region-adaptive normalization and structural consistency constraint. IEEE Trans. Med. Imaging **42**(10), 2974–2987 (2023)
10. Jiang, C., et al.: S2DGAN: generating dual-energy CT from single-energy CT for real-time determination of intracerebral hemorrhage. In: International Conference on Information Processing in Medical Imaging, pp. 375–387 (2023)
11. Jiang, C., Wang, T., Pan, Y., Ding, Z., Shen, D.: Real-time diagnosis of intracerebral hemorrhage by generating dual-energy CT from single-energy CT. Med. Image Anal. **95**, 103194 (2024)
12. Lan, H., D., A., Toga, A., Sepehrband, F.: Three-dimensional self-attention conditional GAN with spectral normalization for multimodal neuroimaging synthesis. Magnet. Reson Med. **86**(3), 1718–1733 (2021)
13. Ma, L., Shuai, R., Ran, X., Liu, W., Ye, C.: Combining DC-GAN with ResNet for blood cell image classification. Med. Biol. Eng. Vomput. **58**, 1251–1264 (2020)
14. Maher, T., et al.: Global incidence and prevalence of idiopathic pulmonary fibrosis. Respir. Res. **22**(1), 1–10 (2021)
15. Oktay, O., et al.: Attention U-Net: learning where to look for the pancreas. arXiv preprint arXiv:1804.03999 (2018)
16. Peng, J., et al.: CBCT-based synthetic CT image generation using conditional denoising diffusion probabilistic model. arXiv preprint arXiv:2303.02649 (2023)
17. Radford, A., et al.: Learning transferable visual models from natural language supervision. In: International Conference on Machine Learning, pp. 8748–8763 (2021)
18. Selim, M., Zhang, J., Brooks, M., Wang, G., Chen, J.: DiffusionCT: latent diffusion model for CT image standardization. arXiv preprint arXiv:2301.08815 (2023)
19. Sohl-Dickstein, J., Weiss, E., Maheswaranathan, N., Ganguli, S.: Deep unsupervised learning using nonequilibrium thermodynamics. In: International Conference on Machine Learning, pp. 2256–2265 (2015)
20. Spagnolo, P., et al.: Idiopathic pulmonary fibrosis: disease mechanisms and drug development. Pharmacol. Therapeut. **222**, 107798 (2021)
21. Torrisi, S., Kahn, N., Vancheri, C., Kreuter, M.: Evolution and treatment of idiopathic pulmonary fibrosis. La Presse Médicale **49**(2), 104025 (2020)

22. Wasserthal, J., et al.: Totalsegmentator: robust segmentation of 104 anatomic structures in CT images. Radiol. Artif. Intell. **5**(5) (2023)
23. Yue, Z., Wang, J., Loy, C.: Resshift: efficient diffusion model for image super-resolution by residual shifting. In: Advances in Neural Information Processing Systems, vol. 36 (2024)

Vision Transformer Model for Automated End-to-End Radiographic Assessment of Joint Damage in Psoriatic Arthritis

Darshana Govind[1], Zijun Gao[2], Chaitanya Parmar[3], Kenneth Broos[4], Nicholas Fountoulakis[5], Lenore Noonan[6], Shinobu Yamamoto[7], Natalia Zemlianskaia[8], Craig S. Meyer[1], Emily Scherer[9], Michael Deman[10], Pablo Damasceno[2], Philip S. Murphy[11], Terence Rooney[12], Elizabeth Hsia[13], Anna Beutler[14], Robert Janiczek[15], Stephen S. F. Yip[2], and Kristopher Standish[3](✉)

[1] Janssen R&D, Data Science, Analytics and Insight, Brisbane, CA, USA
[2] Janssen R&D, Data Science, Analytics and Insight, Cambridge, MA, USA
[3] Janssen R&D, Data Science, Analytics and Insight, La Jolla, CA, USA
kstandis@its.jnj.com
[4] Janssen R&D, Global Development, Janssen Clinical Innovation, Beerse, Belgium
[5] Janssen R&D, Global Regulatory Affairs, Immunology, Spring House, PA, USA
[6] Janssen R&D, Immunology Translational Sciences, Imaging and Digital Health, Spring House, PA, USA
[7] Janssen R&D, Global Development, IDAR, Clinical and Statistical Programming, Raritan, NJ, USA
[8] Janssen R&D, Data Science, Analytics and Insight, New York, NY, USA
[9] Janssen R&D, Data Science, Analytics and Insight, Titusville, NJ, USA
[10] Janssen R&D, Data Science, Immunology, Beerse, Belgium
[11] Janssen R&D, Immunology Translational Sciences, High Wycombe, UK
[12] Janssen R&D, Immunology, Rheum Diseases Area Stronghold, Spring House, PA, USA
[13] Janssen R&D, Data Science, Immunology, Springhouse, PA, USA
[14] Janssen R&D, Immunology, Late Development Rheumatology, Spring House, PA, USA
[15] Janssen R&D, Immunology, Translational Sciences, Spring House, PA, USA

Abstract. Deep learning techniques such as Convolutional Neural Networks (CNNs) and Vision Transformers (ViTs) have demonstrated strong capabilities in region-of-interest (ROI) detection and disease severity scoring, aiming to reduce the manual workload for clinicians. However, studies encompassing an end-to-end method for ROI detection, their subsequent scoring, and aggregation to a patient level score in psoriatic arthritis (PsA) have been limited. To address this gap, we have developed an end-to-end multistep approach for automated radiographic assessment in PsA that detects joints-of-interest from X-rays, assigns structural damage scores to the identified joints, and aggregates them to an extremity level, and subsequently to a patient level score. Furthermore, we compare our approach with state-of-the-art methods and human experts. Our results demonstrate the strong performance of our joint detection models, achieving an average intersection over union (IoU) value of 0.88 for foot joints and 0.76 for hand joints. The

---

D. Govind and Z. Gao—Equal contribution.

subsequent scoring models for structural damage show excellent (intra-class correlation coefficient (ICC) > 0.90) agreement with an ICC of 0.94, 0.98, and 0.97 for joint level erosion, JSN, and patient level scores, respectively, when compared to human readers.

Keywords: Deep Learning · Self-supervised Learning · Psoriatic Arthritis · Automated Van der Heijde–Sharp Scoring

1 Introduction

In recent years, the field of medical imaging has experienced advancements fueled by the progress in deep learning techniques [1, 2], which can be particularly beneficial for clinicians who frequently encounter the challenge of manually identifying regions-of-interest (ROIs) to assess disease severity. In particular, in chronic inflammatory diseases such as rheumatoid arthritis (RA) and psoriatic arthritis (PsA) [3–5], radiologists manually evaluate bilateral X-ray radiographs of hands and feet to identify joint-of-interest and subsequently assign severity scores using the Van der Heijde-Sharp (vdHS) scoring method or its modification [6, 7]. This manual scoring process is time-consuming, tedious, and subjective.

For automated ROI detection, You Only Look Once (YOLO) solutions adapted for medical imaging have been widely popular owing to its ability to detect anatomical lesions with high accuracy [8, 9]. Among the YOLO evolutions, YOLOv8 [10] has showcased enhanced detection accuracy and faster processing speed. For the subsequent scoring of the detected ROIs, numerous approaches have been explored. CNNs, particularly variations of DenseNet and ResNet, have shown performances comparable to radiologists in diagnosing diseases such as cancer, using radiographs [11]. An alternative approach using Vision Transformers (ViTs) [12] has been shown to outperform CNNs for classification tasks in medical and non-medical domains [13, 14]. However, studies encompassing an end-to-end method for ROI detection, their subsequent scoring, and aggregation to a patient level score in PsA have been limited.

In vdHS scoring, radiologists assign severity scores for joint space narrowing (JSN) (reflecting cartilage damage) and erosion (reflecting bone destruction) within predefined ranges, which are aggregated to compute the patient vdHS score. Current automated approaches focus only on hand or foot radiographs, or on either erosion or JSN scoring, potentially due to the lack of high-quality datasets with both hand and foot X-rays from a large dataset of patients along with multiple human expert scores, thereby unable to automate vdHS scoring at the patient level [15–18]. Further, current methods have targeted RA [15–20], but not PsA, like Li and Guan [19], an approach that excelled in the DREAM Rheumatoid Arthritis Challenge. Given the distinct features of PsA [21], such as the scoring of additional joints (distal inter-phalangeal) in the hands, tailored approaches may be required for accurate and reliable assessment of PsA.

To address these challenges, we developed an end-to-end method for PsA vdHS scoring (Fig. 1), using hand and feet X-ray images. Our contributions are three-fold: (1) This is the first comprehensive automated method for vdHS scoring from hand and foot X-rays in PsA. (2) We obtain state-of-the-art performance in vdHS scoring compared to existing approaches. (3) Our results demonstrate excellent (intra-class correlation coefficient (ICC) > 0.90) [22] agreement with ICCs of 0.94, 0.98, and 0.97 for joint level erosion, JSN, and patient level scores, respectively, compared to human readers.

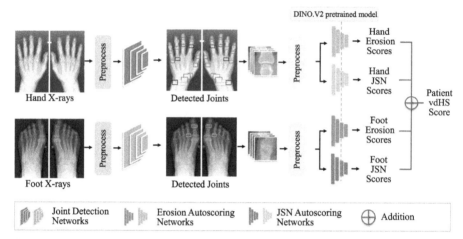

Fig. 1. Schematic diagram of our end-to-end approach. X-ray images are fed to the joint detection networks to locate target joints, which are subsequently assigned an erosion and JSN score through the respective autoscoring networks (each initialized using a single DINO.V2 pretrained model). The joint level scores are aggregated to generate the final vdHS score at the patient level.

2 Methodology

Our approach (Fig. 1) for automated PsA assessment was comprised of two stages. Firstly, joint detection (JD) networks located the joints in X-ray images. Subsequently, the autoscoring (ASC) networks generated erosion and JSN scores for each of the detected joints. Prior to the JD model, the X-rays underwent a preprocessing step which involved removing images with a photometric interpretation of "RGB" (12/17K X-rays (0.07%) in the entire dataset) as they were not scanned but photographed, resulting in lower quality images. Further, images with a photometric interpretation of monochrome 1 were converted to monochrome 2 to ensure that low and high pixel values represent dark and bright areas, respectively. Additionally, the joint level images (crops of the detected joints-of-interest) underwent a preprocessing step before the ASC models were applied, where the images were resized and scaled as detailed in the implementation section.

2.1 Joint Detection Networks

The JD networks utilized the YOLOv8 [10] nano model, pretrained with COCO dataset [23], and were trained separately for foot and hand joint detection to identify 6 joints in each foot (Inter-phalangeal (IP) and Metatarsophalangeal joints (MTPs)), and 26 joints in each hand (IP, Distal Interphalangeal (DIPs), Proximal Interphalangeal (PIPs), and Metacarpophalangeal (MCPs)), respectively. Each JD network was structured with a classification and regression branch for predicting the joint type and location, respectively. The loss function was composed of cross entropy loss for classification and distribution focal loss along with complete intersection over union losses for bounding box regression, with weight configurations of 0.5, 1.5 and 7.5, respectively.

2.2 Self-supervised Pretraining of Joint Level Feature Extractor

To generate a robust ASC network capable of recognizing radiographic features, we first pretrained a ViT model using DINO.V2 [24], a non-contrastive self-supervised learning approach using the DINO.V2 dataset and implementation details described below. **The pretraining** was conducted on both ViT S/14 and ViT B/16 models [25] and the joint level images were resized to $294 \times 294 \times 3$ and $224 \times 224 \times 3$ (reducing the computational complexity), via bicubic interpolation, respectively.

2.3 Autoscoring Tasks with Vision Transformers

Using the pretrained ViT model, four distinct ASC ViT models were trained (all models were finetuned) to generate erosion and JSN scores for hand and foot joints, separately, each constructed using a single-layer regression head and optimized using Mean Squared Error (MSE) loss. The image sizes were adjusted to match pretraining settings and rescaled to the range [0,1] and normalized to adhere to the respective specifications.

2.4 Comparison with Other CNNs in Autoscoring Tasks

We also compared the performance of our approach with CNN-based methods, including ResNet [26] and DenseNet [27]. The implementation details of these models are described in later sections.

3 Experiments

3.1 Dataset

We utilized a total dataset of 1401 patients with 17,442 X-rays ($\approx$8–12 X-rays/pt, spanning 2–3 visits over a 52-week period) from three PsA clinical trials. This dataset was divided into training (738 pts), validation (185 pts), and holdout (478 pts) imaging datasets. Further, to train and optimize the JD and ASC networks, a portion of the training and validation datasets (~3K X-rays from 271 and 72 pts, respectively), and the holdout (580 X-rays from 113 pts) sets were annotated with bounding boxes, hereon referred to as annotated-training, annotated-validation, and annotated-holdout sets, respectively.

Dataset for Joint Detection Networks. The annotated-training and -validation sets were used to build and optimize the JD networks, and the annotated-holdout set was used to test the model performance.

Dataset for DINO.V2 Pretraining. To generate the pretrained DINO.V2 model, the entire training and validation datasets (12,016 X-rays from 923 pts) (to maximize the amount of data for the models) were utilized to extract joint level images for training and optimizing the pretrained model. For images that lacked bounding box annotations, the optimized JD network was used to locate the joints, resulting in 191,879 joint level images (36,042 from feet and 155,837 from hands).

Dataset for Autoscoring Networks. Joint level images were extracted from the annotated-training, -validation, and -holdout sets for training, validating, and testing the ASC models, respectively. As the ground truth for the scoring task, the panel (comprised of two main radiologists (r_1 and r_2) and an adjudicator (r_3)) average scores (generated using Eq. (1)) was utilized for the respective models. For the hands, scores range from 0 to 4 for JSN and 0 to 5 for erosion, while for the feet, scores range from 0 to 4 for JSN and from 0 to 10 for erosion.

$$s = \begin{cases} mean(r_1, r_2), r_3 \text{ not available} \\ mean(r_1, r_3), abs(r_1 - r_3) < abs(r_2 - r_3) \\ mean(r_2, r_3), abs(r_1 - r_3) \geq abs(r_2 - r_3) \end{cases} \quad (1)$$

3.2 Implementation Details

The models listed below were implemented in Python 3.8 and PyTorch 2.0.0.

Joint Detection Networks. The networks were trained for 50 epochs on batch size of 16 with the Adam optimizer, utilizing an initial learning rate (LR) of 10^{-3}, a weight decay of 5×10^{-4}, and a momentum of 0.9. The training was performed on a single NVIDIA A10G GPU with 40 GB memory.

DINO.V2 Pretraining. For the ViT S/14 model, the training process utilized 8 NVIDIA A10G GPUs with a batch size of 512 in total and ran for 125k iterations. During the initial 10% training iterations, the LR underwent a warm-up phase, starting from 0 and gradually increasing to 4×10^{-3}. A cosine decay was scheduled to reduce the LR to 10^{-6} at the end of training. Additionally, global crop scales were set to (0.6, 1.0), while local crop scales were set to (0.2, 0.5) with a crop size of 84. For the ViT B/16 model, training was conducted using a batch size of 256 distributed across 4 NVIDIA A10G GPUs and run for 300k iterations. The LR was initialized at 0, then gradually warmed up to 8×10^{-4} within the initial 25% of iterations, followed by a cosine decay schedule to 10^{-6} for the remaining iterations.

Autoscoring Networks. For ResNet50 and DenseNet121 used for comparison, joint level images were resized to $300 \times 300 \times 3$, and rescaled and normalized the same way as the ViT models for 1:1 comparison. For the CNNs and the ViT models, during training, weighted sampling was executed based on the relevant score associated with each joint to ensure exposing the network to a dynamic range of scores. All models underwent automatic mixed precision training for up to 150 epochs with batch size of 64. An LR scheduler, with a patience of 20 epochs, minimum LR of 10^{-6} and a decay factor of 0.2 based on the validation loss, was applied. To improve robustness, various image augmentations were employed, including contrast adjustments (gamma range: 0.5 to 1.5), random affine transformations (rotation range: up to 0.3; shear range: up to 0.2), and random flips (probability of 0.4). Hyperparameter searches were conducted with LR of $[10^{-5}, 3 \times 10^{-5}, 5 \times 10^{-5}]$ and WD of $[10^{-5}, 10^{-6}, 0]$. The experiments were performed on a single NVIDIA A10G GPU with 40 GB memory.

3.3 Model Evaluation

Performance of JD models was assessed via average Intersection over union (IoU) and proportion of joints with IoU $> = 0.5$ on the annotated-holdout sets. ASC models were assessed by calculating the joint level ICC, also on the annotated-holdout sets. In evaluating the overall approach, the locked JD and ASC models were applied to the entire holdout set. ICCs were computed at the joint-, extremity-, and patient levels. Additionally, the mean squared error (MSE) between the ground truth and predicted scores was determined. In addition, a comparative analysis was conducted between our ViT-based ASC models and other methods such as ResNet [26] and DenseNet [27].

4 Results

4.1 The Joint Detection Models Demonstrated Strong Performances on the Holdout Subset

On the annotated-holdout set, the locked JD models demonstrated strong performance (Fig. 2) with an average IoU of 0.88 ± 0.02 (96.26% of joints ≥ 0.5) and 0.76 ± 0.04 (96.25% of joints ≥ 0.5) for the foot and hand joints, respectively (Fig. 3). Our model outperforms the current range of IoU values reported in literature (~ ICC of 0.72 [17]).

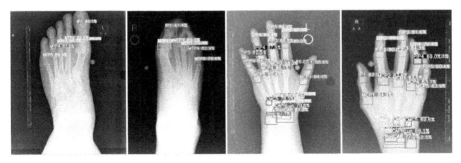

Fig. 2. Joint detection model predictions on the annotated-holdout set. The predictions generated by the models are shown for foot and hand X-rays from patients in the holdout set with varying severities of joint damage.

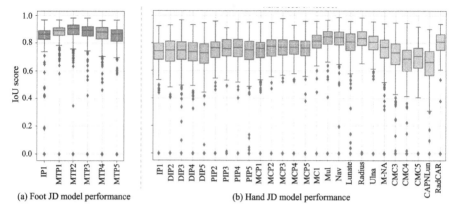

Fig. 3. Box and whisker plots illustrating the performance of JD networks on the annotated-holdout set. Left panel displays results for the foot JD model focusing on foot joints, while the right panel showcases the hand JD model's performance on different hand joints.

4.2 The Autoscoring Models Demonstrated Strong Performances on the Holdout Subset

In the annotated-holdout set, the erosion and JSN ASC models exhibited strong agreement with the ground truth panel average scores at the joint level, achieving ICC values of 0.94, 0.98, 0.95, and 0.97, for hand erosion, hand JSN, foot erosion, and foot JSN respectively. This level of agreement closely aligned with the corresponding inter-rater agreement between human raters, which were ICC values of 0.89, 0.98, 0.86, and 0.97, respectively.

4.3 The Proposed Method Attained Model-Rater Agreement Comparable to that of Inter-Raters on the Holdout Set

To test the performance of the end-to-end pipeline, the complete holdout set was passed through the locked JD and ASC models and the detected joints were subsequently scored and aggregated to extract the final vdHS scores per patient. The values are computed among the detected joints that were scored by both model and human readers.

Our end-to-end approach exhibited consistently lower mean MSE values compared to the inter-rater assessments across all levels (Table 1) and achieved ICC values of 0.94 and 0.98 for bone erosion and JSN at the joint level, respectively, indicating strong agreement between predicted values and panel average. Extremity level assessments for hands and feet also demonstrated substantial ICC values of 0.97 and 0.94, respectively, showing better performances than the inter-rater ICC. Additionally, the patient level evaluation showcased a robust ICC of 0.97, outperforming the agreement observed between human raters (Fig. 4). Further, upon comparison of the joint level MSE values of our approach, with that of the RA dream challenge winner (root mean squared errors of 0.65, 0.49, 1.01, and 0.52 for hand and foot erosion and JSN scores, compared to human readers), our method outperforms the latter.

Table 1. Evaluation of the end-to-end approach on the complete holdout set. Respective ICCs and 95% confidence intervals (CI), and MSE with their standard deviations (SD) are shown.

Evaluation	Model vs. Panel Avg.		R1 vs. R2	
	ICC [95% CI]	MSE (± SD)	ICC [95% CI]	MSE (± SD)
Joints (BE)	0.94 [0.93, 0.94]	0.26 ± 1.25	0.84 [0.83, 0.85]	0.64 ± 2.49
Joints (JSN)	0.98 [0.98, 0.98]	0.1 ± 0.39	0.96 [0.96, 0.97]	0.16 ± 0.59
Hands	0.97 [0.97, 0.97]	87.22 ± 270.35	0.92 [0.91, 0.92]	488.87 ± 1083.73
Feet	0.94 [0.94, 0.95]	35.74 ± 150.77	0.91 [0.90, 0.92]	104.67 ± 375.28
Patient	0.97 [0.96, 0.97]	157.35 ± 444.97	0.91 [0.91, 0.92]	930.24 ± 2221.61

4.4 Autoscoring Model Exhibited Superior Performance Compared to Other Approaches

The DINO.V2 pretrained ViT-based models were the best performing models among the experiments conducted (Table 2). Specifically, ViT B/16 (our approach) achieved an ICC of 0.97, while ViT S/14 obtained an ICC of 0.94 when evaluated against the panel average ground truth. Additionally, the ImageNet pretrained DenseNet achieved an ICC of 0.92 (best CNN model) (Fig. 4), while ResNet achieved an ICC of 0.81. The MSE metric followed a similar trend across all models.

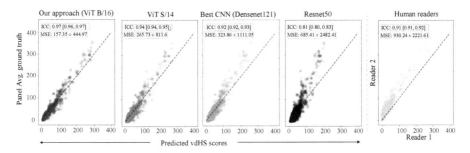

Fig. 4. Comparison of our approach with other methods and human experts. The scatterplots demonstrate the agreement between panel average and predicted scores by our approach (blue), ViT S/14 model (green), DenseNet121 (olive), ResNet50 (black). The agreement between human readers is shown in orange. (Color figure online)

Table 2. Quantitative Comparison of ASC Models on Complete Holdout Set.

Models	ICC [95% CI] (↑)	MSE (± SD) (↓)
ResNet50	0.81 [0.8, 0.83]	685.41 ± 2482.41
DenseNet121	0.92 [0.92, 0.93]	323.86 ± 1111.05
ViT S/14	0.94 [0.94, 0.95]	265.73 ± 811.6
Our model (ViT B/16)	0.97 [0.96, 0.97]	157.35 ± 444.97

5 Conclusion

Our approach demonstrates state-of-the art performance in the radiographic assessment of structural damage in psoriatic arthritis. The vdHS scores obtained with the algorithm align closely with human annotators, showcasing excellent model-rater agreement and are also comparable to the inter-rater consensus. Future work will involve incorporating more radiographs with the entire spectrum of joint damage into the training set to further enhance joint detection and scoring accuracy.

References

1. Rana, M., Bhushan, M.: Machine learning and deep learning approach for medical image analysis: diagnosis to detection. Multimedia Tools Appl. **82**(17), 26731–26769 (2023)
2. Jiang, H., et al.: A review of deep learning-based multiple-lesion recognition from medical images: classification, detection and segmentation. Comput. Biol. Med. 106726 (2023)
3. Imtiaz, M., Shah, S.A.A., ur Rehman, Z.: A review of arthritis diagnosis techniques in artificial intelligence era: current trends and research challenges. Neurosci. Inform. **2**(4), 100079 (2022)
4. Parashar, A., et al.: Medical imaging in rheumatoid arthritis: a review on deep learning approach. Open Life Sci. **18**(1), 20220611 (2023)
5. McMaster, C., et al.: Artificial intelligence and deep learning for rheumatologists. Arthrit. Rheumatol. **74**(12), 1893–1905 (2022)
6. Van der Heijde, D.: How to read radiographs according to the Sharp/van der Heijde method. J. Rheumatol. **27**(1), 261–263 (2000)
7. Rahman, P., et al.: Radiological assessment in psoriatic arthritis. Br. J. Rheumatol. **37**(7), 760–765 (1998)
8. Montalbo, F.J.P.: A computer-aided diagnosis of brain tumors using a fine-tuned YOLO-based model with transfer learning. KSII Trans. Internet Inf. Syst. **14**(12) (2020)
9. Liu, K.: Stbi-yolo: a real-time object detection method for lung nodule recognition. IEEE Access **10**, 75385–75394 (2022)
10. Glenn Jocher, A.C., Qiu, J.: Ultralytics YOLO. 2023 [cited 2024 1]; 8.0.0:[29]. Available from: https://github.com/ultralytics/ultralytics
11. Zhao, X., et al.: A cross-modal 3D deep learning for accurate lymph node metastasis prediction in clinical stage T1 lung adenocarcinoma. Lung Cancer **145**, 10–17 (2020)
12. Dosovitskiy, A., et al.: An image is worth 16x16 words: transformers for image recognition at scale. arXiv preprint arXiv:2010.11929 (2020)
13. Gheflati, B., Rivaz, H.: Vision transformers for classification of breast ultrasound images. In: 2022 44th Annual International Conference of the IEEE Engineering in Medicine & Biology Society (EMBC). IEEE (2022)

14. Matsoukas, C., et al.: Is it time to replace CNNs with transformers for medical images? arXiv preprint arXiv:2108.09038, 2021
15. Huang, Y.-J., et al.: Automatic joint space assessment in hand radiographs with deep learning among patients with rheumatoid arthritis. In: Arthritis & Rheumatology. 2020. Wiley 111 RIVER ST, HOBOKEN 07030–5774, NJ USA
16. Radke, K.L., et al.: Adaptive IoU thresholding for improving small object detection: a proof-of-concept study of hand erosions classification of patients with rheumatic arthritis on X-ray images. Diagnostics **13**(1), 104 (2022)
17. Stolpovsky, A., et al.: RheumaVIT: transformer-based model for Automated Scoring of Hand Joints in Rheumatoid Arthritis. In: Proceedings of the IEEE/CVF International Conference on Computer Vision (2023)
18. Wang, H.-J., et al.: Deep learning-based Computer-Aided diagnosis of rheumatoid arthritis with hand X-ray images Conforming to Modified total Sharp/van der Heijde score. Biomedicines **10**(6), 1355 (2022)
19. Li, H., Guan, Y.: Multilevel modeling of joint damage in rheumatoid arthritis. Adv. Intell. Syst. **4**(11), 2200184 (2022)
20. Rohrbach, J., et al.: Bone erosion scoring for rheumatoid arthritis with deep convolutional neural networks. Comput. Electr. Eng. **78**, 472–481 (2019)
21. Saalfeld, W., et al.: Differentiating psoriatic arthritis from osteoarthritis and rheumatoid arthritis: a narrative review and guide for advanced practice providers. Rheumatol. Therapy **8**, 1–25 (2021)
22. Koo, T.K., Li, M.Y.: A guideline of selecting and reporting intraclass correlation coefficients for reliability research. J. Chiropr. Med. **15**(2), 155–163 (2016)
23. Lin, T.-Y., et al.: Microsoft COCO: common objects in context. In: Fleet, D., Pajdla, T., Schiele, B., Tuytelaars, T. (eds.) ECCV 2014, Part V. LNCS, vol. 8693, pp. 740–755. Springer, Cham (2014). https://doi.org/10.1007/978-3-319-10602-1_48
24. Oquab, M., et al.: Dinov2: Learning robust visual features without supervision. arXiv preprint arXiv:2304.07193 (2023)
25. Khan, S., et al.: Transformers in vision: a survey. ACM Comput. Surv. (CSUR) **54**(10s), 1–41 (2022)
26. He, K., et al.: Deep residual learning for image recognition. In: Proceedings of the IEEE Conference on Computer Vision and Pattern Recognition (2016)
27. Huang, G., et al.: Densely connected convolutional networks. In: Proceedings of the IEEE Conference on Computer Vision and Pattern Recognition (2017)

CorticalEvolve: Age-Conditioned Ordinary Differential Equation Model for Cortical Surface Reconstruction

Wenxuan Wu[1,2], Tong Xiong[1], Dongzi Shi[1], Ruowen Qu[1], Xiangmin Xu[1,3], Xiaofen Xing[1(✉)], and Xin Zhang[1(✉)]

[1] The School of Electronic and Information Engineering, South China University of Technology, Guangzhou 510640, China
{xfxing,eexinzhang}@scut.edu.cn
[2] BGI Research, Hangzhou 310030, China
[3] Institute of Artificial Intelligence, Hefei Comprehensive National Science Center, Hefei 230088, China

Abstract. Cortical surface reconstruction utilizing Neural Ordinary Differential Equation (NODE) stands as a prominent method, renowned for generating surfaces of accuracy and robustness. However, these methodologies, tailored predominantly for the adult brain, fall short in capturing the substantial morphological differences in the cortical surfaces arising from the rapid early development of the neonatal brain. Hence, post menstrual age (PMA) is an essential factor which should not be ignored. To address this, we propose the CorticalEvolve, a diffeomorphic transformation based progressive framework, for neonatal cortical surface reconstruction. We design a neonatal deformation block (DB) to preform deformation with developmental characteristics and a dynamic deformation block to refine the generated surface. Specifically, the proposed Neonatal DB firstly incorporates time-activated dynamic convolution to transform the stationary velocity field into a time-varying dynamic velocity field. Secondly, Age Conditioned ODE (Age-CDE) is proposed to depict the brain development nature, replacing the fixed step integration scheme in NODE with age-related one. Then, each discrete state of intermediate step represents the neonatal cortical surface at the corresponding PMA. The experimental results on the dHCP dataset effectively demonstrate the reconstruction capabilities of the proposed method, surpassing several state-of-the-art methods by a significant margin across various metrics. Visualization results underscore CorticalEvolve's ability to generate more robust cortical surfaces.

Keywords: Neonatal brain · Cortical surface reconstruction · Neural ODE · Brain development

---

W. Wu and T. Xiong—Equal Contribution.

1 Introduction

Cerebral cortex-based analysis holds a pivotal role in the realm of neuroimaging studies, providing valuable insights into the brain's topology for diagnosing psychiatric disorders. Consequently, there exists a substantial demand for reliable cortical surface reconstruction techniques aimed at extracting the 3D structure of the human brain from given Magnetic Resonance Imaging (MRI) scans. Specifically, cortical surface reconstruction strives to precisely retrieve triangular meshes for the inner surface (white matter surface) and outer surface (pial surface) from the provided MRI scans. In this domain, numerous challenges impede accurate reconstruction, including the highly folded nature of the cerebral cortex and partial volume effects in MRI scans. Furthermore, due to the heterogeneity in cortical surfaces, particularly among newborns of different ages, it is challenging for a single model to perform well on MRI scans from various groups. Despite extensive research conducted by predecessors, cortical surface reconstruction remains an open issue and is still underexplored.

In the early developmental era, the cortical surface reconstruction pipeline typically involved sequential image processing modules, with a representative framework being the renowned FreeSurfer [4,7,8]. Although these empirically designed modules effectively ensure the quality of reconstruction, the substantial processing time and the considerable effort required for fine-tuning parameters limit their practical application.

With the rapid advancement of deep learning, numerous studies have proposed integrating deep neural networks into cortical surface reconstruction, such as DeepCSR [3], FastSurfer [9], Vox2Cortex [1]. Given their end-to-end processing mode and universal approximation ability, these deep learning approaches showcase significant potential in both reconstruction efficiency and quality. Among all these deep learning methods, research [12–14,18], based on diffeomorphic transformation has witnessed growing interest due to its capability in reducing self-interaction on the generated surface. These approaches typically involve learning the flow field of diffeomorphic mapping from white matter surface to pial surface. Once the flow field is estimated, the 3D shape can be easily reconstructed by continuously solving the ordinary differential equation (ODE) under the initial mesh state-white matter surface template.

While prevalent diffeomorphic transformation-based methods have demonstrated superior performance on several public adult datasets [11,20], challenges arise in neonatal cortical surface reconstruction. Since cortex varies smoothly among adults with different age, a fixed step integration scheme is sufficient to cover all age groups. However, in the first few postnatal periods, the cortex undergoes rapid development with significant growth in terms of many morphology measurements such as sulcal depth [5,6,16,19]. This leads to evident structural differences among newborns at different post-menstrual ages (PMA). Therefore, existing methods utilizing a fixed step deformation is unable to capture these structure difference in newborns. Furthermore, a time-varying flow field can also benefit the surface deformation. Few studies consider this crucial factor in the flow field estimation process. Although CoTAN [13] introduce the time and age prior information in the estimation of flow field, they simply utilize attention computed by age and time to fuse flow fields under different resolutions. Given the above discussions, it is crucial to emphasize the age effect in both flow field estimation and the approximation of the flow ODE.

Contribution: In this study, we present CorticalEvolve, a diffeomorphic transformation based framework tailored for neonatal cortical surface reconstruction. CorticalEvolve improves the typical deformation module with two distinct deformation blocks (Neonatal DB and Dynamic DB). They both contain a dynamic flow field estimator to introduce diversity in the learning process, transforming the flow field into a time-varying one. For the Neonatal DB, we relax the constraint of a fixed step in the approximation process of solving the flow ODE. We propose an alternate Age Conditioned ODE (Age-CDE), where the approximation step is proportional to age, effectively optimizing the solving process. The Age-CDE is interpreted as early brain development, in which each discrete state of intermediate step represent the neonatal cortical surface at the corresponding PMA. The experimental results on the dHCP dataset effectively demonstrate the reconstruction capabilities of the proposed method, surpassing state-of-the-art methods by a significant margin across various metrics.

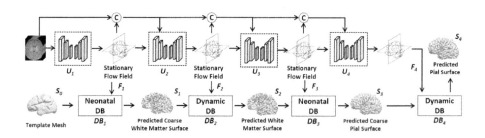

Fig. 1. The overall flowchart of the proposed CorticalEvolve. The model gradually utilizes UNet-3D to estimate the flow field, and then uses the proposed deformation block to generate the reconstructed surface.

2 Method

2.1 Model Architecture

In this section, we will elaborate the proposed CorticalEvolve in detail. As an diffeomorphic transformation-based method, it primarily involves the learning of the flow filed and mesh deformation as demonstrated in Fig. 1. Specifically, CorticalEvolve can be decomposed into 4 progressively ordered similar sub-modules. The sub-module firstly estimates the stationary flow field with UNet-3D [17] U_i and then incorporates the result to model the deformation process using the Deformation Block (DB_i). The deformation process from S_0 to S_1 and S_2 to S_3

model the meaningful characteristic of brain development, while the deformation from S_1 to S_2 and S_3 to S_4 simply serve as surface refinement. Therefore, CorticalEvolve introduces two distinct types of Deformation Blocks (Neonatal DB and Dynamic DB) to adjust to these two different situations.

With these 4 sequential sub-module, CorticalEvolve progressively transform the template mesh S_0 into the coarse white matter surface S_1, followed by the fine-grained white matter surface S_2, the coarse pial surface S_3, and finally the fine-grained pial surface S_4. The computing process in each sub-module can be formalized as:

$$S_1 = DB_1(F_1, S_0), \quad F_1 = U_1(I),$$
$$S_i = DB_i(F_i, S_{i-1}), \quad F_i = U_i(C(I, F_{i-1})), \quad \text{for } i > 1, \tag{1}$$

where $S_i \in \mathbb{R}^{N \times 3}$ represents the mesh state, and $C(I, F_{i-1})$ denote the channel-wise concatenation of the image I and flow field F_{i-1}.

2.2 Neonatal Deformation Block

Compared with previous diffeomorphic transformation-based method, the key improvement of CorticalEvolve lies in mesh deformation. The proposed Neonatal DB initially conduct post-processing on the stationary field to integrate time information with dynamic flow field estimator. Subsequently, it employs an Age Conditioned ODE to facilitate surface deformation from the input to the predicted surface.

Dynamic Flow Field Estimator: In the generation of velocity fields, we follow the CoTAN [13] to incorporate temporal information into the deformation process through an attention mechanism. As illustrated in Fig. 2, the dynamic flow field operates as a time-activated dynamic convolution with K kernels. We initially convert the input time prior information t into an attention vector $v_t \in \mathbb{R}^{1 \times K}$ using a fully connected layer. Each value in v_t signifies the weight of the corresponding K_{th} kernel in dynamic convolution. At different time points t, the dynamic convolution assigns distinct weights to the stationary flow field F_i produced by U_i, resulting in a time-varying dynamic flow field crucial for surface deformation.

Age Conditioned ODE: In the context of deformation from template mesh S_0 to coarse white matter surface S_1, we argue this process is a manifestation of brain development, implying abundant temporal information. Therefore, in this deformation stage, we specifically design an Neonatal DB with an Age-Conditioned ODE (Age-CDE) to depict the dynamic nature. Within Age-CDE, this brain development process is modeled as an flow ODE over developmental period $t \in [0, T]$. Then the coarse white matter surface can be naturally interpreted as the integration result under the initial mesh state–template mesh. Concerning the cortex's complexity is highly correlated to age, we introduce

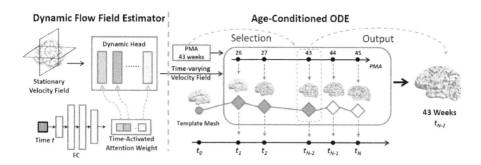

Fig. 2. The proposed Neonatal DB post-process the input stationary field with the dynamic flow field estimator at first. It then progressively solve the Age-Conditioned ODE to generate the surface at given time point.

adaptive integration step setting for different age groups. Given that the PMA of newborns varies from 26 to 45 weeks, we uniformly select $N = 20$ time points in $[0, T]$, satisfying $0 < t_1 < t_2 < \ldots < t_N < T$. By solving the flow ODE at discrete time point t_i, we can theoretically obtain the cortical surface ϕ_{t_i} at $i + 25$ weeks. In summary, we formalize the Age-CDE as an autonomous ODE with variable integral upper bound as:

$$\frac{\partial}{\partial t}\phi_t = V(\boldsymbol{F}_1, t), \ \phi_0 = \boldsymbol{S}_0, \ t \in [0, T], \tag{2}$$

$$\boldsymbol{S}_1 = \phi_{t_i}(\boldsymbol{S}_0) = \boldsymbol{S}_0 + \int_0^{t_i} V(\boldsymbol{F}_1, s)\mathrm{d}s, \ i = a - 25 \tag{3}$$

Here, a is the PMA and ϕ_t signifies the deformation trajectory of $\boldsymbol{S}_0$ over time t. $V(\boldsymbol{F}_1, s)$ represents the dynamic flow field estimator introduced in Sect. 2.3. At various time points t, it transforms $\boldsymbol{F}_1$ into a time-varying dynamic flow field.

Notably, Equation (2) can be interpreted as the continuous evolving process of surface from time t_i to time t_{i-1}. In the context of brain development, for a given subject, the cerebral cortex with PMA at a originates from the cortex with PMA at $a - 1$.

As for the reconstruction from the white matter surface S_2 to the coarse pial surface S_3, we posit that cortical thickness consistently increases with PMA. From a surface perspective, the escalating PMA in neonates corresponds to a greater distance between the white matter surface and pial surface. Thus, we also utilize Neonatal DB to perform the deformation. The variable step size property of Age-CDE facilitates more iterations over the extended surface distance, resulting in enhanced deformation stability.

2.3 Dynamic Deformation Block

The Dynamic DB serves to refine the surface, applied in deformation from S_1 to S_2 and S_3 to S_4. Aligning with the Neonatal DB, Dynamic DB also post-process the flow field with dynamic flow field estimator. However, we assert that such refinement is exclusive to individual variances, devoid of the aforementioned collective attributes. Thus, the predefined Age-CDE with individual specified integrating upper bond can not provide additional enhancement to the deformation process. Consequently, we simply employ a fixed-step ODE for the following effecting corrective surface deformation. Experimental results demonstrated in Sect. 3.3 effectively suggest the rationality of this setting.

2.4 Training Strategy

Our training process is divided into 4 stages, individually training each 3D Unet and respective DB in CorticalEvolve at each stage, while keeping the remaining model parameters frozen. Following previous work, we apply consistent loss functions to all surfaces, including Chamfer distance loss $\mathcal{L}_c$, mesh Laplacian loss $\mathcal{L}_{lap}$, and normal consistency loss $\mathcal{L}_n$. The total loss is expressed as follows:

$$\mathcal{L} = \mathcal{L}_c + \lambda_{lap}\mathcal{L}_{lap} + \lambda_n\mathcal{L}_n \tag{4}$$

Herein, we compute the loss for S_0 and S_1 in relation to the pseudo ground truth white matter surface, while assessing the loss for S_2 and S_3 against the pseudo ground truth pial surface. This training strategy ensures the attainment of accurate and stable cortical surface reconstruction.

3 Experiments

3.1 Setup

Dataset and Evaluation Metrics. In our study, we utilized 877 T2-weighted brain MRI images from the third release of the developing Human Connectome Project (dHCP) [2,10,15]. These scans, obtained from newborns with a PMA between 26 and 45 weeks, were divided into training (70%), validation (10%), and test sets (20%). Geometric accuracy was evaluated using Chamfer Distance (CD), average symmetric surface distance (ASSD), and Hausdorff Distance (HD). These metrics were calculated on point clouds, each containing 100k uniformly sampled points from both predicted and pseudo ground-truth surfaces. Furthermore, the ratio of self-intersecting (SIF) faces is employed as a metric for assessing the geometric quality of the mesh.

Implementation Details. We sequentially trained the deformation blocks along with their corresponding 3D Unet. DB_1 and DB_3 underwent a 200-epoch training, with the initial 100 epochs employing a learning rate of 0.0001, a λ_{lap} value of 0.5, and a λ_n value of 0.0005. Subsequently, for the next 100 epochs, the learning rate remained constant, and both λ_{lap} and λ_n experienced decay with a weight of 0.5. DB_2 and DB_4 are trained for 100 epochs with a learning rate of 0.0001, and both λ_{lap} and λ_n are set to the same decayed value. The Adam optimizer was employed for updating the model parameters. Throughout the experiments, focus was solely directed towards the left hemi-cortical surface.

Table 1. Comparative results of cortical surface reconstruction on dHCP.

Method	White Matter Surface			Pial Surface		
	ASSD (mm)	HD (mm)	SIF (%)	ASSD (mm)	HD (mm)	SIF (%)
Corticalflow	0.122 (±0.020)	0.252 (±0.060)	0.048 (±0.032)	0.149 (±0.025)	0.331 (±0.063)	9.798 (±2.902)
CFPP	0.119 (±0.021)	0.246 (±0.064)	0.075 (±0.057)	0.126 (±0.023)	0.269 (±0.057)	2.457 (±1.003)
CoTAN	0.109 (±0.018)	0.220 (±0.048)	0.001 (±0.004)	0.120 (±0.022)	0.254 (±0.050)	0.071 (±0.034)
Proposed	**0.103 (±0.017)**	**0.209 (±0.047)**	**0.001 (±0.001)**	**0.117 (±0.021)**	**0.245 (±0.051)**	**0.035 (±0.016)**

3.2 Comparative Results

To assess the efficacy of the proposed method, we conducted a comparative analysis with existing cortical surface reconstruction models (Corticalflow [12], CFPP [18], and CoTAN [13]). Given the absence of these methods being evaluated on the dHCP dataset or providing a dataset partitioning scheme, we retrained and evaluated each model using their default parameters. As depicted in Table 1, the proposed method surpasses all existing methods in geometric accuracy and demonstrates the minimal standard deviation. Furthermore, as shown in Fig. 3, the visualization results indicate that the proposed method exhibits a smoother error map compared to other methods, demonstrating better performance on vertex-wise level. These results underscore the accurate and robust performance of our proposed model.

3.3 Ablation Study

As shown in the Fig. 1, the proposed method comprises four sequentially connected deformation blocks. This section offers a detailed comparative analysis of the impact of standard DB, Dynamic DB, and neonatal DB on the reconstruction performance of the left white matter surface. Reiterating, the standard DB utilizes a fixed-step ODE for integrating the stationary flow field. The Dynamic DB extends the standard DB by incorporating a time-varying dynamic flow field predicted through the dynamic flow field estimator. Subsequently, the Neonatal DB builds upon the Dynamic DB, integrating a variable-step Age-CDE inspired by brain development to handle the dynamic flow field.

For template-to-white-matter surface reconstruction, position1 corresponds to DB_1, while position2 corresponds to DB_2. As illustrated in Table 2, at position1, the proposed Neonatal DB achieves superior geometric accuracy. In comparison to the Dynamic DB, it demonstrates reductions of 1.50%, 1.76%, and 2.07% in CD, ASSD, and HD, respectively. When compared to the Standard DB, the Neonatal DB achieves reductions of 2.72%, 4.50%, and 4.32% in CD, ASSD,

Table 2. Comparative Results of White Matter Surface Reconstruction Quality under Different DB Configurations.

Position1	Position2	White Matter Surface Reconstruction		
		CD (mm)	ASSD (mm)	HD (mm)
Standard DB	–	0.282 ± 0.074	0.145 ± 0.025	0.309 ± 0.078
	Standard DB	0.262 ± 0.064	0.125 ± 0.022	0.260 ± 0.067
	Dynamic DB	0.255 ± 0.060	0.116 ± 0.020	0.238 ± 0.063
Dynamic DB	–	0.240 ± 0.076	0.119 ± 0.020	0.241 ± 0.058
	Standard DB	0.230 ± 0.073	0.109 ± 0.019	0.219 ± 0.052
	Dynamic DB	0.228 ± 0.075	0.105 ± 0.019	0.210 ± 0.053
Neonatal DB	–	0.230 ± 0.040	0.110 ± 0.017	0.222 ± 0.046
	Standard DB	0.225 ± 0.040	0.105 ± 0.017	0.210 ± 0.046
	Dynamic DB	**0.224 ± 0.040**	**0.103 ± 0.017**	**0.209 ± 0.047**

and HD, respectively. Moreover, the Neonatal DB exhibits enhanced stability, as depicted in Fig. 2. The surface deformations at each time point during the Age-CDE process for different PMA white matter surfaces show smaller margins. At position2, the Dynamic DB surpasses the Standard DB in reconstruction results, affirming the effectiveness of the dynamic flow field estimator. We also conduct experiment on the reconstruction from the white matter surface to the pial surface, we obtain the same conclusion that the employment of Neonatal DB and Dynamic DB as DB_3 and DB_4 achieves the superior result.

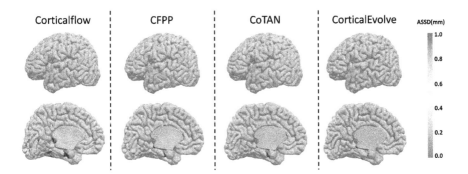

Fig. 3. Visualization of the distance between the predicted pial surface and the corresponding pseudo-ground-truth surface.

4 Conclusion

In this paper, we propose CorticalEvolve, a diffeomorphic transformation based model specifically designed for infant cortical reconstruction. We interpret the deformation process of template mesh to cortical surface as brain development and therefore develop a age-conditioned ODE to capture the diverse developmental of different age groups. And the introduce of dynamic flow field estimator further enhance model's capability to capture the time information. Since this study focuses solely on cortical surface reconstruction from brain MRI, we did not fully exploit the model's ability to generate surfaces at any given PMA. However, Age-CDE essentially depict the brain developmental process, possessing the potential to evolve into cortical surface at various time point. Our future work will focus on extending the generation capability of CorticalEvolve to datasets encompassing paired data from multiple time points, facilitating the dual task of cerebral cortex reconstruction and generation.

Acknowledgments. This work is supported by Key-Area Research and Development Program of Guangdong Province (2023B0303040001), Guangdong Basic and Applied Basic Research Foundation (2024A1515010180) and Guangdong Provincial Key Laboratory of Human Digital Twin (2022B1212010004).

Disclosure of Interests. The authors have no competing interests to declare that are relevant to the content of this article.

References

1. Bongratz, F., Rickmann, A.-M., Pölsterl, S., Wachinger, C.: Vox2cortex: fast explicit reconstruction of cortical surfaces from 3D MRI scans with geometric deep neural networks. In: Proceedings of the IEEE/CVF Conference on Computer Vision and Pattern Recognition, pp. 20773–20783 (2022)
2. Bozek, J., et al.: Construction of a neonatal cortical surface atlas using multimodal surface matching in the developing human connectome project. Neuroimage **179**, 11–29 (2018)
3. Cruz, R.S., Lebrat, L., Bourgeat, P., Fookes, C., Fripp, J., Salvado, O.: Deepcsr: a 3D deep learning approach for cortical surface reconstruction. In: Proceedings of the IEEE/CVF Winter Conference on Applications of Computer Vision, pp. 806–815 (2021)
4. Dale, A.M., Fischl, B., Sereno, M.I.: Cortical surface-based analysis: I segmentation and surface reconstruction. Neuroimage **9**(2), 179–194 (1999)
5. Dubois, J., et al.: Primary cortical folding in the human newborn: an early marker of later functional development. Brain **131**(8), 2028–2041 (2008)
6. Dubois, J., et al.: Mapping the early cortical folding process in the preterm newborn brain. Cereb. Cortex **18**(6), 1444–1454 (2008)
7. Fischl, B.: Freesurfer. Neuroimage **62**(2), 774–781 (2012)
8. Fischl, B., Sereno, M.I., Dale, A.M.: Cortical surface-based analysis: Ii: inflation, flattening, and a surface-based coordinate system. Neuroimage **9**(2), 195–207 (1999)
9. Henschel, L., Conjeti, S., Estrada, S., Diers, K., Fischl, B., Reuter, M.: Fastsurfer-a fast and accurate deep learning based neuroimaging pipeline. Neuroimage **219**, 117012 (2020)
10. Hughes, E.J., et al.: A dedicated neonatal brain imaging system. Magn. Reson. Med. **78**(2), 794–804 (2017)

11. Jack, C.R., Jr., et al.: The Alzheimer's disease neuroimaging initiative (ADNI): MRI methods. J. Magnet. Reson. Imaging Offic. J. Int. Soc. Magnet. Reson. Med. **27**(4), 685–691 (2008)
12. Lebrat, L., et al.: Corticalflow: a diffeomorphic mesh transformer network for cortical surface reconstruction. Adv. Neural. Inf. Process. Syst. **34**, 29491–29505 (2021)
13. Ma, Q., et al.: Conditional temporal attention networks for neonatal cortical surface reconstruction. In: Greenspan, H., et al. (eds.) MICCAI 2023, vol. 14223, pp. 312–322. Springer, Cham (2023). https://doi.org/10.1007/978-3-031-43901-8_30
14. Ma, Q., Li, L., Robinson, E.C., Kainz, B., Rueckert, D., Alansary, A.: Cortexode: learning cortical surface reconstruction by neural odes. IEEE Trans. Med. Imaging **42**(2), 430–443 (2022)
15. Makropoulos, A., et al.: The developing human connectome project: a minimal processing pipeline for neonatal cortical surface reconstruction. Neuroimage **173**, 88–112 (2018)
16. Orasanu, E., et al.: Cortical folding of the preterm brain: a longitudinal analysis of extremely preterm born neonates using spectral matching. Brain Behav. **6**, e00488 (2016)
17. Ronneberger, O., Fischer, P., Brox, T.: U-Net: convolutional networks for biomedical image segmentation. In: Navab, N., Hornegger, J., Wells, W.M., Frangi, A.F. (eds.) MICCAI 2015, Part III. LNCS, vol. 9351, pp. 234–241. Springer, Cham (2015). https://doi.org/10.1007/978-3-319-24574-4_28
18. Santa Cruz, R., et al.: Corticalflow++: boosting cortical surface reconstruction accuracy, regularity, and interoperability. In: Wang, L., Dou, Q., Fletcher, P.T., Speidel, S., Li, S. (eds.) MICCAI 2022. LNCS, vol. 13435, pp. 496–505. Springer, Cham (2022). https://doi.org/10.1007/978-3-031-16443-9_48
19. Studholme, C.: Mapping fetal brain development in utero using magnetic resonance imaging: the big bang of brain mapping. Annu. Rev. Biomed. Eng. **13**, 345–368 (2011)
20. Van Essen, D.C., et al.: The WU-MINN human connectome project: an overview. Neuroimage **80**, 62–79 (2013)

CSR-dMRI: Continuous Super-Resolution of Diffusion MRI with Anatomical Structure-Assisted Implicit Neural Representation Learning

Ruoyou Wu[1,2,3], Jian Cheng[4], Cheng Li[1], Juan Zou[5], Jing Yang[1,3], Wenxin Fan[1,3], Yong Liang[2], and Shanshan Wang[1(✉)]

[1] Paul C. Lauterbur Research Center for Biomedical Imaging, Shenzhen Institute of Advanced Technology, Chinese Academy of Sciences, Shenzhen 518055, China
ss.wang@siat.ac.cn
[2] Pengcheng Laboratory, Shenzhen 518055, China
[3] University of Chinese Academy of Sciences, Beijing 100049, China
[4] School of Computer Science and Engineering, Beihang University, Beijing 100191, China
[5] School of Physics and Electronic Science, Changsha University of Science and Technology, Changsha 410114, China

Abstract. Deep learning-based dMRI super-resolution methods can effectively enhance image resolution by leveraging the learning capabilities of neural networks on large datasets. However, these methods tend to learn a fixed scale mapping between low-resolution (LR) and high-resolution (HR) images, overlooking the need for radiologists to scale the images at arbitrary resolutions. Moreover, the pixel-wise loss in the image domain tends to generate over-smoothed results, losing fine textures and edge information. To address these issues, we propose a novel continuous super-resolution method for dMRI, called CSR-dMRI, which utilizes an anatomical structure-assisted implicit neural representation learning approach. Specifically, the CSR-dMRI model consists of two components. The first is the latent feature extractor, which primarily extracts latent space feature maps from LR dMRI and anatomical images while learning structural prior information from the anatomical images. The second is the implicit function network, which utilizes voxel coordinates and latent feature vectors to generate voxel intensities at corresponding positions. Additionally, a frequency-domain-based loss is introduced to preserve the structural and texture information, further enhancing the image quality. Extensive experiments on the publicly available HCP dataset validate the effectiveness of our approach. Furthermore, our method demonstrates superior generalization capability and can be applied to arbitrary-scale super-resolution, including non-integer scale factors, expanding its applicability beyond conventional approaches.

Keywords: Diffusion MRI · Continuous super-resolution · Implicit neural representation

1 Introduction

Diffusion magnetic resonance imaging (dMRI) can reflect early changes in the microstructure of brain tissue in neurological diseases by measuring the diffusion displacement distribution of water molecules in the brain tissue [5,11]. As a noninvasive method, dMRI is extensively utilized for the diagnosis of brain diseases. Moreover, it provides a distinctive avenue for exploring the neural foundations of human cognitive behavior. However, to accurately estimate quantitative evaluation parameters for diffusion tensor imaging (DTI) or diffusion kurtosis imaging (DKI), it is typically necessary to acquire HR data, resulting in prolonged acquisition time for dMRI. This can lead to motion artifacts and patient discomfort. In clinical practice, reducing acquisition time is often achieved by sacrificing the spatial resolution of dMRI images, which can affect the estimation accuracy of DTI or DKI parameters. Therefore, the reconstruction of HR dMRI images from LR dMRI images holds significant clinical value.

Image post-processing methods are one of the feasible solutions to address the issue. There are some traditional interpolation methods, such as nearest-neighbor interpolation and linear interpolation, that can be utilized. However, Van et al., [17] pointed out that interpolation methods often lead to blurred image edges and are unable to recover fine details. On the other hand, super-resolution reconstruction (SR) methods are an interesting alternative solution that can generate HR images from LR images [10,16,18,19,21]. The existing SR methods can be roughly divided into two categories: model-based methods [8,9,12–15,20] and data-driven methods [3,4,7]. Model-based methods rely on mathematical models to build connections between LR and HR images, while data-driven methods utilize neural networks to learn the nonlinear mapping relationship between LR and HR images. With the fast development of deep learning (DL), SR methods based on deep learning have been widely investigated [6,22,24]. For example, Lim et al., [6] proposed EDSR, which improves super-resolution image reconstruction quality by stacking deeper or wider networks to extract more information within the same computational resources. Zhang et al., [24] introduced the RDN network, which utilizes residual dense blocks with densely connected convolutional layers to extract rich local features. It also allows direct connections from all layers of the previous residual dense block (RDB) to the current RDB, making full use of hierarchical features extracted from the original LR image. These methods have shown promising performance in enhancing the quality of the reconstructed super-resolution images.

Recently, some researchers have started applying SR techniques to diffusion MRI [1,7], primarily using learning-based methods. For instance, Chatterjee et al., [1] proposed a ShuffleUNet architecture to reconstruct super-resolution dMRI data, addressing issues of image blurring and over-smoothing by replacing stridden convolutions with lossless pooling layers. Luo et al., [7] proposed a sub-pixel convolution generative adversarial attentional network (SPC-GAAN) for the reconstruction super-resolution dMRI data, achieving promising results. However, these methods can only reconstruct super-resolution dMRI data with a fixed integer scale factor, limiting their applicability. In practical applications, due to differences in scanning protocols, the resolution of acquired dMRI data is inconsistent. To better facilitate clinical analysis, it is necessary to reconstruct them to a consistent resolution. The aforementioned methods may face challenges in achieving this. Additionally, the pixel-wise loss in the image domain, which was utilized in these methods, tends to generate over-smooth results, leading to the loss of fine textures and edge information.

To address these issues, we propose an anatomical structure-assisted implicit neural representation learning framework for continuous super-resolution of dMRI images. To the best of our knowledge, this is the first attempt of implicit neural representation learning in dMRI super-resolution. Specifically, we introduce implicit neural representation into the super-resolution of dMRI, enabling arbitrary-scale super-resolution for dMRI, and it can be used for transforming dMRI data at different resolutions. Our contributions can be summarized as follows:

1) We propose a novel paradigm for arbitrary-scale super-resolution in diffusion MRI by combining anatomical structure-assisted implicit neural representation learning, called CSR-dMRI. It can be used to transform dMRI data at different resolutions into dMRI data at the same resolution, facilitating clinical analysis.

2) The details and texture information of the image are improved by introducing anatomical image and frequency-domain-based loss.

3) Extensive experiments on the public HCP dataset demonstrate the effectiveness of our approach. Furthermore, our method exhibits better generalization and can be applied to non-integer scale factors, expanding the applicability of this approach.

2 Methods

2.1 Implicit Neural Representation Learning

In implicit neural representation learning, a voxel can be represented by a neural network as a continuous function. the network $\mathcal{F}_\theta$ with parameters θ can be defined as:

$$I = \mathcal{F}_\theta(c),\ c \in [-1,1]^3,\ I \in \mathbb{R}^3 \tag{1}$$

where the input c is the normalized coordinate index in the voxel spatial field, and the output I is the corresponding intensity value in the voxel. The network function $\mathcal{F}$ maps coordinates to voxel intensities, effectively encoding the internal information of the entire voxel into the network parameters. The network $\mathcal{F}$ with parameters θ is also referred to as the neural representation of the voxel. Our voxel coordinates are constructed based on the size of the dMRI data and normalized along each dimension to the range of $[-1,1]$. This network is only applicable to the super-resolution of individual dMRI images and is not suitable for multiple dMRI images. Additional auxiliary information needs to be provided.

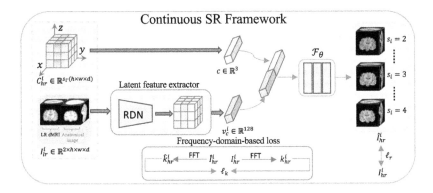

Fig. 1. The overall structure of the proposed CSR-dMRI model. The latent feature extractor is used to extract latent space feature maps from LR dMRI and anatomical images. Implicit function network $\mathcal{F}_\theta$ is used to predict voxel intensity at spatial coordinate c. Frequency-domain-based loss is used to improve the details and texture information of the image.

2.2 Model Overview

The overall architecture of the proposed CSR-dMRI model is illustrated in Fig. 1. Due to the primary focus of dMRI data on tissue microstructure and diffusion information, which provides limited anatomical information. So, we introduce T1-weighted data to provide anatomical prior information. Additionally, by incorporating frequency-domain-based loss to preserve texture and structural information in the image, the quality of the image is enhanced. The entire network consists of a latent feature extractor and an implicit function network. The latent feature extractor is used to extract features from LR dMRI and anatomical images, while the implicit function network predicts voxel intensities at arbitrary coordinates by integrating coordinates and corresponding latent feature vectors, thereby achieving super-resolution. Specifically, for a given pair of LR dMRI and anatomical images from the dataset $D = \{I_{lr}^i \in \mathbb{R}^{2 \times h \times w \times d}, I_{hr}^i \in \mathbb{R}^{s_i \cdot (h \times w \times d)}\}_{i=1}^{N}$, along with their corresponding HR dMRI, where N represents the total number of samples, and s_i represents the upsampling scale factor for the i-th sample pair, dimension 2 indicates concatenating DWI images and anatomical images along the channel. We first utilize a latent feature extractor to transform the LR dMRI and anatomical images into feature maps $v_{lr}^i \in \mathbb{R}^{h \times w \times d \times 128}$, where each element $v_c^i \in \mathbb{R}^{128}$ corresponding to the feature vector of the LR dMRI at coordinate c. For any voxel coordinate c in HR dMRI, we generate the corresponding feature map $v_{hr}^i \in \mathbb{R}^{s_i \cdot (h \times w \times d)}$ by bilinear interpolation of v_{lr}^i. Subsequently, the queried coordinate c and the corresponding feature vector v_c^i are concatenated as input to $\mathcal{F}_\theta$, and the output of the $\mathcal{F}_\theta$ is the predicted voxel intensity $\hat{I}_{hr}^i(c)$ at spatial coordinate c. By minimizing the difference between the predicted voxel intensity $\hat{I}_{hr}^i(c)$ and the real voxel intensity $I_{hr}^i(c)$, both the latent feature extractor and implicit function network are simultaneously optimized.

Latent Feature Extractor. For the latent feature extractor, inspired by the works [2,23], Residual Dense Network (RDN) [24] is used to extract latent space feature vectors from LR dMRI and anatomical image data. The latent feature extractor takes the LR data $I_{lr}^i \in \mathbb{R}^{2 \times h \times w \times d}$ as input and outputs the feature map $v_{lr}^i \in \mathbb{R}^{h \times d \times w \times 128}$. This latent space feature extraction approach helps the implicit function network $\mathcal{F}_\theta$ effectively integrate local information of the image, enabling the recovery of details in HR images even at large upsampling scale factors. Additionally, to maintain the scale of the extracted features uncharged, we removed the upsampling operation from the last layer of the original RDN network and expanded all 2D convolutional layers to 3D convolutional layers. To ensure extracting a sufficient amount of features for each coordinate position, the output channel number of the last layer is set to 128.

Implicit Function Network. The implicit function network $\mathcal{F}_\theta$ consists of a sequence of 8 consecutive fully connected layers, each followed by a *ReLU* activation layer. A residual connection is established between the input of the network and the output of the fourth *ReLU* activation layer. The goal of $\mathcal{F}_\theta$ is to predict the voxel intensity at any spatial coordinate c. The specific process is as follows:

$$\hat{I}_{hr}^i = \mathcal{F}_\theta(c, v_c^i) \tag{2}$$

where c represents the spatial coordinate position to be predicted, v_c^i represents the specific latent feature vector of the voxel at spatial coordinate c. Instead of using spatial coordinates alone as inputs to the implicit function network, we combine the spatial coordinates with their corresponding latent feature vectors. This approach effectively integrates the local semantic information from the image, thereby enhancing the ability of $\mathcal{F}_\theta$ to recover image details.

Loss Function. In image super-resolution tasks, the pixel-wise loss is commonly used as a loss function to improve the model's performance and convergence [25]. However, since pixel-wise loss does not take into account the perceptual quality of the image, the generated results tend to be smooth and lack high-frequency details. To address this issue, we introduce a frequency-domain-based loss that better captures the frequency information of the image, preserving both the structure and texture details in the image, thereby further improving the image quality. The specific process is as follows:

$$\ell_k = \frac{1}{N} \sum_{i=1}^{N} \left| \hat{k}_{hr}^i - k_{hr}^i \right| \tag{3}$$

where $\hat{k}_{hr}^i$ and k_{hr}^i represent the fast Fourier transforms of $\hat{I}_{hr}^i$ and I_{hr}^i, respectively, as shown in Fig. 1. N denotes the total number of samples. Finally, the objective function of the CSR-dMRI model is defined as:

$$\ell_f = \ell_r + \lambda_k \ell_k \tag{4}$$

we set λ_k=0.01 empirically to balance the two losses, ℓ_r denotes the L1 loss between $\hat{I}_{hr}^i$ and I_{hr}^i.

3 Experiments

3.1 Datasets

In this paper, we utilize data from the Human Connectome Project (HCP) WU-Minn-Ox Consortium public dataset[1]. We select 100 pre-processed diffusion and T1-weighted MRI data. Out of these, 70 were used for training, 10 for validation, and 20 for testing. The whole-brain diffusion MRI data were acquired with an isotropic resolution of 1.25 mm, featuring four b-values (0, 1000, 2000, 3000). For diffusion MRI data, we only utilize the data from b1000, normalized by dividing the b1000 data by the average value of 18 b0 data. Firstly, we extract 9 patches of 40^3 dimension size from each sample. Subsequently, these patches were cropped to generate HR patches of $(10 \times s)^3$ dimension size. Finally, the HR patches were downsampled to obtain LR patches of 10^3 dimension size using bicubic interpolation. The scale s is randomly sampled from a uniform distribution $\mathcal{U}(2,3)$.

3.2 Experimental Setup

All networks are trained using the Pytorch framework with one NVIDIA RTX A6000 GPU (with 48 GB memory). The Adam optimizer is employed for model training with an initial learning rate of 10^{-4}, and the batch size is set to 9. Every 200 epochs, the learning rate is multiplied by 0.5. The model is trained for a total of 1000 epochs. We compare our CSR-dMRI model with some state-of-the-art methods, including traditional interpolation method: Bicubic, deep learning-based fixed-scale super-resolution methods: DCSRN [3], ResCNN3D [4], and an arbitrary scale super-resolution method: ArSSR [23]. PSNR and SSIM are used for quantitative evaluation.

3.3 Results

To comprehensively assess the effectiveness of our approach, we conduct experiments in three aspects: the first involves in-distribution experiments, including 2× and 3× experiments. The second is out-of-distribution experiments where we tested the results under the 4×. The third involves experiments with non-integer scale factors, testing the results under the 2.4×.

---

[1] https://www.humanconnectome.org.

Table 1. Quantitative results of different methods on the HCP dataset. Bold numbers indicate the best results.

method	In-distribution				Out-of-distribution		Non-integer scale	
	2×		3×		4×		2.4×	
	PSNR	SSIM	PSNR	SSIM	PSNR	SSIM	PSNR	SSIM
Bicubic	25.8880	0.9075	23.6657	0.8360	22.6657	0.7774	24.6268	0.8778
DCSRN	26.4039	0.8901	24.3236	0.8479	23.1425	0.7974	25.1044	0.8797
ResCNN3D	26.7900	0.9291	24.7841	0.8667	23.3626	0.8135	25.4493	0.8927
ArSSR	26.6123	0.9343	25.3040	0.9059	23.8770	0.8554	26.2907	0.9266
CSR-dMRI	**27.3611**	**0.9458**	**26.0762**	**0.9235**	**24.5061**	**0.8752**	**27.1196**	**0.9410**

Quantitative Results. Table 1 shows the quantitative results of different methods under in-distribution, out-of-distribution, and non-integer scale factor conditions. Since the training scales for the DCSRN and ResCNN3D models are fixed at 2× and 3×, respectively, when testing with the non-integer scale factor of 2.4×, we first use the 3× model for super-resolution and then perform downsampling to obtain the 2.4× results. For testing with the out-of-distribution scale factor of 4×, we first use the 3× model for super-resolution and then upsampling to obtain the 4× results. The arbitrary-scale super-resolution method ArSSR and our CSR-dMRI model are both trained at scale factors of 2× to 3×. Overall, our method achieves the best results across all scales. In out-of-distribution scenarios, our method outperforms existing approaches and maintains better performance even with non-integer scale factors. The experimental results confirm that our method excels in performance and generalization compared to existing methods, and it can also accommodate non-integer scale factors effectively, meeting the diverse imaging resolution needs of medical professionals.

Qualitative Results. To further compare the performance of the models, Fig. 2 displays the reconstruction results of various methods at a non-integer scale (2.4×) in the axial, sagittal, and coronal directions, along with corresponding local zoomed-in results for two areas. From a visual perspective, our method can preserve more detailed information. Compared to ArSSR, our method CSR-dMRI performs better in terms of image details, benefiting from the introduced anatomical images and frequency-domain-based loss constraints. Additionally, locally magnified images also indicate that our method can eliminate block artifacts and preserve better texture details. Overall, our method achieves the best results.

3.4 Ablation Study

To evaluate the effectiveness of key components in the CSR-dMRI model, we conduct ablation experiments on different components, as shown in Table 2. INR

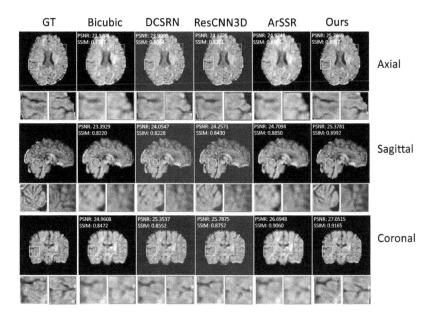

Fig. 2. Qualitative results of different methods on the HCP dataset and corresponding local zoomed-in results. Including the axial, sagittal, and coronal directions. The corresponding quantitative metrics are presented in the top-left corner.

indicates whether implicit neural representations are used, as ArSSR [23], T1w indicates the inclusion of anatomical image, ℓ_k indicates the introduction of frequency-domain-based loss. The quantitative results in Table 2 indicate that different components have positively contributed to the results. Furthermore, the structural prior information provided by anatomical images effectively enhances the image quality.

Table 2. Ablation experiments of different components on the HCP dataset. Bold numbers indicate the best results.

Setting	INR	T1w	ℓ_k	In-distribution				out-of-distribution		Non-integer scale	
				2×		3×		4×		2.4×	
				PSNR	SSIM	PSNR	SSIM	PSNR	SSIM	PSNR	SSIM
Baseline	×	×	×	26.9584	0.9352	24.8094	0.8789	23.3321	0.8240	25.4377	0.9015
M1	√	×	×	26.6123	0.9343	25.3040	0.9059	23.8770	0.8554	26.2907	0.9266
M2	×	√	×	27.0454	0.9444	25.8534	0.9222	24.2769	0.8734	26.9627	0.9408
M3	×	×	√	26.8047	0.9354	25.5623	0.9089	24.0327	0.8575	26.5996	0.9294
CSR-dMRI	√	√	√	**27.3611**	**0.9458**	**26.0762**	**0.9235**	**24.5061**	**0.8752**	**27.1196**	**0.9410**

4 Conclusion

In this paper, we propose a CSR-dMRI method for continuous super-resolution of diffusion MRI with anatomical structure-assisted implicit neural representation learning. Details and texture information in the image are preserved by incorporating anatomical images and frequency-domain-based loss during training. Extensive experiments on the HCP dataset indicate superior performance and generalization of our CSR-dMRI model, showing applicability across non-integer scale factors. This contributes to addressing the clinical demand for images at different resolutions.

Acknowledgements. This research was partly supported by the National Natural Science Foundation of China (62222118, U22A2040), Guangdong Provincial Key Laboratory of Artificial Intelligence in Medical Image Analysis and Application (2022B1212010011), Shenzhen Science and Technology Program (RCYX20210706092104034, JCYJ20220531100213029), and Youth Innovation Promotion Association CAS.

References

1. Chatterjee, S., et al.: ShuffleUNet: super resolution of diffusion-weighted MRIs using deep learning. In: 2021 29th European Signal Processing Conference (EUSIPCO), pp. 940–944. IEEE (2021)
2. Chen, Y., Liu, S., Wang, X.: Learning continuous image representation with local implicit image function. In: Proceedings of the IEEE/CVF Conference on Computer Vision and Pattern Recognition, pp. 8628–8638 (2021)
3. Chen, Y., Xie, Y., Zhou, Z., Shi, F., Christodoulou, A.G., Li, D.: Brain MRI super resolution using 3D deep densely connected neural networks. In: 2018 IEEE 15th International Symposium on Biomedical Imaging (ISBI 2018), pp. 739–742. IEEE (2018)
4. Du, J., et al.: Super-resolution reconstruction of single anisotropic 3D MR images using residual convolutional neural network. Neurocomputing **392**, 209–220 (2020)
5. Li, C., Li, W., Liu, C., Zheng, H., Cai, J., Wang, S.: Artificial intelligence in multiparametric magnetic resonance imaging: a review. Med. Phys. **49**(10), e1024–e1054 (2022)
6. Lim, B., Son, S., Kim, H., Nah, S., Mu Lee, K.: Enhanced deep residual networks for single image super-resolution. In: Proceedings of the IEEE Conference on Computer Vision and Pattern Recognition Workshops, pp. 136–144 (2017)
7. Luo, S., Zhou, J., Yang, Z., Wei, H., Fu, Y.: Diffusion MRI super-resolution reconstruction via sub-pixel convolution generative adversarial network. Magn. Reson. Imaging **88**, 101–107 (2022)
8. Nedjati-Gilani, S., Alexander, D.C., Parker, G.J.: Regularized super-resolution for diffusion MRI. In: 2008 5th IEEE International Symposium on Biomedical Imaging: From Nano to Macro, pp. 875–878. IEEE (2008)
9. Ning, L., Setsompop, K., Michailovich, O., Makris, N., Westin, C.-F., Rathi, Y.: A compressed-sensing approach for super-resolution reconstruction of diffusion MRI. In: Ourselin, S., Alexander, D.C., Westin, C.-F., Cardoso, M.J. (eds.) IPMI 2015. LNCS, vol. 9123, pp. 57–68. Springer, Cham (2015). https://doi.org/10.1007/978-3-319-19992-4_5

10. Park, S.C., Park, M.K., Kang, M.G.: Super-resolution image reconstruction: a technical overview. IEEE Signal Process. Mag. **20**(3), 21–36 (2003)
11. Razek, A.A.K.A., Ashmalla, G.A.: Assessment of paraspinal neurogenic tumors with diffusion-weighted MR imaging. Eur. Spine J. **27**, 841–846 (2018)
12. Scherrer, B., Gholipour, A., Warfield, S.K.: Super-resolution in diffusion-weighted imaging. In: Fichtinger, G., Martel, A., Peters, T. (eds.) MICCAI 2011. LNCS, vol. 6892, pp. 124–132. Springer, Heidelberg (2011). https://doi.org/10.1007/978-3-642-23629-7_16
13. Shi, F., Cheng, J., Wang, L., Yap, P.T., Shen, D.: LRTV: MR image super-resolution with low-rank and total variation regularizations. IEEE Trans. Med. Imaging **34**(12), 2459–2466 (2015)
14. Shi, F., Cheng, J., Wang, L., Yap, P.-T., Shen, D.: Super-resolution reconstruction of diffusion-weighted images using 4D low-rank and total variation. In: Fuster, A., Ghosh, A., Kaden, E., Rathi, Y., Reisert, M. (eds.) Computational Diffusion MRI. MV, pp. 15–25. Springer, Cham (2016). https://doi.org/10.1007/978-3-319-28588-7_2
15. Tobisch, A., Neher, P.F., Rowe, M.C., Maier-Hein, K.H., Zhang, H.: Model-based super-resolution of diffusion MRI. In: Schultz, T., Nedjati-Gilani, G., Venkataraman, A., O'Donnell, L., Panagiotaki, E. (eds.) Computational Diffusion MRI and Brain Connectivity. MV, pp. 25–34. Springer, Cham (2014). https://doi.org/10.1007/978-3-319-02475-2_3
16. Umirzakova, S., Ahmed, S., Khan, L.U., Whangbo, T.: Medical image super-resolution for smart healthcare applications: a comprehensive survey. Inform. Fusion **103**, 102075 (2023)
17. Van Ouwerkerk, J.: Image super-resolution survey. Image Vis. Comput. **24**(10), 1039–1052 (2006)
18. Wang, S., et al.: Review and prospect: artificial intelligence in advanced medical imaging. Front. Radiol. **1**, 781868 (2021)
19. Wang, S., et al.: Knowledge-driven deep learning for fast MR imaging: undersampled MR image reconstruction from supervised to un-supervised learning. Magn. Reson. Med. **92**(2), 496–518 (2024)
20. Wang, S., et al.: PARCEL: physics-based unsupervised contrastive representation learning for multi-coil MR imaging. IEEE/ACM Trans. Comput. Biol. Bioinf. **20**(5), 2659–2670 (2022)
21. Wang, S., Xiao, T., Liu, Q., Zheng, H.: Deep learning for fast MR imaging: a review for learning reconstruction from incomplete k-space data. Biomed. Signal Process. Control **68**, 102579 (2021)
22. Wang, Z., Chen, J., Hoi, S.C.: Deep learning for image super-resolution: a survey. IEEE Trans. Pattern Anal. Mach. Intell. **43**(10), 3365–3387 (2020)
23. Wu, Q., et al.: An arbitrary scale super-resolution approach for 3D MR images via implicit neural representation. IEEE J. Biomed. Health Inform. **27**(2), 1004–1015 (2022)
24. Zhang, Y., Tian, Y., Kong, Y., Zhong, B., Fu, Y.: Residual dense network for image super-resolution. In: Proceedings of the IEEE Conference on Computer Vision and Pattern Recognition, pp. 2472–2481 (2018)
25. Zhao, H., Gallo, O., Frosio, I., Kautz, J.: Loss functions for image restoration with neural networks. IEEE Trans. Comput. Imaging **3**(1), 47–57 (2016)

Atherosclerotic Plaque Stability Prediction from Longitudinal Ultrasound Images

Jan Kybic[1]((✉)), David Pakizer[2], Jiří Kozel[2], Patricie Michalčová[2], František Charvát[3], and David Školoudík[2]

[1] Faculty of Electrical Engineering, Czech Technical University in Prague, Prague, Czech Republic
kybic@fel.cvut.cz
[2] Center for Health Research, Faculty of Medicine, University of Ostrava, Ostrava, Czech Republic
[3] Military University Hospital, Prague, Czech Republic

Abstract. We aim to predict the stability of carotid artery plaques from longitudinal ultrasound images. This is important since atherosclerosis is the primary cause of heart disease and stroke. Accurately predicting plaque stability would allow for more targeted follow-up and treatment, saving healthcare costs.

We analyze data from over 400 patients followed for 3 years, exceeding the size of previous studies. We first localize the carotid artery and segment the plaque within the images. A self-supervised learning approach was used for plaque segmentation, leveraging the power of unlabeled data. The plaque stability predictor uses three image channels derived from the ultrasound image and its segmentation. As an auxiliary task, we predict the plaque width, which helps to prevent overfitting. The balance between the criteria is maintained automatically.

Our estimate of the plaque width correlated well with expert measurements ($\rho = 0.56$). We confirmed that there is a relationship between the plaque ultrasound appearance in longitudinal images and their stability. However, the future width correlation and the plaque stability prediction performance remained modest (AUC $= 0.61$), similar to previous studies.

Keywords: carotid · stability · segmentation · regression · deep learning · self-supervision

1 Introduction

Atherosclerosis, characterized by plaque buildup in blood vessels, poses a significant risk by restricting blood flow. It can lead to a stroke or blockage of smaller arteries of the brain or heart.

Although carotid artery plaques are easily detectable by ultrasound [5,13] and can be treated by medication or surgery, the mechanism underlying their

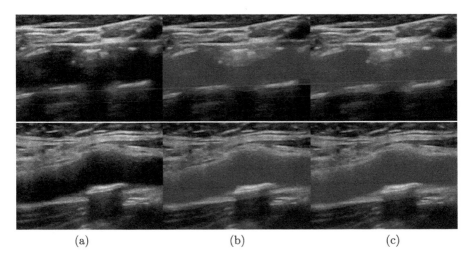

Fig. 1. Input image *(a)*, expert segmentation *(b)*, and automatic segmentation *(c)*.

stability remains poorly understood. While many risk factors are known, there is a lack of reliable tools for predicting how stable a specific plaque is.

While atherosclerotic plaques are common in older adults [19], most of them are *stable* and cause no adverse effects. However, these patients nevertheless require monitoring and potentially medication.

It this work we attempt to predict the plaque stability from standard longitudinal B-mode ultrasound images of the carotid artery. Previous studies have linked plaque stability to image features such as the plaque size, surface irregularity, echogenicity or texture [1,4,6,10,18], However, the associations are weak, limiting their use in clinical decision-making. Additionally, studies have yielded conflicting results, some finding no correlation [7].

Unlike most methods exploring artificial intelligence for carotid artery image analysis [3,12,17,20], we are not concerned with classifying the current state, but we aim to *predict* the future plaque stability over several years. Our advantage lies in having access to one of the largest datasets of this kind reported in the literature.

1.1 Data

Our data comes from the clinical study ANTIQUE [21] approved by the Ethics Board of Ethics Board of the Faculty Hospital Ostrava (approval number 605/2014), which followed a large group of subjects. We used images from two examinations approximately 3 years apart for each of the two carotids (left and right) of each subject, that had an observable atherosclerotic plaque in both examinations, obtaining 6931 longitudinal ultrasound images (see Fig. 1, left column) from 466 subjects with a median age of 69 years. For each examination and each carotid, our experts estimated the atherosclerotic plaque width d. Furthermore, the experts manually segmented 148 longitudinal images and marked the vessel wall, lumen, and atherosclerotic plaque (Fig. 1b).

A plaque from the first examination is considered *unstable*, if its width increase between examinations $D = d_2 - d_1$ is larger than a threshold $\tau = 0.1$ mm.

2 Methods

The pipeline consist of segmenting the images, predicting the parameters of interest, and optionally aggregating the results.

2.1 Segmentation

A region of interest of size 400×568 containing the carotid artery is automatically extracted by a Faster R-CNN architecture [14] trained on 300 manually annotated images [1]. The segmentation into $K = 4$ classes (background, vessel lumen, vessel wall, plaque) is performed by a U-Net [16] f with a ResNet-18 [8] backbone [9]. To take advantage of the large set of unlabeled images available, we use the idea of self-supervised augmentation consistency (SAC), a simplified version of [2]. Both the supervised and self-supervised learning use the Dice loss:

$$\ell_{\text{Dice}}(p, q) = 1 - \frac{2}{K} \sum_{k=1}^{K} \frac{\sum_i p_k(i) q_k(i)}{\epsilon + \sum_i p_k(i) + q_k(i)} \quad (1)$$

where $p_k(i)$, $q_k(i)$ are probabilities of class k at pixel i in the two images and ϵ is a small constant. The Dice loss is symmetric and has complexity $O(K)$. We use the 'macro' version, averaging over classes to give more weight to small but important classes, such as the plaque.

We optimize the sum of two losses, the first being the supervised loss

$$L_{\text{sup}} = \mathbb{E}_{(I,S) \sim \Omega_s} \left[\ell_{\text{Dice}}\left(f(\phi_I(I)), \phi_S(S)\right) \right] \quad (2)$$

where the image I and its reference segmentation S are sampled from the training part of the supervised dataset Ω_s. To emphasize the analogy with self-supervised learning below, we explicitly write the augmentation operators for images and segmentations, ϕ_I and ϕ_S. In our case, geometrical augmentations applied to both I and S are horizontal flip and random rotation and cropping, while brightness and contrast transforms and multiplicative noise are only applied to the image I.

The main idea of self-supervised augmentation consistency (Fig. 2) is that a geometrically transformed image should lead to a segmentation transformed in the same way (equivariance), while applying an intensity transformation or adding noise should not change the segmentation (invariance). This leads to a loss function

$$L_{\text{SAC}} = \mathbb{E}_{I \sim \Omega_u} \left[\ell_{\text{Dice}}\left(f(\phi_I(I)), \phi_S(f(I))\right) \right] \quad (3)$$

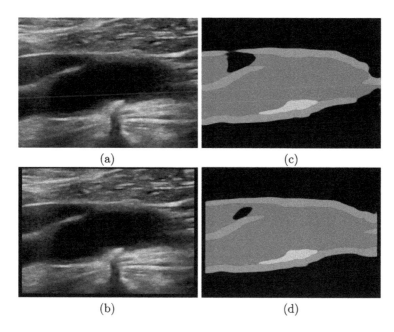

Fig. 2. Illustration of self-supervised augmentation consistency (SAC). Image *(a)* is transformed to *(b)* and segmented *(c)*. Segmenting the image *(a)* and transforming it should give the same segmentation *(d)*, except at the boundaries.

where the images are taken from the large unlabeled dataset Ω_u. Boundary regions with invalid data are ignored. We use the Kornia library [15] to implement the augmentations, as they need to be differentiable.

At training time, batches from the labeled and unlabeled datasets are alternatively fed into the network. We use 119 labeled images and 6781 unlabeled images for training and 29 labeled images for validation. We use the ADAM optimizer, reducing the learning rate when the validation loss stops decreasing. Resulting segmentations can be seen in Fig. 1 (right).

Using SAC leads to a relatively modest increase in the Dice score on the validation set from 0.790 to 0.794. Subjectively, in many cases the segmentations look more plausible.

2.2 Regression

The text step is to predict the parameters of interest from the first examination images (Sect. 1.1). We use a ResNet 18 [8] network g to get an estimate of the current plaque width d and the expected increase D

$$\hat{d} = c_d \sigma(g_1(J)) \tag{4}$$
$$\hat{D} = c_D \sigma(g_2(J)) \tag{5}$$

with a sigmoid transformation σ and scaling constants $c_d = 8$ mm and $c_D = 2$ mm. The network takes as input a 3 channel image J consisting of the original grayscale input image I, I masked with the thresholded plaque segmentation and the plaque mask itself (Fig. 3). This way the network can use the precalculated segmentation and focus on the plaque but it can also look at the shape from the segmentation and the original image. The network is encouraged to use all three input channels by setting them randomly to zero at training time with probability 0.5, similar to dropout.

The regression network g trains to optimize a criterion

$$L_{\text{reg}} = \underset{(J,d,D)\sim\Omega_r}{\mathrm{E}}\left[w_1\left(\hat{d}(J)-d\right)^2 + w_2\left(\hat{D}(J)-D\right)^2 + w_3 B\left(\hat{D}(J),D\right)\right]$$

$$-\frac{1}{2}\sum_{i=1}^{3}\log w_i \qquad (6)$$

where the training part of the regression dataset $\Omega_r \subseteq \Omega_u$ contains 2532 images from the first examination, for which the annotations d and D are available. The remaining 648 images are used for testing. There is no intersection between the subjects in the training and testing part.

The first two terms of L_{reg} are standard mean square errors, while the third is a binary cross entropy between probabilities $\pi(D)$ and $\pi(\hat{D})$ that the plaque is unstable.

$$B(\hat{D},D) = \pi(D)\log\pi\left(\hat{D}(J)\right) + \left(1-\pi(D)\right)\log\left(1-\pi\left(\hat{D}(J)\right)\right) \qquad (7)$$

$$\text{where} \quad \pi(D) = c\,\sigma(D-\tau), \quad \text{with} \quad c = 50 \qquad (8)$$

The motivation for the cross entropy term is that we are more interested whether the plaque is stable or unstable than in the precise increment $\hat{D}$. The weights of the three terms are adjusted automatically using logarithmic penalties [11], downweighting difficult parts of the loss function. We use small amplitude affine transformations and horizontal flip as augmentations and the same optimizer configuration as for the segmentation.

2.3 Aggregation

In the previous section, we have treated d and D as per image properties. However, there is just one prediction per carotid, i.e. the group of images. We average the probabilities $\pi(\hat{D})$ for each carotid C

$$\bar{\pi}_C = \underset{J\in C}{\text{mean}}\,\pi\left(\hat{D}(J)\right) \qquad (9)$$

3 Results

All results are evaluated on the test part of the regression dataset Ω_r containing 648 longitudinal images from 113 subjects with 225 carotids, each with an expert estimate of the plaque width d and plaque width increase D.

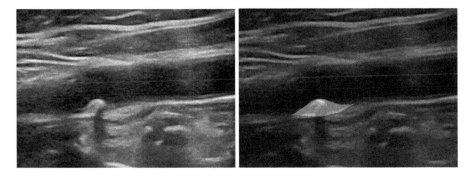

Fig. 3. An input image *(a)* and a corresponding fused image *(b)* composed of the input image (red channel), the input image multiplied by the thresholded plaque mask (green channel) and the mask itself (blue channel). (Color figure online)

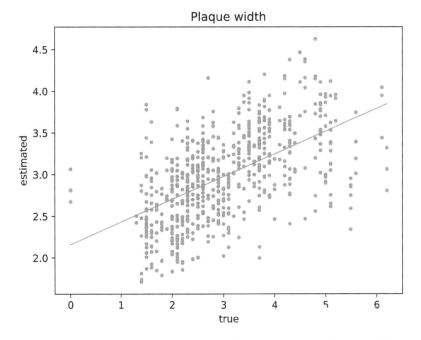

Fig. 4. The dependency between the expert indicated plaque widths d and their estimate $\hat{d}$ on the test dataset. The red line indicates a linear fit. All values are in mm. (Color figure online)

The current plaque width d can be estimated well (Fig. 4) with Pearson's correlation coefficient $\rho = 0.56$ and Spearman's rank correlation coefficient $r_s = 0.58$, which is statistically significant ($p < 10^{-50}$). The residual variability can be explained by the fact that 'plaque width' is interpreted somewhat subjectively by each expert and that the experts actually used different images.

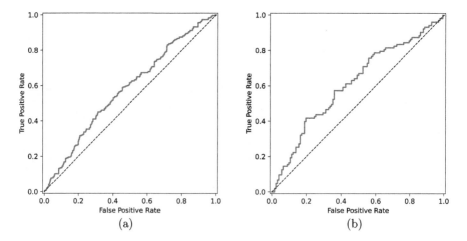

Fig. 5. The ROC curves for the unstable/stable classification at the image *(a)* and carotid *(b)* level. The dashed diagonal line is the performance of a random classifier.

Table 1. Confusion matrices and performance measures of the binary classifier to distinguish between unstable and stable plaques. The threshold was chosen to maximize F_1.

Image level results

true	predicted	
	stable	unstable
stable	62	289
unstable	31	266

AUC 0.598
accuracy 0.506
sensitivity 0.896
precision 0.479
specificity 0.177
F_1 0.624

(threshold $\tau = 0.06$ [mm])

(a)

Carotid level results

true	predicted	
	stable	unstable
stable	12	110
unstable	7	96

AUC 0.614
accuracy 0.480
sensitivity 0.932
precision 0.466
specificity 0.098
F_1 0.621

(threshold $p = 0.5$)

(b)

The continuous regression results for the plaque width increase $\hat{D}$ are not statistically significantly correlated with the true increase D, so we only present the results of the binary classification to stable and unstable plaques. Both the ROC curve (Fig. 5a) and the numerical results in Table 1a show that the classification performance is better than random, so there is certainly some relevant information hidden in the images. However, the accuracy is too low for clinical use. The results aggregated per carotid (Sect. 2.3, Fig. 5b, Table 1b) are a little better in terms of AUC, but not significantly so (AUC = 0.61, F_1 = 0.62).

4 Conclusions

We assembled and processed the largest reported dataset of longitudinal carotid artery ultrasound images for studying atherosclerosis. We created a method for automatic segmentation of the atherosclerotic plaque and vessel wall using self-supervised augmentation consistency to train on unlabeled images. The segmentation appears to work well. We also succeeded in automatically estimating the current size of the plaque. Both these tasks are directly relevant for clinical use.

We were less successful in predicting the future development of the plaque. We showed that there is indeed a relationship between the longitudinal image appearance and the plaque stability but the prediction is not good enough for clinical use. The performance of our method is similar to previously published results using other approaches on this or other types of ultrasound images [1, 7, 18] We are inclined to believe that ultrasound plaque development depends on many factors, not all of which can be estimated from the ultrasound images.

Acknowledgments. The authors acknowledge the support of the Czech Health Research Council project NV19-08-00362.

Disclosure of Interests. The authors have no competing interests to declare that are relevant to the content of this article.

References

1. Anonymous: Anonymous
2. Araslanov, N., Roth, S.: Self-supervised augmentation consistency for adapting semantic segmentation. In: 2021 IEEE/CVF Conference on Computer Vision and Pattern Recognition (CVPR), pp. 15379–15389 (2021). https://doi.org/10.1109/CVPR46437.2021.01513, https://api.semanticscholar.org/CorpusID:233481857
3. Brinjikji, W., Huston, J., Rabinstein, A.A., Kim, G.M., Lerman, A., Lanzino, G.: Contemporary carotid imaging: from degree of stenosis to plaque vulnerability. J. Neurosurg. **124**, 27–42 (2016). https://doi.org/10.3171/2015.1.JNS142452
4. Brinjikji, W., et al.: Ultrasound characteristics of symptomatic carotid plaques: a systematic review and meta-analysis. Cerebrovasc. Dis. **40**, 165–174 (2015). https://doi.org/10.1159/000437339
5. Chen, X., Kong, Z., Wei, S., Liang, F., Feng, T., Wang, S., Gao, J.: Ultrasound lmaging-vulnerable plaque diagnostics: automatic carotid plaque segmentation based on deep learning. J. Radiat. Res. Appl. Sci. **16**, 100598 (2023). https://doi.org/10.1016/j.jrras.2023.100598
6. Doonan, R.J., et al.: Plaque echodensity and textural features are associated with histologic carotid plaque instability. J. Vasc. Surg. **64**, 671–677 (2016)

7. D'Oria, M., et al.: Contrast Enhanced Ultrasound (CEUS) is not able to identify vulnerable plaques in asymptomatic carotid atherosclerotic disease. Eur. J. Vasc. Endovasc. Surg. **56**, 632–642 (2018). https://doi.org/10.1016/j.ejvs.2018.07.024
8. He, K., Zhang, X., Ren, S., Sun, J.: Deep residual learning for image recognition. In: CVPR: IEEE Conference on Computer Vision and Patter Recognition (2106)
9. Iakubovskii, P.: Segmentation models PyTorch (2019). https://github.com/qubvel/segmentation_models.pytorch
10. Kakkos, S., et al.: Computerized texture analysis of carotid plaque ultrasonic images can identify unstable plaques associated with ipsilateral neurological symptoms. Angiology **62**(4), 317–328 (2011)
11. Kendall, A., Gal, Y., Cipolla, R.: Multi-task learning using uncertainty to weigh losses for scene geometry and semantics. In: Proceedings of the IEEE Conference on Computer Vision and Pattern Recognition (CVPR) (2018)
12. Miceli, G., et al.: Artificial intelligence in symptomatic carotid plaque detection: a narrative review. Appl. Sci. **13**, 4321 (2023). https://doi.org/10.3390/app13074321
13. Nicolaides, A., Beach, K.W., Kyriacou, E., Pattichis, C.S.: Ultrasound and Carotid Bifurcation Atherosclerosis. Springer, London (2012). https://doi.org/10.1007/978-1-84882-688-5
14. Ren, S., He, K., Girshick, R.B., Sun, J.: Faster R-CNN: towards real-time object detection with region proposal networks. In: NIPS: Neural Information Processing Systems Conference, pp. 91–99 (2015)
15. Riba, E., Mishkin, D., Ponsa, D., Rublee, E., Bradski, G.: Kornia: an open source differentiable computer vision library for PyTorch. In: Winter Conference on Applications of Computer Vision (2020). https://arxiv.org/pdf/1910.02190.pdf
16. Ronneberger, O., Fischer, P., Brox, T.: U-Net: convolutional networks for biomedical image segmentation. In: MICCAI: Medical Image Computing and Computer-Assisted Intervention (2015)
17. Saba, L., et al.: Multimodality carotid plaque tissue characterization and classification in the artificial intelligence paradigm: a narrative review for stroke application. Ann. Transl. Med. 9, 1206–1206 (2021). https://doi.org/10.21037/atm-20-7676
18. Salem, M., et al.: Identification of patients with a histologically unstable carotid plaque using ultrasonic plaque image analysis. Eur. J. Vasc. Endovasc. Surg. **48**, 118–125 (2014). https://doi.org/10.1016/j.ejvs.2014.05.015
19. Salonen, R., Seppänen, K., Rauramaa, R., et al.: Prevalence of carotid atherosclerosis and serum cholesterol levels in eastern Finland. Arteriosclerosis **6**(8), 788–792 (1988)
20. Skandha, S.S., et al.: 3-D optimized classification and characterization artificial intelligence paradigm for cardiovascular/stroke risk stratification using carotid ultrasound-based delineated plaque: Atheromatic^{TM} 2.0. Comput. Biol. Med. **125**, 103958 (2020). https://doi.org/10.1016/j.compbiomed.2020.103958
21. Školoudík, D.: Atherosclerotic plaque characteristics associated with a progression rate of the plaque in carotids and a risk of stroke. (2015). https://clinicaltrials.gov/ct2/show/NCT02360137, clinical trial NCT02360137

Leveraging IHC Staining to Prompt HER2 Status Prediction from HE-Stained Histopathology Whole Slide Images

Yuping Wang[1], Dongdong Sun[2], Jun Shi[2], Wei Wang[4], Zhiguo Jiang[3], Haibo Wu[4], and Yushan Zheng[1(✉)]

[1] School of Engineering Medicine, Beijing Advanced Innovation Center on Biomedical Engineering, Beihang University, Beijing 100191, China
yszheng@buaa.edu.cn
[2] School of Software, Hefei University of Technology, Hefei 230601, China
[3] Image Processing Center, School of Astronautics, Beihang University, Beijing 102206, China
[4] The First Affiliated Hospital of USTC, Division of Life Sciences and Medicine, University of Science and Technology of China, Hefei 230036, China

Abstract. The development of artificial intelligence has significantly impacted the predictive analysis of molecular biomarkers, which is crucial for targeted cancer therapy. Traditional assessment of HER2 in breast cancer utilizes both Hematoxylin and Eosin (H&E) and Immunohistochemistry (IHC) stained slides. Recent models have sought to predict HER2 status using H&E-stained slides to reduce reliance on the costly and time-consuming IHC staining. However, these models overlook the information from IHC staining. In this paper, we proposes a novel framework that integrates IHC-stained WSIs during the training phase to enhance the HER2 prediction capabilities based on the H&E-stained WSIs. This framework uses IHC-predicted HER2 status as a proxy task, embedding the learned relevant information as prompts into the encoder for H&E slides. Meanwhile, our model only requires H&E slides during inference, which maintains the data-efficiency of the HER2 prediction system. Experimental results show that our method achieves an AUC of 0.860 and a F1 score of 0.652 in the tasks of HER2 0/1+/2+/3+ status grading for breast cancer, which significantly outperforms state-of-the-art models.

Keywords: HER2 · Whole slide image · Multi-modal learning

1 Introduction

With the rapid development of artificial intelligence, the prediction of molecular biomarkers from pathology whole slide images (WSIs) has gained significant attention due to its implications for targeted therapy. Human Epidermal Growth Factor Receptor 2 (HER2) overexpression serves as a critical biomarker for diagnosing breast cancer [10,12]. In routine breast cancer diagnostics, hematoxylin and eosin (H&E) stained slides are initially employed to identify cancerous tissue regions. Following this, Immunohistochemistry (IHC) staining and In Situ Hybridization (ISH) techniques are used to confirm the presence of HER2 overexpression [17]. Accurately predicting HER2 status using WSIs can significantly influence treatment strategies, thereby improve patient outcomes [6].

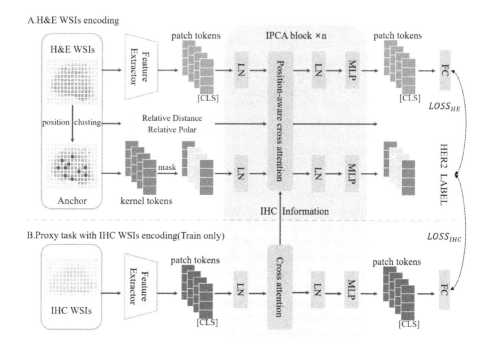

Fig. 1. The overview of the proposed method. (A) illustrates the H&E-stained WSIs encoding process, where the H&E patch tokens are updated by the Position-aware cross attention module with H&E information and through H&E-IHC interaction with IHC information. (B) illustrates the IHC-stained WSIs encoding process, where the IHC patch tokens are updated through cross attention. This process, as a proxy task, is used only during the training phase.

In early research, manually designed image features coupled with classifiers were employed to predict HER2 status from Immunohistochemistry (IHC) stained slides [3,9]. Recently, deep learning techniques [11,13,14,20] have shown substantial benefits in the analysis of IHC-stained slides, significantly enhancing the accuracy and efficiency of HER2 status prediction. Then, various deep learning frameworks [1,16] have been developed to predict HER2 status using H&E stained slides, aiming to reduce reliance on IHC staining which suffers from longer processing times, higher costs, and increased complexity. Notably, frameworks such as SlideGraph+ [8] and HEAT [2] utilize WSI-level graphs to effectively capture spatial and structural information. These advanced graph-based models significantly improve diagnostic accuracy with H&E-stained slides.

However, these models do not harness the extensive data available from IHC staining, which provides critical insights into HER2 expression. This omission of detailed IHC data during model training could limit the predictive capabilities and accuracy of systems relying solely on H&E-stained slides.

In this paper, we proposed a novel framework for HER2 status prediction based on cooperative training of H&E-stained and IHC-stained WSIs. As illustrated in Fig. 1, our framework fully utilizes IHC-stained slides during the training phase to guide the learning of the H&E encoder. This allows more explicit information from the IHC to be embedded into the encoding process of H&E-stained slides. During the inference phase, the framework only requires H&E-stained slides as input and does not depend on IHC slides. Experimental results on a breast cancer dataset demonstrate that our method significantly outperforms baseline models trained solely on H&E-stained slides, with a 2.0% improvement in AUC and an 6.8% improvement in F1 score. The contribution of this paper can be summarized into two key aspects:

1) During the training phase of the H&E encoder, we utilize the prediction of HER2 status from IHC-stained slides as a proxy task. This proxy task allows the model to capitalize on the detailed information specific to HER2 from IHC slides, thereby enhancing its capability to identify similar patterns in H&E stains. In the inference phase, the model exclusively relies on H&E-stained slides, ensuring a high data efficiency.

2) We designed a novel module named IHC-prompted cross-attention (IPCA), as illustrated in Fig. 2. The core idea of this module is to establish a set of learnable prompts that are shaped under the dual guidance of both H&E and IHC branch. This design allows the IPCA module to integrate IHC-specific patterns into the H&E encoder, thus improving the model's predictive accuracy and robustness.

2 Methods

2.1 Problem Formulation

The flowchart of the proposed work is illustrated in Fig. 1. After WSIs segmentation and patch features extraction, we formulate the features extracted from H&E and IHC slides as $\mathbf{X}_{HE} \in \mathbb{R}^{n_p^h \times d_f}$ and $\mathbf{X}_{IHC} \in \mathbb{R}^{n_p^i \times d_f}$, where n_p^h and n_p^i are the numbers of patches segmented from H&E and IHC slides, and d_f is the feature dimension. Simultaneously, by clustering the position coordinates of the H&E patches, multiple position anchors are extracted and represented as $\mathbf{K}_{HE} \in \mathbb{R}^{n_k^h \times d_f}$, where n_k^h is the number of anchors.

2.2 H&E-Stained WSI Encoding

Figure 2 illustrates the process of H&E encoding. We utilized the Position-aware Cross-Attention (PACA) module in PAMA [18] to build the encoder for H&E-stained WSIs. PACA can extract the local and global features for WSIs by

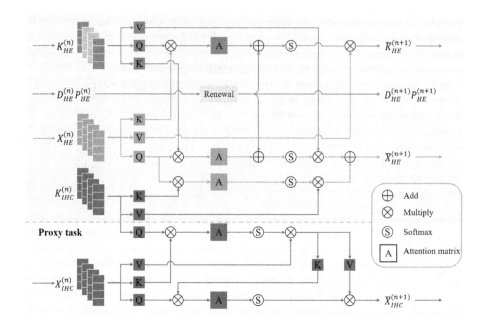

Fig. 2. Detailed View of the proposed IPCA module. The module comprises two main branches: the H&E branch (top) and the IHC branch (bottom). The two branches interact through a cross-attention mechanism to enhance the predictive capability for HER2 status. Each branch includes transformations for queries, keys, and values, along with an attention matrix and Softmax operation, facilitating feature updating and information transmission.

performing cross-attention operations between anchor tokens and patch tokens. Specifically, a relative spatial distance matrix $\mathbf{D}_{HE} \in \mathbb{N}^{n_k^h \times n_p^h}$ and a relative polar angle matrix $\mathbf{P}_{HE} \in \mathbb{N}^{n_k^h \times n_p^h}$ are calculated between the anchors and patches, to describe the structural information of WSI.

In PACA, the anchors first collect local information from the patches by the formula:

$$\bar{\mathbf{K}}_{HE}^{(n+1)} = \sigma \left(\frac{\mathbf{X}_{HE}^{(n)} \mathbf{W}_q^{(n)} \cdot (\mathbf{X}_{HE}^{(n)} \mathbf{W}_k^{(n)})^\top + \mathbf{\Phi}^{(n)}}{\sqrt{d_e}} \right) \cdot (\mathbf{X}_{HE}^{(n)} \mathbf{W}_v^{(n)}), \quad (1)$$

where $\mathbf{W}_l \in \mathbb{R}^{d_f \times d_e}$ are the learnable parameters for the projection matrices, where $l = q, k, v$ stands for query, key, and value projections, $\bar{\mathbf{K}}_{HE}^{(n+1)}$ is passed to the next block after undergoing layer normalization and MLP operations to obtain $\mathbf{K}_{HE}^{(n+1)}$, and

$$\mathbf{\Phi}^{(n)} = \varphi_d(\mathbf{D}_{HE}^{(n)}) + \varphi_p(\mathbf{P}_{HE}^{(n)}), \quad (2)$$

where φ_d and φ_p are the embedding functions that map the distance and polar angle information into corresponding embedding values.

Symmetrically, each patch token captures the information of all anchors into their own local representations using the equation:

$$\bar{\mathbf{X}}_{HE \leftarrow HE}^{(n)} = \sigma \left(\frac{\mathbf{X}_{HE}^{(n)} \mathbf{W}_q^{(n)} \cdot (\mathbf{K}_{HE}^{(n)} \mathbf{W}_k^{(n)})^\top + \mathbf{\Phi}^{(n)\top}}{\sqrt{d_e}} \right) \cdot (\mathbf{K}_{HE}^{(n)} \mathbf{W}_v^{(n)}). \quad (3)$$

This allows anchor tokens to update patch tokens based on valuable information obtained from the kernels.

2.3 Proxy HER2 Prediction Task with IHC-Stained WSIs

As illustrated in Fig. 2, we constructed an additional branch alongside the H&E encoder to serve as a proxy task. This branch processes IHC-stained WSIs and symmetrically predicts the HER2 status. In each IPCA block, a set of prompts, $\mathbf{K}_{IHC} \in \mathbb{R}^{n_k \times d_f}$, are built to describe IHC-related patterns and will be optimized throughout the training process. The module first takes learnable kernel tokens $\mathbf{K}_{IHC}$ and patch feature $\mathbf{X}_{IHC}$ as input. Specifically, *Query* is derived from $\mathbf{K}_{IHC}$, and the *Key* and *Value* are derived from $\mathbf{X}_{IHC}$, all of which are achieved through linear transformations. Afterwards, we get the kernel tokens $\bar{\mathbf{K}}_{IHC}^{(n)}$ with IHC information:

$$\bar{\mathbf{K}}_{IHC}^{(n)} = \sigma \left(\frac{\mathbf{K}_{IHC}^{(n)} \mathbf{W}_q^{(n)} \cdot (\mathbf{X}_{IHC}^{(n)} \mathbf{W}_k^{(n)})^\top}{\sqrt{d_e}} \right) \cdot (\mathbf{X}_{IHC}^{(n)} \mathbf{W}_v^{(n)}). \quad (4)$$

Following this, we utilize $\bar{\mathbf{K}}_{IHC}^{(n)}$ with IHC information to perform another cross-attention with $\mathbf{X}_{IHC}$. In this process, The *Query* is derived from $\mathbf{X}_{IHC}$, and the *Key* and *Value* are derived from $\bar{\mathbf{K}}_{IHC}^{(n)}$. This process can be represented as follows:

$$\bar{\mathbf{X}}_{IHC}^{(n+1)} = \sigma \left(\frac{\mathbf{X}_{IHC}^{(n)} \mathbf{W}_q^{(n)} \cdot (\bar{\mathbf{K}}_{IHC}^{(n)} \mathbf{W}_k^{(n)})^\top}{\sqrt{d_e}} \right) \cdot (\bar{\mathbf{K}}_{IHC}^{(n)} \mathbf{W}_v^{(n)}). \quad (5)$$

As HER2 prediction with IHC-stained WSIs is a proxy task, Eq. 4 and Eq. 5 is only used in the training phase.

2.4 IHC Staining Information Embedding

Based on Eq. 4 and Eq. 5, $\mathbf{K}_{IHC}^{(n)}$ is expected to learn the prototypes that are crucial to recognize IHC features. Therefore, we can use $\mathbf{K}_{IHC}^{(n)}$ to assist the H&E encoder to extract IHC-related features. To achieve this, we utilize an cross attention, where the *Query* is derived from $\mathbf{X}_{HE}$, while the *Key* and *Value* are derived from $\mathbf{K}_{IHC}$ that contains the IHC information. The result of this cross attention operation is an updated $\bar{\mathbf{X}}_{HE \leftarrow IHC}$, which is then combined with the $\bar{\mathbf{X}}_{HE \leftarrow HE}$ to produce the final updated $\mathbf{X}_{HE}$. The mathematical description of this process is as follows:

$$\bar{\mathbf{X}}_{HE \leftarrow IHC}^{(n)} = \sigma \left(\frac{\mathbf{X}_{HE}^{(n)} \mathbf{W}_q^{(n)} \cdot (\mathbf{K}_{IHC}^{(n)} \mathbf{W}_k^{(n)})^\top}{\sqrt{d_e}} \right) \cdot (\mathbf{K}_{IHC}^{(n)} \mathbf{W}_v^{(n)}), \qquad (6)$$

$$\bar{\mathbf{X}}_{HE}^{(n+1)} = \bar{\mathbf{X}}_{HE \leftarrow HE}^{(n)} + \bar{\mathbf{X}}_{HE \leftarrow IHC}^{(n)}. \qquad (7)$$

These designs enable the IPCA module to effectively utilize the IHC staining information in H&E WSI encoding even without the input of the IHC slides. It remains data-efficient during inference by relying solely on H&E slides.

2.5 Objective and Optimization

Before the model inference, we add [CLS] tokens to the patch tokens of both H&E and IHC slides. After passing through n IPCA blocks, the [CLS] tokens are unfolded and passed through a fully connected layer for HER2 status prediction, categorizing into HER2 0, 1+, 2+, and 3+. Finally, we applied the cross-entropy loss function to both [CLS] tokens for training.

Additionally, to promote the integration of IHC information into the H&E encoder during training, we introduce a kernel masking (KM) strategy for a certain proportion of $\mathbf{K}_{HE}$. This masking strategy encourages the encoder to more effectively integrate IHC information into the H&E WSI encoding.

3 Experiments and Results

3.1 Implementation Details

The dataset used in this study is a breast cancer dataset, collected from the First Affiliated Hospital of USTC (University of Science and Technology of China). This dataset comprises 358 cases, with each case including a H&E-stained slide and a paired IHC-stained slide. The HER2 status of each case has been affirmed by expert pathologists. The dataset is split by a ratio of 7:3. Five-fold cross-validation is performed on the training set to determine the model hyper-parameters. Then, the determined model was used for testing.

The slides are segmented into 256×256 patches at a 20× magnification using sliding window strategy where the background region of each slide was removed by a threshold. Then, a pre-trained Vision Transformer (ViT-base) from PLIP [4] was applied to extracting features for the patches.

During the training phase, the Adam optimizer [5] is used with a learning rate of 1e−4, employing a cosine decay strategy. The batch size is set to 16. The evaluation metrics include Accuracy (ACC), Macro-AUC, Weighted-AUC, Macro-F1, and Weighted-F1. All experiments are conducted on two GPUs of NVIDIA GeForce RTX 4090.

3.2 Hyper-Parameter Settings

We tuned the main hyper-parameters of our model through cross-validation. The results are detailed in Table. 1. Initially, we focused on the kernel masking strategy by tuning the mask-ratio in the range of [0, 0.25, 0.5 0.75]. The results showed that a kernel mask-ratio of 0.75 provided the best performance. Based on this experiment, we finally set the mask-ratio as 0.75. Subsequently, we tuned the number of prompts n_k of the shared $\mathbf{K}_{IHC}$ from 32 to 256. The results show that the performance improves as the number of shared prompts increase, and $n_k = 256$ delivered the best performance. Considering that a larger number of prompts would bring greater computation to the model, we finally set $n_k = 256$.

Table 1. Hyper-parameter Experiments.

Settings	Accuracy	Macro-AUC	Weighted-AUC	Macro-F1	Weighted-F1
$r = 0.75$	0.617 ± 0.003	0.837 ± 0.024	0.847 ± 0.024	0.586 ± 0.034	0.617 ± 0.028
$r = 0.5$	0.616 ± 0.030	0.827 ± 0.027	0.839 ± 0.020	0.589 ± 0.052	0.613 ± 0.047
$r = 0.25$	0.616 ± 0.010	0.833 ± 0.026	0.842 ± 0.021	0.572 ± 0.030	0.603 ± 0.018
$r = 0$	0.600 ± 0.010	0.819 ± 0.034	$0.831 v 0.029$	0.564 ± 0.028	0.600 ± 0.019
$n_k = 32$	0.617 ± 0.003	0.837 ± 0.024	0.847 ± 0.024	0.586 ± 0.034	0.610 ± 0.028
$n_k = 64$	0.621 ± 0.003	0.825 ± 0.024	0.832 ± 0.019	0.592 ± 0.028	0.614 ± 0.014
$n_k = 128$	0.623 ± 0.002	0.841 ± 0.022	0.851 ± 0.015	0.568 ± 0.016	0.606 ± 0.014
$n_k = 256$	0.622 ± 0.002	0.847 ± 0.018	0.855 ± 0.010	0.569 ± 0.036	0.604 ± 0.022

3.3 Comparison with SOTA Methods

Then, we compared our model with six methods, inlcuding CLAM [7], Nystromformer [19], TransMIL [15], SlideGraph+ [8], HEAT [2], and PAMA [18]. These methods are used for training and inference with H&E inputs. The results are presented in Table. 2. Overall, our method demonstrates the best performance across all evaluation metrics. In comparison to the second-best methods, our method demonstrates a performance increase of 1.9% in Macro-AUC and 2.0% in Weighted-AUC. Additionally, it shows improvements of 6.5% in Macro-F1 and 6.8% in Weighted-F1, respectively.

SlideGraph+ and HEAT, both based on graph-based approaches, are considered state-of-the-art methods for predicting HER2 status using H&E inputs. PAMA can more effectively capture the context of WSIs based on its position-aware and anchor cross-attention mechanisms compared to SlideGraph+ and HEAT. It enables PAMA to achieve superior performance through enhanced spatial and semantic representation capabilities. Our proposed method significantly improves upon PAMA by leveraging an anchor structure combined with an IHC-prompted mechanism. We plotted the Macro-average ROC curve for

PAMA and our method in Fig. 3. The AUC values for our method are consistently higher than those for PAMA, particularly in predicting HER2 1+ and HER2 2+ status, where we observed increases of 2.3% and 2.0%, respectively. These improvements are crucial, as distinguishing these HER2 statuses using only H&E inputs is challenging due to the subtle morphological features in H&E slides. Our integration of IHC information enhances the discriminative power of the H&E encoder, enabling it to more effectively differentiate these subtle distinctions.

Table 2. Comparison with the state-of-the-art methods.

Method	Train	Inference	ACC	M-AUC	W-AUC	M-F1	W-F1
CLAM [7]	H&E	H&E	0.572	0.823	0.825	0.543	0.567
Nystromformer [19]	H&E	H&E	0.554	0.806	0.811	0.510	0.536
TransMIL [15]	H&E	H&E	0.576	0.789	0.798	0.545	0.571
SlideGraph+ [8]	H&E	H&E	0.589	0.822	0.825	0.546	0.577
HEAT [2]	H&E	H&E	0.548	0.800	0.803	0.519	0.541
PAMA [18]	H&E	H&E	0.583	0.839	0.840	0.567	0.584
Ours w/o proxy	H&E	H&E	0.601	0.845	0.847	0.578	0.606
Ours	H&E+IHC	H&E	**0.657**	**0.858**	**0.860**	**0.632**	**0.652**

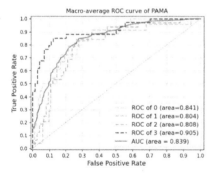
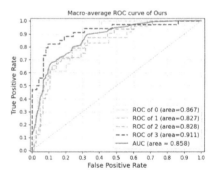

Fig. 3. Comparison of ROC curves between PAMA and our proposed method.

3.4 Ablation Study

We conducted ablation studies to evaluate the impact of embedding IHC information in Table. 2. In *Ours w/o proxy*, we removed the IHC input in both the training and inference stages but retained $\mathbf{K}_{IHC}$ along with the kernel tokens of

H&E to update the patch tokens of H&E. Here, $\mathbf{K}_{IHC}$ serves as prompts trained to encapsulate information about H&E, which can be utilized during the inference phase. The results demonstrate that *Ours w/o proxy* outperforms PAMA across all evaluation metrics, indicating the prompt enable leverages the learned information. Furthermore, our method perform better than *Ours w/o proxy* in AUC of 1.3% and F1 of 5.4%, underscoring the importance of embedding IHC information for enhancing the H&E encoder's capabilities.

4 Conclusion

In this paper, we proposed a novel framework for HER2 prediction, which effectively incorporates the information from IHC into the H&E encoder. This model demonstrates data efficiency by utilizing both H&E and IHC inputs during training, while only requiring H&E input during inference. Ablation experiments validate the effective integration of IHC information into the H&E encoder. Our results exhibit superiority over existing HER2 prediction methods. Future work will focus on training the framework using datasets across multiple organs to enhance its generalization ability for HER2 status prediction.

Acknowledgements. This work was partly supported by Beijing Natural Science Foundation (Grant No. 7242270), partly supported by the National Natural Science Foundation of China (Grant No. 62171007, 61901018, and 61906058), partly supported by the Fundamental Research Funds for the Central Universities of China (grant No. YWF-23-Q-1075), and partly supported by Joint Fund for Medical Artificial Intelligence (Grant No. MAI2023C014).

References

1. Anand, D., et al.: Deep learning to estimate human epidermal growth factor receptor 2 status from hematoxylin and eosin-stained breast tissue images. J. Pathol. Inform. **11**(1), 19 (2020)
2. Chan, T.H., Cendra, F.J., Ma, L., Yin, G., Yu, L.: Histopathology whole slide image analysis with heterogeneous graph representation learning. In: Proceedings of the IEEE/CVF Conference on Computer Vision and Pattern Recognition (CVPR), pp. 15661–15670 (2023)
3. Cordeiro, C.Q., Ioshii, S.O., Alves, J.H., Oliveira, L.F.: An automatic patch-based approach for HER-2 scoring in immunohistochemical breast cancer images using color features. CoRR abs/1805.05392 (2018)
4. Huang, Z., Bianchi, F., Yuksekgonul, M., Montine, T.J., Zou, J.: A visual-language foundation model for pathology image analysis using medical twitter. Nat. Med. **29**, 2307–2316 (2023)
5. Kingma, D.P., Ba, J.: Adam: a method for stochastic optimization (2017)
6. van der Laak, J., Litjens, G., Ciompi, F.: Deep learning in histopathology: the path to the clinic. Nat. Med. **27**, 775–784 (2021)
7. Lu, M.Y., Williamson, D.F., Chen, T.Y., Chen, R.J., Barbieri, M., Mahmood, F.: Data-efficient and weakly supervised computational pathology on whole-slide images. Nat. Biomed. Eng. **5**(6), 555–570 (2021)

8. Lu, W., Toss, M., Dawood, M., Rakha, E., Rajpoot, N., Minhas, F.: SlideGraph+: whole slide image level graphs to predict HER2 status in breast cancer. Med. Image Anal. **80**, 102486 (2022)
9. Mukundan, R.: Analysis of image feature characteristics for automated scoring of HER2 in histology slides. J. Imaging **5**, 35 (2019)
10. Piccart, M., Lohrisch, C., Di Leo, A., Larsimont, D.: The predictive value of HER2 in breast cancer. Oncology **61**(Suppl.2), 73–82 (2001)
11. Qaiser, T., Rajpoot, N.M.: Learning where to see: a novel attention model for automated immunohistochemical scoring. CoRR abs/1903.10762 (2019)
12. Ross, J.S., Slodkowska, E.A., Symmans, W.F., Pusztai, L., Ravdin, P.M., Hortobagyi, G.N.: The HER-2 receptor and breast cancer: ten years of targeted anti-HER-2 therapy and personalized medicine. Oncologist **14**(4), 320–368 (2009)
13. Saha, M., Chakraborty, C.: Her2Net: a deep framework for semantic segmentation and classification of cell membranes and nuclei in breast cancer evaluation. IEEE Trans. Image Process. **27**(5), 2189–2200 (2018)
14. Selcuk, S.Y., et al.: Automated HER2 scoring in breast cancer images using deep learning and pyramid sampling (2024)
15. Shao, Z., Bian, H., Chen, Y., Wang, Y., Zhang, J., Ji, X., et al.: TransMIL: transformer based correlated multiple instance learning for whole slide image classification. Adv. Neural. Inf. Process. Syst. **34**, 2136–2147 (2021)
16. Wang, J., Zhu, X., Chen, K., Hao, L., Liu, Y.: HaHNet: a convolutional neural network for HER2 status classification of breast cancer (2023)
17. Wolff, A.C., et al.: Human epidermal growth factor receptor 2 testing in breast cancer: asco-college of american pathologists guideline update. J. Clin. Oncol. **41**(22), 3867–3872 (2023). pMID: 37284804
18. Wu, K., Zheng, Y., Shi, J., Xie, F., Jiang, Z.: Position-aware masked autoencoder for histopathology wsi representation learning. In: Greenspan, H., et al. (eds.) Medical Image Computing and Computer Assisted Intervention – MICCAI 2023, vol. 14225. pp. 714–724. Springer Nature Switzerland, Cham (2023). https://doi.org/10.1007/978-3-031-43987-2_69
19. Xiong, Y., et al.: Nyströmformer: A nyström-based algorithm for approximating self-attention. In: Proceedings of the AAAI Conference on Artificial Intelligence **35**, 14138–14148 (2021)
20. Yao, Q., et al.: Using whole slide gray value map to predict HER2 expression and fish status in breast cancer. Cancers **14**(24), 6233 (2022)

VIMs: Virtual Immunohistochemistry Multiplex Staining via Text-to-Stain Diffusion Trained on Uniplex Stains

Shikha Dubey[1,2](✉), Yosep Chong[3,4], Beatrice Knudsen[4,5], and Shireen Y. Elhabian[1,2]

[1] Kahlert School of Computing, University of Utah, Salt Lake City, USA
{shikha.d,shireen}@sci.utah.edu
[2] Scientific Computing and Imaging Institute, University of Utah, Salt Lake City, USA
[3] The Catholic University of Korea College of Medicine, Seoul, South Korea
ychong@catholic.ac.kr
[4] Huntsman Cancer Institute, University of Utah Health, Salt Lake City, USA
[5] Department of Pathology, University of Utah, Salt Lake City, USA
beatrice.knudsen@path.utah.edu

Abstract. This paper introduces a Virtual Immunohistochemistry Multiplex staining (VIMs) model designed to generate multiple immunohistochemistry (IHC) stains from a single hematoxylin and eosin (H&E) stained tissue section. IHC stains are crucial in pathology practice for resolving complex diagnostic questions and guiding patient treatment decisions. While commercial laboratories offer a wide array of up to 400 different antibody-based IHC stains, small biopsies often lack sufficient tissue for multiple stains while preserving material for subsequent molecular testing. This highlights the need for virtual IHC staining. Notably, VIMs is the first model to address this need, leveraging a large vision-language single-step diffusion model for virtual IHC multiplexing through text prompts for each IHC marker. VIMs is trained on uniplex paired H&E and IHC images, employing an adversarial training module. Testing of VIMs includes both paired and unpaired image sets. To enhance computational efficiency, VIMs utilizes a pre-trained large latent diffusion model fine-tuned with small, trainable weights through the Low-Rank Adapter (LoRA) approach. Experiments on nuclear and cytoplasmic IHC markers demonstrate that VIMs outperforms the base diffusion model and achieves performance comparable to Pix2Pix, a standard generative model for paired image translation. Multiple evaluation methods, including assessments by two pathologists, are used to determine the performance of VIMs. Additionally, experiments with different prompts highlight the impact of text conditioning. This paper represents the first attempt to accelerate histopathology research by demonstrating the generation of multiple IHC stains from a single H&E input using a

Supplementary Information The online version contains supplementary material available at https://doi.org/10.1007/978-3-031-73284-3_15.

single model trained solely on uniplex data. This approach relaxes the traditional need for multiplex training sets, significantly broadening the applicability and accessibility of virtual IHC staining techniques.

Keywords: Virtual Immunohistochemistry Multiplex · Histopathology Images · Generative model · Virtual Staining · Diffusion Stainer

1 Introduction

Pathologists begin their microscopic assessments of tissues using hematoxylin and eosin (H&E) stained tissue sections. However, challenging cases require more precise information that cannot be delivered solely by H&E staining. This additional information is provided through the staining of tissues with antibodies using a method called immunohistochemistry (IHC). Antibodies label specific proteins in the nucleus, cytoplasm, or membrane of cells, and their binding is visualized by a brown color in the tissue. The antibody stain is easily distinguished from the blue Hematoxylin stain, which highlights DNA and RNA to reveal cellular organization. IHC staining, though informative, is time-consuming and expensive, often taking hours or days. Standard IHC methods rely on staining with one antibody at a time (uniplex stain). However, multiple stains are needed to arrive at the correct diagnosis, requiring multiple tissue sections [17]. Small biopsies contain insufficient tissue for multiple IHC assays and cannot provide sufficient material for proper diagnostic workup and subsequent molecular analysis for treatment selection. Standard multiplexed IHC, which consists of staining the same tissue section with multiple antibodies, would solve the problem of limited tissue availability. However, slides stained by multiplexed IHC cannot be easily analyzed by regular light microscopy. In addition, the number of IHC assays that can be performed in the same tissue section is limited by overlapping staining patterns of antibodies. Each antibody is labeled with a unique color, necessitating color deconvolution of multiplexed IHC images. While a seemingly simple task, algorithmic color unmixing of IHC images constitutes a problem [6]. Altogether, generating multiple virtual IHC stains from an H&E image would provide a benefit to patients who have only limited tissue samples for diagnostic workup and treatment planning or who live in countries where IHC resources are not available.

Advancements in vision-based image generation using generative adversarial network (GAN) models, such as Pix2Pix [9] and CycleGAN [33], have inspired their application in the medical domain, specifically in histopathology [5,10,13,14,16,20,21,27,28,30,34]. These models have been employed for tasks such as virtual H&E and Masson's Trichrome staining [34] and converting autofluorescence images to virtual HER2 [21], stain normalization [10,27], and virtual IHC generation from H&E images [5,13,16,28,30]. So far, virtual IHC staining models, trained on paired or unpaired data, typically generate a uniplex or single IHC stain from each H&E image. Paired IHC and H&E data are difficult to generate due to the need for pixel-wise registration of whole-slide images (WSIs) H&E to IHC, whereas unpaired data do not require such registration. Generating multiple IHC markers from a single H&E sample remains relatively unexplored and challenging, though a few recent GAN-based models [1,3,15] have attempted to address this issue. While models in [3,15] are trained independently for different IHC markers without learning proper associations among IHCs, Multi-VSTAIN [1] requires multiplexed paired data for training, which are difficult to obtain. Therefore, GAN-based models reduces their practical application. The proposed model, Virtual Immunohistochemistry Multiplex staining (VIMs), does not need multiplex paired training data and trains the model in an end-to-end manner, taking advantage of associations among IHCs.

Recently, conditional diffusion models (DMs) [18,22,24–26,31], a class of generative models, have demonstrated remarkable success in image synthesis tasks. These models operate by learning to reverse a gradual noising process, enabling the generation of high-quality images. However, traditional DMs [22,24,31] require multiple denoising steps during inference, which can be time-consuming. Despite this, DMs produce higher fidelity images compared to conventional GAN-based models. This is crucial in the medical domain, where detailed cellular-level morphological information is vital for accurate disease diagnosis. Large vision-language diffusion models (LVDMs) are generally challenging to train from scratch, and fine-tuning their billions of parameters requires extensive computational power and large datasets. Recent advancements, such as adapters [8] and fine-tuning methods [31], have improved their training efficiency. Additionally, optimized DMs [18,26] developed to reduce the number of steps needed during inference. However, the potential of these optimized DMs for virtual IHC staining has not yet been explored. This paper aims to fill this gap by investigating the use of such advanced DMs for multiplex IHC marker generation. Inspired by recent advances incorporating adversarial objectives into conditional DMs [18,26], this paper introduces VIMs. VIMs leverages a text-conditioned single-step DM to generate multiple paired virtual IHC images from the same H&E image, allowing multiple IHC markers to be visualized in the exact same cells, which is difficult to obtain from slides used in clinical practice. Additionally, evaluating generative models, especially in medical imaging, is challenging [5]; thus, this study uses multiple evaluation methods, including a manual assessment by two pathologists.

The primary contributions of this study are as follows:

1. Introduces VIMs[1], a method for virtual multiplexed IHC that generates multiple IHC stains from a single H&E image using a text-conditioned single-step DM. VIMs is the first to adapt such a model for virtual IHC staining, incorporating adversarial learning objectives to preserve tissue structures during translation.
2. VIMs is trained on uniplex paired (pixel-level registered H&E to IHC) data with pathologist-validated prompts, using a large latent DM with small trainable weights via the Low-Rank Adapter (LoRA) for efficient end-to-end training.

---

[1] We will release all the code related to the paper at a future date for public usage.

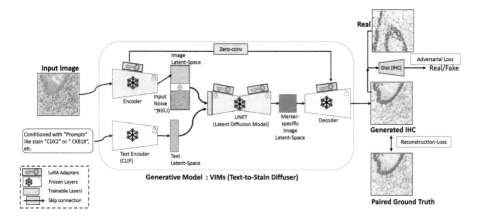

Fig. 1. VIMs: Proposed Multiplex IHC Staining Model. The pre-trained LDM, one of the Large Language Models (LLMs), is optimized for the virtual multiplex IHC generation task with a minimal number of trainable parameters.

3. Extensive experiments are conducted on two gland markers, CDX2 and CK8/18, using both paired and unpaired test sets. Two pathologists confirmed the high performance of VIMs in both settings. Different prompts were tested to highlight the impact of text conditioning. For the first time, DICE and mIoU metrics were used to evaluate whether VIMs accurately stains the correct cells and places the staining correctly in the nucleus or cytoplasm.

The paper covers VIMs methodology in Sect. 2, experimental and results in Sect. 3, and conclusions and future directions in Sect. 4.

2 Methods: Virtual Immunohistochemistry Multiplex Staining (VIMs)

The proposed model, VIMs (Fig. 1), is adapted from recent advancements in conditional DMs [18,22,26,31] and applies an LVDM to the IHC image generation task. VIMs consists of multiple pre-trained and trainable modules: a pre-trained text encoder, Contrastive Language-Image Pretraining (CLIP) [19] to obtain the text latent space; an encoder to obtain the input H&E image latent space; a denoising UNet block [22]; and a decoder to generate the IHC stain image from the marker-specific image latent space (the output of the denoising UNet block). Gaussian noise is added before passing the features from the latent space of the input H&E image through the denoising UNet block of the latent diffusion model (LDM). This UNet also receives input from the encoded prompt/text. The adapter LoRA [8] and techniques from [31] are incorporated into the framework of the model (refer to Fig. 1) for efficient training. Additionally, adversarial learning [9], combined with reconstruction loss, ensures high-fidelity image reconstruction and accurate domain translation. The details of each module and the training of VIMs are provided in the following subsections.

2.1 Latent Diffusion Model (LDM): Encoder-UNet-Decoder

VIMs utilizes a pre-trained image encoder, denoising UNet, and decoder from LDM [22]. Unlike [31] (comparison shown in Figs. 2 and 3, Table 1), the embeddings of the input image are directly combined with the noise maps, rather than being conditioned through the decoder of the denoising UNet. This approach avoids conflicts between the noise map and the input image domain, preserving input details for generating associated IHC stains [18]. Most layers of the LDM-UNet are frozen, except for the first layer, which remains trainable for feature preservation during fine-tuning. VIMs uses the pre-trained encoder (E) to encode the H&E input image into the latent space ($x = E(H)$) and the decoder (D) to decode the denoised marker-specific latent space to generate the IHC image ($I = D(y)$). The diffusion process progressively denoises the combination of the input H&E image latent space and a noise variable (z_t), sampled from a Gaussian distribution $\mathcal{N}(0, I)$. Additionally, the denoising is conditioned on text prompts until a clean, denoised IHC image latent space is generated. This process is represented as:

$$y_i = Q(x_i + z_t, c_i) \tag{1}$$

where Q maps the i_{th} input domain H&E sample's latent space (x_i), combined with the noise map (z_t), to the associated specific IHC latent space (y_i) based on the input prompt (c_i).

2.2 Text/Prompt Encoder

The text/prompt encoder plays a vital role in VIMs, enabling the generation of multiple IHC images that represent different antibody IHC stains from a single H&E image through text conditioning. VIMs maps the H&E latent space to marker-specific IHC latent spaces and learns correlations among different markers based on the input prompt. As an end-to-end model, VIMs is trained on a uniplex H&E and IHC dataset using the prompt encoder, thereby eliminating the need for multiplex paired data (input images stained and registered at the pixel level with multiplex IHC). We evaluated the impact of different types of prompts (refer Suppl. Figure 4) on VIMs performance (refer Suppl. Table 3, Fig. 3) during training. Additionally, VIMs was compared with a base model [22], where IHC generation is conditioned only on the input H&E image with an empty prompt to assess the general effect of text conditioning (refer Tables 1 and 4, Figs. 2, 3). VIMs utilizes the pre-trained CLIP model [19] as the text/prompt encoder, which encodes textual descriptions p into feature vectors t_p that guide the image generation process. This enables VIMs to generate specific types of IHC stains based on user-provided text prompts, ensuring that the output aligns with the desired staining characteristics.

2.3 LoRA Adapter and Skip Connections

To efficiently adapt the pre-trained LDM [18,22,26] for the virtual IHC staining task and integrate new functionalities without overfitting, VIMs employs the LoRA [8] adapter. LoRA introduces a small number of trainable parameters into the image encoder, denoising UNet, and image decoder of the LDM model (refer Fig. 1). This enhances its adaptability while significantly reducing the need for extensive computational resources (refer to the original work [8]). This approach allows VIMs to avoid training from scratch and leverage the pre-trained model, which was trained on a very large vision-language task, and supports single-step inference. To preserve high-resolution details, VIMs incorporates zero-conv skip connection layers [31] between the image encoder and decoder, facilitating the flow of information across different layers of the network [18,31]. These connections improve gradient propagation and maintain the integrity of histopathological structures during the image generation process.

2.4 Losses and Adversarial Training

VIMs utilizes adversarial loss [9] along with other losses and employs a CLIP-based discriminator [12,18], similar to [18,26]. The training objective for VIMs involves three key losses: the reconstruction loss $\mathcal{L}_{rec}$, the adversarial (GAN) loss $\mathcal{L}_{adv}$, and the CLIP text-image alignment loss $\mathcal{L}_{clip}$ [19]. The reconstruction loss $\mathcal{L}_{rec}$, composed of L_2 and $\mathcal{L}_{lpips}$ losses [32], ensures that the generated IHC images closely match the ground truth. The GAN loss $\mathcal{L}_{adv}$, facilitated by the CLIP-based discriminator, helps produce realistic IHC images by guiding the generator through feedback. The text-image alignment loss $\mathcal{L}_{clip}$ ensures that the output matches the desired IHC marker by aligning the generated images with the provided text prompts. The overall training objective of VIMs is defined as:

$$\mathcal{L}_{total} = \mathcal{L}_{rec} + w_{clip}\mathcal{L}_{clip} + w_{adv}\mathcal{L}_{adv} \tag{2}$$

where w_{adv} and w_{clip} are the weights for the GAN and CLIP losses, respectively. The objective is to minimize $\mathcal{L}_{total}$ as: $\arg\min_G \mathcal{L}_{total}$. This adversarial training setup enables VIMs to generate high-fidelity IHC stains from H&E images while maintaining consistency with the text prompts, allowing for end-to-end training.

2.5 Inference

During inference, the VIMs model generates multiplex IHC images from an input H&E image and a text prompt. The process involves encoding the H&E image, integrating it with the encoded text prompt, and decoding the generated latent representation to produce the final IHC image. This allows for the generation of images for multiple IHC markers from the same H&E image. Our approach addresses the lack of paired H&E and multiplex IHC data by using conditional text prompts for each marker and incorporating negative samples (images lacking cells positive for CDX2 or CK8/18) in the training dataset. This reduces false positive rates and offers a scalable solution for virtual IHC staining (see Sect. 3).

3 Experimentation and Discussion

3.1 Dataset and Training Details

This study utilized a uniplex paired dataset, where each H&E image is pixel-wise registered with a single IHC marker, to train our model. We focused on two gland IHC markers: CDX2, a nuclear marker expressed in the epithelial cells of the colon, and CK8/18, a cytoplasmic marker in epithelial cells. The goal was to generate and evaluate multiplex IHC markers from the same H&E patch. The model's capability was assessed using a partially paired dataset: paired test data where H&E is pixel-wise registered to one of the ground truth (GT) IHC markers, and an unpaired test set where the second marker was unavailable. The dataset comprised H&E WSIs from surveillance colonoscopies of five ulcerative colitis patients. Slides were stained with H&E, scanned at 0.25 μm/pixel at 40x magnification, restained by IHC with CDX2 or CK8/18 antibodies, and rescanned. Pixel-level alignment was achieved using ANTSpY [2], as detailed in [11]. For model training, 16,000 patches for CDX2-H&E and CK8/18-H&E uniplex pairs were sampled from four patients, and 4,000 patches each for paired testing from a fifth patient. The unpaired test set included H&E images without the second marker (e.g., CK8/18 stains absent from CDX2-H&E pairs). Extracted RGB patches were 512×512. Each training sample was paired with prompts, as detailed in Sect. 2.2 and Fig. 4. Mixed/hybrid prompts (MxP) were used for VIMs (MxP) model training. Additionally, the dataset for training the UNet [23] gland segmentation model was created using the pipeline from [11] (refer to GT images Suppl. Figure 7). VIMs was trained on an Nvidia A30 GPU with a batch size of 1 for 200,000 steps, with weights $w_{adv} = 0.4$ and $w_{clip} = 4$ as per Eq. 2 [18]. VIMs was effectively fine-tuned with few epochs due to the techniques, including the LoRA adapter, skip connections, and adversarial training. Further details on the evaluation are provided in Sect. 3.2.

3.2 Evaluation Methods

This study employed three comprehensive quantitative evaluation methods to assess the performance of VIMs, as illustrated in Suppl. Figure 5, for both paired and unpaired settings of H&E and IHC tiles. We aimed to determine if the model (1) correctly reconstructed the glands in the colon and (2) correctly colored the nuclei of epithelial cells when prompted for the CDX2 marker and the cytoplasm of epithelial cells when prompted for the CK8/18 marker. In virtually stained images, unstained glandular epithelial cells were counted as false negatives (FNs) and staining of cells outside of glands as false positives (FPs). Gland segmentation and masking of brown color pixels (DAB-mask) allowed the calculation of DICE, IoU, and Hausdorff distance (Haus. Dist.) [4] as the metrics of staining accuracy. The evaluation of VIMs performance included:

1. **Downstream Task (Gland Segmentation) Evaluation**: We used a UNet model [23] trained on gland segmentation tasks (examples in Suppl. Figure 7)

to evaluate both paired and unpaired test sets. IoU, DICE, and Haus. Dist. [4] were used to determine the overlap in gland outlines. In the paired setting, the UNet was trained on IHC images with corresponding GT gland segmentation [11]. In the unpaired test set, the UNet was trained on H&E images with corresponding GT gland segmentation. Examples of this evaluation are shown in Fig. 6.

2. **Quantitative Metrics**: Standard metrics were used to measure the quality and accuracy of generated IHC images, calculated only for the paired test data. Metrics included DICE, IoU, and Haus. Dist. on DAB-channel masks (Suppl. Figure 6), as well as Mean Squared Error (MSE), Structural Similarity Index (SSIM) [29], and Fréchet Inception Distance (FID) [7] calculated on the GT and generated IHC images.

3. **Qualitative Assessment by Pathologists**: Two board-certified study pathologists visually inspected the generated images to evaluate their quality and accuracy. This involved assessing the overall image quality, identifying FP and FN cells, and ranking the models based on their accuracy. This evaluation was performed for both paired and unpaired test datasets.

In addition to these, qualitative assessments were included for evaluating the model. These evaluation methods collectively provide a robust framework for analyzing and validating the effectiveness of the VIMs model.

Table 1. Quantitative analysis of dual IHC stain generation for CDX2 and CK8/18 in the paired test set. Pix2Pix and LDM are uniplex stain generation models, trained separately for CDX2 and CK8/18 markers on a fully paired dataset. VIMs are compared only with models trained on paired data. The ControlNet inference step is 25, while others are 1 step. SP: Small Prompt, MxP: Mixed (Hybrid) Prompt. Bold numbers represent the best scores.

Markers (Paired Test Set)	Models	MSE(%) ↓	SSIM(%) ↑	FID ↓	Gland Segmentation			DAB-Channel Mask		
					DICE (%) ↑	IoU (%) ↑	Haus. Dist. ↓	DICE (%) ↑	IoU (%) ↑	Haus. Dist. ↓
CDX2	Pix2Pix(CDX2)	12.01	63.15	17.86	86.66	83.64	96.80	81.19	74.63	120.54
	LDM*(CDX2)	11.51	67.34	18.73	92.81	89.67	63.57	84.44	76.83	106.36
	ControlNet	15.63	41.05	94.34	51.75	45.95	348.91	27.56	19.81	457.84
	VIMs (SP) (Ours)	12.47	66.00	22.76	87.57	84.01	99.16	69.80	66.73	210.04
	VIMs (MxP) (Ours)	12.32	**68.11**	19.21	87.83	84.79	99.66	**85.10**	**76.90**	**103.45**
CK8/18	Pix2Pix(CK8/18)	12.69	64.75	18.25	91.96	89.55	62.35	45.46	37.72	178.87
	LDM*(CK8/18)	12.18	67.46	17.32	91.45	88.50	131.95	36.28	27.49	216.70
	ControlNet	15.68	32.66	92.19	57.47	53.56	315.63	20.65	13.53	314.85
	VIMs (SP) (Ours)	12.72	66.46	**17.22**	89.83	88.02	90.03	28.10	19.55	275.51
	VIMs (MxP) (Ours)	12.15	**68.36**	18.52	**93.43**	**91.03**	60.53	**45.67**	**37.73**	**171.66**

3.3 Results

The proposed model VIMs (MxP) was evaluated on paired and unpaired test data for markers CDX2 and CK8/18, using the methods outlined in Sect. 3.2. It was compared with Pix2Pix [9], LDM* [22], ControlNet [31], and VIMs (SP) (using less informative prompts, Small Prompts). LDM* shares a similar structure to VIMs, including LoRA adapters and adversarial loss, but without prompts. ControlNet used image and prompt conditioning separately and was trained from scratch with 50 steps and inference with 25 steps, lacking image-prompt direct association, pretrained LVDM, and adversarial training. Unlike VIMs, which is a multiplex staining model, Pix2Pix and LDM* are trained separately for each marker.

Quantitative Analysis: **(i) Downstream Task Analysis**: CDX2 and CK8/18 were evaluated for gland segmentation. Table 1 (paired test set) and Table 4 (unpaired test set, Suppl.) show VIMs (MxP) outperforming other methods in CK8/18 segmentation. LDM showed better CDX2 segmentation, suggesting benefits in separate marker training. ControlNet struggled due to lack of pretrained LVDM, necessitating further optimization and data. Pix2Pix gives comparable results in all cases. **(ii) Quantitative Metrics Analysis**: Metrics (MSE, SSIM, FID scores, and DAB-Mask scores) were evaluated on paired data (Table 1). VIMs (MxP) exhibited superior SSIM scores for both markers, indicating better structural fidelity than Pix2Pix, LDM, and ControlNet. For MSE and FID scores, VIMs (MxP) showed comparable results to Pix2Pix and LDM models, demonstrating its capability in generating high-quality images similar to specialized models trained separately for each marker. VIMs also outperformed in DAB-Mask scores, despite challenges with the less prominent brown color in CK8/18. **(iii) Qualitative and Quantitative Assessment by Pathologists**: This evaluation is particularly significant for cases absent of GT for the generated markers. Two pathologists evaluated 50 challenging H&E images with generated CDX2 and CK8/18 (refer Table 2) for both unpaired and paired test sets. Models M1, M2, M3, and M4, are Pix2Pix [9], LDM [22], VIMs (SP), and VIMs (MxP), respectively. Pix2Pix and LDM are trained separately on CDX2 and CK8/18 markers. M4 achieved top rankings for image fidelity and marker localization accuracy, showing fewer FN compared to other models. M1 performed well in minimizing FP for CDX2, while M4 also showed competitive results. Overall, pathologists rated VIMs (M4) with the highest rank for both paired (only CDX2) and unpaired scenarios.

Qualitative Analysis: Qualitative analysis (Fig. 2, Suppl. Fig. 3) supported pathologists' assessments. VIMs (MxP) accurately highlighted CDX2 and CK8/18 markers, even with small prompts (VIMs SP). In negative H&E samples, where glandular structures were absent, VIMs effectively generated negative IHCs (Fig. 3), whereas other models struggled. Similar performance for positive samples is shown in the second example in Fig. 2. This robust performance underscores VIMs' efficacy in generating multiplex IHC images from a single trained model.

Overall, these evaluations demonstrate VIMs (MxP)'s capability in generating multiple markers with high fidelity, closely matching GT feature spaces, exhibiting fewer pixel-wise variations, and accurately positioning structures in H&E samples.

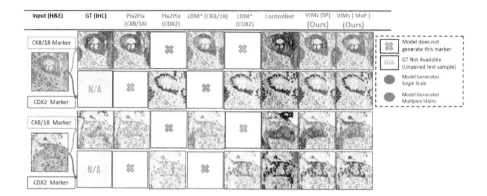

Fig. 2. Visualization of Multiplex IHC stain generation on the test set with the CK8/18 GT marker. VIMs generates visually realistic images, accurately highlighting both markers, and performs well across various cases, including difficult samples like the 2nd H&E input example.

Table 2. Pathologists' evaluation on the unpaired and paired test sets with a single inference step. Model rankings (Range: {1, 2, 3, 4}). P1 and P2 are Pathologists 1 and 2, respectively; Avg.: Average; M1: Pix2Pix [9]; M2: LDM [22]; M3: VIMs (SP) (Ours); **M4: *VIMs (MxP) (Ours)***. Pix2Pix and LDM are trained separately on CDX2 and CK8/18 markers.

Markers	Pathologists	False Positive (FP)(%) ↓				False Negative (FN)(%) ↓				Model Ranking ↓			
		M1	M2	M3	**M4**	M1	M2	M3	**M4**	M1	M2	M3	**M4**
Unpaired CDX2 Test	P1	10.00	11.00	11.00	10.80	12.00	10.20	10.00	10.40	3	2	4	1
	P2	10.20	10.60	11.40	10.80	13.10	10.20	10.00	10.00	2	3	4	1
	Ave. Score	**10.10**	10.80	11.20	10.80	12.55	10.20	**10.00**	10.20	2	2	3	1
Unpaired CK8/18 Test	P1	11.40	11.80	11.60	10.80	16.40	12.00	12.60	10.80	3	4	2	1
	P2	10.00	10.20	10.80	10.20	16.60	13.40	11.80	11.20	4	3	2	1
	Ave. Score	10.70	11.00	11.20	**10.50**	16.50	12.70	12.20	**11.00**	3	3	2	1
Paired CDX2 Test	P1	10.00	10.00	10.20	10.20	17.70	16.00	14.10	11.80	3	2	4	1
	P2	10.20	10.40	11.40	10.20	21.10	17.20	14.30	11.00	2	3	4	1
	Ave. Score	**10.10**	10.20	10.80	10.20	19.40	16.60	14.20	**11.40**	2	2	3	1

3.4 Ablation Study

In addition to the evaluations in Sect. 3.3, this study conducted an ablation analysis on various prompts used in the VIMs models for both markers. We examined VIMs on different data types, such as training without negative samples on a single marker (CDX2), Single-Marker Positive Prompts (SMPP), and combined positive and negative samples on a single marker (SMP). We also assessed the impact of different prompt conditions, including number prompts (Num) and varying lengths (small (SP), medium (MP), long (LP), mixed/hybrid (MxP)). Results from Suppl. Table 3 and Fig. 3 demonstrate that VIMs (MxP) outperforms other variations, including VIMs (SP)(Tables 1 and 4). SMPP generated more FP than SMP, highlighting the importance of including negative samples for better accuracy and visual quality. The comparison of VIMs (MxP) to VIMs (Num) shows the importance of informative prompt conditioning. VIMs (MP) performed similarly to VIMs (MxP), but the latter benefits from requiring less informative prompts during inference. VIMs (LP) needs further optimization of the prompt encoder. Overall, VIMs (MxP) consistently delivered high-quality results for both markers, leveraging hybrid prompts for robust performance, highlighting its efficiency in generating multiplex IHC images from a single trained model.

4 Conclusion and Future Work

This study introduced Virtual Immunohistochemistry Multiplex staining (VIMs), using a text-conditioned single-step DM to generate multiple IHC stains from a single H&E sample. VIMs, the first to adapt a LVDM for virtual IHC multiplexing with adversarial learning. Our approach addresses the challenge of unavailable paired H&E and multiplex IHC data, providing a scalable solution. Extensive experiments showed VIMs outperforming traditional GAN-based models, achieving high fidelity in IHC stain generation. Pathologists' evaluation confirmed its ability to produce diagnostically relevant IHC images. Efficient training with the LoRA and incorporating negative samples to reduce false positives further enhanced it's performance. Future work includes expanding VIMs to more challenging IHC markers like CD3 and CD20, improving text conditioning with advanced NLP techniques, and integrating VIMs into other clinical uses. Investigating robustness and generalizability across diverse datasets and developing automated evaluation metrics aligned with pathologist assessments are crucial next steps. Enhancing model efficiency for real-time applications and eliminating the need for uniplex paired data will also be explored. VIMs advances virtual IHC staining, offering a scalable method for generating multiplex IHC stains from a single H&E sample.

Acknowledgements. We thank the Department of Pathology and the Kahlert School of Computing at the University of Utah for their support of this project. This work was supported in part by the U.S. National Science Foundation (NSF) through award 2217154.

References

1. Andani, S., Chen, B., Ficek-Pascual, J., et al.: Multi-V-Stain: Multiplexed virtual staining of histopathology whole-slide images. medRxiv (2024)
2. Avants, B.B., Tustison, N., Song, G., et al.: Advanced normalization tools (ants). Insight J. **2**(365), 1–35 (2009)
3. Berijanian, M., Schaadt, N.S., Huang, B., Lotz, J., et al.: Unsupervised many-to-many stain translation for histological image augmentation to improve classification accuracy. J. Pathol. Inform. **14**, 100195 (2023)

4. Huttenlocher, D.P., Klanderman, G.A., Rucklidge, W.J.: Comparing images using the hausdorff distance. IEEE Trans. Pattern Anal. Mach. Intell. **15**(9), 850–863 (1993)
5. Dubey, S., Kataria, T., Knudsen, B., Elhabian, S.Y.: Structural cycle GAN for virtual immunohistochemistry staining of gland markers in the colon. In: Cao, X., Xu, X., Rekik, I., Cui, Z., Ouyang, X. (eds.) Medical Image Computing and Computer-Assisted Intervention-Workshop (MICCAI-W), pp. 447–456. Springer, Cham (2023). https://doi.org/10.1007/978-3-031-45676-3_45
6. Fassler, D.J., Abousamra, S., Gupta, R., et al.: Deep learning-based image analysis methods for brightfield-acquired multiplex immunohistochemistry images. Diagn Pathol. **15**(1), 1–11 (2020)
7. Heusel, M., Ramsauer, H., Unterthiner, T., Nessler, B., Hochreiter, S.: GANs trained by a two time-scale update rule converge to a local NASH equilibrium. In: Conference on Neural Information Processing Systems (NeurIPS) (2017)
8. Hu, E.J., Shen, Y., Wallis, P., Allen-Zhu, Z., et al.: LORA: low-rank adaptation of large language models. In: International Conference on Learning Representations (ICLR) (2022)
9. Isola, P., Zhu, J.Y., Zhou, T., Efros, et al.: Image-to-image translation with conditional adversarial networks. In: Proceedings of the IEEE Conference on Computer Vision and Pattern Recognition, pp. 1125–1134 (2017)
10. Kang, H., Luo, D., Feng, W., Zeng, S., Quan, T., et al.: StainNet: a fast and robust stain normalization network. Front. Med. **8**, 746307 (2021)
11. Kataria, T., et al.: Automating ground truth annotations for gland segmentation through immunohistochemistry. Mod. Pathol. **36**, 100331 (2023)
12. Kumari, N., Zhang, R., Shechtman, E., et al.: Ensembling off-the-shelf models for GAN training. In: IEEE Conference on Computer Vision and Pattern Recognition (CVPR) (2022)
13. Li, F., Hu, Z., Chen, W., Kak, A.: Adaptive supervised patching loss for learning H&E to IHC stain translation with inconsistent ground truth image pairs. In: Medical Image Computing and Computer-Assisted Intervention (MICCAI) (2023)
14. Li, J., Garfinkel, J., Zhang, X., Wu, D., et al.: Biopsy-free in vivo virtual histology of skin using deep learning. Light Sci. Appl. **10**, 233 (2021)
15. Lin, Y., Zeng, B., Wang, Y., et al.: Unpaired multi-domain stain transfer for kidney histopathological image. In: AAAI Conference on Artificial Intelligence (2022)
16. Liu, S., Zhang, B., Liu, Y., Han, A., et al.: Unpaired stain transfer using pathology-consistent constrained generative adversarial networks. IEEE Trans. Med. Imaging **40**, 1977–1989 (2021)
17. Magaki, S., Hojat, S.A., Wei, B., So, A., Yong, W.H.: An introduction to the performance of immunohistochemistry. In: Yong, W.H. (ed.) Biobanking. MMB, vol. 1897, pp. 289–298. Springer, New York (2019). https://doi.org/10.1007/978-1-4939-8935-5_25
18. Parmar, G., Park, T., Narasimhan, S., Zhu, J.Y.: One-step image translation with text-to-image models. arXiv preprint arXiv:2403.12036 (2024)
19. Radford, A., Kim, J.W., Hallacy, C., Ramesh, A., Goh, G., et al.: Learning transferable visual models from natural language supervision. In: International Conference on Machine Learning (ICML) (2021)
20. Rivenson, Y., Liu, T., Wei, Z., Zhang, Y., de Haan, K., Ozcan, A.: PhaseStain: the digital staining of label-free quantitative phase microscopy images using deep learning. Light Sci. Appl. **8**(1), 23 (2019)

21. Rivenson, Y., Wang, H., Wei, Z., de Haan, K., Zhang, Y., et al.: Virtual histological staining of unlabelled tissue-autofluorescence images via deep learning. Nat. Biomed. Eng. **3**(6), 466–477 (2019)
22. Rombach, R., Blattmann, A., Lorenz, D., et al.: High-resolution image synthesis with latent diffusion model. In: IEEE Conference on Computer Vision and Pattern Recognition (CVPR) (2022)
23. Ronneberger, O., Fischer, P., Brox, T.: U-Net: convolutional networks for biomedical image segmentation. In: Navab, N., Hornegger, J., Wells, W.M., Frangi, A.F. (eds.) MICCAI 2015. LNCS, vol. 9351, pp. 234–241. Springer, Cham (2015). https://doi.org/10.1007/978-3-319-24574-4_28
24. Saharia, C., Chan, W., Chang, H., et al.: Palette: Image-to-image diffusion model. In: ACM SIGGRAPH (2022)
25. Saharia, C., Chan, W., Saxena, S., et al.: Photorealistic text- to-image diffusion models with deep language understanding. In: Conference on Neural Information Processing Systems (NeurIPS) (2022)
26. Sauer, A., Lorenz, D., Blattmann, A., Rombach, R.: Adversarial diffusion distillation. arXiv preprint arXiv:2311.17042 (2023)
27. Shaban, M.T., Baur, C., Navab, N., Albarqouni, S.: StainGAN: stain style transfer for digital histological images. In: 2019 IEEE 16th International Symposium on Biomedical Imaging (ISBI 2019), pp. 953–956. IEEE (2019)
28. Shengjie, L., Chuang, Z., Feng, X., Xinyu, J., Shi, et al.: BCI: breast cancer immunohistochemical image generation through pyramid pix2pix. In: Proceedings of the IEEE/CVF Conference on Computer Vision and Pattern Recognition, pp. 1815–1824 (2022)
29. Wang, Z., Bovik, A., Sheikh, H., Simoncelli, E.: Image quality assessment: from error visibility to structural similarity. IEEE Trans. Image Process. **13**(4), 600–612 (2004)
30. Xu, Z., Huang, X., Moro, C.F., Bozóky, B., Zhang, Q.: Gan-based virtual re-staining: a promising solution for whole slide image analysis. arXiv preprint arXiv:1901.04059 (2019)
31. Zhang, L., Rao, A., Agrawala, M.: Adding conditional control to text-to-image diffusion model. In: IEEE International Conference on Computer Vision (ICCV) (2023)
32. Zhang, R., Isola, P., Efros, A.A., Shechtman, E., Wang, O.: The unreasonable effectiveness of deep features as a perceptual metric. In: IEEE Conference on Computer Vision and Pattern Recognition (CVPR) (2018)
33. Zhu, J.Y., Park, T., Isola, P., Efros, A.A.: Unpaired image-to-image translation using cycle-consistent adversarial networks. In: Proceedings of the IEEE International Conference on Computer Vision, pp. 2223–2232 (2017)
34. Zingman, I., Frayle, S., Tankoyeu, I., Sukhanov, S., Heinemann, F.: A comparative evaluation of image-to-image translation methods for stain transfer in histopathology. arXiv preprint arXiv:2303.17009 (2023)

Structural-Connectivity-Guided Functional Connectivity Representation for Multi-modal Brain Disease Classification

Zhaoxiang Wu, Biao Jie$^{(\boxtimes)}$, Wen Li, Wentao Jiang, Yang Yang, and Tongchun Du

School of Computer and Information, Anhui Normal University, Anhui 241003, Wuhu, China
jbiao@ahnu.edu.cn

Abstract. Multi-modal brain network (*e.g.*, structural and functional connectivity networks) analysis utilizes complementary information of neuroimaging data with multiple modalities, thus help to gain a more comprehensive understanding of the brain's function and structure. However, it is still a challenge task to effectively fuse the representation of structural and functional connectivity networks, with considering the intrinsic complementary of multi-modal data and unique characteristics of each modality data. To solve this problem, in this paper, we propose a multi-modal brain connectivity network (called M2BCN) learning framework that integrates the structural and functional connectivity networks for brain disease classification. Different with existing methods, we use structural connectivity from diffusion tensor imaging to induce more representative features of functional connectivity, using graph attention network method. Then we develop a novel deep learning framework to fuse features of two modality data for classification of brain diseases. The experiment results on an real epilepsy dataset demonstrate the effectiveness of the proposed M2BCN method.

Keywords: Multimodal brain network · graph attention network · classification · brain disease

1 Introduction

Epilepsy is a chronic neurological disorder characterized by sudden abnormal discharges of neurons in the brain, resulting in transient brain dysfunction [1]. Accurate diagnosis of epilepsy and its subtypes, *i.e.*, frontal lobe epilepsy (FLE) and temporal lobe epilepsy (TLE), is crucial as patients with epilepsy that may experience mental anxiety and physical safety hazards. Recently, functional magnetic resonance imaging (fMRI) and diffusion tensor imaging (DTI) have been successfully applied to computer-aided epilepsy diagnosis.

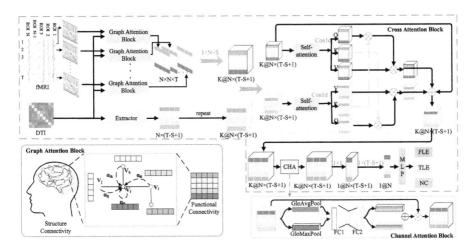

Fig. 1. Illustrates of the proposed Graph attention-based multimodal feature fusion network framework, which consists of (a) Structural Connectivity-based Graph Attention Network; (b) Multimodal Feature Fusion Module; (c) Classification.

Functional connectivity (FC) built from fMRI data primarily focusing on functional association patterns within short periods [2]. In contrast, structural connectivity (SC) constructed from DTI data remains stable over extended periods, describing the anatomical connectivity patterns between brain regions with an emphasis on topological stability [3]. Studies have demonstrated that incorporating SC information into FC network can improve the classification performance of brain disease classification. For example, Kim et al. [4] employed a regression model with graph-constrained elastic networks to predict brain cognitive performance, using FC networks(FCNs) enriched by SC networks (SCNs). Similarly, Liu et al. [5] extracted significant features from structural graph networks and subsequently adjusted the edge weights of FC networks for classification of neuropsychiatric disorders. However, these studies mainly focus on fusion strategies of multi-modal brain network data, and do not take account of the inherent complementary information between multi-modal data and the specific characteristics of each modality data. For example, studies have indicated that SC between specific brain regions often exhibit stronger functional signal expression capabilities [6], and the FC in these brain regions may be effected first in the early stages of brain diseases. Identifying these changes can further enhance the sensitivity and specificity of disease prediction. Also, studies have suggested that there exists an "imperfect" matching characteristic between FC and SC of specific brain regions [7], where the FC of certain brain regions may increase as their SC weakens, possibly as a compensatory mechanism [8]. These important information is ignored in existing studies.

To address these issues, we propose a multi-modal brain connectivity network (called M2BCN) learning framework that integrates the structural and functional connectivity networks for brain disease classification. Different with existing methods, we use SC from DTI to induce more representative features of FC, using graph attention network (GAT) method. Then we develop a novel deep learning framework to fuse features of two modality data for classification of brain diseases. Specifically, For the attributes of fMRI sequences, we first construct dynamic FCNs (dFCN), using non-overlapping time windows approach with the Pearson correlation coefficient (PCC) as measure of FC, and build SCN from DTI data using the PANDA suite to process the DTI data. Then we extract significant FC features guided by SC networks with GAT method. Here, the attention mechanism is used to enhance the connectivity of specific brain regions with strong FC and weak SC. To further alleviate the "imperfect" matching, we subsequently construct second-order SCNs to better match FC and SC relation. Furthermore, it could further alleviate the effect of noise by incorporating SCN information into dFCNs. Finally, we construct a cross-modal feature fusing framework for brain disease classification, by using channel attention to concatenate features from different channels of multi-modal data. Experimental results on an real epilepsy dataset demonstrate the effectiveness of the proposed M2BCN method.

2 Method

2.1 Structural Connectivity-Based Graph Attention Network

Currently, most multimodal research is still limited to methods that first extract modality features separately and then design fusion strategies. This approach mainly focuses on the characteristics of the modalities themselves, overlooking the inherent complementarity between different modalities. Therefore, we combine the dependency relationships between functional and structural information to extract more representative functional features for brain disease research.

As shown in Fig. 1, we construct all Regions of Interest (ROIs) as nodes $V = \{v_1, v_2, ..., v_N\}$, with DTI data used to build the connectivity matrix $A \in R^{N \times N}$. Each row of the FCN at time t is constructed as the node feature of the corresponding ROI $H = \{h_1, h_2, ...h_N\} \in R^{N \times F}$. h_i represents the node feature of the $i - th$ brain region, and F is its feature dimension. To extract functional information between ROIs on the SCNs, we will use the connectivity matrix generated by the SCN to guide the extraction of functional features. Additionally, to highlight brain regions that are not directly connected on the SCNs but exhibit active functional information, we adopt an attention mechanism to assign weight values to the PCC between pairs of brain regions. This ensures that brain regions with active functional information but less prominent structural information receive additional attention. Specifically, the operation is implemented using the GAT:

$$\alpha_{ij} = \frac{\exp(Leak\text{ReLU}(\vec{\mathbf{a}}[Wh_i||Wh_j]))}{\sum_{k \in N_i} \exp(Leak\text{ReLU}(\vec{\mathbf{a}}[Wh_i||Wh_k]))} \quad (1)$$

Here, **a** is a trainable weight vector,W is a shared weight matrix used to linearly transform node features to obtain higher-order representations. The weight relationship between two node features is calculated by $a[Wh_i||Wh_j])$,$||$ represents the concatenation operation, and N_i denotes the adjacent nodes corresponding to the central node. LeakyReLU is an activation function with a negative slope. Finally, the weight vector obtained through normalization assigns a weight coefficient to each node.

To address the issue of "imperfect" matching, we further consider matching structural and functional information between non-adjacent brain regions to extract richer functional feature information. By constructing a second-order SCNs and performing the same operations as mentioned above, we multiply and fuse the first-order features with the second-order features to enhance the representational capacity of the fMRI features. Additionally, guided by the SCNs, this approach can avoid the accumulation of noise connections and weak connections in long-sequence fMRI data.

2.2 Multimodal Feature Fusion Module

Although FC has been optimised by SC, the original SC features still contain important information about the physical connections of the brain network. By integrating these two features, the physical basis and functional performance of the brain network can be comprehensively considered, further revealing the complexity and dynamics of the brain network.

We constructed the graph convolution of DTI data to extract features with a feature dimension of N×(T−S+1). Then, we replicated it K times to obtain a feature dimension of K×N×(T−S+1). For the enhanced fMRI features, we use 3D convolution with a convolution kernel size of 1×N×S and a channel size of K. Where 1×N represents the spatial size of the FC and S is the temporal size. From this, we perform aggregation of spatio-temporal features within adjacent windows of fMRI. After that, we fused the DTI features with the fMRI features in each channel to obtain the connected features in different modes of the transmembrane state. Specifically, we perform a self-attention mechanism operation on each of them, taking the feature data N×(T−S+1) in each channel as input. The formula is shown below:

$$Q_A = X^A W_q^A, K_A = X^A W_k^A, V_A = X W_v^A$$
$$Q_B = X^B W_q^B, K_B = X^B W_k^B, V_B = X W_v^B$$
$$A = soft\max(\frac{Q_A K_A^T}{\sqrt{D_k}})V_A, B = soft\max(\frac{Q_B K_B^T}{\sqrt{D_k}})V_B$$
(2)

Where W_q, W_k, W_v is the weight matrix, Q query vector, K key vector, and V value vector pass to, after which their attention weights are computed by dot product, normalized and multiplied with V to get the respective attention matrices. For DTI data, we can further obtain the relationship between unpaired ROIs and get the local-global structural topology information. For fMRI data, we extracted the sequence features to make the model adaptively focus on the connectivity between brain regions with different importance at different moments. The DTI features are then fused with the fMRI features through a cross-attention mechanism. The specific operations are as follows:

$$\begin{aligned}
&Q_A = AW_q^{A'}, K_A = AW_k^{A'}, V_A = AW_v^{A'}\\
&Q_B = BW_q^{B'}, K_B = BW_k^{B'}, V_B = BW_v^{B'}\\
&f_A = soft\max(\tfrac{Q_A K_B^T}{\sqrt{D_k}})V_A, f_B = soft\max(\tfrac{Q_B K_A^T}{\sqrt{D_k}})V_B
\end{aligned} \quad (3)$$

f_A and f_B are the fMRI and DTI features after we fused them through the cross-attention mechanism, and we subsequently summed these features with the original fMRI features to obtain the modal fusion features in each channel H_{out}^i. The above operations are implemented separately in different channels, so we obtain the cross-modal fusion features in other channels. We use channel attention operations to calculate different weights to different channels further to distinguish the importance of information in each channel and obtain more representative feature information.

2.3 Classification

Before classification, we perform convolution operations with two convolutional kernels of sizes 1×1 and 1×(N−S+1) respectively, to reduce the dimensionality of the features to 1×N. Subsequently, a three-layer perceptron is used for classification, with the number of neurons set to 90, 32, and 2, respectively. ReLU is used as the activation function, and the cross-entropy loss function is employed in this method, with a dropout regularization of 0.35.

3 Experiment

3.1 Data Acquisition and Processsing

Our epilepsy dataset was obtained from Jinling Hospital of Nanjing University School of Medicine [9]. It included 114 normal controls (NC), 103 patients with frontal lobe epilepsy (FLE), and 89 patients with temporal lobe epilepsy (TLE). The data were collected using a Siemens Trio 3T scanner. The scanning parameters of Rs-fMRI were as follows: repetition time = 2000 ms, echo time = 30 ms, flip angle = 90°, voxel size = 3.75 × 3.75 ×3.75 mm³. The DTI scanning parameters were repetition time = 6100 ms, echo time = 93 ms, flip angle = 90°, and voxel size was 0.94 × 0.94 × 3 mm³. We used SPM8 implemented in the DPARSF toolbox [36] for all rs fMRI data preprocessing. This was done using EPI templates for the split sequences to subject the data to subsequent head correction, line band-pass filtering, and regression of white matter, cerebrospinal fluid, and motor parameters. The rs-fMRI scans of the subjects were divided into 90 ROIs with 240-time points using a predefined automatic anatomical labeling (AAL) template and a non-parametric alignment method. For the preprocessing of the DTI data, we used the PANDA tool, corrected for DTI distortions using the FSL toolbox, TrackVis to generate the fiber images, and, finally, the number of fibers could be considered as a measure of SC.

3.2 Experimental Setup

In this study, we validate the method's effectiveness using five-fold cross-validation covering four binary classification tasks (TLE vs. AD, FLE vs. AD, TLE vs. FLE, and TLE&FLE vs. AD). We divide all sample data into five equal parts, one used as the test set and the rest as the training set in each cross-validation. During each validation, 20% of the training set was divided into the validation set, and the rest was used for training to determine the optimal parameters of the model. Each sample is considered a new individual during the training process for each data fold to utilize the data to fully enhance the model generalization capability. We used the AdamW optimizer with a cross-entropy loss function for model training. A total of 500 epochs were performed, and the size of each batch was set to 16. The parameters of this model were set as follows: N = 90, K = 4, S = 2.

Table 1. Performance of nine methods on FLE vs.NC and TLE vs. NC classifications. ACC = Accuracy.

Method	FLE vs. NC(%)			TLE vs. NC(%)		
	ACC	SEN	SPE	ACC	SEN	SPE
SCP-GCN	59.93	43.48	71.29	63.62	61.29	65.00
Siamese-GCN	73.56	74.28	68.60	76.52	79.34	72.91
DCNN	76.58	74.67	78.02	80.49	78.76	83.14
MPCA	72.46	75.60	68.64	78.66	83.30	73.58
BrainNetCNN	71.95	66.77	77.73	74.09	63.88	85.76
M2BCN_f	78.52	83.52	71.82	78.31	76.17	83.62
M2BCN_ws2	81.69	86.71	77.22	84.31	89.64	76.90
M2BCN	**86.62**	**89.37**	**85.56**	**88.12**	**91.94**	**86.23**

3.3 Comparison Methods

We compared our proposed method with the following approaches: (1) **BrainNetCNN** [9]: Three special convolutional kernels are designed for extracting the structural information of the brain network and combined with fMRI data for attention computation to obtain new multimodal feature representations for classification. (2) **DCNN** [10]: Combining fMRI and DTI data, a diffusion convolution operation obtains multi-scale feature representations of brain regions. Setting the network parameter H = 2, direct and indirect connections are considered, and the dimensionality of the brain region representation vectors is reduced through the fully connected layer to achieve the prediction of the sample categories. (3) **Siamese-GCN**: Aims to measure the similarity between brain networks using fMRI to describe ROI features and DTI to describe brain map structure. Low-dimensional representation vectors are generated by graph convolutional networks and finally fused to learn the representation vectors and use KNN classifiers for sample category prediction. (4) **MPCA** [11]: Brain network analysis using multilinear principal component analysis (MPCA) method to integrate fMRI and DTI features into a three-dimensional tensor. The MPCA method is used to learn the low-dimensional feature representation of each sample; all samples are classified, and the support vector machine predicts labels. (5) **SCP-GCN**: Pairwise similarity is modeled using a Siamese network structure, where DTI is considered as the graph's adjacency matrix. The Laplace matrix of the normalized graph is used as the input to the network, and the classification task is performed through a dense layer, following the approach of the related papers.

Table 2. Performance of nine methods on FLE&TLE vs. NC and FLE vs. TLE classifications. ACC = Accuracy.

Method	FLE&TLE vs. NC(%)			TLE vs. FLE(%)		
	ACC	SEN	SPE	ACC	SEN	SPE
SCP-GCN	67.95	72.39	61.26	57.81	40.66	63.64
Siamese-GCN	70.92	72.18	66.42	69.36	70.88	65.94
DCNN	77.66	78.82	75.99	74.94	75.43	73.28
MPCA	73.25	78.22	70.49	75.40	**78.67**	71.36
BrainNetCNN	69.04	71.61	61.33	71.84	68.52	74.45
M2BCN_f	78.64	74.5	81.85	70.1	68.34	72.26
M2BCN_ws	82.58	**73.5**	87.40	72.07	71.33	72.06
M2BCN	**82.68**	71.50	**90.57**	**77.76**	68.33	**86.67**

3.4 Classification Performance

As shown in Table 1 and Table 2, our method has achieved optimal performance in four binary classification tasks (TLE vs. NC, FLE vs. NC, TLE vs. FLE and FLE & TLE vs. NC), which is enough to prove the effectiveness of our method.

In addition, we added two sets of ablation experiments. The first set of M2BCN_f is where we remove the structural connection part in the graph attention module. Replace it with the connection matrix constructed after thresholding the functional network to verify the guiding role of SCN on FCN. We can observe that with SCN guidance, the accuracy will stay high. This is because there is a lack of guidance from structural information, and there will be noisy connections and weak connections in the functional connections, which will affect the model's performance. In addition, in the second set of ablation experiments, we removed the DTI data module that participated in the fusion as additional supplementary information and removed the cross-attention module. The experimental results are at M2BCN_ws. We can observe that even without this step, our model Performance is still superior. This further proves that DTI and fMRI data characteristics have been reasonably used in our method. Compared with ablation experiments, DTI provides some implicit structural connection information to further improve the performance of our model (Fig. 2).

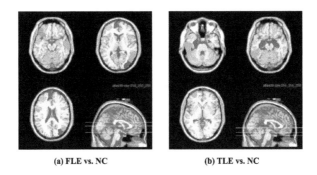

(a) FLE vs. NC   (b) TLE vs. NC

Fig. 2. Discriminative brain regions identified by the ours in 2(a) FLE vs. NC classification and 2(b) TLE vs. NC classification. With colors representing brain regions.

3.5 Visual Illustration of Discriminative Brain Regions

In this section, we further identified experimentally the top ten brain regions that contribute most to a given classification task, which are considered discriminative brain regions. The fused multimodal features were analyzed, and to simplify the complexity of the manipulation, we averaged the channels and used standard t-tests for all samples (p-value < 0.05). Brain regions with the highest discrimination in the identification of TLE vs. NC and FLE vs. NC were measured by calculating the frequency of brain region occurrences in cross-validation.

For FLE vs. NC classification, we obtained the same brain regions as in previous studies, including Middle frontal gyrus left, Inferior frontal gyrus (opercular) left, Inferior frontal gyrus (opercular) right, Orbitofrontal cortex (opercular) left, and Superior frontal gyrus (media) left [12]. In addition, most of the selected brain regions were located in the frontal part of the brain, and damage to the frontal part of the brain was associated with loss of behavior, attention, language, and thinking skills in patients, consistent with the idea that frontal lobe epilepsy originates in the frontal lobe [13].

For the TLE vs. NC classification, the brain regions we identified were also confirmed in previous work and include Superior frontal gyrus (dorsal) left, Hippocampus left, Hippocampus right, ParaHippocampal gyrus left, ParaHippocampal gyrus right, Amygdala right, Temporal pole (superior) left [14]. The hippocampal region is closely associated with long-term memory, and some studies have reported that parts of the Hippocampus are closely associated with TLE

[15], which is one of the primary sources of seizures. We also identified brain regions associated with the temporal lobe, which is related to human recognition, perception, and language comprehension, and one of the hallmarks of temporal lobe epilepsy is damage to these brain regions [16].

These results demonstrate the potential of our proposed multimodal approach in FLE vs. NC and TLE vs. NC classification tasks and highlight the critical role that specific brain regions play in classifying these disorders.

4 Summary

This paper proposes a M2BCN learning framework.Using the structural characteristics of DTI to guide the extraction of fMRI functional features, we obtain multi-modal features with more representational capabilities. We make up for the shortcomings in existing work that need to fully utilize the characteristics of the modality itself and the complementarity of data between modalities. Experiments have shown that our method performs better on the epilepsy classification task, proving its superiority.

Acknowledgment. Z. Zhang, B. Jie, Z. Wang, and J. Zhou are supported in part by NSFC (Nos. 61976006, 61573023, 61902003), Anhui NSFC (Nos. 1708085MF145, 1808085MF171) and AHNU FOYHE (No. gxyqZD2017010).

References

1. Thijs, R.D., Surges, R., O'Brien, T.J., Sander, J.W.: Epilepsy in adults. Lancet **393**(10172), 689–701 (2019)
2. Logothetis, N.K.: What we can do and what we cannot do with FMRI. Nature, **453**(7197), 869–878 (2008)
3. Zalesky, A., Fornito, A.: A DTI-derived measure of cortico-cortical connectivity. IEEE Trans. Med. Imaging **28**(7), 1023–1036 (2009)
4. Kim, M., et al.: A structural enriched functional network: an application to predict brain cognitive performance. Med. Image Anal. **71**, 102026 (2021)
5. Liu, L., Wang, Y.-P., Wang, Y., Zhang, P., Xiong, S.: An enhanced multi-modal brain graph network for classifying neuropsychiatric disorders. Med. Image Anal. **81**, 102550 (2022)
6. Gu, Z., Jamison, K.W., Sabuncu, M.R., Kuceyeski, A. et al.: Heritability and interindividual variability of regional structure-function coupling. Nat. Commun. **12**(1), 4894 (2021)
7. Suárez, L.E., Markello, R.D., Betzel, R.F., Misic, B.: Linking structure and function in macroscale brain networks. Trends Cogn. Sci. **24**(4), 302–315 (2020)
8. Lei, B., et al.: Self-calibrated brain network estimation and joint non-convex multi-task learning for identification of early Alzheimer's disease. Med. Image Anal. **61**, 101652 (2020)
9. Xu, R., Zhu, Q., Li, S., Hou, Z., Shao, W., Zhang, D.: MSTGC: Multi-channel spatio-temporal graph convolution network for multi-modal brain networks fusion. IEEE Trans. Neural Syst. Rehabil. Eng. **31**, 2359–2369 (2023)

10. Zhu, Q., Wang, H., Bingliang, X., Zhang, Z., Shao, W., Zhang, D.: Multimodal triplet attention network for brain disease diagnosis. IEEE Trans. Med. Imaging **41**(12), 3884–3894 (2022)
11. Huang, J., Zhou, L., Wang, L., Zhang, D.: Attention-diffusion-bilinear neural network for brain network analysis. IEEE Trans. Med. Imaging **39**(7), 2541–2552 (2020)
12. Xintong, W., et al.: Altered intrinsic brain activity associated with outcome in frontal lobe epilepsy. Sci. Rep. **9**(1), 8989 (2019)
13. McGonigal, A., Chauvel, P.: Frontal lobe epilepsy: seizure semiology and presurgical evaluation. Pract. Neurol. **4**(5), 260–273 (2004)
14. Stéphan Chabardès, S., et al.: The temporopolar cortex plays a pivotal role in temporal lobe seizures. Brain **128**(8), 1818–1831 (2005)
15. Doucet, G.E., et al.: Presurgery resting-state local graph-theory measures predict neurocognitive outcomes after brain surgery in temporal lobe epilepsy. Epilepsia **56**(4), 517–526 (2015)
16. Zhao, F., Kang, H., You, L., Rastogi, P., Venkatesh, D.: Neuropsychological deficits in temporal lobe epilepsy: a comprehensive review. Ann. Indian Acad. Neurol. **17**(4), 374–382 (2014)

Clinical Brain MRI Super-Resolution with 2D Slice-Wise Diffusion Model

Runqi Wang[1,2], Zehong Cao[2], Yichu He[2], Jiameng Liu[1], Feng Shi[2(✉)], and Dinggang Shen[1,2,3(✉)]

[1] School of Biomedical Engineering and State Key Laboratory of Advanced Medical Materials and Devices, ShanghaiTech University, Shanghai 201210, China
Dinggang.Shen@gmail.com
[2] Department of Research and Development, United Imaging Intelligence Co., Ltd., Shanghai 200232, China
feng.shi@uii-ai.com
[3] Shanghai Clinical Research and Trial Center, Shanghai 201210, China

Abstract. Magnetic resonance imaging (MRI) plays a vital role in brain imaging, offering exceptional soft tissue contrast without the use of ionizing radiation, ensuring safe and effective medical diagnosis. In clinic settings, 2D acquisitions are preferred by physicians due to fewer slices, large spacing, and high in-plane resolution, balancing spatial resolution, signal-to-noise ratio (SNR), 0 and acquisition time. However, these MR images may lack through-plane resolution, which may hinder lesion detection, tissue segmentation, accurate volumetric measurements, and cortical reconstruction. Most existing deep learning methods are built with purely synthetic data by collecting only high-resolution images, creating a gap between synthetic data and real-world paired data. To address these issues, we propose a slice-wise framework using a diffusion model for inter-slice super-resolution of brain MR images: 1) Employ a real-world coarse super-resolution model for initial prediction; 2) Use a score-based diffusion model for detailed iterative refinement; 3) Leverage total variation (TV) penalty with a plug-and-play (PnP) optimization module for enhanced consistency. We validate our method on over 450 real paired cases, demonstrating that our method could generate realistic images with satisfactory 3D consistency and significantly reduce over-smooth problems, thereby improving current data quality. This 2D slice-wise diffusion model also provides an effective solution for improving the quality of brain MRI images in real-world scenarios.

Keywords: Brain MRI · Super Resolution · Diffusion Model · Plug and Play · Clinical Imaging

---

Supplementary Information The online version contains supplementary material available at https://doi.org/10.1007/978-3-031-73284-3_17.

1 Introduction

Magnetic resonance imaging is an essential medical imaging technology [20,21], because of its superior ability to reflect soft tissue contrast compared to CT and other modalities, without the use of ionizing radiation. Globally, hospitals acquire millions of brain MRI scans annually, holding immense potential for advancing our understanding of various neurological diseases. However, their morphometric analysis has been hindered by their anisotropic resolution, characterized by thick-slice in the clinical settings. So, inter-slice super-resolution is likely to build a bridge between clinical and research-level brain MRI scans. The goal of super-resolution is to reconstruct a high-resolution (HR) image from its low-resolution (LR) counterpart. While deep neural network approaches have made significant progress in this area [1,5,19,22,23], most methods assume an ideal bicubic downsampling kernel, which does not accurately reflect real-world degradations. This limitation makes these approaches impractical for real-world clinical scenarios. Although some blind super-resolution methods [18,24] have been proposed for natural images, they have not yet gained traction in medical imaging, particularly for MR images. Moreover, most methods rely solely on the synthetic data. Liu *et al.* introduce a structure constrained super-resolution network on real paired dataset [11], but PSNR-oriented models (POM) often produce over-smooth results that deviate from the distribution of high-quality MR images. This makes the SR result untruthfulness. While diffusion models have achieved competitive performance in natural image SR [2,3,9,13–15,25], they need long timesteps for the reverse process, resulting in memory-heavy 3D representation and a dependency on extensive 3D data. Some traditional methods [16] have been presented for SR using known priors. Leveraging the slow reverse process and incorporating priors on each step may lead to improved performance.

In this paper, we intend to develop a framework for enhancing low-resolution MR images, to address the aforementioned issues and the challenge of low inter-slice resolution in clinical MR images, where a limited number of slices hinder accurate volumetric measurements and cortical reconstruction. Our main contributions are summarized as follows:

- We collect a large set of real paired LR and HR images in the clinic and use a high-order degradation process on the high-quality public data to generate more realistic training pairs. This effectively deals with the gap between predefined degradation and real-world degradation. Then a PSNR-oriented model is applied for a coarse prediction.
- We present a score-based generative model for detailed iterative refinement. This model builds a stochastic differential equation (SDE) between smooth images and realistic images, effectively reducing the over-smoothing effects.
- We design a plug-and-play optimization module for enhanced through-plane consistency, which alleviates the 3D data hungry and memory-heavy problems by efficiently using a 2D slice-wise diffusion model.

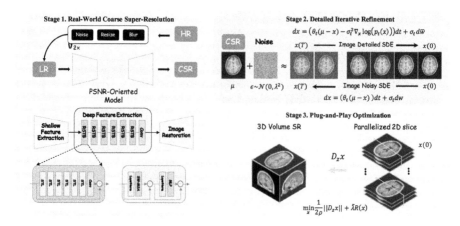

Fig. 1. The pipeline of our method contains three stages: 1) Employ real-world coarse super-resolution for initial prediction; 2) Use a score-based generative model for detailed iterative refinement; 3) Leverage total variation penalty to achieve better consistency with a plug-and-play optimization module

2 Method

We proposed a real-world MRI super-resolution framework to generate realistic HR images from LR images, which contains three parts: 1) real-world coarse super-resolution, 2) detailed iterative refinement and 3) a plug-and-play optimization module. These three stages will be introduced in this section.

2.1 Real-World Coarse Super-Resolution

This section aims to achieve coarse super-resolution (CSR) and to minimize the gap between predefined degradation and real-world degradation. We introduce both a degradation model and a PSNR-oriented model for the initial prediction stage.

Degradation Model. Here, we adopt a high-order degradation model to enhance robustness across various MR image domains with differing contrast and resolution. Traditional degradation models often simplify the representation to introduce only predetermined basic degradations such as the bicubic downsampling kernel. These basic degradations could include common types of noise, blurring, compression artifacts, or other types of distortion. However, the complexity of degradation process encountered in real-world scenarios cannot be adequately captured by these traditional first-order models. To address this, we introduce a high-order degradation model.

$$x = \mathcal{D}^n(y) = (\mathcal{D}_n \circ \cdots \circ \mathcal{D}_2 \circ \mathcal{D}_1)(y) \qquad (1)$$

This model consists of n iterations of the standard degradation process. Each iteration applies the same degradation procedure but with different hyperparameters. To encompass a broader range of degradation effects, both diverse and intricate, we implement a random selection from the available degradation methods. The degradation model contains **Resize** (i.e., given that the nearest-neighbor interpolation can lead to misalignment problems, we have chosen to disregard this operation, focusing solely on area, bilinear, trilinear and bicubic interpolation operations instead), **Blur** (i.e., we set anisotropic Gaussian and isotropic Gaussian kernels), and **Noise** (i.e., we add Gaussian noise and Poisson noise randomly). We adopt this high-order degradation model with the same procedure but different hyperparameters on MR images.

PSNR-Oriented Model. To achieve robust distortion performance and a general coarse result, a strong backbone is essential in this phase. The Swin transformer has innovatively incorporated hierarchical attention with shifted windows, enabling the fusion of contextual information while mitigating the computational burden. Among its notable applications in the domain of low-level vision, SwinIR [10] particularly stands out. Figure 1 visually illustrates the architecture, comprising multiple convolutional blocks and Swin transformer modules. Directly employing a vision transformer for super-resolution or other image restoration, it can result in patch artifacts due to the patch splitting. So a 3×3 convolution layer is built for shallow feature extraction. The patch splitting operation is performed on the feature map after the initial convolution layer, ensuring avoiding patch artifacts in the final images. For deep feature extraction, we incorporate multiple Residual Swin Transformer Blocks (RSTB), each containing several Swin Transformer Layers (STL). The combination of shallow and deep features ensures preservation of both low-frequency and high-frequency information.

2.2 Detailed Iterative Refining Network

Even though the initial stage of our method effectively addresses the majority of degradations, the images often exhibit over smooth ing. This over-smoothing effect implies that the CSR lacks intricate textures and sharpness which are characteristics of high-quality MR images. Consequently, there remains a significant gap between CSR and the distribution of high-quality MR images. We aim to bridge this gap by introducing perturbation and denoising techniques to transition from smooth to realistic representations.

Image Noisy SDE. In our approach to image detailed modeling, we construct a specialized stochastic differential equation that encapsulates the essence of the detailed process. This forward SDE denoted as in Eq. 2 is designed to translate a realistic image towards a noisy over-smoothing image by adding fixed Gaussian noise. The SDE is formulated as follows:

Fig. 2. The process of converting smooth results to realistic results with perturbing and denoising.

Algorithm 1. Plug-and-Play Optimization for Better 3D Consistency
---
1: **Input**: Low-resolution MR image μ (Volume), $\hat{\epsilon}_\theta$, T, ρ, $\{\sigma_t\}_{t=1}^T$, $\tilde{\lambda}$.
2: **Output**: Optimized super-resolution MR image $\mathbf{x}_0$ (Volume).
3: **for** $t = T$ to 1 **do**
4:     $\epsilon_t \sim \mathcal{N}(\mathbf{0}, \mathbf{I})$
5:     $\mathbf{z}_{t-1} = \frac{1}{\sqrt{\alpha_t}}\left(\mathbf{x}_t - \frac{\beta_t}{\sqrt{1-\bar{\alpha}_t}}\hat{\epsilon}_\theta(x(t),\mu,t)\right) + \sqrt{\beta_t}\epsilon_t$ // slice-wise as a network output
6:     $\mathbf{x}_{t-1} = \mathbf{z}_{t-1} - \frac{\sigma_t^2}{2\tilde{\lambda}\rho}\nabla_{\mathbf{z}_{t-1}}\|D_z\mathbf{z}_{t-1}\|^2$ //slice-wise as through-plane constraint
7: **end for**
8: **return** $\mathbf{x}_0$

$$dx = \theta_t(\mu - x)\, dt + \sigma_t\, dw, \tag{2}$$

where x represents the state of the image at time t, and μ is the mean state corresponding to the over-smoothing image. θ_t and σ_t are time-dependent parameters used to govern the speed of mean reversion and the stochastic volatility, respectively, and w is a standard Wiener process.

Image Detailed SDE. To facilitate the restoration of over-smoothing images, we employ a reverse-time stochastic differential equation that inverts the noisy process modeled by the forward SDE. This reverse SDE, derived from the forward process, is instrumental in guiding the recovery of the realistic image from its smooth state. It is formulated as:

$$dx = \left[\theta_t(\mu - x) - \sigma_t^2 \nabla_x \log p_t(x)\right] dt + \sigma_t\, d\hat{w}, \tag{3}$$

where $\nabla_x \log p_t(x)$ represents the score function of the marginal distribution at time t. The score function, which is typically intractable, is approximated by training a neural network $s_\theta(x, t)$ under a score-matching objective.

2.3 Plug-and-Play Optimization Module

To efficiently utilize the coarse-to-fine framework, we apply our score-based generative model slice by slice across the axial plane. However, this approach may overlook the relationship along z-axis, potentially resulting in slice gap artifacts. To solve this problem, we introduce a plug-and-play optimization module aimed at enhancing through-plane consistency. The main idea of this plug-and-play module is to separate the TV penalty across the z-axis and implicit regularization. Unlike traditional methods that incorporate a determined data term with a known degradation kernel, which may be too simplistic for real-world clinical scenarios, we opt for an implicit regularization approach. We assume the implicit regularization contains the data item, as we use the score function to implicitly learn the distribution of data. Here's the optimization problem:

$$\hat{\mathbf{x}} = \arg\min_{\mathbf{x}} \frac{1}{2\rho} \|D_z x\|^2 + \tilde{\lambda}\mathcal{R}(\mathbf{x}), \tag{4}$$

We can use Half-Quadratic-Splitting tricks to solve above optimization problem as the same reverse process of DDPM (proof in the supplementary material). Our sampling method is described in Algorithm 1. This optimization term will be inserted in each reverse step of our detailed iterative refinement network as

$$\hat{\mathbf{x}}_t \approx \hat{\mathbf{z}}_t - \frac{\sigma_t^2}{2\tilde{\lambda}\rho} \nabla_{\mathbf{z}_t} \|D_z \mathbf{z}_t\|^2. \tag{5}$$

2.4 Loss Function

The loss function in the first stage is composed of a pixel-wise loss:

$$I_{\text{CSR}} = \text{SwinIR}(I_{LQ}), \quad \mathcal{L}_{\text{POM}} = \|I_{\text{CSR}} - I_{HQ}\|_2^2. \tag{6}$$

Specifically, the Swin Transformer is optimized by minimizing the L2 pixel-wise loss without any perceptual optimization guidance. So, the coarse prediction has mediocre perceptual quality (over-smoothing) like the results of other PSNR-oriented models but has good distortion performance. Then we can train the condition time-dependant neural network $\hat{\epsilon}_\theta(x(t), \mu, t)$ using the following objective like DDPM [6]:

$$\mathcal{L}_{Diff} = \sum_{i=1}^{T} \gamma_i \, \mathbb{E}\Big[\|\hat{\epsilon}_\theta(x_i, \mu, i) - \epsilon_i\|\Big], \tag{7}$$

where $\gamma_1, \ldots, \gamma_T$ are positive weights. The obtained result in this stage is denoted as $\mathcal{L}_{Diff}$. We can use this neural network $\hat{\epsilon}$ to generate realistic images by perturbing the I_{CSR} with noise to sample a noisy state x_T and iteratively denoising it.

3 Experiments

Datasets. We use both a large public dataset, the Human Connectome Project (HCP), and a large in-house clinical brain MRI dataset (called as clinical dataset

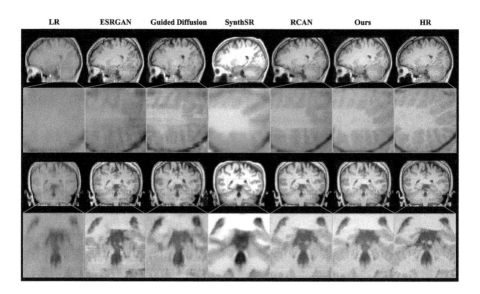

Fig. 3. Visual super-resolution results on clinical dataset. We compare GAN-based model, Diffusion-based model, CNN-based model and a general SR model with our methods. It shows that our results perform well with good perceptual quality.

Table 1. Quantitative evaluation results of different methods on clinical datasets.

	Clinical Dataset A			Clinical Dataset B		
	PSNR↑	SSIM↑	LPIPS↓	PSNR↑	SSIM↑	LPIPS↓
Trilinear	14.98	0.5163	0.1085	22.18	0.8060	0.0816
RCAN [23]	27.35	0.8690	0.0513	24.01	0.8166	0.0637
SRGAN [8]	24.76	0.8132	0.0451	22.65	0.7489	0.0522
ESRGAN [19]	25.71	0.8474	0.0421	22.31	0.7561	0.0501
Guided Diffusion [4]	26.75	0.8563	0.0382	23.49	0.8017	0.0467
SynthSR [7]	27.11	0.8692	0.0563	23.27	0.8021	0.0628
Ours	28.03	0.8891	0.0314	24.20	0.8179	0.0458

A). We evaluate our model on clinical dataset B from another center. The HCP-1200 dataset is an extensive and publicly accessible collection of high-resolution, isotropic, 3D T1-weighted magnetic resonance (MR) images of the human brain, encompassing a total of 1,113 individual scans. In the clinical routine, two-dimensional thick-slice images were obtained for each subject within approximately 90 s using a spin-echo T1-weighted imaging (T1WI) sequence, with the following parameters: repetition time (TR) = 350 ms, echo time (TE) = 2.46 ms, field of view (FOV) = 24 × 24 cm^2, matrix size = 320 × 320, slice thickness/gap = 4/0.4 mm, voxel size = 0.75 × 0.75 × 4.4 mm^3, and a total of 30 slices. Additionally, high-resolution three-dimensional T1-weighted (3D-T1w) images were acquired in approximately 450 s per subject using a magnetization-prepared rapid gradient-echo (MPRAGE) sequence, with the following acquisition parameters: TR = 2300 ms, TE = 2.98 ms, FOV = 256 × 256, matrix size = 256 × 256, slice thickness = 1 mm, voxel size = 0.5 × 0.5 × 1 mm³, and a total of 176 slices. We compare our method with five single image SR methods (trilinear interpolation, RCAN [23], SRGAN [8], ESRGAN [19] and Guided Diffusion [4]) and one general SR method (SynthSR [7]).

Experiment Settings. All models are run with PyTorch. Here are the settings: We use 700 subjects in HCP and 230 subjects in clinical dataset A for training and evaluate on about 100 subjects from clinical dataset A and clinical dataset B. All real paired clinical images are preprocessed to the same resolution in axial plane and we use trilinear to interpolate them before feed images into our model. The number of high-order degradation is six. We train the PSNR-oriented model for 120000 iterations. The diffusion model is used with the following hyperparameters: diffusion step $T = 100$, $\sigma_{max} = 30$, cosine schedule, $\rho = 1e5$, $\tilde{\lambda} = 0.01$. The time of restoring a HR image for a subject is about three minutes on a single A100 GPU. Our plug-and-play module could also work when using DDIM [17] and DPM-Solver [12] samplers.

Results and Ablation Study. Fig 3 shows the sagittal and coronal results for each method. Our method can generate more realistic results, outperforming other methods on both distortion and perceptual quality, consistently with their higher performance in PSNR, SSIM and LPIPS. More detailed quantitative measures can be seen in Table 1. We also test the significance of each part in our framework. Table 2 shows that all three models perform worse than ours, which means every module is beneficial to this task. We also show the perceptual difference in Fig 4 to show if the plug-and-play module works, i.e. with significantly fewer gap artifacts using our PnP module.

Table 2. Ablation study for different parts of our method.

	PSNR↑	SSIM↑	LPIPS↓
A) w/o coarse prediction	26.27	0.8402	0.0391
B) w/o detailed refinement	**28.04**	0.8780	0.0499
C) w/o PnP module	27.81	0.8742	0.0320
Ours	28.03	**0.8891**	**0.0314**

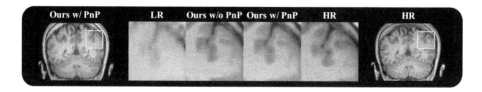

Fig. 4. Visualization of super-resolution images to show the effect of PnP module. We can see that the gap artifacts between slices disappear with our PnP module.

4 Conclusion

We present a novel framework for real-world clinical MR images super-resolution. Our method leverages both the strong distortion performance of the PSNR-oriented model and the excellent perceptual quality of diffusion model for super-resolution in clinical scenarios. Addressing the prevalent 3D data-hungry challenge in medical imaging, our method efficiently uses a 2D slice-wise diffusion model with a plug-and-play module to reduce the inconsistency. The module makes full use of the reverse step in all diffusion models. Experimental results show that our method achieves superior performance on MR images super-resolution. Future work focuses on refining degradation modeling for real-world MR images super-resolution, as well as design of the data item for scenarios involving unknown degradation.

Acknowledgements. This work was supported in part by the National Key Technologies R&D Program of China (Grant No. 82027808), and National Natural Science Foundation of China (grant numbers 62131015, U23A20295).

Disclosure of Interests. Z.C., Y.H., F.S., and D.S. are employees of United Imaging Intelligence. The company has no role in designing and performing the surveillances and analyzing and interpreting the data. All other authors report no conflicts of interest relevant to this article.

References

1. Chen, Y., Shi, F., Christodoulou, A.G., Xie, Y., Zhou, Z., Li, D.: Efficient and accurate MRI super-resolution using a generative adversarial network and 3d multi-level densely connected network. In: Frangi, A.F., Schnabel, J.A., Davatzikos, C., Alberola-López, C., Fichtinger, G. (eds.) MICCAI 2018. LNCS, vol. 11070, pp. 91–99. Springer, Cham (2018). https://doi.org/10.1007/978-3-030-00928-1_11
2. Chung, H., Kim, J., Mccann, M.T., Klasky, M.L., Ye, J.C.: Diffusion posterior sampling for general noisy inverse problems. In: The Eleventh International Conference on Learning Representations (2023)
3. Chung, H., Lee, E.S., Ye, J.C.: MR image denoising and super-resolution using regularized reverse diffusion. IEEE Trans. Med. Imaging **42**(4), 922–934 (2022)

4. Dhariwal, P., Nichol, A.: Diffusion models beat GANs on image synthesis. Adv. Neural. Inf. Process. Syst. **34**, 8780–8794 (2021)
5. Dong, C., Loy, C.C., He, K., Tang, X.: Learning a deep convolutional network for image super-resolution. In: Fleet, D., Pajdla, T., Schiele, B., Tuytelaars, T. (eds.) Computer Vision – ECCV 2014: 13th European Conference, Zurich, Switzerland, September 6-12, 2014, Proceedings, Part IV, pp. 184–199. Springer International Publishing, Cham (2014). https://doi.org/10.1007/978-3-319-10593-2_13
6. Ho, J., Jain, A., Abbeel, P.: Denoising diffusion probabilistic models. Adv. Neural. Inf. Process. Syst. **33**, 6840–6851 (2020)
7. Iglesias, J.E., et al.: Joint super-resolution and synthesis of 1 mm isotropic mp-rage volumes from clinical MRI exams with scans of different orientation, resolution and contrast. In: Neuroimage (2021)
8. Ledig, C., et al.: Photo-realistic single image super-resolution using a generative adversarial network. In: Proceedings of the IEEE Conference on Computer Vision and Pattern Recognition, pp. 4681–4690 (2017)
9. Li, H., et al.: Srdiff: single image super-resolution with diffusion probabilistic models. Neurocomputing **479**, 47–59 (2022)
10. Liang, J., Cao, J., Sun, G., Zhang, K., Van Gool, L., Timofte, R.: Swinir: image restoration using swin transformer. In: Proceedings of the IEEE/CVF International Conference on Computer Vision, pp. 1833–1844 (2021)
11. Liu, G., et al.: Recycling diagnostic MRI for empowering brain morphometric research-critical & practical assessment on learning-based image super-resolution. Neuroimage **245**, 118687 (2021)
12. Lu, C., Zhou, Y., Bao, F., Chen, J., Li, C., Zhu, J.: DPM-solver: a fast ode solver for diffusion probabilistic model sampling in around 10 steps. Adv. Neural. Inf. Process. Syst. **35**, 5775–5787 (2022)
13. Luo, Z., Gustafsson, F.K., Zhao, Z., Sjölund, J., Schön, T.B.: Image restoration with mean-reverting stochastic differential equations. In: International Conference on Machine Learning (2023)
14. Meng, C., et al.: SDEdit: guided image synthesis and editing with stochastic differential equations. In: International Conference on Learning Representations (2022)
15. Rombach, R., Blattmann, A., Lorenz, D., Esser, P., Ommer, B.: High-resolution image synthesis with latent diffusion models. In: Proceedings of the IEEE/CVF Conference on Computer Vision and Pattern Recognition, pp. 10684–10695 (2022)
16. Shi, F., Cheng, J., Wang, L., Yap, P.T., Shen, D.: Lrtv: Mr image super-resolution with low-rank and total variation regularizations. IEEE Trans. Med. Imaging **34**(12), 2459–2466 (2015)
17. Song, J., Meng, C., Ermon, S.: Denoising diffusion implicit models. In: International Conference on Learning Representations (2020)
18. Wang, X., Xie, L., Dong, C., Shan, Y.: Real-esrgan: training real-world blind super-resolution with pure synthetic data. In: Proceedings of the IEEE/CVF International Conference on Computer Vision, pp. 1905–1914 (2021)
19. Wang, X., et al.: Esrgan: enhanced super-resolution generative adversarial networks. In: Proceedings of the European Conference on Computer Vision (ECCV) Workshops, pp. 0–0 (2018)
20. Wei, Y., et al.: Multi-modal learning for predicting the genotype of glioma. IEEE Transactions on Medical Imaging (2023)
21. Wei, Y., et al.: Structural connectome quantifies tumour invasion and predicts survival in glioblastoma patients. Brain (2022)

22. Wu, Q., et al.: An arbitrary scale super-resolution approach for 3D MR images via implicit neural representation. IEEE J. Biomed. Health Inform. **27**(2), 1004–1015 (2022)
23. Zhang, Y., Li, K., Li, K., Wang, L., Zhong, B., Fu, Y.: Image super-resolution using very deep residual channel attention networks. In: Proceedings of the European conference on computer vision (ECCV), pp. 286–301 (2018)
24. Zhou, H., Huang, Y., Li, Y., Zhou, Y., Zheng, Y.: Blind super-resolution of 3D MRI via unsupervised domain transformation. IEEE J. Biomed. Health Inform. **27**(3), 1409–1418 (2022)
25. Zhu, Y., et al.: Denoising diffusion models for plug-and-play image restoration. In: Proceedings of the IEEE/CVF Conference on Computer Vision and Pattern Recognition, pp. 1219–1229 (2023)

Low-to-High Frequency Progressive K-Space Learning for MRI Reconstruction

Xiaohan Xing[1], Liang Qiu[1], Lequan Yu[2], Lingting Zhu[2], Lei Xing[1], and Lianli Liu[1(✉)]

[1] Department of Radiation Oncology, Stanford University, Stanford, USA
{xhxing,llliu}@stanford.edu
[2] Department of Statistics and Actuarial Science, University of Hong Kong, Hong Kong, China

Abstract. Magnetic Resonance Imaging (MRI) is a crucial non-invasive diagnostic tool. The image quality, however, is often limited by k-space under-sampling and noise, which is exacerbated for low-field systems. K-space learning has the potential to support high quality MRI reconstruction by exploiting correlation in the raw data domain and recovering the noise-corrupted or under-sampled measurements. However, the magnitude of the low-frequency (LF) component is usually thousands of times higher than the high-frequency (HF) component in the k-space, thus the network training might be dominated by LF learning while ignoring the recovery of the HF component. To support the effective recovery of all frequency components on the k-space data, we propose a *Low-to-High Frequency Progressive* (LHFP) learning framework, consisting of a cascade of k-space learning networks. In the first round, the model focuses on the learning of the LF component. Starting from the second round, we propose a *High-Frequency Enhancement* (HFE) module to emphasize the HF learning based on a predicted patient-specific low-high frequency boundary. To avoid degradation in LF learning during the subsequent rounds that focus on the HF component, we propose a *Low-Frequency Compensation* (LFC) module that compensates the current prediction of LF component by the last-round prediction with an estimated weight. For the reconstruction of fully-sampled and 4X under-sampled low-field brain MRI on the BraTs dataset, our method demonstrates superior performance than existing k-space learning methods, and surpasses dual-domain learning methods when combined with a simple image domain denoiser. The source codes will be released upon acceptance.

Keywords: Magnetic Resonance Imaging · Low-field MRI reconstruction · Fourier space · Progressive learning

1 Introduction

Magnetic Resonance Imaging (MRI) is a non-invasive and standard method for medical examination and diagnosis. In MRI systems, signal is collected in the k-space (i.e., Fourier space) and then reconstructed into images via Inverse Fourier

Transform (IFT) [1]. Recently, low-field MRI systems (<1T) are gaining popularity for portable imaging and imaging-guided interventional procedures [2]. However, the low-field strength results in decreased signal-to-noise ratio (SNR) and limits the image quality of low-field MRI systems. What's more, the k-space data is usually under-sampled to accelerate the MRI scanning procedure, which will further degrade the image quality [3]. It is highly desirable to reconstruct high quality MRI images from the noisy and under-sampled k-space data.

In recent years, deep learning has dominated the MRI reconstruction field. The mainstream method is image domain learning, which first reconstructs low quality images from the corrupted k-space data, then performs denoising in the image domain (as demonstrated in Fig. 1 (b)) [4,5]. However, the system noise and under-sampling effect in the k-space will significantly degrade the MRI images before the image-domain denoiser takes effect, posing a challenge for these methods to reconstruct high quality images [6]. Another stream of methods recover high quality k-space signal via interpolation and denoising in the Fourier domain, then reconstruct MRI images from the recovered k-space data (as shown in Fig. 1 (a)) [6–9]. These methods improve the quality of k-space data by exploring the correlations among different frequency components, and can be integrated with image domain denoising to further improve the quality of MRI reconstruction [10,11]. Therefore, k-space learning is a crucial step in reconstructing high quality MRI images.

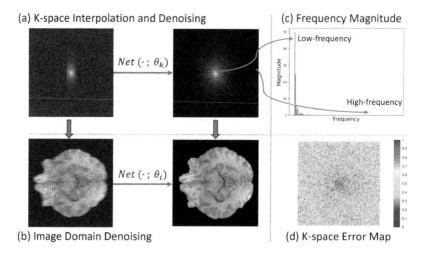

Fig. 1. MRI reconstruction via (a) k-space interpolation and denoising, and (b) image domain denoising. (c) Magnitude of different frequency components, where the magnitude of the low-frequency component (located in the central region of the k-space) is thousands of times higher than the high-frequency component (in the periphery of the k-space). (d) Relative error map of the recovered k-space data supervised by a simple L1 loss across the entire k-space.

In k-space data, low-frequency component corresponds to the overall appearance and basic structure of an image, while high-frequency component is crucial for the sharpness and detailed structure of an image. In k-space learning tasks, accurate recovery of both types of frequency components is essential for obtaining high quality images. However, the magnitudes of different frequency components are highly imbalanced, where the magnitude of the low-frequency component is usually thousands of times higher than the high-frequency component (as demonstrated in Fig. 1 (c)). Therefore, the network training might be dominated by the recovery of high-magnitude low-frequency component while ignoring high-frequency learning, thus yielding sub-optimal reconstruction results. As shown in Fig. 1 (d), the relative reconstruction error of the high-frequency component is much higher than the low-frequency component when optimized using a simple L1 loss across the entire k-space.

To support the effective recovery of both low-frequency and high-frequency components of the k-space data, we propose a *Low-to-High Frequency Progressive* (LHFP) learning framework. Our method includes a cascade of k-space learning networks, each of which progressively refines the k-space data predicted from the preceding round. In the first round, the same weight is imposed across all frequency components, thus the network will naturally focus on the learning of the high-magnitude low-frequency component. Starting from the second round, a *High-Frequency Enhancement* (HFE) module is introduced to emphasize the learning of the high-frequency component by reducing the loss weight of the low-frequency component. The low-high frequency boundary is adaptively estimated by a mask predictor, which is optimized together with the k-space learning network. To avoid degradation of the low-frequency component during the subsequent rounds that focus on high-frequency learning, we propose a *Low-Frequency Compensation* (LFC) module that compensates the current prediction of the low-frequency component by the last-round prediction with an estimated compensation weight w_c. Our main contributions are as follows:

- To tackle the challenge of frequency magnitude imbalance in k-space learning, we propose a novel LHFP framework to progressively recover the low-frequency and high-frequency components.
- We propose a HFE module to emphasize the recovery of the high frequency component and a LFC module to compensate the low-frequency learning during the subsequent rounds. The low-high frequency boundary and compensation weight w_c are adaptively optimized for each image.
- For the reconstruction of fully-sampled and 4X under-sampled low-field MRI on the BraTs dataset, our method demonstrates superior performance than existing k-space learning methods, and surpasses dual-domain learning methods when combined with a simple image domain denoiser.

2 Method

During MRI acquisition, the underlying image subject and the corresponding k-space signals can be represented as $I_H \in R^{h \times w}$ and $k_H \in R^{h \times w \times 2}$ respectively, where h and w denote the height and width of the image, and the k_H has two

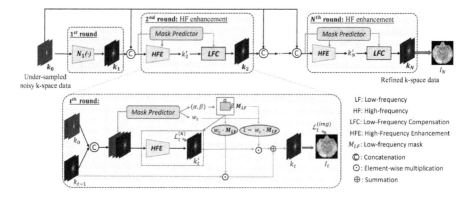

Fig. 2. Illustration of our proposed *Low-to-High Frequency Progressive* (LHFP) learning framework, which includes N networks trained in successive rounds. The 1st round focuses on low-frequency learning while the subsequent rounds emphasize the recovery of the high-frequency component. The obtained high quality k-space data k_N is used to reconstruct high quality MRI images.

channels that correspond to the real and imaginary parts of the k-space data. Due to system noise and under-sampling, the acquired k-space signal $k_0 \in R^{h \times w \times 2}$ is corrupted by noise ϵ and the under-sampling mask **M**, i.e., $k_0 = \mathbf{M} \odot (k_H + \epsilon)$. In this work, we aim to reconstruct the high quality image I_H by recovering the high quality k-space signal k_H from the corrupted k_0.

Inspired by the iterative unrolling-based MRI reconstruction methods [12,13], we design a k-space progressive learning framework, where a cascade of networks are trained in several successive rounds. The framework is demonstrated in Fig. 2. In the t-th round, the acquired low quality k-space data k_0 and last-round prediction k_{t-1} are concatenated as the model input, and the model prediction is denoted as $k_t = N_t([k_0, k_{t-1}]; \theta_t)$, where N_t is the network of the t-th round. The network parameters θ_t are optimized by jointly minimizing the k-space loss $\mathcal{L}_t^{(k)}$ and the image domain loss $\mathcal{L}_t^{(img)}$.

2.1 Low-to-High Frequency Progressive Learning

For k-space learning, the learning pace and difficulties of recovering low-frequency and high-frequency components are different due to the following reasons. First, the magnitude of the low-frequency component is thousands times higher than the high-frequency component, thus the network optimization might be dominated by the recovery of high-magnitude low-frequency component. Second, since the low-frequency signal is stronger than the high-frequency component, it is less affected by the system noise and has a higher SNR. Therefore, the recovery of low-frequency component is easier and will serve as a good basis for the high-frequency learning. Based on these observations, we propose a *Low-to-High Frequency Progressive* (LHFP) Learning strategy that first recovers low-frequency component, then progressively emphasizes the recovery of high-frequency component in the subsequent rounds.

Specifically, in the first round, the same loss weight is assigned to all frequency components across the space and the k-space loss is defined as

$$\mathcal{L}_1^{(k)} = ||k_1 - k_H||_1. \qquad (1)$$

Due to the imbalance between the amplitudes of the low-frequency and high-frequency components, the network will naturally focus on the learning of the low-frequency component. Starting from the second round, we introduce the *High-Frequency Enhancement* (HFE) module by reducing the weight of the low-frequency component during loss computation. The k-space loss is defined as:

$$\mathcal{L}_t^{(k)} = \lambda \cdot \mathbf{M}_{LF} \cdot ||k_t' - k_H||_1 + (1 - \mathbf{M}_{LF}) \cdot ||k_t' - k_H||_1, \qquad (2)$$

where k_t' is the k-space data predicted by the HFE module in the t-th round, $\mathbf{M}_{LF}$ is the mask encompassing the low-frequency component, and λ is set as 0.01 to avoid the domination of the low-frequency component loss.

Due to the different learning paces for individual cases, the use of a fixed low-frequency mask may be suboptimal for the progressive learning process. Instead, a patient-specific low-high frequency boundary is optimized together with the network weights for each round. Specifically, the input k-space data is fed into a Mask Predictor to estimate the width α and height β of the low-frequency mask. By setting the low-frequency and high-frequency components as 1 and 0 respectively, the mask $\mathbf{M}_{LF}$ is defined as:

$$\mathbf{M}_{LF}[\frac{1-\alpha}{2}h : \frac{1+\alpha}{2}h, \frac{1-\beta}{2}w : \frac{1+\beta}{2}w] = 1. \qquad (3)$$

2.2 Low-Frequency Compensation

As the network progressively focuses on high-frequency learning in the subsequent rounds, learning of the low-frequency component may be degraded, which will negatively affect the overall image reconstruction quality. To compensate for the insufficient learning of the low-frequency component during the subsequent rounds, we propose a *Low-Frequency Compensation* (LFC) layer.

As shown in Fig. 2, we combine the last-round prediction k_{t-1} and the current prediction k_t' inside the predicted low-frequency mask $\mathbf{M}_{LF}$ with a compensation weight w_c, which is also optimized together with the network parameters. Specifically, the k-space data predicted at the t-th round is defined as:

$$k_t = w_c \cdot \mathbf{M}_{LF} \odot k_{t-1} + (1 - w_c \cdot \mathbf{M}_{LF}) \odot k_t'. \qquad (4)$$

In this way, the high-frequency component is predicted solely based on results from the current round. In contrast, the low-frequency component is predicted by integrating results from the preceding round, to compensate for the insufficient low-frequency learning of the current round. Besides the k-space loss, an image

Table 1. For the T1WI MRI of the BraTs dataset, comparison of our method with state-of-the-art methods.

Domain	Method	Low-field		4X, Low-field	
		PSNR ↑	SSIM ↑	PSNR ↑	SSIM ↑
K-space only	Zero filling	22.27 ± 2.08	0.226 ± 0.020	21.61 ± 2.06	0.169 ± 0.021
	K-UNet [7]	32.29 ± 3.38	0.931 ± 0.014	29.07 ± 2.65	0.881 ± 0.018
	K-Net [14]	34.61 ± 2.03	0.890 ± 0.023	30.43 ± 2.01	0.829 ± 0.030
	K-Trans. [8]	34.45 ± 2.40	0.945 ± 0.018	30.25 ± 1.98	0.892 ± 0.014
	Ours	**37.60 ± 1.88**	**0.963 ± 0.008**	**31.27 ± 1.92**	**0.907 ± 0.014**
Dual Domain	Ding et al. [9]	37.26 ± 2.02	0.967 ± 0.008	31.95 ± 2.03	0.924 ± 0.012
	KV-Net [14]	36.50 ± 2.20	0.969 ± 0.006	31.80 ± 1.97	0.926 ± 0.011
	DudoRNet [10]	37.92 ± 1.97	0.970 ± 0.006	32.93 ± 2.00	0.934 ± 0.011
	Ours*	**39.76 ± 1.88**	**0.981 ± 0.004**	**33.66 ± 1.97**	**0.941 ± 0.011**

domain loss is also employed to refine the detailed structural information and optimize the compensation weight w_c. Specifically, the predicted k-space data k_t is transformed into the image I_t, and is supervised by L1 loss in the image domain:

$$\mathcal{L}_t^{(img)} = ||I_t - I_H||_1. \quad (5)$$

In each training round, the k-space loss in Eq. (2) and image domain loss in Eq. (5) are optimized simultaneously. The width α and height β of the low-frequency mask $\mathbf{M}_{LF}$, and the compensation weight w_c are optimized together with the network parameters to achieve the optimal reconstruction results.

3 Experiments

3.1 Dataset and Implementation

Dataset Description. We evaluate our method on the BraTs dataset [15], which contains both T1-weighted (T1WI) and T2-weighted (T2WI) brain MRI volumes. The dataset is split into training and testing sets with a ratio of 3:1, including 75 subjects (3,621 slices) for training and 25 subjects (1,088 slices) for validation. We conduct experiments on two settings: 1) low-field MRI denoising, where Gaussian noise is added to the k-space data to simulate degraded SNR for a low-field system; 2) low-field MRI acceleration, where the k-space data is corrupted by Gaussian noise and 4× under-sampling (with the central 10% data retained and the remaining signals randomly sampled). The model performance is measured by Structural Similarity Index Measure (SSIM) and Peak Signal-to-Noise Ratio (PSNR) on the testing set.

Implementation Details. Our method and all comparison methods are implemented on an NVIDIA RTX 2080ti GPU using PyTorch [16]. For the k-space data prediction network in each round depicted in Fig. 2, we employ the UNet

Table 2. For the T2WI MRI of the BraTs dataset, comparison of our method with state-of-the-art methods.

Domain	Method	Low-field		4X, Low-field	
		PSNR ↑	SSIM ↑	PSNR ↑	SSIM ↑
K-space only	Zero filling	27.10 ± 1.27	0.287 ± 0.019	25.39 ± 1.15	0.217 ± 0.017
	K-UNet [7]	36.07 ± 2.39	0.959 ± 0.007	30.60 ± 1.58	0.893 ± 0.013
	K-Net [14]	37.35 ± 1.28	0.928 ± 0.011	31.27 ± 1.42	0.846 ± 0.020
	K-Trans. [8]	37.65 ± 1.45	0.965 ± 0.006	31.05 ± 1.40	0.903 ± 0.015
	Ours	**41.05 ± 1.42**	**0.979 ± 0.004**	**32.00 ± 1.44**	**0.919 ± 0.012**
Dual Domain	Ding et al. [9]	40.48 ± 1.53	0.982 ± 0.005	33.02 ± 1.45	0.937 ± 0.012
	KV-Net [14]	40.22 ± 1.71	0.982 ± 0.005	32.72 ± 1.54	0.937 ± 0.012
	DudoRNet [10]	41.76 ± 1.54	0.985 ± 0.005	33.77 ± 1.49	0.944 ± 0.012
	Ours*	**43.35 ± 1.52**	**0.990 ± 0.003**	**34.25 ± 1.52**	**0.948 ± 0.010**

model [17] with 4 encoder and 4 decoder blocks. The Mask Predictor illustrated in Fig. 2 is implemented by 3 convolutional blocks. The training process encompasses a total of 3 rounds. For each round, the model is trained for 100 epochs with an initial learning rate set to 0.0002 and divided by 2 in every 20 epochs. The network is trained by Adam optimizer with $\beta_1 = 0.5$ and $\beta_1 = 0.999$, and batch size of 4.

3.2 Experimental Results

Comparison with State-of-the-Art Methods. We compare our method with the simple zero-filling method [18], and state-of-the-art k-space learning-based MRI reconstruction methods, including the K-UNet [7], K-Net [14], and k-space transformer [8]. As shown in Table 1 and Table 2, our method achieves a PSNR of 37.60 dB and 41.05 dB in the low-field T1WI and T2WI reconstruction, and 31.27 dB and 32.00 dB in the 4× accelerated low-field T1WI and T2WI reconstruction, which consistently outperforms existing k-space learning methods [7,8,14].

We further integrate our proposed k-space learning framework with image domain denoising. Specifically, the MRI image reconstructed by our method is concatenated with the low quality MRI as input of the image domain learning network (which is implemented as UNet [17] in our experiments). As shown in Table 1 and Table 2, our method (i.e., Ours*) outperforms existing dual-domain MRI reconstruction methods [9,10,14], further indicating the effectiveness of our method in refining k-space data and reconstructing high quality MRI images.

Ablation Study. On the T1WI MRI reconstruction task, we conduct an ablation study to analyze the contributions of different components in our method. As shown in Table 3, by training for 3 rounds with the proposed LHFP learning strategy, the PSNR of low-field MRI reconstruction is improved from 32.29

Table 3. Ablation study of different components (i.e., iterative training, HFE and LFC modules) of the proposed method for T1WI MRI reconstruction.

Method	Low-field		4X, Low-field	
	PSNR ↑	SSIM ↑	PSNR ↑	SSIM ↑
Round1	32.29 ± 3.38	0.931 ± 0.014	29.07 ± 2.65	0.881 ± 0.018
Round2	36.61 ± 1.88	0.954 ± 0.010	31.09 ± 1.93	0.901 ± 0.015
Round3 (Ours)	**37.60 ± 1.88**	**0.963 ± 0.008**	**31.27 ± 1.92**	**0.907 ± 0.014**
w/o HFE	34.22 ± 2.45	0.950 ± 0.013	30.50 ± 2.32	0.892 ± 0.016
w/o LFC	35.45 ± 2.23	0.955 ± 0.011	30.76 ± 2.14	0.898 ± 0.016

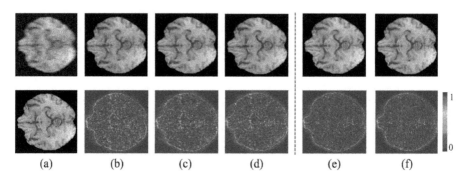

Fig. 3. Reconstructed images and the error maps of different rounds. (a) The 4× accelerated low-field MRI and fully-sampled high quality MRI. The reconstructed image and error map of the (b) Round1, (c) Round2 w/o HFE, (d) Round3 w/o HFE, (e) Round2 w/ HFE, (f) Round3 w/ HFE.

dB to 37.60 dB, and the PSNR of 4× low-field MRI reconstruction is improved from 29.07 dB to 31.27 dB. These results indicate the effectiveness of the proposed iterative LHFP learning framework. Then we remove the HFE module by putting the same emphasis on the low-frequency and high-frequency components in different rounds, and find the PSNR degrades to 34.22 dB and 30.50 dB in the low-field and the 4× low-field MRI reconstruction tasks, respectively. These results demonstrate that the performance gains of our method is not only caused by training multiple rounds, but also caused by emphasizing the high-frequency component in our LHFP leaning strategy. Furthermore, we can observe degraded performance by removing the LFC module, suggesting the effectiveness of compensating for the low-frequency learning during the subsequent rounds.

Figure 3 provides a visualization of the reconstructed images and their error maps comparing to the ground-truth image. By removing the HFE module and putting the same emphasis on all frequency components, the reconstructed images in Fig. 3(b-d) are obscure and some of the detailed information is missing. These results indicate that the high-frequency component is not well recovered since the network is dominated by the learning of the low-frequency component, and this issue cannot be resolved by simply training with more rounds. In comparison, the images (Fig. 3(e, f)) reconstructed by our method recover more detailed information, indicating that the network effectively learns the high-frequency information. This is mainly caused by the emphasis on the high-frequency component in the latter rounds of our method.

4 Conclusion

In this work, we propose a novel LHFP learning framework to progressively recover the low-frequency and high-frequency components in consecutive rounds, which effectively tackles the challenge of frequency magnitude imbalance in k-space learning. An HFE module is proposed to enhance the high-frequency learning by reducing the loss weight of the low-frequency component in subsequent rounds, where the low-high frequency boundary is adaptively estimated by a mask predictor. To avoid degradation of the low-frequency learning during subsequent rounds, an LFC module is proposed to compensate for the low-frequency component by integrating the last-round prediction. Extensive experiments on the BraTs dataset show that our method outperforms existing k-space learning methods, and surpasses dual-domain learning methods when combined with a simple image domain denoiser.

Acknowledgement. The authors would like to thank the support from the National Institutes of Health (NIH) (5R01CA256890 and 1R01CA275772).

References

1. Weishaupt, D., Köchli, V.D., Marincek, B., Froehlich, J.M., Nanz, D., Pruessmann, K.P.: How does MRI work?: An introduction to the physics and function of magnetic resonance imaging, vol. 2. Springer (2006). https://doi.org/10.1007/978-3-540-37845-7
2. Mengye Lyu, et al.: M4Raw: a multi-contrast, multi-repetition, multi-channel MRI k-space dataset for low-field MRI research. Sci. Data **10**(1), 264 (2023)
3. Tsao, J., Kozerke, S.: MRI temporal acceleration techniques. J. Magn. Reson. Imaging **36**(3), 543 560 (2012)
4. Feng, C.-M., Fu, H., Yuan, S., Xu, Y.: Multi-contrast MRI super-resolution via a multi-stage integration network. In: de Bruijne, M., et al. (eds.) MICCAI 2021. LNCS, vol. 12906, pp. 140–149. Springer, Cham (2021). https://doi.org/10.1007/978-3-030-87231-1_14
5. Feng, C.-M., et al.: Multi-modal transformer for accelerated MR imaging. IEEE Trans. Med. Imaging **42**(10), 2804–2816 (2022)
6. Pan, J., et al.: Global k-space interpolation for dynamic MRI reconstruction using masked image modeling. In: Greenspan, H., et al. Medical Image Computing and Computer Assisted Intervention - MICCAI 2023. MICCAI 2023. LNCS, vol. 14229, pp. 228–238. Springer, Cham (2023). https://doi.org/10.1007/978-3-031-43999-5_22
7. Han, Y., Sunwoo, L., Ye, J.C.: k-space deep learning for accelerated MRI. IEEE Trans. Med. Imaging **39**(2), 377–386 (2019)

8. Zhao, Z., Zhang, T., Xie, W., Wang, Y.-F., Zhang, Y.: K-space transformer for undersampled MRI reconstruction. In: BMVC, pp. 473 (2022)
9. Ding, Q., Zhang, X.: MRI reconstruction by completing under-sampled K-space data with learnable Fourier interpolation. In: Wang, L., Dou, Q., Fletcher, P.T., Speidel, S., Li, S. (eds.) Medical Image Computing and Computer Assisted Intervention - MICCAI 2022. MICCAI 2022. LNCS, vol. 13436, pp. 676–685. Springer, Cham (2022). https://doi.org/10.1007/978-3-031-16446-0_64
10. Zhou, B., Zhou, S.K.: DuDoRNet: learning a dual-domain recurrent network for fast MRI reconstruction with deep T1 prior. In: Proceedings of CVPR, pp. 4273–4282 (2020)
11. Lyu, J., Sui, B., Wang, C., Tian, Y., Dou, Q., Qin, J.: DuDoCAF: dual-domain cross-attention fusion with recurrent transformer for fast multi-contrast MR imaging. In: Wang, L., Dou, Q., Fletcher, P.T., Speidel, S., Li, S. (eds.) Medical Image Computing and Computer Assisted Intervention - MICCAI 2022. MICCAI 2022. LNCS, vol. 13436. Springer, Cham (2022). https://doi.org/10.1007/978-3-031-16446-0_45
12. Schlemper, J., Caballero, J., Hajnal, J.V., Price, A.N., Rueckert, D.: A deep cascade of convolutional neural networks for dynamic MR image reconstruction. IEEE Trans. Med. Imaging **37**(2), 491–503 (2017)
13. Fabian, Z., Tinaz, B., Soltanolkotabi, M.: HUMUS-Net: hybrid unrolled multi-scale network architecture for accelerated MRI reconstruction. Proc. NeurIPS **35**, 25306–25319 (2022)
14. Liu, X., Pang, Y., Ruiqi Jin, Yu., Liu, and Zhenchang Wang.: Dual-domain reconstruction network with v-net and k-net for fast mri. Magn. Reson. Med. **88**(6), 2694–2708 (2022)
15. Menze, B.H., et al.: The multimodal brain tumor image segmentation benchmark (BRATS). IEEE Trans. Med. Imaging **34**(10), 1993–2024 (2014)
16. Paszke, A., et al.: PyTorch: an imperative style, high-performance deep learning library. In: Proceedings of NeurIPS, vol. 32 (2019)
17. Ronneberger, O., Fischer, P., Brox, T.: U-Net: Convolutional Networks for Biomedical Image Segmentation. In: Navab, N., Hornegger, J., Wells, W.M., Frangi, A.F. (eds.) MICCAI 2015. LNCS, vol. 9351, pp. 234–241. Springer, Cham (2015). https://doi.org/10.1007/978-3-319-24574-4_28
18. Bernstein, M.A., Fain, S.B., Riederer, S.J.: Effect of windowing and zero-filled reconstruction of MRI data on spatial resolution and acquisition strategy. J. Magn. Reson. Imaging **14**(3), 270–280 (2001)

LSST: Learned Single-Shot Trajectory and Reconstruction Network for MR Imaging

Hemant Kumar Aggarwal[1(✉)], Sudhanya Chatterjee[1], Dattesh Shanbhag[1], Uday Patil[1], and K.V.S. Hari[2]

[1] Advanced Technology Group, AI Organization, GE HealthCare, Bangalore, India
{hemantkumar.aggarwal,sudhanya.chatterjee,dattesh.shanbhag,
uday.patil}@gehealthcare.com
[2] Center for Brain Research, Indian Institute of Science, Bangalore, India
hari@iisc.ac.in

Abstract. Single-shot magnetic resonance (MR) imaging acquires the entire k-space data in a single shot and it has various applications in whole-body imaging. However, the long acquisition time for the entire k-space in single-shot fast spin echo (SSFSE) MR imaging poses a challenge, as it introduces T2-blur in the acquired images. This study aims to enhance the reconstruction quality of SSFSE MR images by (a) optimizing the trajectory for measuring the k-space, (b) acquiring fewer samples to speed up the acquisition process, and (c) reducing the impact of T2-blur. The proposed method adheres to physics constraints due to maximum gradient strength and slew-rate available while optimizing the trajectory within an end-to-end learning framework. Experiments were conducted on publicly available fastMRI multichannel dataset with 8-fold and 16-fold acceleration factors. An experienced radiologist's evaluation on a five-point Likert scale indicates improvements in the reconstruction quality as the ACL fibers are sharper than comparative methods.

Keywords: Trajectory Optimization · Single-Shot MRI · Deep Learning

1 Introduction

Magnetic Resonance (MR) Imaging is a non-invasive technique that offers superior soft-tissue contrast compared to other imaging modalities such as X-ray or CT. The quality of reconstructed images are influenced not only by the number of samples acquired but also by the sampling scheme employed. Therefore,

Supplementary Information The online version contains supplementary material available at https://doi.org/10.1007/978-3-031-73284-3_19.

optimizing the sampling pattern for acquisition is critical to further enhance the reconstruction quality [1–9].

Prior work exist regarding sampling pattern optimization scheme independent of reconstruction algorithm [1,3,4], active sensing techniques [10,11] which optimizes sampling location after each TR, and methods that focus on multi-shot fast spin echo (FSE) sequences such as T2-Shuffling [12]. This study focuses on jointly optimizing sampling pattern and reconstruction network for single-shot spin echo MR imaging sequences.

The process of undersampling k-space in two dimensions to expedite MR acquisition cannot be arbitrary due to practical considerations such as system configuration of gradient and slew-rate. Random undersampling, which necessitates swift gradient switching, is often impractical as it results in high eddy current-related artifacts [13]. When formulating a k-space trajectory, it's crucial to consider the relevant MR system hardware constraints, specifically, the maximum gradient magnitude ($G_{\max}$) and maximum slew-rate ($S_{\max}$). These constraints limit the maximum velocity ($v_{\max} = \gamma G_{\max}$) and acceleration ($a_{\max} = \gamma S_{\max}$) of the trajectory, where γ is the Gyromagnetic ratio [14].

Given the benefits of arbitrary single-shot trajectories, this work focuses on optimizing them for a MR system with specific gradient and slew-rate specifications. This is achieved by determining the shortest path through a random initial set of points, akin to solving a Traveling Salesman Problem (TSP) [15]. A TSP trajectory might not meet the gradient constraints but provides a good initialization for the trajectory in the joint optimization framework.

Our approach improves upon the PILOT [16] study by eliminating the need for solving an explicit TSP during joint optimization. Unlike PILOT, this work explicitly accounts T2-Blur in single-shot acquisition and enhances reconstruction quality by incorporating the Density Compensation Factor (DCF) during training, and initializing the network input with a model-based solution. Unlike the BJORK [17] study, which doesn't account for T2-blur, which is relevant for single-shot MR imaging, our method account for it. We propose a direct-inversion network to account for the unknown T2-Blur. Key contributions of this work include:

- Proposing an approach to accelerate 2D spin echo (SE) MR imaging as single-shot fast spin echo imaging, offering an effective alternative to the 1D undersampling approach that produces considerable aliasing and blurring artifacts for high acceleration rates.
- Development of a framework for accelerated single-shot MR image acquisition and reconstruction with a focus on reducing T2-Blur, noise, and aliasing artifacts occuring in single-shot fast spin echo imaging.
- Handling an unknown forward model in single-shot acquisition due to unknown T2-Blur, using a direct-inversion network. The T2 blur is simulated for the proposed imaging method in a physics aware manner.

2 Proposed Generalized Framework

Figure 1 illustrates the training and inference pipeline of our proposed Learned Single-Shot Trajectory (LSST) optimization framework, which concurrently optimizes both the k-space trajectory and the reconstruction network. Section 2.1 discusses trainable single-shot image acquisition model and Sect. 2.2 discusses proposed joint optimization framework for trajectory optimization.

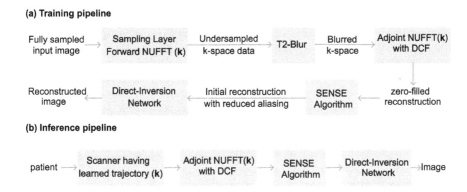

Fig. 1. The training (a) and inference (b) pipeline of the proposed joint optimization framework. Here purple blocks contain trainable parameters.

2.1 Trainable Single-Shot Acquisition Model

Initially a random variable-density sampling mask for a given acceleration factor R is generated, as depicted in Fig. 2(a). An initial trajectory **k** is generated from these random points using a TSP solver, such that the k-space center is sampled as the first point. The trajectory defines the order in which k-space is traversed to acquire the data in a single-shot. The network is initialized with this trajectory shown in Fig. 2(b). A direct TSP-based trajectory is infeasible since it does not satisfy physics constraints based on system hardware. Additionally, this generic trajectory is not necessarily optimal for a given dataset/task. Therefore, optimization of the trajectory is based on a given dataset while enforcing the physics constraints. Thus, trajectory is treated as a trainable parameter **k**. After imposing these constraints, the learned trajectory has relatively smooth curvatures which are evident from Fig. 2(c).

As shown in Fig. 1(a), we input a fully sampled image **x** to a backpropagation-compatible non-uniform Fourier transform (NUFFT) [17] operator $\mathcal{T}_\mathbf{k}$ to generate the undersampled k-space measurements **y** using the forward model

$$\mathbf{y} = \mathcal{T}_\mathbf{k}(\mathbf{x}) + \mathbf{n}, \tag{1}$$

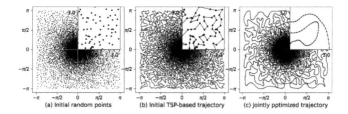

Fig. 2. This figure (a) shows an example of a random set of points acquired to achieve an acceleration factor R=8. (b) shows corresponding TSP-based trajectory that does not satisfy physics constraints on maximum gradient strength and slew-rate. (c) shows optimized trajectory using the proposed LSST framework that has smooth curvature as seen from zoomed portions in range $[2\pi/3, \pi]$ that optimized trajectory has smooth curvature compared to (b) since it satisfy physics constraints.

where $\mathbf{n}$ is an additive Gaussian noise. Here, the operator $\mathcal{T}_\mathbf{k}$ consists of k-space sampling trajectory ($\mathbf{k} \in \mathbb{C}^m$), non-uniform Fourier transform, and coil sensitivity maps (CSM). This forward model is often considered in most image reconstruction and joint optimization frameworks including: [2,16–18,18–20]. However, this model (1) does not account for the T2-blur that is present in the single-shot MR acquisition. We can model this unknown tissue specific T2-blur in k-space as a following modulation operation

$$\mathbf{y}_b = \mathcal{B} \odot \mathbf{y}, \qquad (2)$$

where $\mathcal{B}$ represents the unknown k-space blur modulation function, and $\mathbf{y}_b$ represent modulated, undersampled, and noisy measurements. Combining, (1) and (2), we get relatively accurate forward model for single-shot acquisition as

$$\mathbf{y}_b = \mathcal{B} \odot (\mathcal{T}_\mathbf{k}(\mathbf{x})) + \mathbf{n}. \qquad (3)$$

The next section describes the reconstruction process from these acquired blurred noisy undersampled measurements $\mathbf{y}_b$.

2.2 Proposed Image Reconstruction Pipeline

This section describes the proposed three-step reconstruction process that involves initial iterative Sensitivity Encoding (SENSE) based reconstruction followed by deep learning based reconstruction and artifact reduction step.

Initial SENSE Reconstruction. Given that the trajectory densely samples the center of k-space and sparsely samples the periphery, it can introduce a bias towards the lower frequencies, resulting in a blurred reconstruction from the k-space measurements. Therefore, we explicitly account for density compensation prior to the reconstruction process. This work utilizes a backpropagation compatible implementation of a density compensation algorithm [21] on blurred measurements, $\mathbf{y}_b$, to obtain an approximation of regridding reconstruction, $\widetilde{\mathbf{x}}$ using conjugate of the acquisition operator as $\widetilde{\mathbf{x}} = \mathcal{T}_\mathbf{k}^*(\mathbf{y}_b)$.

The occurrence of blur in single-shot acquisition presents additional challenges due to the pronounced undersampling artifacts and get exacerbated at higher acceleration factors. Additionally, an unknown modulation function contributes to the blur in the reconstructed images (compounding the aliasing artifacts). The density compensated regridding reconstruction $\widetilde{\mathbf{x}}$, obtained from blurred measurements $\mathbf{y}_b$, can be improved by using reconstruction algorithms such as Sensitivity Encoding (SENSE) that solves the following ℓ_2-regularized optimization problem

$$\arg\min_x \|\mathbf{y}_b - \mathcal{T}_\mathbf{k}(\mathbf{x})\|_2^2 + \lambda \|\mathbf{x}\|_2^2. \qquad (4)$$

iteratively using Conjugate-Gradient algorithm, whose solution can be represented as

$$\mathbf{x}_0 = (\mathcal{T}_\mathbf{k}^* \mathcal{T}_\mathbf{k} + \lambda \mathcal{I})^{-1}(\widetilde{\mathbf{x}}). \qquad (5)$$

Although, due to the absence of blur information, single-shot forward model in (3) can not be directly used to obtain the solution $\mathbf{x}_0$, still it acts as a good initial guess to the neural network since the aliasing artifacts are significantly reduced in $\mathbf{x}_0$ as compared to $\widetilde{\mathbf{x}}$.

Direct-Inversion Network. Once we have an initial solution $\mathbf{x}_0$, we can reconstruct the final image using a CNN. Unlike BJORK [17], the single-short forward model (3) has unknown blur therefore a model-based deep learning approach is not directly applicable for the single-shot reconstruction and trajectory optimization. Therefore, in this work, we propose to reduce the artifacts present in $\mathbf{x}_0$ using a direct-inversion network such UNet [22]. It is possible to utilize a fixed feasible trajectory such as [23] and obtain the reconstructed image $\widehat{\mathbf{x}}$ from $\mathbf{x}_0$ using deep learning methods as

$$\widehat{\mathbf{x}} = \mathcal{D}_\theta(\mathbf{x}_0) \qquad (6)$$

Here, $\mathcal{D}_\theta$ is a CNN with trainable parameters θ. This method only optimizes the network parameters once the samples are acquired. However, the joint optimization of the sampling trajectory ($\mathbf{k}$) and reconstruction network parameters (θ) can further improve the reconstruction quality. The joint optimization can be represented as

$$\widehat{\mathbf{x}} = \mathcal{D}_{\theta,\mathbf{k}}(\widetilde{\mathbf{x}}), \qquad (7)$$

where, $\mathcal{D}_{\theta,\mathbf{k}}$ represent the network that jointly optimizes trajectory and reconstruction parameters using the following loss function

$$\{\theta^*, \mathbf{k}^*\} = \arg\min_{\theta,\mathbf{k}} L_{\text{total}}(\widehat{\mathbf{x}}, \mathbf{x}). \qquad (8)$$

Here, $\mathbf{x}$ represents the ground truth data and $L_{\text{total}} = L_{\text{task}} + \beta L_{\text{const}}$ is the total loss consisting of task loss L_{task} and constraint loss L_{const} with β as hyperparameter. L_{task} is a task-based loss function to minimize the error between ground truth and the reconstructed images. The L_task was considered as

$$L_\text{task} = \sum_{i=1}^{N} \left(\|\widehat{\mathbf{x}}_i - \mathbf{X_i}\|_1 + (1 - \text{SSIM}(\widehat{\mathbf{x}}_i, \mathbf{X_i})) \right), \tag{9}$$

where SSIM is the structural similarity index [24]. Here N represents the number of training images. However, solving (9) may make the trajectory infeasible while learning the trajectory for the MRI task. The gradient constraints are enforced by an additional constraint as proposed in [16]

$$L_\text{const} = \lambda_v \sum_{i=1}^{m-1} \max(0, |v_i| - v_\text{max}) + \lambda_a \sum_{i=1}^{m-2} \max(0, |a_i| - a_\text{max}), \tag{10}$$

with v_i and a_i as velocity and acceleration at i^{th} point, calculated using first and second derivatives of the trajectory as $\dot{\mathbf{k}}$ and $\ddot{\mathbf{k}}$, respectively.

The jointly trained network $\mathcal{D}_{\theta,\mathbf{k}}$ using the total loss function results in optimized trajectory $\mathbf{k}$ and network parameters θ for a particular decimation rate (DR). We can acquire undersampled measurements using this optimized trajectory $\mathbf{k}$ and later reconstruct the fully sampled image $\widehat{\mathbf{x}}$ using this same network $\mathcal{D}_{\theta,\mathbf{k}}$ as shown in the inference pipeline in Fig. 1(b).

3 Experiments and Results

We trained and tested our models on a subset of parallel imaging fastMRI knee dataset [25] consisting of 100 volumes for training, 50 volumes for validation, and 94 volumes for testing. We performed network training on complex valued data at actual resolution without cropping. Images were cropped to center 320 × 320 only for display purpose. For simulating the single-shot data, we assumed, echo time (TE) =100 ms, T2 for bone marrow fat=80 ms, and sampling duration= 1 μs. During loss function optimization, we set the value of gradient constraints as $G_\text{max} = 40$ mT/m and $S_\text{max} = 200$ mT/m/ms.

3.1 Improved Input to the Network

Figure 3 highlights the impact of performing SENSE reconstruction using (5) with and without (w/o) considering the T2-Blur in the measurements. As we can see with a lower peak signal to noise ratio (PSNR) value of 19.36 dB in Fig. 3(d), the single-shot reconstruction problem becomes significantly challenging due to the presence of T2-blur. However, we note that the initial solution $\mathbf{x}_0$ (Fig. 3(d)) has reduced artifacts compared to $\widetilde{\mathbf{x}}$ (Fig. 3(c)) and, therefore acts as a good initial input to the neural network.

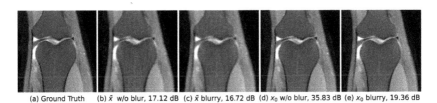

(a) Ground Truth   (b) $\tilde{x}$ w/o blur, 17.12 dB   (c) $\tilde{x}$ blurry, 16.72 dB   (d) x_0 w/o blur, 35.83 dB   (e) x_0 blurry, 19.36 dB

Fig. 3. This figure contrasts analytic and iterative reconstructions used as network inputs. A fully sampled test image shown in (a) when undersampled at 8x and reconstructed using adjoint NUFFT with DCF results in (b). (c) and (d) depict iterative reconstructions via the SENSE algorithm, without and with blur in measurements, respectively. The blurred measurements lower the PSNR from 33.83 dB to 19.36 dB, highlighting the challenge of single-shot reconstruction.

Table 1. Average PSNR (dB) and SSIM values at 8x and 16x acceleration factors shown as mean ± standard deviation for 94 test subjects. Typically, higher average PSNR and SSIM values signify enhanced reconstruction quality. When compared to the classical CSTV reconstruction and the PILOT algorithm, our LSST approach yields superior PSNR and SSIM values for single-shot acquisition.

	PSNR		SSIM × 100	
	8x	16x	8x	16x
CSTV	33.04 ± 1.06	30.91 ± 1.23	86.40 ± 2.32	82.46 ± 2.76
PILOT	36.49 ± 1.58	35.41 ± 1.44	91.88 ± 3.33	90.27 ± 3.78
Proposed	**37.94 ± 1.89**	**36.17 ± 1.48**	**93.92 ± 2.69**	**91.82 ± 3.37**

3.2 Improved Reconstruction by the Network

We performed initial experiment using traditional compressed sensing algorithm [26] (1D undersampling) with a total variation regularization (CSTV) without accounting for the T2-Blur at 8x accelerated data acquired using uniform Cartesian sampling mask having 18 lines in the calibration region. In the second experiment, to ensure fair comparison, we extended the existing PILOT algorithm [16] to the single-shot settings accounting for T2-Blur in the measurements. In the third experiment, we implemented our proposed end-to-end training pipeline LSST as shown in Fig. 1(a).

Table 2. Average Likert scores obtained for 20 volumes on five-point scale with scores as 1 to 5 include 1–non-diagnostic quality, 2–poor diagnostic quality, 3–fair diagnostic quality, 4–good diagnostic quality, 5–excellent diagnostic quality.

	SNR	Artifacts	Resolution	Contrast	Overall
Avg. of 20 volumes	5	5	4	5	5

Table 1 summarizes the PSNR and SSIM metrics on the test dataset at 8x and 16x acceleration factors for the three different methods previously discussed. Table 2 provides the quantitative evaluation results from an experienced radiologist on a five-point Likert scale. The evaluation metrics include SNR, Artifacts, Resolution, Contrast, and Overall Quality. It was observed that sharpness of ACL fibres was improved. The clinical implication for this is improved detection of partial ACL tears involving some of the fibres of the ACL. The proposed method scored highest primarily due to improvements in the ACL fibres being sharper while the meniscus region was found to be uniformly sharp across the different methods.

Figure 4 visually compares the reconstruction quality of CSTV, PILOT and LSST at 8x acceleration on a slice of a test subject. The zoomed regions indicate the improvement in the reconstruction quality. Appendix shows additional results on trajectory evaluation at 8x and 16x acceleration factors. Appendix shows benefits of individual SENSE and DL components of proposed pipeline.

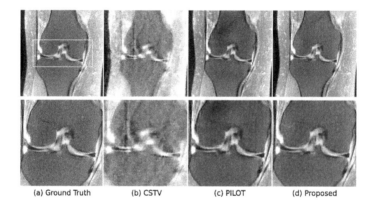

Fig. 4. A comparison of experimental results at 8x acceleration on a test slice. The CSTV output (b) exhibits noticeable artifacts due to high 8x acceleration and the use of a 1D Cartesian sampling mask. The zoomed area shows that ACL fibers are sharper than PILOT (c) whereas meniscus is sharp everywhere in the proposed method (d).

4 Conclusions and Discussions

This study introduces a method for the joint optimization of the k-space trajectory and reconstruction network parameters for single-shot acquisition that accounts for T2-Blur while adhering to MR system constraints. The learned trajectory meets gradient strength and slew-rate constraints and ensures practical trajectory estimation. Initial results and radiologist Likert scores affirm the method's utility.

Acknowledgments. We thank Shriram KS for his help and support with the manuscript.

Disclosure of Interests. The authors have no competing interests to declare that are relevant to the content of this article.

References

1. Senel, L.K.: Statistically segregated k-space sampling for accelerating multiple-acquisition MRI. IEEE Trans. Med. Imaging **38**(7), 1701–1714 (2019)
2. Bahadir, C.D., Dalca, A.V., Sabuncu, M.R.: Learning-based optimization of the under-sampling pattern in MRI. In: Chung, A.C.S., Gee, J.C., Yushkevich, P.A., Bao, S. (eds.) IPMI 2019. LNCS, vol. 11492, pp. 780–792. Springer, Cham (2019). https://doi.org/10.1007/978-3-030-20351-1_61
3. Haldar, J.P., Kim, D.: OEDIPUS: an experiment design framework for sparsity-constrained MRI. IEEE Trans. Med. Imaging **38**(7), 1545–1558 (2019)
4. Gao, Y., Reeves, S.J.: Optimal k-space sampling in MRSI for images with a limited region of support. IEEE Trans. Med. Imaging **19**(12), 1168–1178 (2000)
5. Sherry, F., et al.: Learning the sampling pattern for MRI. arXiv preprint arXiv:1906.08754 (2019)
6. Gözcü, B., et al.: Learning-based compressive MRI. IEEE Trans. Med. Imaging **37**(6), 1394–1406 (2018)
7. Liu, F., Samsonov, A., Chen, L., Kijowski, R., Feng, L.: SANTIS: sampling-augmented neural network with incoherent structure for MR image reconstruction. Magn. Reson. Med. **82**(5), 1890–1904 (2019)
8. Kim, T.H., Bilgic, B., Polak, D., Setsompop, K., Haldar, J.P.: Wave-LORAKS: combining wave encoding with structured low-rank matrix modeling for more highly accelerated 3d imaging. Magn. Reson. Med. **81**(3), 1620–1633 (2019)
9. Aggarwal, H.K., Jacob, M.: J-MoDL: joint model-based deep learning for optimized sampling and reconstruction. IEEE J. Sel. Top. Sig. Process. **14**(6), 1151–1162 (2020)
10. Jin, K.H., Unser, M., Yi, K.M.: Self-supervised deep active accelerated MRI. arXiv preprint arXiv:1901.04547 (2019)
11. Zhang, Z., Romero, A., Muckley, M.J., Vincent, P.L., Drozdzal, M.: Reducing uncertainty in undersampled MRI reconstruction with active acquisition. In: Proceedings of the IEEE Conference on Computer Vision and Pattern Recognition, pp. 2049–2058 (2019)
12. Tamir, J.I., et al.: T2 Shuffling: sharp, multicontrast, volumetric fast spin-echo imaging. Magn. Reson. Med. **77**(1), 180–195 (2017)
13. Geethanath, S., et al.: Compressed sensing MRI: a review. Crit. Rev. Biomed. Eng. **41**(3), 183–204 (2013). https://doi.org/10.1615/critrevbiomedeng.2014008058
14. Buxton, R.B.: Introduction to Functional Magnetic Resonance Imaging: Principles and Techniques. Cambridge University Press, Cambridge (2009)
15. Chauffert, N., Ciuciu, P., Kahn, J., Weiss, P.: Travelling salesman-based variable density sampling. In: Proceedings of the 10th SampTA Conference, pp. 509–512 (2013)
16. Weiss, T., Senouf, O., Vedula, S., Michailovich, O., Zibulevsky, M., Bronstein, A.: Pilot: physics-informed learned optimized trajectories for accelerated MRI. Mach. Learn. Biomed. Imaging **1**, 1–23 (2021)

17. Wang, G., Luo, T., Nielsen, J.F., Noll, D.C., Fessler, J.A.: B-spline parameterized joint optimization of reconstruction and k-space trajectories (BJORK) for accelerated 2d MRI. IEEE Trans. Med. Imaging **41**(9), 2318–2330 (2022)
18. Hammernik, K., et al.: Learning a variational network for reconstruction of accelerated MRI data. Magn. Reson. Med. **79**(6), 3055–3071 (2018)
19. Yang, Y., Sun, J., Li, H., Xu, Z.: Deep ADMM-net for compressive sensing MRI. In: Advances in Neural Information Processing Systems, vol. 29, pp. 10–18 (2016)
20. Yang, G., et al.: DAGAN: deep de-aliasing generative adversarial networks for fast compressed sensing MRI reconstruction. IEEE Trans. Med. Imaging **37**(6), 1310–1321 (2017)
21. Pipe, J.G., Menon, P.: Sampling density compensation in MRI: rationale and an iterative numerical solution. Magn. Reson. Med. Official J. Int. Soc. Magn. Reson. Med. **41**(1), 179–186 (1999)
22. Ronneberger, O., Fischer, P., Brox, T.: U-net: convolutional networks for biomedical image segmentation. In: Navab, N., Hornegger, J., Wells, W.M., Frangi, A.F. (eds.) MICCAI 2015. LNCS, vol. 9351, pp. 234–241. Springer, Cham (2015). https://doi.org/10.1007/978-3-319-24574-4_28
23. Sharma, S., Coutino, M., Chepuri, S.P., Leus, G., Hari, K.: Towards a general framework for fast and feasible k-space trajectories for MRI based on projection methods. Magn. Reson. Imaging **72**, 122–134 (2020)
24. Wang, Z., Bovik, A.C., Sheikh, H.R., Simoncelli, E.P.: Image quality assessment: from error visibility to structural similarity. IEEE Trans. Image Process. **13**(4), 600–612 (2004)
25. Zbontar, J., et al.: fastMRI: an open dataset and benchmarks for accelerated MRI (2018)
26. Lustig, M., Donoho, D., Pauly, J.M.: Sparse MRI: The application of compressed sensing for rapid MR imaging. Magn. Reson. Med. Official J. Int. Soc. Magn. Reson. Med. **58**(6), 1182–1195 (2007)

7T-Like T1-Weighted and TOF MRI Synthesis from 3T MRI with Multi-contrast Complementary Deep Learning

Zheng Zhang[1,2], Zechen Zhou[1(✉)], Lei Xiang[1], Kelei He[3], Zhiqing Zhu[1], Xingang Wang[3], Zhiming Zeng[3], Hongqin Liang[3], and Chen Liu[3(✉)]

[1] Shentou Medical Inc, Shanghai, China
zechen@subtlemedical.com
[2] Centre for Vision, Speech and Signal Processing, University of Surrey, Guildford, UK
[3] Southwest Hospital, Army Medical University, Chongqing, China
liuchen@aifmri.com

Abstract. Ultra-high-field (7T) MR imaging offers superior resolution and exceptional anatomical details when compared to conventional 3T MRI. However, the current scarcity and higher cost of 7T MRI scanners limit their accessibility in both clinical and research field. This study introduces a novel approach for simultaneously synthesizing 7T T1-weighted and Time-of-Flight (TOF) Magnetic Resonance Angiography (MRA) images from corresponding 3T T1-weighted and TOF MR scans. This method emphasizes vessel restoration using a unique loss function that integrates multi-view Maximum Intensity Projection (MIP) of the TOF image volume. The proposed method leverages complementary information between 3D T1-weighted and TOF images to enhance 3T images achieving 7T-like image quality. Extensive experiments demonstrate its superior performance compared to existing single-/multi-contrast image reconstruction methods, highlighting the significance of multi-view MIP loss and multi-contrast restoration in improving clinical 3T MRI scans.

Keywords: TOF MRA · 7T MRI · Image synthesis · Maximum Intensity Projection (MIP) Loss

1 Introduction

Magnetic Resonance Imaging (MRI) is a pivotal, non-invasive medical imaging modality providing superior soft tissue contrast for lesion detection, which has been extensively used for clinical diagnosis and disease staging [19]. However, the inherent limitations associated with the physical and physiological aspects of MRI often result in prolonged scan times for high-quality image acquisition, adversely impacting patient comfort and incurring higher costs [4]. Notably, 7T MRI systems can offer superior imaging resolution and enhanced signal-to-noise ratios (SNR) compared to their 3T counterparts, which have significantly advanced the precision in visualizing tissue structures and pathological progressions for more precise diagnosis [22]. Despite these technical advantages, the limited availability of 7T MRI systems remains a significant barrier, restricting their value in clinical practice and challenging their impact in healthcare settings [6]. In addition, clinical diagnosis often relies on multi-contrast MR image reviews with different contrast weightings that can highlight different physical properties of normal/abnormal tissues, including but not limited to T1-weighted (T1w), T2-weighted (T2w), and proton density-weighted (PDw) images [14]. These sequences are commonly used for various applications, including tumor detection, inflammation identification, soft tissue characterization, and anatomical visualization. Dedicated MR imaging scans will be required in certain MR application to better support disease staging. For example, Time-Of-Flight (TOF) plays a critical role in artery stenosis detection for neurovascular MR Angiography (MRA) imaging. Synthesis of 7T MR image by properly leveraging the multi-contrast images acquired at 3T can be clinically impactful to achieve higher image quality and more 7T-like detailed information for clinical diagnosis. Specifically, we investigate the synthesis of 7T T1w and TOF images from 3T MR acquisitions in this work.

Innovative model-driven deep convolutional neural network (DCNN) techniques have emerged to address a range of image restoration challenges, including but not limited to deraining [23], deblurring [12], and image fusion [13]. In the realm of MRI, various model-based reconstruction solutions have been proposed, leveraging compressed sensing principles [1,20,26], albeit primarily for single-contrast MRI reconstruction without extensively investigating the shareable and complementary information inherent in multi-contrast imaging [15,24,30]. To address the challenge of multi-contrast MRI reconstruction, numerous approaches have been proposed in recent years [3,4,21,25]. Specifically, Dar et al. [3] incorporated an auxiliary contrast as prior knowledge within generative adversarial network (GAN) [7] to enhance MRI acquisition speed. MC-CDic [11] utilized a multi-contrast convolutional dictionary strategy for the super-resolution and reconstruction of multi-contrast MRI. However, existing multi-contrast MR reconstruction methods mainly leverage the single slice in-plane information across multiple contrasts, without considering the shareable information from neighboring through-planes within each contrast and across different contrasts.

For 3T-to-7T MR image synthesis, Bahrami et al. [2] enhanced 3T MRI quality to approximate ultra-high-field 7T-like images using a novel hierarchical reconstruction approach in a multi-level Canonical Correlation Analysis space, significantly improving resolution and contrast for more accurate post-processing tasks like tissue segmentation, while Zhang et al. [29] proposed to synthesize realistic 7T from 3T images with two parallel and interactive multi-stage regression streams based on spatial and frequency domains. To overcome the limitation of extensive paired data, Qu et al. [16] utilized a semi-supervised learning approach with CycleGAN [31] in the spatial-wavelet domain to enable 3T-to-7T MR image mapping. Qu et al. [17] merged spatial and wavelet domain data to upscale 3T MRI images to 7T quality, employing a novel wavelet-based affine transformation layer for enhanced tissue contrast and anatomical detail. However, there are two major limitations to apply existing multi-contrast approaches to the synthesis of 7T-like TOF images: 1) Current methods focus on the translation of the contrast gap between 3T and 7T scans, neglecting the potential vascular information. In the context of MRA applications, TOF images are often paired with Maximum Intensity Projection (MIP) techniques to further enhance vascular signals against the surrounding tissue. This amplification process significantly improves the visibility and discernibility of vascular structures. 2) The calculation of MIP relies on multiple consecutive slices, existing approaches cannot effectively utilize this characteristic to improve the reconstruction. Consequently, the primary challenge in TOF MRA reconstruction extends beyond generating the TOF image itself; it also encompasses the vital task of reconstructing the vascular representation within the volume.

In this study, we introduce an innovative approach for multi-contrast MR image reconstruction, specifically aimed at generating 7T T1w and TOF MRA images from 3T acquisitions. Additionally, we present a unique loss function designed to incorporate multi-view MIP of the TOF image volume. Our key contributions include: 1) pioneering the synthesis of 7T TOF images from 3T MRI scans, 2) developing a novel multi-contrast technique that concurrently produces 7T-like T1-weighted and TOF images, facilitating comprehensive reconstruction of small details by leveraging loss computation with actual 7T images, and 3) introducing an innovative loss function that integrates multi-view MIP of the TOF image volume, enhancing the fidelity and detail of the reconstructed small vessels (e.g. lenticulostriate arteries).

2 Methods

We assume the availability of a multi-contrast MRI training set that contains multiple contrast MRI (i.e., T1-weighted and TOF) of each slice, denoted by $\mathcal{D} = \{x_1^i, x_2^i, y_1^i, y_2^i\}_{i=1}^{|\mathcal{D}|}$, where $x \in \mathcal{X} \subset \mathbb{R}^{H \times W \times R}$ denotes an input of size $H \times W$ and R channels, x_1^i and x_2^i represents R continuous 3T TOF and T1 images, while y_1^i and y_2^i denote R continuous 7T TOF and T1 images. The testing set is similarly defined.

2.1 Architecture

The proposed method is illustrated in Fig. 1. The main concept behind this method is to effectively integrate complementary information from both T1w and TOF images. Specifically, the proposed approach contains a generator $G(.)$ and a discriminator $D(.)$. The generator incorporates the Information Multi-distillation Block (IMDB) [8,9] into a UNet [18] backbone. Similar to the UNet backbone, our proposed method consists of four scales, with each scale having a residual connection between a 2 × 2 strided convolution-based downsampling and a 2 × 2 transposed convolution-based upsampling operation. The number of channels in each layer ranges from 64 on the first scale to 512 on the fourth scale. Within each scale, all original residual blocks are replaced with IMDB blocks. During the loss calculation, in addition to the standard L1 loss and SSIM loss, as well as the GAN loss, we introduce a MIP loss (as elaborated in Sect. 2.2). This addition is intended to emphasize the accurate representation of arteries on the MIP of TOF image that is clinically important for stenosis diagnosis. Furthermore, the discriminator $D(.)$ is a classifier followed by Pixel2Pixel [31], aiming to discriminate the synthesized image belonging to 3T or 7T scans (0 or 1). Consequently, our method is optimized based on the following objectives:

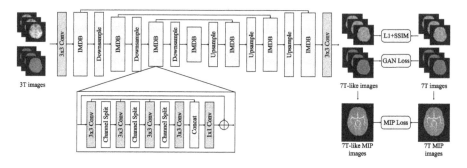

Fig. 1. The architecture of our proposed method.

$$\theta^* = \mathrm{argmin}_\theta \frac{1}{|\mathcal{D}|} \sum_{(x_1^i, x_2^i, y_1^i, y_2^i) \in \mathcal{D}} \lambda_1 \mathcal{L}_{SSIM}(x_1^i, x_2^i, y_1^i, y_2^i) + \lambda_2 \mathcal{L}_{MAE}(x_1^i, x_2^i, y_1^i, y_2^i) + \\ \lambda_3 \mathcal{L}_{GAN}(x_1^i, x_2^i, y_1^i, y_2^i) + \lambda_4 \mathcal{L}_{MIP}(x_1^i, y_1^i), \quad (1)$$

2.2 Multi-view Maximum Intensity Projection (MIP) Loss

MIP (Maximum Intensity Projection) is a widely used technique in medical image processing to enhance the visualization of vascular structures. The mathematical formula for MIP can be expressed as follows:

Given a three-dimensional image denoted as $I(x, y, z)$, where (x, y, z) represents the coordinates in three-dimensional space, the MIP image $MIP(x, y)$ for each pixel (x, y) along the z-axis (depth direction) in a two-dimensional MIP image can be defined as:

$$\mathrm{MIP}_{real}(x, y) = \max_z I(x, y, z), \quad (2)$$

As shown in Eq. 2, the MIP image is obtained by selecting the maximum pixel value for each pixel (x, y) along the z-axis based on $\max_z I(x, y, z)$. However, this operation is non-differentiable, which can pose challenges in optimization. To make the MIP operation differentiable, we replace it with the Gumbel softmax [10] method. The Gumbel softmax produces a continuous distribution that approximates categorical samples, and the gradients of its parameters can be computed using the re-parameterization trick. With this, we can approximate the MIP image for a TOF volume while maintaining differentiability. Hence, the pseudo MIP can be calculated as follows:

$$\text{MIP}_{pseudo}(x,y) = \sum_{i}^{n} \text{GumbelSM}_z I(x,y,z_i), \quad (3)$$

where GumbelSM_z denotes Gumbel Softmax of the z-axis, i and n represent the i_{th} slice and the total number of slices in the input TOF image slices. Thus, according to the real and pseudo MIP images, we can calculate the MIP loss by:

$$\mathcal{L}_{MIP}(x,y) = \mathcal{L}_{MAE}(\text{MIP}_{pseudo}, \text{MIP}_{real}) + \mathcal{L}_{SSIM}(\text{MIP}_{pseudo}, \text{MIP}_{real}). \quad (4)$$

By adjusting the temperature coefficient of the Gumbel Softmax, it is possible to make the weight assigned to the slice where the maximum value is located approach 1. When these weighted slices are summed along the z-axis, x-axis, and y-axis, respectively, it results in approximate MIP results for each axis. This approach enables the emphasis of slices with the maximum intensity value along each axis during the MIP computation.

2.3 Other Loss Functions

Besides MIP loss, the L1 loss, SSIM loss, and GAN loss [31] are used to supervise the reconstruction results, for a single contrast pairs can be represented as:

$$\mathcal{L}_{MAE}(x^i, y^i) = |y^i - x^i|, \quad (5)$$

$$\mathcal{L}_{SSIM}(x^i, y^i) = 1 - \frac{(2\mu_x\mu_y + C_1)(2\sigma_{xy} + C_2)}{(\mu_x^2 + \mu_y^2 + C_1)(\sigma_x^2 + \sigma_y^2 + C_2)}, \quad (6)$$

$$\mathcal{L}_{GAN}(x^i, y^i) = |D(y^i) - D(G(x^i))|^2, \quad (7)$$

where μ_x and μ_y are the average values of x^i and y^i, σ_x and σ_y denote the variance of x^i and y^i, σ_{xy} represents the covariance of x^i and y^i, C_1 and C_2 are used to stabilize division with weak denominators.

3 Experiments and Results

3.1 Dataset

13 subjects were recruited from X hospital, and 3D T1w and TOF images were acquired using both accelerated 3T and standard-of-care (SOC) 7T MRI protocols. The 7T TOF image served as the reference image for cross-contrast and cross-scanner registration. Brain extraction methods [5] were applied to remove the skull. The subjects were divided into two groups: a training set consisting of 10 subjects and a testing set comprising 3 subjects. Noisy slices were removed during dataset preparation. In total, the dataset consisted of 1480 image slice pairs of 3T and 7T T1w-TOF image slices for training and 444 pairs for testing.

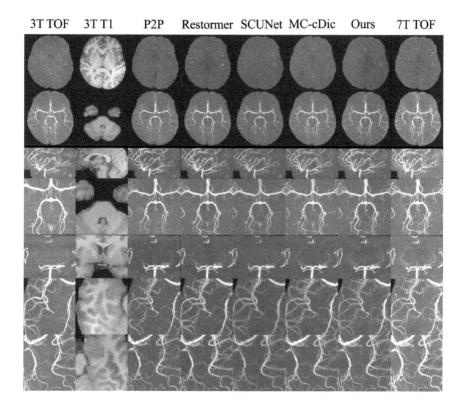

Fig. 2. Visual comparison of TOF/MIP results. The first row denotes the comparison of original images, while the others denote the comparison of MIP results. The brain arteries, particularly those small branches including lenticulastriate arteries in the 5th row, can be more accurately delineated by our proposed method.

3.2 Implementation Details

Our proposed method is implemented in PyTorch with an NVIDIA RTX3090 GPU. We adopt the Adam optimizer with a batch size of 16. λ_1, λ_2, λ_3, λ_4 are 1, 1, 0.01, and 0.1, respectively. The Gumbel softmax temperature parameter is 5. The learning rate is set to 0.0001 and the models are trained by 200 epochs. We selected the last epoch checkpoint to directly evaluate the model performance. We employ the peak-to-noise-ratio (PSNR), structural similarity index (SSIM), and root mean squared error (RMSE) to evaluate the model performance. The higher PSNR and SSIM, the lower RMSE represents the better result.

3.3 Evaluation of Synthesized 7T Images

We compare our proposed model with single/multi-contrast reconstruction methods, including Pixel2Pixel [31], Restormer [27], SCUNet [28] and MC-CDic [11]. Note that all single-contrast reconstruction methods employ multi-contrast

Table 1. Quantitative results of our proposed method and other single/multiple contrast reconstruction methods. Values in bold indicate the best results, and values with underscore denote the secondary best results.

Contrast		Methods					
		3T	P2P	Restormer	SCU-Net	MC-cDic	Ours
TOF-MIP	SSIM	0.8237	0.875	0.8901	0.8616	0.8813	**0.8991**
	PSNR	22.6	26.58	26.97	25.56	26.81	**27.39**
	RMSE	0.0753	0.0476	0.0541	0.0452	0.0456	**0.042**
T1w	SSIM	0.694	0.976	0.9782	0.975	0.9767	**0.9815**
	PSNR	13.87	35.08	35.23	35.45	35.07	**35.57**
	RMSE	0.19	0.0171	0.0167	0.0168	0.0169	**0.016**

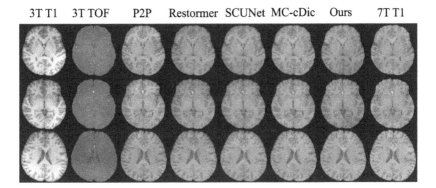

3T T1　　3T TOF　　P2P　　Restormer　SCUNet　MC-cDic　Ours　　7T T1

Fig. 3. Qualitative comparison of T1w results across different methods. Different rows show examples from several different slices. Since the artery information and T1 contrast from TOF are incorporated into our method, our synthesized 7T-like T1w image can be more similar to the acquired 7T MR image in terms of vessel and tissue contrast.

images as input to restore the target contrast images. Quantitative results are presented in Table 1. To emphasize the vascular reconstruction effect in TOF images, we conduct a series of quantitative comparisons of Maximum Intensity Projection (MIP) in TOF images. Specifically, we compute the MIP for each slice by aggregating a volume composed of 64 upper and lower slices, and then calculate the average of the transverse, coronal, and sagittal MIPs as the final result. From the table, it can be observed that our proposed method achieves the best performance for both T1w and TOF cases in the testing sets. An interesting observation is that the performance gap between our method and the suboptimal method is much larger on TOF compared to T1w images, indicating that the introduction of MIP loss mainly improves the TOF image enhancement.

Qualitative comparisons are illustrated in Figs. 2 and 3. In Fig. 2, the first row illustrates one of the TOF image slices, while the subsequent rows display the MIP of the TOF image volume. Our proposed method qualitatively

Table 2. Quantitative results of ablation experiments. Values in bold indicate the best results. MC: Multi-Contrast; MIP: Maximum Intensity Projection.

Contrast	Methods			
	3T	Ours (w/o MIP)	Ours (w/o MC)	Ours
TOF-MIP SSIM	0.8237	0.8783	0.8915	**0.8991**
PSNR	22.6	26.12	26.52	**27.39**
RMSE	0.0753	0.0463	0.044	**0.042**
T1w SSIM	0.694	0.9775	0.9789	**0.9815**
PSNR	13.87	35.51	35.38	**35.57**
RMSE	0.19	0.0162	0.0165	**0.016**

provides more accurate vascular structures (Fig. 2) and tissue structural information (Fig. 3) compared to other methods when considering the acquired 7T TOF and T1w images. Notably, our method integrates T1w structural information to accurately restore more small arteries (Fig. 2), and effectively restores 7T-like T1w contrast when combining TOF information (Fig. 3). Specifically, in the fifth row (coronal view) of Fig. 2, the results of the lenticulostriate artery demonstrate that our proposed method outperforms other methods in restoring blood vessels closer to the 7T effect. Overall, our proposed method offers qualitatively improved vascular and tissue structural information compared to existing methods when considering the acquired 7T TOF and T1w images.

3.4 Ablation Study

In Table 2, we examine how results are affected by the priors added to different components of our methods. We initially assess the model's performance when the MIP loss is removed, observing similar performance in T1w cases but a notable decrease in TOF cases. This highlights the significant influence of the MIP loss. Furthermore, we investigate the impact of multiple contrast embedding by testing two models that only utilize single contrast MRI. Experimental findings indicate that relying solely on a single contrast input leads to a deterioration in overall reconstruction performance, underscoring the importance of leveraging complementary information from different contrasts for reconstruction purposes.

4 Discussion and Conclusion

In this study, we present an innovative approach for multi-contrast MR image synthesis, specifically targeting the generation of 7T T1w and TOF MRA images from 3T scans. To better serve for vessel reconstruction, we introduce a unique loss function designed to integrate multi-view MIP of the TOF image volume. By leveraging the complementary information between T1w and TOF images, our proposed approach achieves 7T-like image quality by enhancing accelerated 3T images. Through extensive experiments, we demonstrate its superior qualitative and quantitative results over other single/multiple contrast reconstruction approaches. Additionally, ablation studies highlight the importance of the MIP loss and the introduction of multi-contrast information for better synthesis of 7T-like T1w and TOF images. Our future work will delve into performance evaluation with more clinical cases and the restoration of more smaller vessels that are only visible on 7T TOF. Such 3T clinical MR scans with enhanced efficiency and quality as 7T will better support neurovascular disease diagnosis.

References

1. Aggarwal, H.K., Mani, M.P., Jacob, M.: MoDL: model-based deep learning architecture for inverse problems. IEEE Trans. Med. Imaging **38**(2), 394–405 (2019). https://doi.org/10.1109/TMI.2018.2865356
2. Bahrami, K., Shi, F., Zong, X., Shin, H.W., An, H., Shen, D.: Reconstruction of 7T-like images from 3T MRI. IEEE Trans. Med. Imaging **35**(9), 2085–2097 (2016)
3. Dar, S.U., Yurt, M., Shahdloo, M., Ildız, M.E., Tınaz, B., Cukur, T.: Prior-guided image reconstruction for accelerated multi-contrast MRI via generative adversarial networks. IEEE J. Sel. Top. Sig. Process. **14**(6), 1072–1087 (2020)
4. Feng, C.M., et al.: Multi-modal transformer for accelerated MR imaging. IEEE Trans. Med. Imaging **42**(10), 2804–2816 (2022)
5. Fisch, L., et al.: Deepbet: Fast brain extraction of T1-weighted MRI using convolutional neural networks (2023). arXiv preprint arXiv:2308.07003
6. Forstmann, B.U., Isaacs, B.R., Temel, Y.: Ultra high field MRI-guided deep brain stimulation. Trends Biotechnol. **35**(10), 904–907 (2017)
7. Goodfellow, I., et al.: Generative adversarial nets. In: Advances in Neural Information Processing Systems, vol. 27 (2014)
8. Hui, Z., Gao, X., Yang, Y., Wang, X.: Lightweight image super-resolution with information multi-distillation network. In: Proceedings of the 27th ACM International Conference on Multimedia, pp. 2024–2032 (2019)
9. Hui, Z., Wang, X., Gao, X.: Fast and accurate single image super-resolution via information distillation network. In: Proceedings of the IEEE Conference on Computer Vision and Pattern Recognition, pp. 723–731 (2018)
10. Jang, E., Gu, S., Poole, B.: Categorical reparameterization with Gumbel-Softmax (2016). arXiv preprint arXiv:1611.01144
11. Lei, P., Fang, F., Zhang, G., Xu, M.: Deep unfolding convolutional dictionary model for multi-contrast MRI super-resolution and reconstruction. In: Proceedings of the Thirty-Second International Joint Conference on Artificial Intelligence, pp. 1008–1016 (2023)
12. Liu, R., Cheng, S., Ma, L., Fan, X., Luo, Z.: Deep proximal unrolling: algorithmic framework, convergence analysis and applications. IEEE Trans. Image Process. **28**(10), 5013–5026 (2019). https://doi.org/10.1109/TIP.2019.2913536
13. Liu, R., Liu, J., Jiang, Z., Fan, X., Luo, Z.: A bilevel integrated model with data-driven layer ensemble for multi-modality image fusion. IEEE Trans. Image Process. **30**, 1261–1274 (2021). https://doi.org/10.1109/TIP.2020.3043125
14. Lyu, Q., et al.: Multi-contrast super-resolution MRI through a progressive network. IEEE Trans. Med. Imaging **39**(9), 2738–2749 (2020)

15. Nitski, O., Nag, S., McIntosh, C., Wang, B.: CDF-Net: cross-domain fusion network for accelerated MRI reconstruction. In: Martel, A.L., et al. (eds.) MICCAI 2020. LNCS, vol. 12262, pp. 421–430. Springer, Cham (2020). https://doi.org/10.1007/978-3-030-59713-9_41
16. Qu, L., Wang, S., Yap, P.-T., Shen, D.: Wavelet-based semi-supervised adversarial learning for synthesizing realistic 7T from 3T MRI. In: Shen, D., et al. (eds.) MICCAI 2019. LNCS, vol. 11767, pp. 786–794. Springer, Cham (2019). https://doi.org/10.1007/978-3-030-32251-9_86
17. Qu, L., Zhang, Y., Wang, S., Yap, P.T., Shen, D.: Synthesized 7T MRI from 3T MRI via deep learning in spatial and wavelet domains. Med. Image Anal. **62**, 101663 (2020)
18. Ronneberger, O., Fischer, P., Brox, T.: U-Net: convolutional networks for biomedical image segmentation. In: Navab, N., Hornegger, J., Wells, W.M., Frangi, A.F. (eds.) MICCAI 2015. LNCS, vol. 9351, pp. 234–241. Springer, Cham (2015). https://doi.org/10.1007/978-3-319-24574-4_28
19. Song, P., Weizman, L., Mota, J.F., Eldar, Y.C., Rodrigues, M.R.: Coupled dictionary learning for multi-contrast MRI reconstruction. IEEE Trans. Med. Imaging **39**(3), 621–633 (2019)
20. Sriram, A., et al.: End-to-end variational networks for accelerated MRI reconstruction (2020)
21. Sun, L., Fan, Z., Fu, X., Huang, Y., Ding, X., Paisley, J.: A deep information sharing network for multi-contrast compressed sensing MRI reconstruction. IEEE Trans. Image Process. **28**(12), 6141–6153 (2019)
22. Uğurbil, K.: Magnetic resonance imaging at ultrahigh fields. IEEE Trans. Biomed. Eng. **61**(5), 1364–1379 (2014)
23. Wang, H., Xie, Q., Zhao, Q., Meng, D.: A model-driven deep neural network for single image rain removal. In: Proceedings of the IEEE/CVF Conference on Computer Vision and Pattern Recognition (CVPR) (2020)
24. Wang, S., et al.: Accelerating magnetic resonance imaging via deep learning. In: 2016 IEEE 13th International Symposium on Biomedical Imaging (ISBI), pp. 514–517. IEEE (2016)
25. Xiang, L., et al.: Deep-learning-based multi-modal fusion for fast MR reconstruction. IEEE Trans. Biomed. Eng. **66**(7), 2105–2114 (2018)
26. Yang, Y., Sun, J., Li, H., Xu, Z.: ADMM-CSNet: a deep learning approach for image compressive sensing. IEEE Trans. Pattern Anal. Mach. Intell. **42**(3), 521–538 (2020). https://doi.org/10.1109/TPAMI.2018.2883941
27. Zamir, S.W., Arora, A., Khan, S., Hayat, M., Khan, F.S., Yang, M.H.: Restormer: efficient transformer for high-resolution image restoration. In: Proceedings of the IEEE/CVF Conference on Computer Vision and Pattern Recognition, pp. 5728–5739 (2022)
28. Zhang, K., et al.: Practical blind image denoising via Swin-Conv-UNet and data synthesis. Mach. Intell. Res. **20**(6), 822–836 (2023)
29. Zhang, Y., Cheng, J.-Z., Xiang, L., Yap, P.-T., Shen, D.: Dual-domain cascaded regression for synthesizing 7T from 3T MRI. In: Frangi, A.F., Schnabel, J.A., Davatzikos, C., Alberola-López, C., Fichtinger, G. (eds.) MICCAI 2018. LNCS, vol. 11070, pp. 410–417. Springer, Cham (2018). https://doi.org/10.1007/978-3-030-00928-1_47

30. Zheng, H., Fang, F., Zhang, G.: Cascaded dilated dense network with two-step data consistency for MRI reconstruction. In: Advances in Neural Information Processing Systems, vol. 32 (2019)
31. Zhu, J.Y., Park, T., Isola, P., Efros, A.A.: Unpaired image-to-image translation using cycle-consistent adversarial networks. In: Proceedings of the IEEE International Conference on Computer Vision, pp. 2223–2232 (2017)

A Probabilistic Hadamard U-Net for MRI Bias Field Correction

Xin Zhu[1,2(✉)], Hongyi Pan[2], Batuhan Gundogdu[3], Debesh Jha[2], Yury Velichko[2], Adam B. Murphy[2], Ashley Ross[2], Baris Turkbey[4], Ahmet Enis Cetin[1], and Ulas Bagci[2]

[1] Department of ECE, University of Illinois Chicago, Chicago, USA
[2] Machine and Hybrid Imaging Lab, Northwestern University, Chicago, USA
xzhu61@uic.edu
[3] Department of Radiology, University of Chicago, Chicago, USA
[4] Molecular Imaging Branch, NCI, National Institutes of Health, Bethesda, MD, USA

Abstract. Magnetic field inhomogeneity correction remains a challenging task in MRI analysis. Most established techniques are designed for brain MRI by supposing that image intensities in the identical tissue follow a uniform distribution. Such an assumption cannot be easily applied to other organs, especially those that are small in size and heterogeneous in texture (large variations in intensity) such as the prostate. To address this problem, this paper proposes a probabilistic Hadamard U-Net (PHU-Net) for prostate MRI bias field correction. First, a novel Hadamard U-Net (HU-Net) is introduced to extract the low-frequency scalar field. HU-Net converts the input image from the time domain into the frequency domain via Hadamard transform. In the frequency domain, high-frequency components are eliminated using the trainable filter (scaling layer), hard-thresholding layer, and sparsity penalty. Next, a conditional variational autoencoder is used to encode possible bias field-corrected variants into a low-dimensional latent space. Random samples drawn from latent space are then incorporated with a prototypical corrected image to generate multiple plausible images. Experimental results demonstrate the effectiveness of PHU-Net in correcting bias-field in prostate MRI with a fast inference speed. It has also been shown that prostate MRI segmentation accuracy improves with the high-quality corrected images from PHU-Net. The codes are available at https://github.com/Holmes696/Probabilistic-Hadamard-U-Net.

Keywords: Hadamard transform · Probabilistic U-Net · MRI bias field correction · Prostate MRI segmentation

1 Introduction

The main magnetic field in an Magnetic Resonance Imaging (MRI) scanner is not perfectly uniform across the imaging volume, making the quantification tasks

---

This work is supported by the NIH funding: R01-CA246704, R01-CA240639, U01-DK127384-02S1, and U01-CA268808, and NSF IDEAL 2217023.

(detection, diagnosis, segmentation) challenging and sub-optimal. To address this non-uniformity problem in MRI, one critical step is to correct bias field [8,25]. The bias field refers to a low-frequency multiplicative field arising from various sources, such as coil sensitivity variations and radio-frequency inhomogeneities. It distorts intensity distributions and impacts the accuracy of subsequent downstream tasks, most importantly image segmentation algorithms.

Numerous approaches have been proposed for bias field removal. Among them, the N4 bias field correction algorithm (N4ITK) has gained popularity for its effectiveness in rectifying inhomogeneities within brain MRI by assuming uniform intensity across distinct types of brain tissues [3,22,23]. However, this condition is not valid for prostate MRI because there exist large intensity variations within the same prostate tissue. Additionally, N4ITK corrects the bias field with a high computation cost, leading to a low inference speed. To improve inference speed, convolutional layers-based frameworks [19,21] were proposed. Yet, these methods are limited by their accuracy, efficiency, and generalization ability. Alternatively, adversarial learning [3,6] has emerged as a promising avenue for addressing non-uniformities in medical imaging. This technique contains two models: while a *generator* corrects bias fields, a *discriminator* distinguishes corrected images from ground truth images. Although adversarial networks enhance correction accuracy, they still face challenges such as model collapse, sensitivity to hyperparameters, and longer training times.

The most crucial step of bias field correction is to extract the low-frequency multiplicative field; therefore, proper frequency analysis can improve the quality of the corrected images. At present, orthogonal transforms including discrete Fourier transform [4], discrete cosine transform [2] and Hadamard transform [18] have been widely employed in neural networks to extract low-frequency features hidden in the transform domain. In this family, the Hadamard transform is the most implementation-efficient because its transform matrix elements take only $+1$ and -1 values. To the best of our knowledge, there is no literature study combining U-Nets (or other neural networks) with orthogonal transforms for MRI bias field correction.

In this study, we propose a probabilistic Hadamard U-Net (PHU-Net) to correct the bias field of MRI and we demonstrate its effectiveness in prostate MRI due to its unique challenges of large intensity variations and complexity of tissue interactions. More specifically, our contributions are summarized as follows: (1) We develop a novel Hadamard U-Net based on the trainable filters and hard-thresholding layers to obtain an efficient representation of the underlying MR scalar field. (2) We incorporate a conditional variational autoencoder (CVAE) to model the joint distribution of all intensity values in ground truth and generate extensive plausible bias field corrected images. (3) We design a hybrid loss optimization composed of Kullback Leibler divergence (KLD) loss [16], total variation loss, and mean squared error (MSE) loss to promote sparsity in the learned representations, resulting in improved robustness and generalization to diverse MRI datasets. (4) Through comprehensive experiments on several benchmark datasets, we validate the superior efficacy of PHU-Net compared to state-of-the-art bias field correction techniques, as well as improved segmentation performance. This demonstrates the potential of PHU-Net in enhancing the quality and reliability of medical image analysis pipelines.

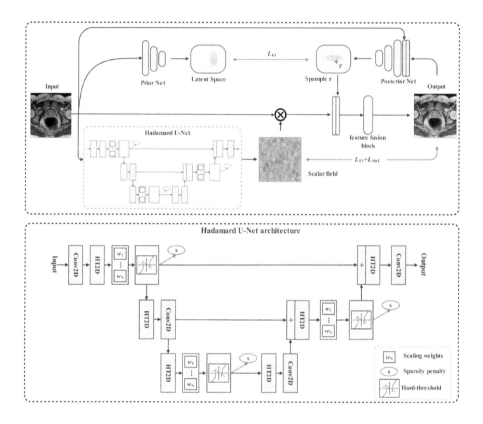

Fig. 1. Probabilistic Hadamard U-Net.

2 Methods

The overall framework of the PHU-Net is presented in Fig. 1. Our supervised pipeline contains two inter-connected modules: a Hadamard U-Net (HU-Net) and a conditional variational autoencoder (CVAE).

2.1 Hadamard U-Net

The Hadamard U-Net (HU-Net) extracts the scalar field, which is the inverse of the bias field. Innovatively, the Hadamard transform (HT) is applied for the frequency analysis. The HT of an $M \times M$ input matrix $\mathbf{X}$ is computed as:

$$\widehat{X}_{a,b} = \frac{1}{M} \sum_{c=0}^{M-1} \sum_{d=0}^{M-1} X_{c,d}(-1)^{\sum_{j=0}^{log_2 M-1} a_j c_j + b_j d_j}, \tag{1}$$

where M is an integer power of 2, and a_j, b_j, c_j and d_j are the j-bits in the binary representations of a, b, c and d, respectively. Next, we take three steps to emphasize energy in the low-frequency band where the bias field lies: **First**, each entry of $\widehat{\mathbf{X}}$ is multiplied with a trainable scaling weight:

$$\widetilde{\mathbf{X}} = \mathbf{W} \circ \widehat{\mathbf{X}}, \tag{2}$$

where $\circ$ denotes the element-wise product and $\mathbf{W}$ is the scaling weights. Applying scaling on the Hadamard coefficients is inspired by the Hadamard convolution theorem [24], stating that the Hadamard-domain element-wise product is similar to the time-domain dyadic convolution:

$$\mathbf{m} *_d \mathbf{n} = \mathcal{T}^{-1}\left(\mathcal{T}(\mathbf{m}) \circ \mathcal{T}(\mathbf{n})\right), \tag{3}$$

where $\mathbf{m}$ and $\mathbf{n}$ are input vectors. $\mathcal{T}(\cdot)$ denotes the HT. $*_d$ is the dyadic convolution. Therefore, the scaling layer in the HT domain works as the convolution filter in the time-domain. On the other hand, the scaling layer requires significantly less computational overhead than the time-domain convolutional filtering.

Second, a trainable hard-thresholding layer $\mathcal{H}_\mathbf{T}(\cdot)$ diminishes the influence of the high-frequency band:

$$\mathbf{Z} = \mathcal{H}_\mathbf{T}(\widetilde{\mathbf{X}}) = \mathcal{C}_\mathbf{T}(\widetilde{\mathbf{X}}) + \text{sign}(\mathcal{C}_\mathbf{T}(\widetilde{\mathbf{X}})) \circ \mathbf{T}, \tag{4}$$

$$\mathcal{C}_\mathbf{T}(\widetilde{\mathbf{X}}) = \text{sign}(\widetilde{\mathbf{X}}) \circ \text{ReLU}(|\widetilde{\mathbf{X}}| - \mathbf{T}), \tag{5}$$

where ReLU$(\cdot)$ stands for the standard ReLU activation function [5] and $\mathbf{T} \in \mathbb{R}^{M \times M}$ is non-negative trainable threshold parameters determined using the back-propagation algorithm. Specially, Eq. (5) can be simplified as:

$$\mathbf{Z} = \mathcal{C}_\mathbf{T}(\widetilde{\mathbf{X}}) - \text{copysign}(|\widetilde{\mathbf{X}}| \quad \mathbf{T}, \widetilde{\mathbf{X}}), \tag{6}$$

where copysign$(\cdot, \cdot)$ replaces the first argument's sign bit with the second argument's sign bit. In this implementation, no multiplication is required. The hard-thresholding layer can constrain the high-frequency band by shifting the small entries to 0s. In the frequency domain, the low-frequency magnitudes are usually much larger than the high-frequency magnitudes. Hence, such a shifting with a proper threshold can remove the high-frequency components while preserving the low-frequency components.

Third, the output of the hard-thresholding layer is penalized using the Kullback Leibler divergence (KLD)-based sparsity constraint. After HU-Net, the original input is element-wise multiplied by the scalar field to achieve the prototypical bias field corrected image.

2.2 Conditional Variational Autoencoder

Since the intensities in the prostate MRI data follow a non-uniform distribution, the conditional variational autoencoder (CVAE) [9] is applied to establish the joint probability of all intensities in the ground truth $\mathbf{Y}$. As shown in Fig. 1, the *prior network* maps the raw image $\mathbf{X}$ to a Gaussian distribution parameterized by its mean $\mathbf{u} \in \mathbb{R}^D$ and covariance $\mathbf{V} \in \mathbb{R}^{D \times D}$ in the latent space, where D is the dimension of the latent space. Next, the raw image and the ground truth are concatenated together as the input of the *posterior network*. This posterior network predicts the distribution of bias field-corrected images conditioned on a raw image. In the training step, the random samples $\mathbf{r}_i \in \mathbb{R}^D$ are drawn from the posterior distribution F:

$$\mathbf{r}_q \sim F(\cdot \mid \mathbf{X}, \mathbf{Y}) = \mathcal{N}(\mathbf{u}(\mathbf{X}, \mathbf{Y}; q), \mathbf{V}(\mathbf{X}, \mathbf{Y}; q)), \tag{7}$$

where q represents the weights of the posterior network. In the testing stage, random samples are drawn from the prior distribution G:

$$\mathbf{r}_p \sim G(\cdot \mid \mathbf{X}) = \mathcal{N}(\mathbf{u}(\mathbf{X}; p), \mathbf{V}(\mathbf{X}; p)), \tag{8}$$

where p stands for the prior network weights. Then, the random samples are incorporated with HU-Net output. Next, the feature fusion block filters all information to obtain the corrected MRI data. Furthermore, a KLD loss function imposes a penalty on differences between the outputs of the prior and posterior networks.

2.3 Loss Function

The overall loss function is composed of three parts: KLD loss $\mathcal{L}_{\mathcal{KL}}$, total variation loss $\mathcal{L}_{\mathcal{TV}}$ [17] and mean squared error (MSE) loss $\mathcal{L}_{\mathcal{MS}}$:

$$\mathcal{L}_{\mathcal{C}} = \lambda_1 \mathcal{L}_{\mathcal{KL}}(F \| G) + \lambda_2 \sum_{i=1}^{L} \mathcal{L}_{\mathcal{KL}}(\delta(\mathbf{Z}_i) \| \beta) + \lambda_3 \mathcal{L}_{\mathcal{TV}}(\mathbf{U}) + \lambda_4 \mathcal{L}_{\mathcal{MS}}(\mathbf{I}, \mathbf{O}), \tag{9}$$

where F and G represent the posterior distribution and the prior distribution respectively as shown in Fig. 1. $\delta(\cdot)$ stands for the sigmoid function. $\mathbf{Z}_i$ is the average output of i-th hard-thresholding layer. L is the number of sparsity constraints applied. β denotes a sparse parameter. $\mathbf{U}$ is the output of Hadamard U-Net. $\mathbf{I}$ and $\mathbf{O}$ represent the input and output respectively. λ_1, λ_2, λ_3 and λ_4 are the constant weights.

We impose sparsity on $\mathbf{Z}_i$ by minimizing the term $\mathcal{L}_{\mathcal{KL}}(\delta(\mathbf{Z}_i) \| \beta)$. Additionally, variance between F and G is penalized using $\mathcal{L}_{\mathcal{KL}}(F \| G)$. Moreover, we employ $\mathcal{L}_{\mathcal{TV}}(\mathbf{U})$ to keep spatial gradients of $\mathbf{U}$ sparse. Furthermore, the difference between output and ground truth is minimized by decreasing $\mathcal{L}_{\mathcal{MS}}(\mathbf{I}, \mathbf{O})$.

3 Experiments

3.1 Datasets and Implementation

Datasets. We use four distinct publicly available MRI prostate datasets (all T2-weighted) in our experiments: HK dataset [13], UCL dataset [13], HCRUDB dataset [11] and AWS dataset [1]. Since N4ITK is one of the most prevalent MRI bias field correction methods in the last decades, we employ it to clean all scans in these datasets and utilize processed scans as a reference baseline that serves as a substitute for ground truth. To validate the generalization performance of different methods, all models are trained on the UCL dataset and then tested on the HK, HCRUDB, and AWS datasets. In summary, the training set contains 318 scans, while the three testing sets contain 288, 1,216, and 904 scans, respectively. All these scans are reshaped into 256 × 256 in-plane resolution while they have different sizes of slices.

Implementation Details. The HU-Net includes 4 convolutional layers. The kernel sizes are 16 × 16, 7 × 7, 7 × 7 and 16 × 16, respectively. In CVAE, the prior and the posterior nets share a similar encoder structure. This encoder structure is comprised of two 3 × 3 convolutional blocks with 32 and 64 filters. Each convolutional block consists of 4 convolutional layers activated by the ReLU function. After the first block, a 2×2 average pooling layer with stride 2 downsamples the image tensors. Moreover, the feature fusion block shown in Fig. 1 contains three 1×1 convolutional layers. Each of the first 2 convolutional layers contains 32 filters, and they are activated by the ReLU function. The third convolutional layer contains 1 filter. Our model is trained using the AdamW optimizer [14] with a batch size of 128 and a learning rate of 0.0001 under 100 training epochs. λ_1, λ_2, λ_3 and λ_4 in Eq. (9) are set to 10, 0.1, 1, and 1, respectively.

Performance Metrics. Models are evaluated on the following metrics: Coefficient of variation (CV) [12,15,20], signal-to-noise ratio (SNR) [27] to evaluate the bias field correction performance, and Dice [10], positive predictive value (PPV) [26], and intersection over union (IoU) [7] to assess prostate segmentation performance. In general, a lower CV and higher other metrics indicate a better model performance.

3.2 Experimental Results

We compare the proposed PHU-Net with three up-to-date bias field correction methods, including N4ITK [23], convolutional autoencoder (CAE) [21] and implicitly trained CNN (ITCNN) [19].

Table 1 compares the CV results of our proposed PHU-Net versus other state-of-the-art methods on three test datasets. Compared with N4ITK, PHU-Net reduces CV from 78.24 to 65.95 (15.71%), from 56.76 to 54.14 (4.62%), and

Table 1. Comparison experiments on CV and SNR.

Metrics	CV ↓			SNR↑
Dataset	HK	AWS	HCRUDB	HK
Original	78.24	59.13	53.78	24.64
N4ITK	78.24	56.76	52.41	24.64
CAE	66.97	62.51	51.39	13.18
ITCNN	78.25	59.02	53.69	23.50
PHU-Net	**65.95**	**54.14**	**43.39**	**27.05**

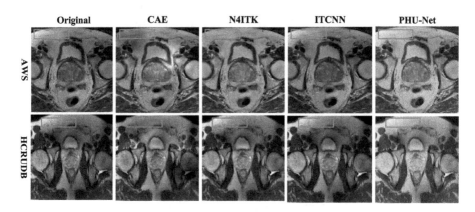

Fig. 2. Comparison between the original and the corrected MRIs on the AWS and HCRUDB dataset. Last column is our proposed method.

from 52.41 to 43.39 (17.21%) on the HK, AWS and HCRUDB datasets, respectively. PHU-Net also outperforms CAE and ITCNN, indicating that PHU-Net has a better capability for removing intensity variations regardless of scanner types. Additionally, Table 1 also presents SNR values for the MRI dataset corrected by different methods. In this comparison, models are only evaluated on the HK dataset because other datasets do not contain a clear background as the HK dataset, essential for SNR calculation. As a result, PHU-Net achieves the highest SNR. Furthermore, only PHU-Net improves SNR compared to the original images, which demonstrates our proposed method introduces no additional noise to the region of interest during the bias field correction stage.

Experimental results are visualized in Fig. 2. Images processed by PHU-Net exhibit superior intensity uniformity compared to those processed by other baseline methods.

Ablation Study. To verify each module in PHU-Net, ablation studies are carried out in Table 2. When the scaling layer, sparsity, hard-thresholding layer, TV loss, or CVAE are removed, SNR is reduced by 2.96% at least and 5.99% at most. Although CV improves slightly in the HCRUDB dataset when the scaling

Table 2. Ablation experiments on CV and SNR.

Metrics	CV↓			SNR↑
Dataset	HK	AWS	HCRUDB	HK
No scaling	67.38	54.92	**43.12**	25.69
No sparsity	71.36	58.60	49.19	25.76
No threshold	67.44	54.83	43.13	25.43
No TV loss	67.21	55.07	43.77	25.73
No CVAE	68.27	56.15	44.09	26.25
PHU-Net	**65.95**	**54.14**	43.39	**27.05**

layer is not present, it degrades significantly in the HK and AWS datasets. When there is no CVAE, PHU-Net only contains HU-Net. Its results are still better than those of other models in Table 1.

Table 3. Segmentation experiment on the HK dataset.

Metrics	Dice(%)	IoU(%)	PPV(%)
Original	71.43	66.22	76.22
N4ITK	72.38	67.38	75.74
CAE	68.20	63.04	74.07
ITCNN	75.51	70.27	80.06
PHU-Net	**76.86**	**71.72**	**82.34**

Segmentation Experiment. Corrected images are further utilized in the segmentation experiments. Here, a traditional probabilistic U-Net [9] is used as the segmentation model, containing a standard U-Net and a CVAE. The top 50% of the corrected dataset is utilized as the training dataset while the remaining is employed as the testing dataset. The MSE loss function is applied. Other implementation details are the same as in [9]. As shown in Table 3, it achieves the highest Dice, IoU, and PPV on the images corrected by PHU-Net. Therefore, PHU-Net ensures more accurate and reliable segmentation of the underlying tissue structures than other correction models.

Runtime Analyses. Another benefit of PHU-Net lies in its inference efficiency. For one single NIfTI image file ($\tilde{1}5$ slices), the bias field correction runtime of N4ITK, CAE, ITCNN, and PHU-Net are around 44.26, 1.06, 1.23, and 1.13 s, respectively. These tests are executed on a computer with an Intel Core i7-12700H CPU. Compared to N4ITK, our proposed model exhibits nearly 40 times faster processing speed, and similar inference times with ITCNN and CAE while better in performance than them.

4 Conclusion

In this paper, we developed a novel framework named PHU-Net for MRI bias field correction. The key idea of PHU-Net is to combine a U-Net structure with transform domain methods, trainable filters, and hard-thresholding layers to extract biased field information. Experimental results showed that PHU-Net achieved a better bias field correction performance than other state-of-the-art methods with a fast running speed. What is more, PHU-Net significantly improved the segmentation accuracy of the probabilistic U-Net.

References

1. Antonelli, M., et al.: The medical segmentation decathlon. Nat. Commun. **13**(1), 4128 (2022)
2. Badawi, D., Agambayev, A., Ozev, S., Cetin, A.E.: Discrete cosine transform based causal convolutional neural network for drift compensation in chemical sensors. In: ICASSP 2021-2021 IEEE International Conference on Acoustics, Speech and Signal Processing (ICASSP), pp. 8012–8016. IEEE (2021)
3. Chen, L., et al.: ABCnet: adversarial bias correction network for infant brain MR images. Med. Image Anal. **72**, 102133 (2021)
4. Chi, L., Jiang, B., Mu, Y.: Fast fourier convolution. Adv. Neural. Inf. Process. Syst. **33**, 4479–4488 (2020)
5. Fukushima, K.: Visual feature extraction by a multilayered network of analog threshold elements. IEEE Trans. Syst. Sci. Cybern. **5**(4), 322–333 (1969)
6. Goldfryd, T., Gordon, S., Raviv, T.R.: Deep semi-supervised bias field correction of MR images. In: 2021 IEEE 18th International Symposium on Biomedical Imaging (ISBI), pp. 1836–1840. IEEE (2021)
7. Jiang, Y., Ye, M., Huang, D., Lu, X.: AIU-Net: an efficient deep convolutional neural network for brain tumor segmentation. Math. Probl. Eng. **2021**, 1–8 (2021)
8. Juntu, J., Sijbers, J., Van Dyck, D., Gielen, J.: Bias field correction for MRI images. In: Kurzyński, M., Puchala, E., Woźniak, M., zolnierek, A. (eds.) Computer Recognition Systems. Advances in Soft Computing, vol. 30, pp. 543–551. Springer, Berlin, Heidelberg (2005). https://doi.org/10.1007/3-540-32390-2_64
9. Kohl, S., et al.: A probabilistic U-net for segmentation of ambiguous images. In: Advances in Neural Information Processing Systems, vol. 31 (2018)
10. Laradji, I., et al: A weakly supervised consistency-based learning method for COVID-19 segmentation in CT images. In: Proceedings of the IEEE/CVF Winter Conference on Applications of Computer Vision, pp. 2453–2462 (2021)
11. Lemaître, G., Martí, R., Freixenet, J., Vilanova, J.C., Walker, P.M., Meriaudeau, F.: Computer-aided detection and diagnosis for prostate cancer based on mono and multi-parametric MRI: a review. Comput. Biol. Med. **60**, 8–31 (2015)
12. Likar, B., Viergever, M.A., Pernus, F.: Retrospective correction of MR intensity inhomogeneity by information minimization. IEEE Trans. Med. Imaging **20**(12), 1398–1410 (2001)
13. Litjens, G., et al.: Evaluation of prostate segmentation algorithms for MRI: the PROMISE12 challenge. Med. Image Anal. **18**(2), 359–373 (2014)
14. Loshchilov, I., Hutter, F.: Decoupled weight decay regularization (2017). arXiv preprint arXiv:1711.05101

15. Madabhushi, A., Udupa, J.K.: Interplay between intensity standardization and inhomogeneity correction in MR image processing. IEEE Trans. Med. Imaging **24**(5), 561–576 (2005)
16. Ng, A., et al.: Sparse autoencoder. CS294A Lecture Notes **72**(2011), 1–19 (2011)
17. Osher, S., Burger, M., Goldfarb, D., Xu, J., Yin, W.: An iterative regularization method for total variation-based image restoration. Multiscale Model. Simul. **4**(2), 460–489 (2005)
18. Pan, H., Zhu, X., Atici, S.F., Cetin, A.: A hybrid quantum-classical approach based on the hadamard transform for the convolutional layer. In: International Conference on Machine Learning, pp. 26891–26903. PMLR (2023)
19. Simkó, A., Löfstedt, T., Garpebring, A., Nyholm, T., Jonsson, J.: MRI bias field correction with an implicitly trained CNN. In: International Conference on Medical Imaging with Deep Learning, pp. 1125–1138. PMLR (2022)
20. Sled, J.G., Zijdenbos, A.P., Evans, A.C.: A nonparametric method for automatic correction of intensity nonuniformity in MRI data. IEEE Trans. Med. Imaging **17**(1), 87–97 (1998)
21. Sridhara, S.N., Akrami, H., Krishnamurthy, V., Joshi, A.A.: Bias field correction in 3D-MRIs using convolutional autoencoders. In: Medical Imaging 2021: Image Processing, vol. 11596, pp. 671–676. SPIE (2021)
22. Tustison, N., Gee, J.: N4ITK: nick's N3 ITK implementation for MRI bias field correction. Insight J. **9**, 1–22 (2010)
23. Tustiso, N.J., et al.: N4ITK: improved N3 bias correction. IEEE Trans. Med. Imaging **29**(6), 1310–1320 (2010)
24. Ušáková, A., Kotuliaková, J., Zajac, M.: Walsh-hadamard transformation of a convolution. Radioengineering **11**(3), 40–42 (2002)
25. Van Leemput, K., Maes, F., Vandermeulen, D., Suetens, P.: Automated model-based bias field correction of MR images of the brain. IEEE Trans. Med. Imaging **18**(10), 885–896 (1999)
26. Wang, H., Gu, H., Qin, P., Wang, J.: CheXLocNet: automatic localization of pneumothorax in chest radiographs using deep convolutional neural networks. PLoS ONE **15**(11), e0242013 (2020)
27. Welvaert, M., Rosseel, Y.: On the definition of signal-to-noise ratio and contrast-to-noise ratio for FMRI data. PLoS ONE **8**(11), e77089 (2013)

Structure-Preserving Diffusion Model for Unpaired Medical Image Translation

Haoshen Wang[1], Xiaodong Wang[2], and Zhiming Cui[1(✉)]

[1] School of Biomedical Engineering, ShanghaiTech University, Shanghai, China
{wanghsh2022,cuizhm}@shanghaitech.edu.cn
[2] United Imaging Healthcare, Shanghai, China

Abstract. Multi-modality imaging plays a crucial role in clinical diagnosis. Reconstructing missing modality images, such as CT-to-MR, is quite important when only one modality is available. Previous works either fall short in preserving the anatomical structures during translation or require paired data, leaving significant challenges unaddressed in the realm of unpaired medical image translation. This study introduces a novel structure-preserving diffusion model specifically designed for unpaired medical image translation, leveraging edge information to represent common anatomical structures across different modalities. To bridge the domain gap effectively, we further propose a novel Interleaved Sampling Refinement (ISR) mechanism that dynamically alternates the use of edge information. This approach not only generates high-quality images but also preserves structural integrity across modalities. Our experiments conducted on two public datasets have achieved the state-of-the-art performance, demonstrating the advantage of our method on unpaired medical image translation. The code of our implementation is available at GitHub.

Keywords: Unpaired image translation · Diffusion model · Anatomical structures preserving

1 Introduction

Multi-modality medical images, captured using distinct physical principles, play essential and complementary roles in clinical diagnosis. However, the simultaneous acquisition of multi-modality images, such as CT and MR scans for a single patient, is often constrained by practical limitations. Consequently, medical image translation, transforming images from one modality to another while retaining clinically significant information, is an attractive yet challenging task.

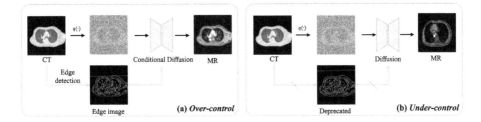

Fig. 1. The main challenges of conditional diffusion model for unpaired medical image translation, including the over-control and under-control problem.

Supervised learning-based approaches offer straightforward solutions to the medical image translation task. However, obtaining paired data in different modalities for training is a significant obstacle. To overcome this, many strategies have been developed, mainly exploiting techniques for unpaired image-to-image translation. Among these, CycleGAN [1] and its variants [2–4] have been recognized as leading frameworks. Unfortunately, these methods often suffer from model collapse and training instability, requiring careful hyperparameters tuning and frequently producing outputs lacking realism [5].

In recent years, the Denoising Diffusion Probabilistic Model (DDPM) [6] has garnered significant attention for its ability to generate high-quality images. Although it has been applied to address medical image translation problem, controlling the translation process while preserving anatomical structures is still challenging, especially for the unpaired data. An effort to tackle this challenge is the Adversarial Diffusion Models [7] by combing the adversarial training and diffusion process. However, it still suffers from drawbacks similar to those of GAN-based methods. Concurrently, Li et al. [8] proposed an unsupervised CBCT-to-CT translation using frequency-guided diffusion models. This method shares the similar idea that utilizes edge information extracted from source image as the high-frequency guidance for denoising process. However, its effectiveness for more intricate translations, such as CT-to-MR, is restricted due to significant differences in image appearance between CT and MR images.

To tackle these limitations, we introduce a structure-preserving diffusion model for unpaired medical image translation. Our framework leverages the inherent high-quality image generation properties of the diffusion model, and more importantly, maintains the anatomical structures throughout the image translation process. Specifically, we train a conditional diffusion model to generate target domain images based on the edge information, e.g., canny edge, extracted from source domain images. However, excessive reliance on edge information can lead to *over-control*, generating images with unrealistic edges for the target modality (see Fig. 1(a)), even if the style closely resembles the target domain. On the other hand, completely ignoring edge information during the diffusion reverse process can result in *under-control*, where the structures are not preserved at all (see Fig. 1(b)). To tackle this challenge, we introduce an Interleaved Sampling Refinement (ISR) mechanism to use or drop the edge

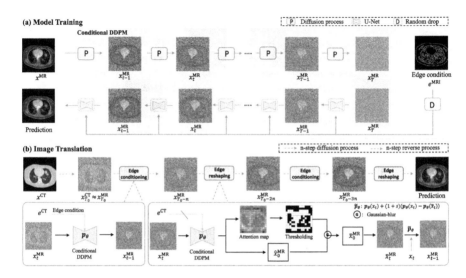

Fig. 2. The overview illustration of our proposed framework for unpaired medical image translation. The model training stage (a) enables the diffusion model have the ability to generate target domain (MR) images with structural edge guidance, while in the image translation stage (b) we use or drop the edge information from the source domain (CT) in interleaved manner to reduce the domain gap between different modalities.

information in an interleaved manner. This strategy aims to strike a balance between *over-control* and *under-control*, facilitating the generation of high-quality images that preserves anatomical structures. Experimental results are validated on two public datasets, and we further utilize a segmentation model as a downstream task to evaluate the quality of the generated images. The results indicate that our proposed method achieves the state-of-the-art performance for unpaired medical image translation task.

2 Method

Figure 2 illustrates our structure-preserving diffusion model for unpaired medical image translation. We first employ the edge information, i.e., canny edge, of the image as a condition to guide the generation in diffusion model training. After training, in the image translation process, we introduce an Interleaved Sampling Refinement (ISR) mechanism, including *edge conditioning* and *edge reshaping*, to reduce the domain gap between different modalities.

2.1 Conditional Diffusion Network

In the diffusion network [6], two fundamental processes are involved: the forward diffusion process $q(\cdot)$ and the reverse denoising process $p_\theta(\cdot)$. During the forward diffusion process, Gaussian noise is incrementally introduced to an image. During the reverse process, a trained U-Net [9] model is utilized to denoise step-by-step, restoring the clear image from Gaussian noise. The Conditional diffusion network extends the traditional diffusion network by incorporating an additional input as a condition to guide the model in generating images with specific content. However, the conditional information should be paired with generating images, making it much more difficult for unpaired image-to-image translation task.

2.2 Structure-Preserving DDPM

In our unpaired medical image translation task, we suppose to have unpaired CT images $x^{\text{CT}} \in X^{\text{CT}}$ and MR images $x^{\text{MR}} \in X^{\text{MR}}$, our goal is to train a diffusion model $p_\theta(\cdot)$ that generates corresponding MR images from the source domain CT images. A key property is that medical images of different modalities are captured to reveal the same human anatomical structure information, thus explicitly leveraging the common anatomical structures across different modalities may benefit the translation. Hence, in this study, using the edge information, i.e., canny edge, to represent the anatomical structures and guide the generation process is an intuitive idea, since it does not incorporate any style information. However, owing to difference in imaging principles, tissue edges across different modality images, such as CT and MR, exhibit variations to some content. For example, CT imaging exhibits enhanced clarity in hard tissues, whereas MR provides superior resolution for soft tissues. Relying excessively on edges may result in *over-control*, leading to the generated images with unrealistic edges for the target modality (refers to Fig. 1(a)), even if the style is similar to the target domain. Conversely, disregarding edge information entirely during the diffusion reverse process may lead to *under-control*, where the structures are not preserved (refers to Fig. 1(b)).

Model Training. To address this, a feasible solution is to strike a balance between *over-control* and *under-control*, generating fidelity target images with reasonable tissue edges. To achieve the balance, as shown in Fig. 2, our diffusion model should have flexible control over edges, such that 1) it can recover clear images with corresponding structural edge guidance, for example, recovering MR images guided by the canny edges on itself; and 2) it can generate realistic target images from pure Gaussian noise. Specifically, during training process, we first provide the canny edge e^{MR} extracted from the MR image x^{MR} to guide the diffusion model, enabling it to recover clear target images, denoted as $p_\theta(x_{1:T}^{\text{MR}}, e = e^{\text{MR}})$. Then, we randomly exclude the edge condition during training by setting $e = 0$, allowing the model to generate images from pure noise without any condition, denoted as $p_\theta(x_{1:T}^{\text{MR}}, e = 0)$. In this way, our model can generate target MR images from Gaussian noise with or without edge guidance.

Image Translation. With such flexibility, we further propose a novel Interleaved Sampling and Refinement (ISR) mechanism in our CT-to-MR translation, as shown in Fig. 2, including *edge conditioning* and *edge reshaping*, to interleave the use or drop of canny edge information from the source images to achieve a balance between *over-control* and *under-control*.

The *edge conditioning* aims to guide the denoising process using edge information, which is the key to preserve the source domain anatomical structures during translation. Formally, the diffusion model p_θ generates target domain images with the source domain anatomical structures e, denoted as:

$$x_{t-1}^{MR} = p_\theta(x_t^{MR}, e = e^{CT}). \tag{1}$$

The *edge reshaping* aims to alleviate unrealistic edges in the generated target images. Specifically, it involves two stages: unrealistic edges elimination and realistic edges regeneration. During this phase, we drop the edge condition. This allows the diffusion model to generate images relying the results of previous steps $x_{T:t}^{MR}$ rather than the edge condition e^{CT}. In the first stage, unrealistic edges elimination, we directly predict $\hat{x}_0^{MR}$ from x_t^{MR}. And we obtain the attention map A_t by multiplying the queries Q and keys K in the attention layer of the diffusion U-Net.

$$\hat{x}_0^{MR}, A_t = p_\theta(x_t^{MR}, e = 0). \tag{2}$$

In the attention map, regions with high attention values indicate where the network focuses and contain the primary anatomical structures. Thus, we perform the Gaussian blur to regions of $\hat{x}_0^{MR}$ where the attention map shows high values, denoted as $M_t = \mathbb{1}(A_t > \xi)$, resulting in $\bar{x}_0^{MR}$. This approach helps to mitigate the unrealistic edges produced during the *edge conditioning* phase. Finally, the forward diffusion process $q(\cdot)$ is applied to reintroduce noise and obtain $\bar{x}_t^{MR}$.

$$\begin{aligned}\bar{x}_0^{MR} &= (1 - M_t) \odot \hat{x}_0^{MR} + M_t \odot \text{Gaussian-blur}(\hat{x}_0^{MR}), \\ \bar{x}_t^{MR} &= q\left(\bar{x}_t^{MR} | \bar{x}_0^{MR}\right).\end{aligned} \tag{3}$$

In the second stage, realistic edges regeneration, the network denoises $\bar{x}_t^{MR}$, regenerating the edges eliminated in the first step and making them more realistic. To enhance the level of detail in the generated image, we guide the sampling process using details guidance, as suggested by [10].

$$\tilde{p}_\theta(x_t^{MR}) = p_\theta(\bar{x}_t^{MR}) + (1 + s)\left(p_\theta(x_t^{MR}) - p_\theta\left(\bar{x}_t^{MR}\right)\right). \tag{4}$$

In this way, the diffusion model takes the noisy blurred image $\bar{x}_t$ as input, with the details guidance determined by the difference between the denoising result of the unblurred image, $p_\theta(x_t^{MR})$, and the denoising result of the blurred image, $p_\theta\left(\bar{x}_t^{MR}\right)$. This approach guides the generation process from insufficient to sufficient details, where s represents the scaling coefficient.

2.3 Implementation Details

In our task, we apply perturbation to the source image over $T_0 = 500$ steps to preserve some information from the source image. To maintain the anatomical structures during translation, we set T_α. When $t > T_\alpha$, only *edge conditioning* is applied, ensuring that the anatomical structure condition is sufficiently incorporated. Conversely, when $t < T_\alpha$, *edge conditioning* and *edge reshaping* are performed in an interleaved manner, each for n steps. And we set $T_\alpha = 250$ and $n = 25$ based on 4-fold cross-validation results. During the *edge reshaping* phase, the Gaussian blur operation is not applied to the entire image but rather to the areas with high values in the attention map, specifically where $A_t > \xi$. We set $\xi = 0.9$ based on experimental results from 4-fold cross-validation. Similarly, we determined the scaling coefficient $s = 0.5$ through 4-fold cross-validation. The coefficient regulates the influence of details guidance in the sampling process.

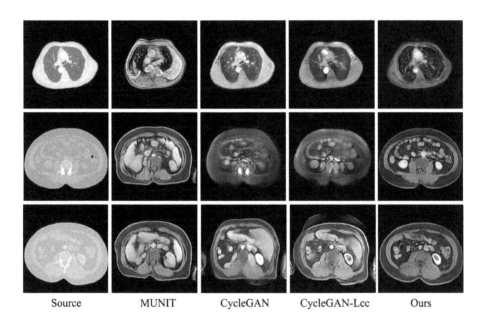

Fig. 3. Qualitative results on AMOS dataset.

3 Experiments and Results

3.1 Datasets and Evaluation Metrics

The performance of our proposed method is evaluated on two multi-modality datasets: AMOS [11] and the Gold Atlas Pelvic dataset [12]. The AMOS dataset, aiming at multi-modality abdominal multi-organ segmentation, consists of 500 CT and 100 MR scans of the abdominal region. Meanwhile, the Gold Atlas Pelvic dataset contains CT, T1-weighted MR, and T2-weighted MR images from the male pelvic region, collected from 19 patients. Note that the CT and MR images in the Gold Atlas Pelvic dataset have undergone pairwise deformation registration. In this study, we specifically evaluated the performance of our method on the CT-to-MR (T2-weighted) translation task due to its high-level of difficulty compared to other translation scenarios, such as MR-to-CT and T1-to-T2.

While our method is tailored for unsupervised medical image translation and does not require paired images, evaluating its performance often requires paired source-target images for certain metrics. Therefore, we compute the Peak Signal-to-Noise Ratio (PSNR) and Structure Similarity Index Measure (SSIM) for the registered pairwise CT-MR images in the Gold Atlas Pelvic dataset. However, as the CT and MR images in the AMOS dataset are not paired, ground-truth based metrics cannot be directly calculated. To address this, we train a segmentation network [13] on the target domain for assessment purposes. For the segmentation results on generated images, we evaluate the Dice coefficient [14] and Hausdorff distance for four organs: spleen, right kidney, left kidney, and liver.

3.2 Comparative Results

We compared proposed method with four typical unpaired image-to-image translation methods, including CycleGAN [1], MUNIT [15], CycleGAN-Lcc [16] and FGDM [8]. This section details the competitive results of their performance.

Qualitative Results. The qualitative results of CT-to-MR translation are presented in Fig. 3 and Fig. 4. Across the translation of the abdominal region, all compared methods fail to deliver high-quality images due to limitations inherent to GAN-based approaches. It can be observed that the anatomical structure information is severely compromised in MUNIT. While CycleGAN and CycleGAN-Lcc methods showing marginal improvement due to the cycle consistent scheme, a substantial gap still persists when compared to our proposed method, particularly in terms of image quality, structural preservation, and fine image details. In the image translation of the pelvic region, all three competing methods achieve improved visual performance, likely attributable to the less complex anatomical structures of the pelvic area compared to those of the abdominal regions. Despite these improvements, our proposed method consistently outperform the others, especially in capturing the nuanced details of the translations.

Quantitative Results. The quantitative results are provided in Table 1. Our proposed method achieves the highest performance in terms of the Dice coefficient and Hausdorff distance on the Amos dataset. In contrast, MUNIT shows the lowest segmentation accuracy, primarily due to its lack of structural constraints compared to other approaches. Visualization of the segmentation results can be found in Supplementary Fig. S1. Regarding the pelvic dataset, our method not only achieves superior performance in PSNR and SSIM, but also showcases a notable advancement in structural similarity (SSIM) over GAN-based counterparts. Additionally, it outperforms FGDM, another diffusion-based approach that incorporates edge conditions, underscoring our method's effectiveness in

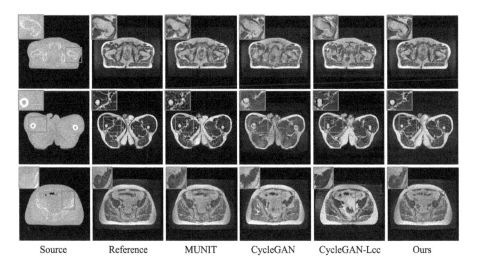

Fig. 4. Qualitative results on Gold Atlas Pelvic dataset.

Table 1. Quantitative comparison on the two public dataset.

Method	AMOS		Pelvic	
	Dice (%)↑	HD95↓	PSNR↑	SSIM (%)↑
MUNIT [15]	78.23 ± 7.51	4.03 ± 0.84	24.72 ± 1.01	80.01 ± 5.17
CycleGAN [1]	84.37 ± 4.37	3.93 ± 0.52	22.77 ± 1.42	78.78 ± 6.47
CycleGANLcc [16]	85.91 ± 5.24	2.51 ± 0.44	22.06 ± 2.19	79.13 ± 6.11
FGDM [8]	86.89 ± 6.32	2.49 ± 0.58	25.13 ± 1.31	82.51 ± 4.64
Ours wo/edge	/	/	12.89 ± 1.05	35.09 ± 13.47
Ours wo/ISR	86.61 ± 5.49	2.53 ± 0.61	25.41 ± 1.53	82.69 ± 5.21
Ours	**87.52 ± 6.17**	**2.47 ± 0.77**	**25.45 ± 1.58**	**83.31 ± 5.39**

reducing the domain gap existing in the edge representation. Our method outperformed the second-place method across all metrics, with a p-value < 0.03.

3.3 Ablation Study

To validate the effectiveness of our proposed strategies, we conduct experiments by implementing two variants: 1) removing the edge condition, equivalent to the under-control scenario; 2) removing the ISR mechanism, equivalent to the over-control scenario. The results shown in Table 1 indicate a consistent degradation in all metrics across both datasets, highlighting the significance of both the edge condition and ISR enhancement. Specifically, upon removal of the edge condition, the diffusion model struggles in generating images with anatomical structures similar to those in the source image. This lead to significant degradations in both the PSNR and SSIM, making the segmentation metrics unnecessary for reporting. Furthermore, with the removal of ISR mechanism, the Dice coefficient decreased by nearly 1%, indicating the misinterpretation of some unrealistic structures by the segmentation model. Despite a decrease in the SSIM metric, it maintains its rank as the second-highest, indicating that the edge information is crucial to support the model in preserving organ structures, and the ISR mechanism finely refines the boundaries of generated images by effectively bridging the domain gap. Some visual results are shown in Fig. S2.

4 Conclusion

In our study, we present a structure-preserving diffusion model for unpaired medical image translation. By leveraging edge information from source images as condition, our approach ensures the preservation of anatomical structures throughout the translation process. To mitigate unrealistic edges resulted by domain gap, we propose a novel Interleaved Sampling Refinement (ISR) mechanism. This mechanism effectively adjusts the control of edge information during sampling. Experimental results demonstrate that our method achieves the state-of-the-art performance on unpaired medical image translation task.

Acknowledgement. This work was supported in part by NSFC grants (No. 6230012077) and Shanghai Municipal Central Guided Local Science and Technology Development Fund Project (No: YDZX20233100001001).

References

1. Zhu, J.-Y., Park, T., Isola, P., Efros, A.A.: Unpaired image-to-image translation using cycle-consistent adversarial networks. In: Proceedings of the IEEE International Conference on Computer Vision, pp. 2223–2232 (2017)
2. Fu, H., Gong, M., Wang, C., Batmanghelich, K., Zhang, K., Tao, D.: Geometry-consistent generative adversarial networks for one-sided unsupervised domain mapping. In: Proceedings of the IEEE/CVF Conference on Computer Vision and Pattern Recognition, pp. 2427–2436 (2019)
3. Yang, H., et al.: Unsupervised MR-to-CT synthesis using structure-constrained cycleGAN. IEEE Trans. Med. Imaging **39**(12), 4249–4261 (2020)
4. Wolterink, J.M., Dinkla, A.M., Savenije, M.H.F., Seevinck, P.R., van den Berg, C.A.T., Išgum, I.: Deep MR to CT synthesis using unpaired data. In: Tsaftaris, S., Gooya, A., Frangi, A., Prince, J. (eds.) Simulation and Synthesis in Medical Imaging: Second International Workshop, SASHIMI 2017, Held in Conjunction with MICCAI 2017, Québec City, QC, Canada, 10 September 2017, Proceedings 2, pp. 14–23. Springer, Cham (2017). https://doi.org/10.1007/978-3-319-68127-6_2
5. Creswell, A., White, T., Dumoulin, V., Arulkumaran, K., Sengupta, B., Bharath, A.A.: Generative adversarial networks: an overview. IEEE Signal Process. Mag. **35**(1), 53–65 (2018)

6. Ho, J., Jain, A., Abbeel, P.: Denoising diffusion probabilistic models. Adv. Neural. Inf. Process. Syst. **33**, 6840–6851 (2020)
7. Özbey, M., et al.: Unsupervised medical image translation with adversarial diffusion models. IEEE Trans. Med. Imaging (2023)
8. Li, Y., et al.: Zero-shot medical image translation via frequency-guided diffusion models. arXiv preprint arXiv:2304.02742 (2023)
9. Ronneberger, O., Fischer, P., Brox, T.: U-Net: convolutional networks for biomedical image segmentation. In: Navab, N., Hornegger, J., Wells, W., Frangi, A. (eds.) Medical Image Computing and Computer-Assisted Intervention–MICCAI 2015: 18th International Conference, Munich, Germany, 5–9 October 2015, Proceedings, Part III 18, pp. 234–241. Springer, Cham (2015). https://doi.org/10.1007/978-3-319-24574-4_28
10. Hong, S., Lee, G., Jang, W., Kim, S.: Improving sample quality of diffusion models using self-attention guidance. In: Proceedings of the IEEE/CVF International Conference on Computer Vision, pp. 7462–7471 (2023)
11. Ji, Y., et al.: AMOS: a large-scale abdominal multi-organ benchmark for versatile medical image segmentation. arXiv preprint arXiv:2206.08023 (2022)
12. Nyholm, T., et al.: MR and CT data with multiobserver delineations of organs in the pelvic area-part of the gold atlas project. Med. Phys. **45**(3), 1295–1300 (2018)
13. Isensee, F., Jaeger, P.F., Kohl, S.A.A., Petersen, J., Maier-Hein, K.H.: nnU-Net: a self-configuring method for deep learning-based biomedical image segmentation. Nat. Methods **18**(2), 203–211 (2021)
14. Milletari, F., Navab, N., Ahmadi, S.-A.: V-Net: fully convolutional neural networks for volumetric medical image segmentation. In: 2016 Fourth International Conference on 3D Vision (3DV), pp. 565–571. IEEE (2016)
15. Huang, X., Liu, M.-Y., Belongie, S., Kautz, J.: Multimodal unsupervised image-to-image translation. In: Ferrari, V., Hebert, M., Sminchisescu, C., Weiss, Y. (eds.) ECCV 2018. LNCS, vol. 11207, pp. 179–196. Springer, Cham (2018). https://doi.org/10.1007/978-3-030-01219-9_11
16. Ge, Y., Xue, Z., Cao, T., Liao, S.: Unpaired whole-body MR to CT synthesis with correlation coefficient constrained adversarial learning. In: Medical Imaging 2019: Image Processing, vol. 10949, pp. 28–35. SPIE (2019)

Simultaneous Image Quality Improvement and Artefacts Correction in Accelerated MRI

Georgia Kanli[1,2,3]($\boxtimes$), Daniele Perlo[1], Selma Boudissa[1,2], Radovan Jiřík[4], and Olivier Keunen[1,2]

[1] Translational Radiomics, Luxembourg Institute of Health, Strassen, Luxembourg
Georgia.Kanli@lih.lu
[2] In-Vivo Imaging Platform, Luxembourg Institute of Health, Strassen, Luxembourg
[3] Faculty of Electrical Engineering and Communication, Brno University of Technology, Brno, Czech Republic
[4] Institute of Scientific Instruments of the Czech Academy of Sciences, Brno, Czech Republic

Abstract. MR data are acquired in the frequency domain, known as k-space. Acquiring high-quality and high-resolution MR images can be time-consuming, posing a significant challenge when multiple sequences providing complementary contrast information are needed or when the patient is unable to remain in the scanner for an extended period of time. Reducing k-space measurements is a strategy to speed up acquisition, but often leads to reduced quality in reconstructed images. Additionally, in real-world MRI, both under-sampled and full-sampled images are prone to artefacts, and correcting these artefacts is crucial for maintaining diagnostic accuracy. Deep learning methods have been proposed to restore image quality from under-sampled data, while others focused on the correction of artefacts that result from the noise or motion. No approach has however been proposed so far that addresses both acceleration and artefacts correction, limiting the performance of these models when these degradation factors occur simultaneously. To address this gap, we present a method for recovering high-quality images from under-sampled data with simultaneously correction for noise and motion artefact called USArt (Under-Sampling and Artifact correction model). Customized for 2D brain anatomical images acquired with Cartesian sampling, USArt employs a dual sub-model approach. The results demonstrate remarkable increase of signal-to-noise ratio (SNR) and contrast in the images restored. Various under-sampling strategies and degradation levels were explored, with the gradient under-sampling strategy yielding the best outcomes. We achieved up to 5× acceleration and simultaneously artefacts correction without significant degradation, showcasing the model's robustness in real-world settings.

Keywords: magnetic resonance imaging · acceleration · under-sampling · artefact/noise correction · deep learning

1 Introduction

Magnetic Resonance Imaging (MRI) provides detailed anatomical and functional information on soft tissues by collecting raw data in k-space [1]. The MRI scan time is influenced by the number of phase encoding steps required to reconstruct an image. Increasing the resolution or quality of an image typically requires more phase encoding steps, leading to longer scan times. This poses challenges, especially for patients who struggle to remain still, such as children and people with claustrophobia or uncontrolled movements disorder. Reducing scan time improves patient comfort and lowers medical costs by increasing throughput, but it can also reduce image quality [2–4].

Accelerating MRI acquisition has been a major focus in the field [5–12], leveraging both physics-driven and data-driven strategies to reduce scan times and improve image quality. Physics-driven methods like parallel imaging (SENSE, GRAPPA), compressed sensing (CS), and Echo Planar Imaging (EPI) exploit physical principles to decrease acquisition time but can introduce artefacts like noise amplification, residual aliasing, and Nyquist ghosting. Data-driven approaches, particularly deep learning with convolutional neural networks [13] (CNNs) or autoencoders, such as U-net [14], have also been proposed to predict missing data resulting from various under-sampled k-space data acquisition strategies, offering robust image reconstruction.

In real-world MRI, signal acquisition is also subject to various degradation factors that cause artefacts, which are undesired and unreal information that appear in the reconstructed image. Although full sampling may theoretically provide the most complete data, it is still susceptible to motion, noise and other imperfections inherent in the imaging process. These issues distort anatomical structures, introduce false information or cause signal loss, compromising diagnostic accuracy. Therefore, in the pursuit of improving image quality through acceleration techniques, it's crucial to not only address artefacts arising from under-sampling but also to carefully manage and mitigate additional artefacts inherent to imperfect acquisitions settings that may be further amplified by the under-sampling process.

In the present paper, we present a new method to restore quality in images reconstructed from under-sampled MRI data acquisitions. We propose a neural network model called USArt (Under-Sampling and Artifact correction model) that restores missing k-space data and simultaneously corrects for motion and noise related artefacts. By addressing both under-sampling and artefacts correction, we aim to enhance the overall quality and accuracy of real-life fast MRI methods, thus contributing to more reliable and effective image reconstruction techniques. Our approach involves a dual-model framework, featuring one model operating in the k-space domain and another in the image domain, inspired by prior work that exploits the different characteristics of these domains [7,8,12]. We examined different under-sampling strategies, acceleration factors and artefacts. This project focuses on single-channel coils and Cartesian sampling for simplicity; hence, parallel MRI is not discussed.

2 Methods

2.1 Dataset Description

In vivo 2D T2w anatomical images of mouse brain with tumors were acquired according to established protocols [15,16]. Images acquisitions used a Cartesian sampling trajectory and were performed on a preclinical 3T MRI system (MRSolutions, UK) equipped with a quadrature head coil. The dataset consists of 5649 complex-valued images from 224 different mice. The train/validation dataset includes 5000/449 images from 204/30 subjects. The remaining 200 images from 10 different subjects were used to test the performance of the trained networks. There was no overlapping of the same subjects or images in the different datasets.

2.2 Under-Sampling

Under-sampling was achieved by retrospectively dropping lines in fully acquired Cartesian k-spaces, corresponding to phase-encoding steps. Masks were for this purpose applied to selectively zero out lines in the k-space using one of three strategies: gradient, random, and uniform under-sampling. The gradient under-sampling mask progressively reduces the number of lines acquired as the trajectory moves away from the k-space center. This favors low frequency k-space information that determines contrast, brightness, and general shapes, over high frequencies that pertain to edges and sharp transitions. Random under-sampling selects the retained lines randomly. Uniform under-sampling uniformly discard lines in the k-space, without targeting specific regions or frequencies. Three under-sampling acceleration factors were used: 2×, 5×, and 10×. For all under-sampling strategies, we additionally retained some low-frequency lines in order to reduce the aliasing artefact; 25%, 10%, and 4% of the k-space's lines for 2×, 5×, and 10× accelerations respectively.

2.3 Artefacts

Artefacts in real-world acquisitions were simulated in the k-space domain using a custom developed library [17], accessible through GitHub[1]. We explored the most common artefacts in anatomical images: Gaussian noise and motion artefact.

Gaussian Noise. The noise present in MRI usually originates from electronic sources involved in the data acquisition process. In the k-space domain, Gaussian noise is typically observed, resulting in Rician noise in the reconstructed images [18,19]. In this study, the noise was applied in the k-space as an additive complex Gaussian random signal, i.e. with the real and imaginary parts being independent random variables with zero mean, and variance set according to the simulated SNR. The maximum noise level caused a reduction of the original image SNR by half.

---

[1] https://github.com/TransRad/MRArt.git.

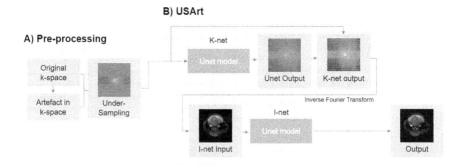

Fig. 1. The preprocessing pipeline and USArt. A) Artifacts and noise are added to full k-space, before under-sampling is performed using specific masks and acceleration factors. This degraded k-space dataset is used as input for the USArt model. B) USArt utilizes two U-Net based components: K-net and I-net. K-net operates in the k-space domain to fill missing lines, and its output is transformed to the image domain via an inverse Fourier Transform. I-net then refines this output, focusing on artifacts correction and image consistency.

Motion Artefacts. Motion artefacts (MA) in MRI typically originate from patient motion; this motion can lead to inconsistencies in the magnetic field interactions, resulting in deviations from the expected signal patterns. MA primarily affect the acquired data (k-space data) and manifest as discrepancies or distortions in the k-space lines corresponding to the moment of motion. The appearance of MA depend on the k-space scanning strategy, particularly whether the acquisition is conducted in 2D or 3D. Movements that occur when the acquisition trajectory corresponds to low frequencies (central k-space lines) often result in ghosting artefacts, while movements occurring during high-frequency acquisitions often result in ringing and blurring artefacts. In this work, MA were implemented as rotation or shift in the image domain as described by [20] and its effect on k-space data reflected by the Fourier Transformation of the artefacted images.

2.4 Metrics

To assess the quality of the images, we used the well-established reference-based Structural SIMilarity (SSIM) index focused on the subject only by removing background (SSIMf) [21] and Peak Signal-to-Noise Ratio (PSNR). These metrics allow an objective comparison between USArt reconstructed images and the ground truth images. Since ground truth images themselves are not perfect and always contain some level of degradation, we also used reference-free metrics, including SNR and Contrast to evaluate an absolute image quality on an individual basis. Contrast was measured as the standard deviation of tissue signal.

Table 1. Performance of our proposed model with various under-sampling strategies, acceleration factors, and artefacts. The first three lines compare USArt performance with different under-sampling strategies. The next three lines show USArt performance with different acceleration factors. The next 4 lines show the robustness of USArt to the presence of artefacts. The bottom part of the table provides benchmark values for the reference KIKI [8] model using a 5× acceleration factor and gradient under-sampling in the presence of artifacts, showing the superiority of our model in real-world acquisitions.

Acc.	US masks	Artifact	SSIMf	PSNR	SNR	Contrast
	Original	No			15.249 ± 2.170	0.671 ± 0.141
USArt						
5x	Gradient	No	**0.971 ± 0.008**	**77.219 ± 1.436**	**78.976 ± 11.125**	0.710 ± 0.148
5x	Random	No	0.969 ± 0.009	76.586 ± 1.467	78.946 ± 11.868	0.709 ± 0.148
5x	Uniform	No	0.969 ± 0.009	76.541 ± 1.541	75.712 ± 11.558	**0.721 ± 0.152**
2x	Gradient	No	0.979 ± 0.006	78.477 ± 1.457	51.656 ± 4.290	0.708 ± 0.137
5x	Gradient	No	0.971 ± 0.008	77.219 ± 1.436	75.712 ± 11.558	0.710 ± 0.148
10x	Gradient	No	0.962 ± 0.011	74.934 ± 1.798	107.222 ± 16.006	0.746 ± 0.148
5x	Gradient	No	0.971 ± 0.008	**77.219 ± 1.436**	75.712 ± 11.558	0.710 ± 0.148
5x	Gradient	Noise	0.966 ± 0.010	76.126 ± 1.747	85.638 ± 14.070	0.735 ± 0.149
5x	Gradient	MA	0.964 ± 0.011	75.141 ± 1.816	55.759 ± 08.380	0.705 ± 0.143
5x	Gradient	N+MA	0.960 ± 0.013	74.557 ± 1.607	84.251 ± 16.347	0.743 ± 0.143
KIKI [8]						
5x	Gradient	No	0.971 ± 0.007	76.496 ± 1.175	77.125 ± 11.824	0.735 ± 0.154
5x	Gradient	Noise	0.954 ± 0.049	76.091 ± 1.303	44.297 ± 19.010	0.677 ± 0.140
5x	Gradient	MA	0.952 ± 0.013	74.935 ± 1.952	**57.293 ± 20.947**	0.642 ± 0.162
5x	Gradient	N+MA	0.951 ± 0.017	73.179 ± 3.408	40.172 ± 19.683	0.612 ± 0.164

2.5 Pre-processing

The pre-processing pipeline (Fig. 1-A) begins by introducing artefacts, such as MA and/or noise, into the k-space domain. Subsequently, the data experience under-sampling, a process that zeros specific lines in the k-space. To achieve this, one of the three under-sampling masks and acceleration factors is applied. These steps collectively result in a k-space dataset degraded by both artefacts and under-sampling, which acts as the input for our USArt.

2.6 Proposed Model

We address the correction of image degradation caused by under-sampling in the presence of artefacts by using a dual model strategy, that has recently showed successes in the context of MR images [7,8,12]. Considering the complex nature of the MR k-space, we divide MR complex data into real and imaginary parts and use them as two input channels in the model. In USArt, the first part (K-net) focuses solely on the under-sampled k-space, while the second part (I-net) concentrates on improving image quality and consistency (Fig. 1-B). Both parts (K-net and I-net) are based on the U-Net architecture; U-Net was proposed

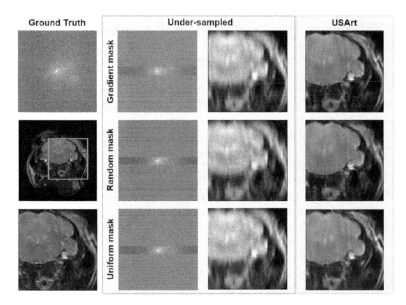

Fig. 2. Under-sampling strategies with acceleration factor 5×: Left column from top to bottom) k-space, reconstructed, and zoom images (blue frame) for the original image. Grey panel) Under-sampled k-space and the zoom corresponding reconstructed images for gradient, random and uniform under-sampling. Blue-light panel) the corresponding USArt's output. (Color figure online)

for biomedical image segmentation [22] and has since become a popular choice also for image-to-image translation tasks. In detail, K-net operates within the k-space domain, addressing the task of filling the missing lines. The output of the K-net is modified by inserting back the acquired k-space lines while keeping the lines predicted by the K-net model at the k-space positions with no data acquired. Transformation from the k-space domain to the image domain is achieved by an inverse Fourier Transform operation. In the last step, I-net operates within the image domain, focusing on artefacts correction and image consistency improvement. The K-net and I-net networks were trained sequentially; after K-net training was completed, I-net was trained using the output from K-net. We compare USArt[2] with the KIKI model [8]; KIKI has focused on utilizing strategy for reconstructing under-sampled image data, our work extends its application to concurrently tackle both under-sampling and artefacts.

The image input size is $2 \times 256 \times 256$. U-Net with 4 layers is trained for 150 epochs with batch size 16 and an initial learning rate of 0.0001. We use AdamW optimization as a stochastic gradient descent method. A learning rate decay with 20 epochs of patience is applied with 10% drop. For the activation function, a leaky rectified linear unit (Leaky-ReLU) is used with negative slope 0.2. L2 regularization penalties are applied on a per-layer basis. We used a focus Multi-

---

[2] https://github.com/TransRad/USArt.git.

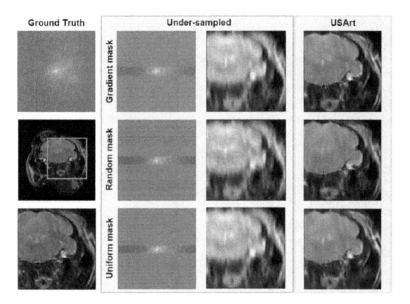

Fig. 3. Different acceleration factors: Left column) from top to bottom: k-space, reconstructed, and zoom images (blue frame) from the original image. Grey panel) Under-sampled k-space and the zoom corresponding reconstructed images for acceleration factors 2×, 5×, and 10×. Blue panel) the corresponding USArt's output. (Color figure online)

Scale Structural Similarity (MS-SSIM) [23] loss function of the missing lines of the k-space for the K-net network and MS-SSIM of the whole reconstructed image for the I-net network in order to minimize the loss error. The networks are implemented using the TensorFlow framework. The training and inference used a GPU NVIDIA GeForce RTX 3080Ti graphics card with 12GB RAM.

3 Results

In preparation for combining the acceleration and artefacts correction, we first established the optimal under-sampling (US) strategy and the acceleration factor. The MRI details showcased in Fig. 2 present variations under different under-sampling strategies. The highest SSIMf and PSNR were achieved with the gradient under-sampling mask. For the reference free metrics, the gradient under-sampling provided the best results in SNR, while the uniform under-sampling mask provided the best result for Contrast (Table 1). The gradient under-sampling strategy was thus retained for the subsequent tasks.

We then evaluated USArt using various acceleration factors, and were able to efficiently reconstruct quality images for all cases (Fig. 3). We found that 5× acceleration was possible without significant degradation (Table 1), with SSIMf to 0.971 and PSNR to 77.219. Subsequently, the acceleration factor 5× was retained for the ensuing tests.

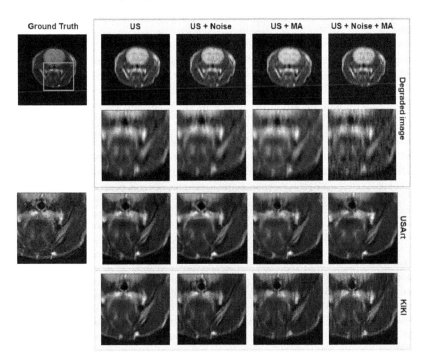

Fig. 4. Illustration of artefacts correction in accelerated images: First column) Original and zoom images (blue frame). Grey box) 5×under-sampled full and zoomed images with no artefacts, noise, motion artefact and their combination. Blue box) Corresponding images with quality restored by USArt and KIKI model's. US: Under-sampled, MA: Motion Artifact. (Color figure online)

Using the gradient under-sampling strategy and an acceleration factor of 5× we then evaluated the simultaneous correction of under-sampled data and artefacts, with the model detailed earlier. USArt effectively restored quality in under-sampled acquisitions, despite the presence of artefacts (Fig. 4). The results were confirmed by the metrics in Table 1, where only a limited reduction was observed in the reconstructed under-sampled images in the presence of artefacts, compared to those obtained in the absence of artefacts, with reductions of less than 1.2% in SSIM and 3.5% in PSNR, respectively. The results from USArt outperformed those from previous well-established models used for image restoration from under-sampled data, notably the KIKI model [8], establishing the superiority of our model in real-word settings.

4 Discussion and Perspectives

Acquiring high-quality MR images can be time-consuming. Reducing k-space sampling saves time but typically lowers image quality. The USArt model introduces a novel approach that simultaneously improves image quality and corrects artifacts in accelerated MR imaging. The K-net and I-net sub-models work together to enhance image details, restore contrast, and ensure image consistency. Moreover, the model demonstrates robustness against real-world degradations such as noise and motion artifacts, even when applied to under-sampled data.

Having established confidence in the model's capacity to manage simultaneous under-sampling and artefact correction, future research could explore alternative model architectures such as Vision Transformers. These architectures, proven effective in image-to-image translation tasks, hold promise for enhancing our USArt. Moreover, our approach is also applicable to accelerating clinical data and other types of trajectories and more advanced protocols, where various types of artefacts may be observed.

References

1. Moratal, D., Vallés-Luch, A., Martí-Bonmati, L., Brummers, M.E.: k-Space tutorial: an MRI educational tool for a better understanding of k-space. Biomed. Imaging Interv. J. **4**, 1 (2008)
2. Wang, S., Xiao, T., Liu, Q., Zheng, H.: Deep learning for fast MR imaging: a review for learning reconstruction from incomplete k-space data. Biomed. Signal Process. Control **68**, 102579 (2021)
3. Hollingsworth, K.G.: Reducing acquisition time in clinical MRI by data undersampling and compressed sensing reconstruction. Phys. Med. Biol. **60**, R297–R322 (2015)
4. Huang, J., et al.: Data and physics driven learning models for fast MRI – fundamentals and methodologies from CNN, GAN to attention and transformers. IEEE Signal Process. Mag. (2022)
5. Sriram, A., et al.: End-to-end variational networks for accelerated MRI reconstruction. In: Martel, A.L., et al. (eds.) MICCAI 2020. LNCS, vol. 12262, pp. 64–73. Springer, Cham (2020). https://doi.org/10.1007/978-3-030-59713-9_7
6. Ramzi, Z., Ciuciu, P., Starck, J.L.: Benchmarking MRI reconstruction neural networks on large public datasets. Appl. Sci. (Switzerland) **10**, 2020 (1816)
7. Haji-Valizadeh, H., et al.: Comparison of complex k-Space data and magnitude-only for training of deep learning-based artifact suppression for real-time cine MRI. Front. Phys. **9**, 9 (2021)
8. Eo, T., Jun, Y., Kim, T., Jang, J., Lee, H.J., Hwang, D.: KIKI-net: cross-domain convolutional neural networks for reconstructing undersampled magnetic resonance images. Magn. Reson. Med. **80**, 2188–2201 (2018)
9. Hyun, C.M., Kim, H.P., Lee, S.M., Lee, S., Seo, J.K.: Deep learning for undersampled MRI reconstruction. Phys. Med. Biol. **63**, 135007 (2018)
10. Muckley, M.J., et al.: Results of the 2020 fastMRI challenge for machine learning MR image reconstruction. IEEE Trans. Med. Imaging **40**, 2306–2317 (2021)
11. Zbontar, J., et al.: fastMRI: an open dataset and benchmarks for accelerated MRI. Radiol. Artif. Intell. **2**, 1–35 (2018)
12. Cheng, J., Wang, H., Ying, L., Liang, D.: Model learning: primal dual networks for fast MR imaging. In: Shen, D., et al. (eds.) MICCAI 2019. LNCS, vol. 11766, pp. 21–29. Springer, Cham (2019). https://doi.org/10.1007/978-3-030-32248-9_3

13. O'Shea, K., Nash, R.: An introduction to convolutional neural networks. arXiv:1511.08458, November 2015
14. Ronneberger, O., Fischer, P., Brox, T.: U-Net: convolutional networks for biomedical image segmentation. In: Navab, N., Hornegger, J., Wells, W.M., Frangi, A.F. (eds.) MICCAI 2015. LNCS, vol. 9351, pp. 234–241. Springer, Cham (2015). https://doi.org/10.1007/978-3-319-24574-4_28
15. Oudin, A., et al.: Protocol for derivation of organoids and patient-derived orthotopic xenografts from glioma patient tumors. STAR Protocols **2**, 6 (2021)
16. Golebiewska, A., et al.: Patient-derived organoids and orthotopic xenografts of primary and recurrent gliomas represent relevant patient avatars for precision oncology. Acta Neuropathol. **140**, 919–949 (2020)
17. Boudissa, S., Kanli, G., Perlo, D., Jaquet, T., Keunen, O.: Addressing artefacts in anatomical MR images: a k-space-based approach. In: IEEE International Symposium on Biomedical Imaging (2024)
18. Constantinides, C.D., Atalar, E., McVeigh, E.R.: Signal-to-noise measurements in magnitude images from NMR phased arrays. Magn. Reson. Med. **38**, 852–857 (1997)
19. Dietrich, O., Raya, J.G., Reeder, S.B., Reiser, M.F., Schoenberg, S.O.: Measurement of signal-to-noise ratios in MR images: influence of multichannel coils, parallel imaging, and reconstruction filters. J. Magn. Reson. Imaging **26**, 375–385 (2007)
20. Shaw, R., Sudre, C.H., Varsavsky, T., Ourselin, S., Cardoso, M.J.: A k-Space model of movement artefacts: application to segmentation augmentation and artefact removal. IEEE Trans. Med. Imaging **39**, 2881–2892 (2020)
21. Qiu, W., Li, D., Jin, X., Liu, F., Sun, B.: Deep neural network inspired by iterative shrinkage-thresholding algorithm with data consistency (NISTAD) for fast Undersampled MRI reconstruction. Magn. Reson. Imaging **70**, 134–144 (2020)
22. Krithika, M., Alias AnbuDevi, Suganthi, K.: Review of semantic segmentation of medical images using modified architectures of UNet. Diagnostics **12**, 3064 (2022)
23. Wang, Z., Simoncelli, E.P., Bovik, A.C.: Multi-scale structural similarity for image quality assessment. In: Conference Record of the Asilomar Conference on Signals, Systems and Computers, vol. 2, pp. 1398–1402 (2003)

Full-TrSUN: A Full-Resolution Transformer UNet for High Quality PET Image Synthesis

Boyuan Tan[1], Yuxin Xue[1], Lei Bi[1,2], and Jinman Kim[1(✉)]

[1] School of Computer Science, The University of Sydney, Sydney, NSW, Australia
jinman.kim@sydney.edu.au
[2] Institute of Translational Medicine, National Center for Translational Medicine, Shanghai Jiao Tong University, Shanghai, China

Abstract. Positron Emission Tomography (PET) is an established functional imaging modality integral to clinical practices. Despite its widespread utility, the attendant radiation exposure from PET scans has raised substantial health concerns. To address these challenges, numerous CNN-based methodologies have been developed to reconstruct standard-dose PET (SPET) images by using low-dose PET (LPET). These reconstructions are generally via image synthesizes using CNNs which by design, predominantly capture localized features, and thus struggle to encapsulate the long-range global feature correlations that are essential for fine-grained image synthesis. Transformers have demonstrated an inherent strength in capturing these extensive dependencies. However, the high computational and memory demands constrain its use to images at reduced resolutions, leading to potential loss of essential textural information for accurate PET synthesis. Our research proposes a new Full-resolution Transformer based model, named as Full-TrSUN, that applies a 3D transformer block at full image resolution designed to discern fine-grained, long-range dependencies. We also integrate a CNN-based encoder and decoder process in a U-Net architecture for PET image synthesis. The Full-TrSUN framework is designed to preserve the vital texture nuances at full resolution, enhancing the functional detail captured in PET synthesis. Our experimental results with public benchmark datasets show that our method outperformed the state-of-the-art methods with high efficiency.

Keywords: Transformer · Image Synthesis · Low-dose PET

1 Introduction

Positron Emission Tomography (PET) imaging is widely acknowledged for its crucial role in oncology for diagnosis and management of various cancer diseases [1]. However, standard-dose PET (SPET) images will take a high cumulative radiation exposure, which raises concerns about potential health risks [2]. Specifically, employment of radioactive tracers such as 18F-FDG results in a radiation dose of 25 mSv [3], markedly exceeding that of X-rays [4] and CT scans [5]. To mitigate the risks associated with radiation exposure, a common approach is to reduce the injected dosage, i.e., low-dose PET (LPET). However, when compared to the PET images acquired under the standard protocol (SPET), LPET suffers from lower signal-to-noise-ratio (SNR). In light of these challenges, adopting deep learning strategies within SPET synthesis from LPET has been identified as a promising alternative.

With the recent advances of Convolutional Neural Networks (CNNs) and generative adversarial network (GAN) [6], methods have been developed for SPET synthesis [7–9]. Wang et al. [10] utilized a 3D conditional generative adversarial networks (3D c-GANs) to reconstruct high-quality SPET images, with a 3D U-net-like architecture to ensure consistent information between both imaging types. Furthermore, it introduced concatenated 3D c-GANs (stacked-GAN)to enhance the quality of the estimated images. Zhou et al. [11] and Zhao et al. [12] utilized CycleGAN for SPET synthesis combined with cycle-consistency loss, Wasserstein distance loss, and a supervised learning loss. Luo et al. [13] introduced AR-GAN model, which used an adaptive rectification network (AR-Net) and a spectral regularization term to address discrepancies and high-frequency distortions in synthesized PET images. Xue et al. [14] introduced SS-AEGAN model, which employed an adaptive residual estimation mechanism to dynamically rectify preliminary synthesized PET images. Additionally, this model incorporated a self-supervised pre-training strategy to enhance feature representation, effectively addressing discrepancies in texture and structure between synthesized and real images. In a later work, Xue et al. [15] proposed a CG-3DSRGAN, a Classification-Guided Generative Adversarial Network enhanced with Super Resolution Refinement. This model included a multi-tasking coarse generator and an auxiliary super-resolution network. Both methods demonstrated superior performances across various dose reduction factors when compared to the existing state-of-the-art methods.

However, CNN-based methods tend to enlarge the receptive field by systematically down-sampling the image features and also to aggregate multiple convolutional layers [16]. These methods constrain the receptive field when dealing with high-resolution inputs and, consequently, struggles to capture the fine-grained, long-range visual correlations. Such long-distance relationships are crucial in dense prediction tasks, offering contextual insight necessary for a subtle semantic interpretation [17]. Transformer [18] models, particularly hierarchical architectures like the Swin Transformer [19], due to their capabilities in capturing long-range dependencies via self-attention mechanisms, have recently outperformed CNN-based methods. Unfortunately, self-attention process within the Transformer model is computationally and memory-intensive and therefore are often limited to its deployment at lower image resolutions. This limitation, as a result, restricts the detailed textural representation available at full image resolutions. To address this challenge, researchers have proposed channel-wise self-attention methods [20–22], aimed to balance computational efficiency and detailed feature representation, to enhance performance in high-resolution tasks. Jang et al. combined the spatial-wise with channel-wise to do the self-attention [22]. However, the local content capture capabilities of the pure transformer architectures also face limitations, especially for the small dataset [23]. Additionally, these methods, along with the transformer encoder and decoder, are memory-intensive, posing constraints on model efficiency and necessitating further research and optimization to achieve desired performance across diverse applications.

In this study, we introduce Full-TrSUN—a full-resolution hierarchical transformer framework. We innovate in employing hierarchical transformer blocks to distil multi-scale features, starting at the full image resolution. The process integrates XCIT Transformer Blocks [20] with a CNN encoder-decoder architecture and UNet connections [24], collectively enabling the model to capture a comprehensive range of textural nuances. The use of the CNN structure also captures spatial relationships and enhances efficiency, resulting in higher performance.

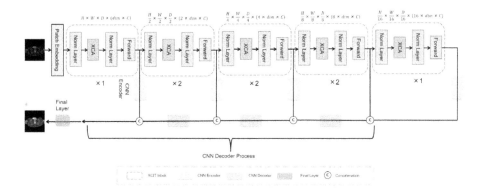

Fig. 1. Architecture of the Full-TrSUN.

2 Method

The objective of this study is to generate standard-dose PET images I_S from low-dose PET images I_L. Both I_L and I_S are defined as three-dimensional (3D) volumes situated within a spatial domain $\Omega \subseteq \mathbb{R}^3$. Initially, the process utilizes Transformer encoder as a foundational hidden block. Assuming the presence of t distinct stages within this block, each stage produces a downsampled output. At the ith level $i \in \{1, 2, \cdots, t\}$, the resultant output is denoted as I_L^i which employed as inputs for a CNN-encoder block, denoted as I_E^i. The output is then directed through a CNN decoder block, facilitated by skip connections at various resolutions to effectively maintain and integrate multi-scale features. Upon completion of the decoding process through the final layer, an estimated standard-dose PET image I_S' is produced.

The architecture of the Full-TrSUN is illustrated in Fig. 1, which consists of a transformer encoder stage combined with the CNN encoder process followed by the CNN decoder using the hidden state and UNet connection.

2.1 Transformer Stage

Starting with the low-dose PET image $I_L \in \mathbb{R}^{H \times W \times D \times C}$ as the input, the transformer stage initially employs a $3 \times 3 \times 3$ convolution layer to obtain the low-level feature embeddings. Following the patch embedding, the input image retains its full resolution, ensuring the extraction of complete textural information at the tissue level. The entire encoder process comprises five Transformer Stages, each with a different number of transformer blocks arranged as (1,2,2,2,1). A critical precursor in this methodology is the initial convolutional transformation of an LPET input image $I_L \in \mathbb{R}^{H \times W \times D \times C}$ to extract foundational feature representations $F_0 \in \mathbb{R}^{H \times W \times D \times (dim \times C)}$, where H, W, and D delineates the spatial dimensions, and $dim \times C$ denotes the channel count. This extraction lays the groundwork for subsequent transformations across five distinct levels, thereby evolving F_0 into enriched feature embeddings $F_d \in \mathbb{R}^{H \times W \times D \times (d \times dim \times C)}$.

Two components of the Transformer block are Cross-Covariance Attention (XCA) and Forward process [20]. The XCA mechanism is articulated as,

$$\hat{\mathbf{X}} = W Attention(\mathbf{Q}, \mathbf{K}, \mathbf{V}) + \mathbf{X} \quad (1)$$

$$Attention(\mathbf{Q}, \mathbf{K}, \mathbf{V}) = \mathbf{V} Attention(\mathbf{Q}, \mathbf{K}) = \mathbf{V} \cdot Softmax(\frac{\mathbf{K}^T \mathbf{Q}}{\alpha}) \quad (2)$$

where $\mathbf{X}$ and $\hat{\mathbf{X}}$ are the input and output features map and $\mathbf{Q}(query)$, $\mathbf{K}(key)$, $\mathbf{V}(value)$ are obtained after reshaping tensors from the original size. Here $\mathbf{Q} = \mathbf{X}W_q$, $\mathbf{K} = \mathbf{X}W_k$ and $\mathbf{V} = \mathbf{X}W_v$, which W_q, W_k, W_v are the weights. It shifts the computational emphasis from the spatial domain to the channel domain, thereby achieving a linear complexity. Depth-wise convolutions complement this shift to underscore local context before the computation of the global attention. Here α is the scaling factor.

After the XCA block, we use a Forward block to enhance the communication between each XCA block by two $3 \times 3 \times 3$ convolution layers with Batch Normalization and GELU non-linearity.

2.2 CNN Encoder and CNN Decoder

Our model extracts a sequence of representations F_i ($i \in \{1, 2, \ldots, 5\}$), each with dimensions $\frac{H}{P} \times \frac{W}{P} \times \frac{D}{P} \times C$, where $P = 2^i$, $C = dim \times i$. We choose the feature size and the special dim with 24 and 3 for the CNN-Encoder and CNN-Decoder blocks. For the Encoder block, we use these representations as the input of the encoder blocks with two $3 \times 3 \times 3$ convolutional layers, each followed by a normalization layer. For the Decoder block, the spatial resolution of these feature maps is subsequently amplified by a factor of 2 via a deconvolutional layer. Utilizing these convolutional layers enables the effective projection of complex, high-dimensional data into a more interpretable and spatially relevant form, laying the groundwork for enhanced feature synthesis and integration within the decoder through skip connections at corresponding levels. The synthesized features are further processed in another decoder block, adhering to the previously described configuration and including an upsampling phase. The final outputs are then generated by applying a $1 \times 1 \times 1$ convolutional layer, followed by a sigmoid activation function, thus enabling the effective synthesis task.

We use the hidden state output as the output of the Transformer Block to connect the Transformer Block and the CNN-encoder block. Then, we use the concatenation of the Encoder output as the Decoder input like UNet [24].

Upon obtaining the final estimated standard-dose PET (SPET) images I'_S, we proceed to juxtapose these with the target SPET images I_S for loss computation and subsequent model optimization. The commonly used L1 Loss function was used to estimate the difference between the predicted and the target images. The optimization of the model is based on using the Adam optimizer [25].

3 Experiment Setup

3.1 Dataset

In this study, we use the data from Ultra-low Dose PET Imaging Challenge [26]. 387 patient data were acquired from the Siemens Biograph Vision Quadra scanner. Images are reconstructed with OSEM to be $440 \times 440 \times 644$ voxels at a voxel spacing of $1.65 \, \text{mm}^3$. The dataset allocation includes 75 patients for testing, 38 for validation, and 284 for training. All acquired data was in 'list mode', which can be used to reconfigure the acquisition duration to simulate various low-dose scenarios. Each low-dose PET was characterized by a dose reduction factor (DRF) derived from reconstructed counts over a reduced acquisition time frame centered around the midpoint of the original acquisition duration. The generation of low-dose images employed DRFs of 4, 10, 20, 50, and 100, in conjunction with a corresponding full-dose image, to span a comprehensive range of dose levels. Our research focuses on minimizing radiation exposure by synthesizing high-quality standard-dose images from the lowest dose inputs; hence, we exclusively utilize the DRF of 100, representing a dose reduction of 100%, for our training input.

3.2 Implementation Details

All experiments were conducted using PyTorch with TensorBoard for visual analytics, using i7-5930K CPU and 24 GB NVIDIA RTX A3090 GPU. The learning rate was set to 0.0002 and batch size was set to 1. We trained our method for more than 300,000 iterations.

3.3 Experimental Settings

The study focused on the ultra-low-dose synthesis and therefore only used DRF100. After compared different model parameters (in Sect. 4, Tables 2 and 3), the dimensionality (dim) of the Transformer block was set to 24. The architecture of the Transformer stage was set to $(1, 2, 2, 2, 1)$, and the number of attention heads was set to $(1, 2, 4, 8, 16)$. Additionally, for the hidden states encoder and the UNet skip-connection for the decoding processes, the $feature\_size$ was set to 24, and the $spatial\_dims$ parameter was set to 3.

We evaluate the effectiveness of the synthesized results using the commonly used evaluation metrics, including Peak Signal-to-Noise Ratio (PSNR), Structural Similarity Index Measure (SSIM), Normalized Root Mean Squared Error (NRMSE) and Mean Squared Error (MSE).

3.4 Comparison Methods

The Full-TrSUN model was evaluated against several state-of-the-art synthesis methods. For comparison, we included the traditional 3D-UNet [27], four GAN-based methods specifically designed for PET synthesis-cGAN [10], Cycle-GAN [11,12], AR-GAN [13], and SS-AEGAN [14]-as well as two transformer-based methods, Restormer [21] and SPACH-Transformer [22]. To ensure a fair comparison, all models were implemented using PyTorch. Due to GPU memory limitations, the dimensional settings for Restormer and SPACH-Transformer were adjusted to 6 and 10, respectively.

4 Results and Discussion

4.1 Comparison with the State of the Art Models

The results of our Full-TrSUN model and various comparative methods are summarized in Table 1. The visualisation results in Fig. 2 further demonstrate that the Full-TrSUN model performs the best among the comparison methods. The Full-TrSUN method consistently surpasses other approaches across all evaluation metrics, establishing it as the most effective technique for synthesizing SPET images from LPET images in this study. The Full-TrSUN model achieves the highest PSNR of 52.7788 dB, indicating that the synthesized images exhibit higher fidelity and closer resemblance to the ground truth compared to those produced by other models. Additionally, with an SSIM of 0.9927, the Full-TrSUN model demonstrates superior structural similarity to the original SPET images, reflecting its capability to accurately preserve image textures and structures. Furthermore, the model records the lowest MSE of 0.0296, suggesting minimal errors in image reconstruction and high precision in replicating the original images. The lowest NRMSE of 0.2733% achieved by the Full-TrSUN model further underscores its effectiveness in minimizing the discrepancy between the synthesized and original images.

4.2 Comparison of Model Parameters

We further evaluated the impact of varying the *number of blocks* and the parameter *dim* on the model's performance.

Table 1. Comparison with the State of the Art models

	PSNR (dB) (↑)	SSIM (↑)	MSE (↓)	NRMSE (%)(↓)	Memory (GB)
raw	42.6927	0.9141	0.3235	0.8992	–
unet	49.8706	0.9887	0.0665	0.3770	4.0
argan	47.4489	0.9782	0.1046	0.4809	5.1
cgan	49.1875	0.9876	0.0636	0.4265	4.3
cyclegan	50.6001	0.9907	0.0471	0.3509	10.5
SS-AEGAN	52.0535	0.9918	0.0340	0.2943	4.5
Restormer	52.0575	0.9874	0.0346	0.2958	19.7
SPACH-Transformer	52.1662	0.9857	0.0339	0.2926	20.2
Full Tr-SUN	**52.7788**	**0.9927**	**0.0296**	**0.2733**	21.7

Table 2. Comparison of parameter num_block with ten epochs training using dim = 12

num_blocks	PSNR (dB) (↑)	SSIM (↑)	NRMSE (%) (↓)	MSE (↓)
[4, 6, 6, 6, 4]	51.6793	0.9916	0.3058	0.0489
[2, 3, 3, 3, 2]	51.8052	0.9909	0.3024	0.0423
[2, 2, 2, 2, 2]	51.6094	0.9910	0.3082	0.0437
[1, 2, 2, 2, 1]	**51.9578**	**0.9915**	**0.2980**	**0.0381**
[1, 1, 1, 1, 1]	51.8097	0.9908	0.3027	0.0424

Table 3. Comparison of parameter dim with ten epochs training using num_block = (1,2,2,2,1)

dim	PSNR (dB) (↑)	SSIM (↑)	NRMSE (%) (↓)	MSE (↓)
6	51.2461	0.9882	0.3191	0.0520
12	51.9578	**0.9915**	0.2980	0.0381
24	**52.0575**	0.9874	**0.2958**	**0.0346**

Evaluation of Block Configurations. Table 2 presents the results for different block configurations with *dim* set to 12. The configuration [1, 2, 2, 2, 1] demonstrated the best performance, achieving a PSNR of 51.9578 dB, an SSIM of 0.9915, an NRMSE of 0.2980%, and an MSE of 0.0381. These metrics suggest that this configuration effectively balances computational efficiency and model performance.

Evaluation of Dimension Values. We also assessed the effect of varying *dim* values, starting with the optimal block configuration [1, 2, 2, 2, 1]. Table 3 shows that under limited GPU memory constraints, a *dim* value of 24 yielded the best results, achieving a PSNR of 52.0575 dB, an SSIM of 0.9874, an NRMSE of 0.2958%, and an MSE of 0.0346. This larger dim value enhances the model's capacity to capture more complex spatial relationships, thus improving image quality and reconstruction accuracy. In the context of channel-wise transformer methods, the parameter *dim* is particularly critical. Under the same GPU constraints, our Full-TrSUN model can utilize a larger *dim* value. Using a CNN encoder-decoder structure to replace the transformer decoder process can help save memory, enabling the use of better parameter settings to achieve higher results. This approach leverages the spatial relationship capabilities of CNNs, enhancing computational efficiency and performance.

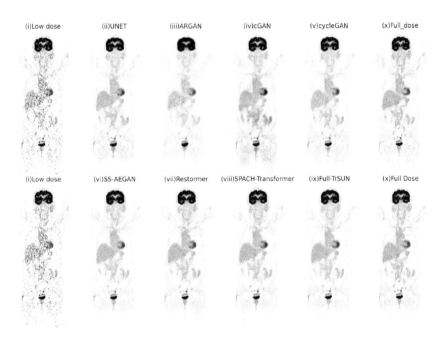

Fig. 2. Visualization of different models.

5 Conclusion

In this study, we presented Full-TrSUN - a Full-Resolution Transformer that amalgamates the XCIT transformer block and UNet architectures for synthesising Standard-Dose PET (SPET) images from Low-Dose PET (LPET) sources. Distinct from prevailing traditional CNN and GAN-based approaches, our method is able to be more effectively in leveraging textural information at multiple levels e.g., tissue level, within the PET images. Full-TrSUN demonstrates superior efficiency when compared to other transformer-based methods. The preliminary experimental findings corroborate that Full-TrSUN surpasses existing synthesis techniques in terms of evaluation metrics and in optimizing training memory. Despite its strong performance in evaluation metrics, the visual output appears to be smoother than the target full-dose image. Future work will focus on addressing this issue. We aim to refine the model to better capture and reproduce high-frequency textural details, ensuring that the synthesized images more closely resemble the sharpness and details of SPET images.

Acknowledgement. This project was funded in part by Tour De Cure RSP-581-2024.

References

1. Maisey, M.N.: Positron emission tomography in clinical medicine. In: Positron Emission Tomography: Basic Sciences, pp. 1–12. Springer, Cham (2005)
2. Voss, S.D., Reaman, G.H., Kaste, S.C., Slovis, T.L.: The ALARA concept in pediatric oncology. Pediatr. Radiol. **39**, 1142–1146 (2009). https://doi.org/10.1007/s00247-009-1404-5
3. Brix, G., et al.: Radiation exposure of patients undergoing whole-body dual-modality 18F-FDG PET/CT examinations. J. Nucl. Med. **46**(4), 608–613 (2005)
4. Huda, W., Sandison, G., Palser, R., Savoie, D.: Radiation doses and detriment from chest X-ray examinations. Phys. Med. Biol. **34**(10), 1477 (1989). https://doi.org/10.1088/0031-9155/34/10/010
5. Donadieu, J., Roudier, C., Saguintaah, M., Maccia, C., Chiron, R.: Estimation of the radiation dose from thoracic CT scans in a cystic fibrosis population. Chest **132**(4), 1233–1238 (2007). https://doi.org/10.1378/chest.07-0221
6. Goodfellow, I., et al.: Generative adversarial nets. In: Advances in Neural Information Processing Systems, vol. 27 (2014)
7. Xiang, L., et al.: Deep auto-context convolutional neural networks for standard-dose PET image estimation from low-dose PET/MRI. Neurocomputing **267**, 406–416 (2017). https://doi.org/10.1016/j.neucom.2017.06.048
8. Bi, L., Kim, J., Kumar, A., Feng, D., Fulham, M.: Synthesis of positron emission tomography (PET) images via multi-channel generative adversarial networks (GANs). In: Cardoso, M., et al. (eds.) Molecular Imaging, Reconstruction and Analysis of Moving Body Organs, and Stroke Imaging and Treatment: Fifth International Workshop, CMMI 2017, Second International Workshop, RAMBO 2017, and First International Workshop, SWITCH 2017, Held in Conjunction with MICCAI 2017, QuÉBec City, QC, Canada, 14 September 2017, Proceedings 5. pp. 43–51. Springer, Cham (2017). https://doi.org/10.1007/978-3-319-67564-0_5
9. Spuhler, K., Serrano-Sosa, M., Cattell, R., DeLorenzo, C., Huang, C.: Full-count pet recovery from low-count image using a dilated convolutional neural network. Med. Phys. **47**(10), 4928–4938 (2020). https://doi.org/10.1002/mp.14402
10. Wang, Y., et al.: 3D conditional generative adversarial networks for high-quality pet image estimation at low dose. Neuroimage **174**, 550–562 (2018). https://doi.org/10.1016/j.neuroimage.2018.03.045
11. Zhou, L., Schaefferkoetter, J.D., Tham, I.W., Huang, G., Yan, J.: Supervised learning with CycleGAN for low-dose FDG PET image denoising. Med. Image Anal. **65**, 101770 (2020). https://doi.org/10.1016/j.media.2020.101770
12. Zhao, K., et al.: Study of low-dose pet image recovery using supervised learning with CycleGAN. PLoS ONE **15**(9), e0238455 (2020). https://doi.org/10.1371/journal.pone.0238455

13. Luo, Y., et al.: Adaptive rectification based adversarial network with spectrum constraint for high-quality PET image synthesis. Med. Image Anal. **77**, 102335 (2022). https://doi.org/10.1016/j.media.2021.102335
14. Xue, Y., Bi, L., Peng, Y., Fulham, M., Feng, D.D., Kim, J.: Pet synthesis via self-supervised adaptive residual estimation generative adversarial network. IEEE Trans. Radiat. Plasma Med. Sci. (2023). https://doi.org/10.1109/TRPMS.2023.3339173
15. Xue, Y., Peng, Y., Bi, L., Feng, D., Kim, J.: CG-3DSRGAN: a classification guided 3D generative adversarial network for image quality recovery from low-dose PET images. In: 2023 45th Annual International Conference of the IEEE Engineering in Medicine & Biology Society (EMBC), pp. 1–4. IEEE (2023). https://doi.org/10.1109/EMBC40787.2023.10341112
16. Simonyan, K., Zisserman, A.: Very deep convolutional networks for large-scale image recognition. arXiv preprint arXiv:1409.1556 (2014). https://doi.org/10.48550/arXiv.1409.1556
17. Zuo, S., Xiao, Y., Chang, X., Wang, X.: Vision transformers for dense prediction: a survey. Knowl.-Based Syst. **253**, 109552 (2022). https://doi.org/10.1016/j.knosys.2022.109552
18. Vaswani, A., et al.: Attention is all you need. In: Advances in Neural Information Processing Systems, vol. 30 (2017)
19. Liu, Z., et al.: Swin transformer: hierarchical vision transformer using shifted windows. In: 2021 IEEE/CVF International Conference on Computer Vision (ICCV), pp. 9992–10002 (2021). https://doi.org/10.1109/ICCV48922.2021.00986
20. Ali, A., et al.: XCiT: cross-covariance image transformers. Adv. Neural. Inf. Process. Syst. **34**, 20014–20027 (2021)
21. Zamir, S.W., Arora, A., Khan, S., Hayat, M., Khan, F.S., Yang, M.H.: Restormer: efficient transformer for high-resolution image restoration. In: 2022 IEEE/CVF Conference on Computer Vision and Pattern Recognition (CVPR), pp. 5718–5729 (2022). https://doi.org/10.1109/CVPR52688.2022.00564
22. Jang, S.I., et al.: Spach transformer: spatial and channel-wise transformer based on local and global self-attentions for pet image denoising. IEEE Trans. Med. Imaging **43**(6), 2036–2049 (2024). https://doi.org/10.1109/TMI.2023.3336237
23. Shao, R., Bi, X.J.: Transformers meet small datasets. IEEE Access **10**, 118454–118464 (2022). https://doi.org/10.1109/ACCESS.2022.3221138
24. Ronneberger, O., Fischer, P., Brox, T.: U-Net: convolutional networks for biomedical image segmentation. In: Navab, N., Hornegger, J., Wells, W., Frangi, A. (eds.) Medical Image Computing and Computer-Assisted Intervention–MICCAI 2015: 18th International Conference, Munich, Germany, 5–9 October 2015, Proceedings, Part III 18, pp. 234–241. Springer, Cham (2015). https://doi.org/10.1007/978-3-319-24574-4_28
25. Kingma, D.P., Ba, J.: Adam: a method for stochastic optimization. arXiv preprint arXiv:1412.6980 (2014). https://doi.org/10.48550/arXiv.1412.6980
26. Ultra-low dose pet imaging challenge (2023). https://ultra-low-dose-pet.grand-challenge.org/udpet-challenge-2023-announcement/
27. Çiçek, Ö., Abdulkadir, A., Lienkamp, S.S., Brox, T., Ronneberger, O.: 3D U-Net: learning dense volumetric segmentation from sparse annotation. In: Ourselin, S., Joskowicz, L., Sabuncu, M., Unal, G., Wells, W. (eds.) Medical Image Computing and Computer-Assisted Intervention–MICCAI 2016: 19th International Conference, Athens, Greece, 17–21 October 2016, Proceedings, Part II 19, pp. 424–432. Springer, Cham (2016). https://doi.org/10.1007/978-3-319-46723-8_49

TS-SR3: Time-Strided Denoising Diffusion Probabilistic Model for MR Super-Resolution

Zejun Wu[1], Samuel W. Remedios[2](✉), Blake E. Dewey[3], Aaron Carass[1](✉), and Jerry L. Prince[1,2]

[1] Image Analysis and Communications Laboratory, Department of Electrical and Computer Engineering, Johns Hopkins University, Baltimore, MD 21218, USA
aaron_carass@jhu.edu
[2] Department of Computer Science, Johns Hopkins University, Baltimore, MD 21218, USA
sremedi1@jhu.edu
[3] Department of Neurology, Johns Hopkins School of Medicine, Baltimore, MD 21287, USA

Abstract. Iterative refinement based image super-resolution with conditional denoising diffusion probabilistic models (DDPM), such as SR3 [21], has shown promise in the super-resolution of magnetic resonance images (MRIs). However, these methods are dependent on the inference stage of the DDPMs, which can be slow and also require hundreds of iterations to reach the desired denoising level. We address this issue by proposing a time-strided SR3 (TS-SR3) for MRI super-resolution. Traditional DDPM approaches add noise to the high-resolution (HR) image according to small steps of the variance schedule, and the accumulation of the noise results in an image resembling pure Gaussian noise. In contrast, we take larger strides in time across the same variance schedule, leading to less accumulation of noise and diffusing to a different overall noise level. At inference time, this permits us to start denoising not from full noise but with some signal still present. We propose three ways in which to generate the initial estimate where the signal is still present and evaluate the benefits of each. Our experiments show that our TS-SR3 approach is superior to a recently published super-resolution method and that our two alternative initialization approaches further improve results.

Keywords: DDPM · Super-resolution · DDIM · MRI

1 Introduction

Anisotropic low-resolution (LR) magnetic resonance (MR) image volumes are frequently acquired due to the limitations of high-resolution (HR) imaging—i.e. increased scanning time, higher costs, susceptibility to patient movement,

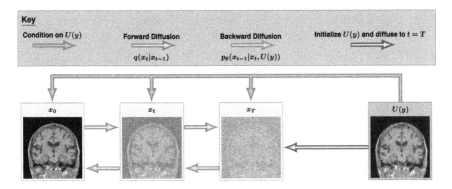

Fig. 1. The figure shows the framework of the proposed approach. Notice how x_T does not result in a pure-noise image in this example, for $T = 160$. The condition $U(y)$ is provided during training and inference for all t.

and hardware constraints [2]. While trained radiologists are able to interpret and make diagnoses from anisotropic acquisitions, automated algorithms typically require isotropic resolutions to perform reliably.

There has been considerable work done in creating HR images from LR input using super-resolution (SR) with various reported methods [4,6,12,16,18–20]. Generative adversarial networks (GANs) have been employed for SR in medical imaging [12,13], but often face challenges such as complex training procedures and mode collapse [1,17]. Alternatively, denoising diffusion probabilistic models (DDPMs), which convert random noise into a structured output over many iterations through a reverse Markov process, have shown promising results in enhancing the quality of super-resolved images [5,14,22,26,27]. SR3 [21], a pioneering DDPM method, incorporates a conditional LR image to guide the reverse Markov process during SR. Despite the use of the conditional LR images, the stability of DDPM training for MRI SR remains challenging; to address this, a residual prediction strategy was proposed [10]. The bottleneck to these conditional DDPMs being more widely used is the time-consuming nature of the inference problem [3,5,26], which we address in this paper.

Several methods have been proposed to expedite the inference stage of DDPMs, including denoising diffusion implicit models (DDIM) [23] and early stopping along the noise trajectory. To the best of our knowledge, early stopping has been implemented in two varieties in the literature: early-stopped DDPMs (ES-DDPMs) [11] and truncated DDPMs [29]. These approaches initiate the reverse denoising process prematurely, starting from non-Gaussian partially noisy images. Both ES-DDPMs and truncated DDPMs use either a variational autoencoder (VAE) [7] or a GAN, the difference between them being that the truncated DDPM generates directly from the VAE or GAN, while ES-DDPM applies some diffusion steps to the VAE or GAN based noisy images.

Inspired by ES-DDPMs, we notice that in the SR problem, relying on a generative model as the initial guess for the early diffusion is not necessary.

The observed LR image can serve as the initial guess and can be injected early in the noise trajectory. Crucially, there exists a time along the noise trajectory where the true HR image with added noise and a noisy version of the LR image are perceptually similar. In this paper, we investigate the feasibility of injecting a transformation of the LR image in the diffusion noise trajectory by changing the *stride* of the timestep. We name our method Time-Strided SR3 (TS-SR3). Experimentally, we show that TS-SR3 not only achieves satisfactory qualitative and quantitative results, but also outperforms Res-SR3 [21], a competitive DDPM approach. Traditional DDPMs train to denoise through 1000 timesteps, and we empirically find that 160 timesteps is sufficient to produce satisfactory results. Additionally, we find that DDIM is still applicable to TS-SR3, further accelerating the sampling.

We also explore which image to inject early during the SR process, with three choices of injected images: 1) bicubic upsampling, 2) pre-super-resolution (pre-SR) with an auxiliary network, and 3) a "cascade" where we iteratively use results of TS-SR3 and inject the estimate from a previous run back into the trajectory.

2 Methodology

2.1 Time-Strided Variance Schedule

The forward process in DDPMs defines how to take a clean image x_0 and produce x_t, a noisy image at time t. While this process is a Markov chain, following Ho et al. [5] it is possible to directly produce x_t from x_0 in one step:

$$q(x_t|x_0) = \mathcal{N}(x_t; \sqrt{\bar{\alpha}_t}x_0, (1-\bar{\alpha}_t)\mathbf{I}). \quad (1)$$

Since $\bar{\alpha}_t = \prod_{s=0}^{t}(1-\beta_s)$, this process is entirely determined by the variance schedule β_t, the amount of noise present at time t. The total number of diffusion timesteps T is conventionally set to 1000 [5] and β_1 and β_T are fixed. This induces a specific choice of Δt on the variance schedule; ideally, Δt is small and the denoising task has an "easier" job.

We propose to *"time-stride"* the diffusion process during both training and inference, as shown in Fig. 1. We set the total number of timesteps T to much smaller numbers, leading to larger Δt. From Fig. 1, we see that by selecting larger Δt, the image x_T does not resemble Gaussian noise and instead has the appearance of a very noisy MR acquisition. Indeed, time-striding a linear variance schedule is sufficient to impact the amount of noise present in x_T. We show this in Fig. 2.

During inference, our proposed TS-SR3 begins from x_T and denoises in steps Δt as in training. Like the base model SR3, our approach is conditioned on the LR image and we estimate the denoised image as

$$\hat{x}_{t-1} = \frac{1}{\sqrt{\alpha_t}}\left(\hat{x}_t - \frac{1-\alpha_t}{\sqrt{1-\bar{\alpha}_t}}\epsilon_\theta^{(t)}(x,\hat{x}_t,\bar{\alpha}_t)\right) + \sqrt{1-\alpha_t}\epsilon, \quad (2)$$

where $\epsilon \sim \mathcal{N}(\mathbf{0},\mathbf{I})$. Similar to SR3, we implement conditioning as concatenation.

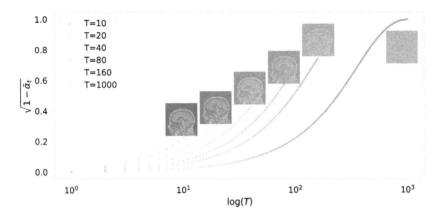

Fig. 2. Visualization of the effect of varying the number of timesteps T and the resulting amount of noise $\sqrt{1-\bar{\alpha}_t}$. Although the variance schedule is linear, due to the cumulative noise application the final amount of noise along the trajectory is nonlinear. The horizontal axis is on a logarithmic scale.

2.2 Initial Estimate for $\hat{x}_0$

Inference with our proposed TS-SR3 framework requires starting from $\hat{x}_T$, and according to the forward diffusion process described in Eq. 1, this requires an initial estimate of the HR image $\hat{x}_0$. We evaluate three choices for $\hat{x}_0$: 1) $U(y)$, the upsampled LR image with bicubic interpolation; 2) $f_\theta(y)$, a super-resolved image using a pretrained auxiliary network f_θ; and 3) $\hat{x}_0^{(i-1)}$, a super-resolved image using the result of TS-SR3 at a previous iteration $(i-1)$.

The second choice requires an external pretrained SR network, similar to SRDiff [10]. We denote this method as Pre-SR TS-SR3. The third choice here begins with $\hat{x}_0^{(0)}$ estimated with TS-SR3 and $\hat{x}_0 = U(y)$. Since this involves a *cascade* of the same network multiple times, we denote it CAS TS-SR3. In our experiments, we show the choice of initial estimate $\hat{x}_0$ matters, and some choices can help improve the performance of TS-SR3.

2.3 Applying a 2D Model to 3D Data

The DDPM we use in our work is a 2D model, taking images of a single channel and two spatial dimensions and producing images of a single channel and two dimensions. However, we are interested in improving anisotropic 3D MR imaging volumes. To address this, we borrow the approach used in Zhao et al. [28] and Remedios et al. [20]: 2D LR through-plane slices are independently super-resolved, then stacked to create a volume. To mitigate slice inconsistency issues, two super-resolved volumes are created in this manner for each of the cardinal through-planes; e.g., for an axially acquired volume, the two cardinal LR through-planes are sagittal and coronal. These two super-resolved volumes are then averaged to produce the final result.

Table 1. Distribution of MRI volumes used for training and testing from the *3D-MR-MS*, *OASIS3*, *IXI*, and *HCP* datasets.

Dataset	Train			Test		
	T_1-w	T_2-w	FLAIR	T_1-w	T_2-w	FLAIR
OASIS3	36	0	0	3	0	0
HCP	7	7	0	2	2	0
IXI	21	0	0	3	0	0
3D-MR-MS	0	0	27	0	0	3
Total	64	7	27	8	2	3

3 Experiments

3.1 Data and Processing

We selected fluid attenuated inversion recovery (FLAIR), T1-weighted (T_1-w), and T2-weighted (T_2-w) MR images from multiple sites for our experiments. Our datasets came from the *3D-MR-MS Dataset* [9], the *OASIS3 Dataset* [8], the *IXI Dataset*[1], and the *Human Connectome Project (HCP) Dataset* [24]. Table 1 shows the breakdown of the training and testing data used.

We simulated LR data from HR data to enable quantitative reference-based metrics. Simulation of LR data was provided by convolving the HR data in the through-plane direction using a Shinnar-Le Roux pulse sequence with a slice-selection profile thickness of 3 mm [15]. The slice separation was set at 4 mm, achieved through cubic B-spline subsampling. These two steps combined result in data with a slice thickness of 3 mm and a gap between slices of 1 mm, and correspondingly we have a SR factor of 4—the inplane data is 1 × 1 mm. All images in our datasets had their backgrounds removed.

3.2 How Large of a Stride?

We wish to know what size of a stride in time is feasible. To test this, we empirically evaluated the number of timesteps $T \in \{10, 20, 40, 80, 160\}$. As a baseline comparison method, we compared to Res-SR3 [21,26], an SR3-based approach which predicts the residual of a bicubic-interpolated image rather than the image itself, since it has been shown that estimating the residual stabilizes training. The Res-SR3 approach used $T = 1000$ timesteps. In all cases, $\beta_1 = 10^{-5}$ and $\beta_T = 10^{-2}$.

Notably, Res-SR3 with $T = 1000$ achieves a PSNR of 33.32 and SSIM of 0.9545, while TS-SR3 with $T = 160$ achieves a PSNR of 33.17 and SSIM of 0.9534. Thus TS-SR3 achieves results similar to Res-SR3 while using only 16% of the inferences time—i.e., a speed-up factor of six. Such performance is consistent with the use of the DDIM sampler as well; this can also be seen in Fig. 3.

---
[1] https://brain-development.org/ixi-dataset/.

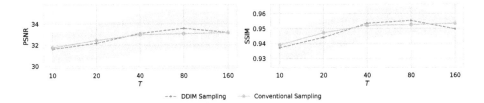

Fig. 3. This figure shows a comparison of the impact of the DDIM sampler under the time-strided training. TS-SR3 with conventional inference sampling is shown as solid orange lines, and TS-SR3 with an 8-step DDIM sampler is shown as dashed blue lines. The horizontal axis denotes the number of training diffusion timesteps T, and the vertical axis is the comparison metric PSNR or SSIM for the left and right subfigures, respectively. Shaded regions represent the standard deviation and the center dot represents the mean. All metrics are computed volume-wise over the 11 test volumes. (Color figure online)

Regardless of the choice of T, DDIM sampling achieves similar performance according to both PSNR and SSIM.

3.3 What to Inject?

As mentioned in Sect. 2.2, we investigated three choices for $\hat{x}_0$. Bicubic interpolation was our first choice. For our second choice, we used $f_\theta(y)$ to pre-super resolve the image before injection. Like SRDiff [10], we used a pre-trained RRDB [25] for f_θ. Our third choice was the cascaded result after a single iteration. Finally, we considered injecting the true HR image x_0 to study the upper bound of our approach. These results are illustrated in Fig. 4. As expected, injecting the true HR image yielded the best results, especially for $T = 10$ for which the image x_T has little noise at all. However, as T increased, all injection methods become more similar in quality after the reverse diffusion process.

For the cascaded model we evaluated two scenarios. The first was to use the same model twice during inference: iteration zero injected the bicubic interpolated image $U(y)$ and super-resolved with TS-SR3, and iteration one injected the output of iteration zero. Second, we considered leveraging the utility of residual predictions. Time-striding is not viable for residual DDPMs conventionally because no such estimate of a noisy residual exists at inference time. However, given a pre-trained model to estimate the residual, we can now cascade these results. We term this method "rTS-SR3" for "residual time-strided SR3" and include the results in Table 2. Both cascaded methods yielded superior results to single-pass methods; the third row requires only one DDPM model and achieves higher PSNR and SSIM compared to the single-pass method of the same model in the same number of DDIM steps, and the fourth row yields the best results in terms of PSNR and SSIM, but requires two independent pre-trained DDPM models.

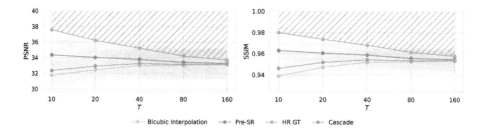

Fig. 4. This figure shows a comparison of injected images. The gray, shaded region shows the true HR ground truth (GT) x_T. This is the upper bound for injection images, since it is unavailable at inference time. In red, we show the pre-SR method; in green, we show the cascade approach; in orange, we show bicubic interpolation. Shaded regions represent standard deviations and the center dots represent the mean. All metrics are computed volume-wise over the 11 test volumes. (Color figure online)

Table 2. This table shows results from the cascade experiment. All methods in the table use a total of 16 denoising steps during inference, but varying T during training (written in parentheses). The first and second rows do not use cascades, and so the estimate of $\hat{x}_0^{(0)}$ is not applicable (N/A). The first row is Res-SR3 trained with $T = 1000$ timesteps, the conventional DDPM with residual prediction, and an aggressive 16 DDIM sampling timesteps. The second row is our proposed TS-SR3 with $T = 80$ training timesteps and using the bicubic interpolated image $U(y)$ to estimate $\hat{x}_0^{(0)}$. The third row is a cascaded approach, where 8 DDIM steps produce the estimate $\hat{x}_0^{(0)}$ and 8 more steps are used to produce $\hat{x}_0^{(1)}$. The final row requires two pre-trained DDPM models, starting the initial guess $\hat{x}_0^{(0)}$ from row 1 (but with half as many DDIM steps) and then the residual TS-SR3 model to estimate $\hat{x}_0^{(1)}$.

$\hat{x}_0^{(0)}$		$\hat{x}_0^{(1)}$		PSNR	SSIM
Model	DDIM Steps	Model	DDIM Steps		
N/A	N/A	Res-SR3 ($T = 1000$)	16	33.60	0.9581
$U(y)$	N/A	TS-SR3 ($T = 80$)	16	33.34	0.9505
TS-SR3 ($T = 80$)	8	TS-SR3 ($T = 80$)	8	33.72	0.9549
Res-SR3 ($T = 1000$)	8	rTS-SR3 ($T = 100$)	8	33.96	0.9604

3.4 Comparison with Existing Methods

In Table 3, we provide a comparison between the state-of-the-art super-resolution in anisotropic MR super-resolution SMORE [20], the residual version of SR3 (Res-SR3), and our proposed method TS-SR3 with both types of cascades. SMORE is a self-supervised SR method for MR images, which means it does not use external training data in its SR process. In contrast the SR3 based methods we present in Table 3 are fully-supervised—i.e. they use external training data. These fully-supervised methods include Res-SR3 ($T = 1000$) with a 16-step DDIM sampler, the cascaded TS-SR3 ($T = 160$), and the pre-SR rTS-SR3 ($T = 100$). Another key difference between these methods is that SMORE is

Table 3. Model comparison for various SR methods. See Fig. 5 for a qualitative comparison of the results. Bolded metrics are the best; regarding inference time, both Res-SR3 and the Cascaded TS-SR3 are bolded since differences in runtime are likely due to environmental inconsistencies. Key: CA - Contrast-agnostic; ED - External Data; AM - Additional Model.

Method	CA	ED	AM	DDIM Steps	Time (s)	PSNR	SSIM
SMORE [20]	✓	✗	✗	N/A	1760	33.10	0.9566
Res-SR3	✗	✓	✗	16	**416**	33.60	0.9581
Cascaded TS-SR3	✗	✓	✗	16	**413**	33.96	0.9604
Pre-SR rTS-SR3	✗	✓	✓	16	462	**34.53**	**0.9643**

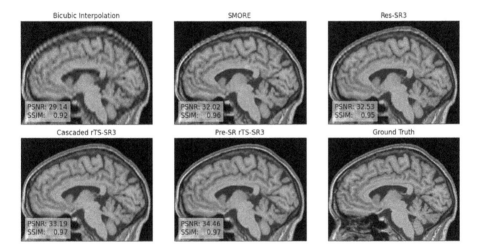

Fig. 5. Qualitative comparison between super-resolution methods of a representative subject from the IXI dataset. The PSNR and SSIM values shown are computed over the entire volume.

contrast and resolution agnostic, while the SR3 based methods are not. All the fully-supervised methods achieve better performance and speed for in-domain inference in Table 3. The difference in speed between Pre-SR rTS-SR3 and the other two comparable methods is due to the inference time for Pre-SR; otherwise the number of DDIM steps is identical for the last three rows of Table 3. The best performing method is pre-SR rTS-SR3, though it requires two independent pre-trained models.

4 Discussion and Conclusion

We proposed a time-striding framework for the super-resolution of anisotropic MR images by training SR3 [21] with fewer timesteps. Three proposed models, TS-SR3, Pre-SR TS-SR3, and a cascaded TS-SR3 are tested in our experiments, where the difference between these methods is the initial estimate $\hat{x}_0$. There are limitations to our work. Since TS-SR3 must train on a dataset, its use is limited to within the training domain. We did not evaluate to what extent TS-SR3 can generalize outside its training domain. In particular we only considered one slice thickness and gap, and more work is needed to evaluate this model at other resolutions, or determine if re-training is necessary.

Our performance speed is relatively fast compared to the baseline SMORE, but slower than other direct, non-iterative SR approaches such as pure CNNs, transformers, or GANs. There are also other works which accelerate the inference process of DDPMs which we intend to explore in the future.

In conclusion our results show that a cascade strategy yields promising results without the reliance on an external model, and the Pre-SR TS-SR3 achieves strong results albeit with an external model. In the future, more experiments could be done to find an optimal strategy that balances sampling time and performance trade-off; also the generalizability of the model to other resolutions, contrasts, and anatomies will be evaluated.

Acknowledgments. This material is supported by the National Science Foundation Graduate Research Fellowship under Grant No. DGE-1746891 (Remedios). This work also received support from NMSS RG-1907-34570 (Pham), FG-2008-36966 (Dewey), and CDMRP W81XWH2010912 (Prince). Data were provided [in part] by the Human Connectome Project, WU-Minn Consortium (Principal Investigators: Van Essen and Ugurbil; 1U54MH091657) funded by the 16 NIH Institutes and Centers that support the NIH Blueprint for Neuroscience Research; and by the McDonnell Center for Systems Neuroscience at Washington University.

References

1. Arjovsky, M., et al.: Wasserstein generative adversarial networks. In: International Conference on Machine Learning, pp. 214–223. PMLR (2017)
2. Du, J., et al.: Super-resolution reconstruction of single anisotropic 3D MR images using residual convolutional neural network. Neurocomputing **392**, 209–220 (2020)
3. Gandikota, R., Brown, N.: Pro-DDPM: progressive growing of variable denoising diffusion probabilistic models for faster convergence. In: BMVC, p. 121 (2022)
4. Guo, L., et al.: Self-supervised super-resolution of 2-D pre-clinical MRI acquisitions. In: Medical Imaging 2024: Clinical and Biomedical Imaging, vol. 12930, pp. 652–658. SPIE (2024)
5. Ho, J., et al.: Denoising diffusion probabilistic models. Adv. Neural. Inf. Process. Syst. **33**, 6840–6851 (2020)
6. Jog, A., Carass, A., Prince, J.L.: Self super-resolution for magnetic resonance images. In: Ourselin, S., Joskowicz, L., Sabuncu, M.R., Unal, G., Wells, W. (eds.) MICCAI 2016. LNCS, vol. 9902, pp. 553–560. Springer, Cham (2016). https://doi.org/10.1007/978-3-319-46726-9_64
7. Kingma, D.P., Welling, M.: Auto-encoding variational Bayes. arXiv preprint arXiv:1312.6114 (2013)
8. LaMontagne, P.J., et al.: OASIS-3: longitudinal neuroimaging, clinical, and cognitive dataset for normal aging and Alzheimer disease. MedRxiv, 2019–12 (2019). https://doi.org/10.1101/2019.12.13.19014902

9. Lesjak, Ž, et al.: A novel public MR image dataset of multiple sclerosis patients with lesion segmentations based on multi-rater consensus. Neuroinformatics **16**, 51–63 (2018)
10. Li, H., et al.: SRDiff: single image super-resolution with diffusion probabilistic models. Neurocomputing **479**, 47–59 (2022)
11. Lyu, Z., et al.: Accelerating diffusion models via early stop of the diffusion process. arXiv preprint arXiv:2205.12524 (2022)
12. Mahapatra, D., Bozorgtabar, B., Hewavitharanage, S., Garnavi, R.: Image super resolution using generative adversarial networks and local saliency maps for retinal image analysis. In: Descoteaux, M., Maier-Hein, L., Franz, A., Jannin, P., Collins, D.L., Duchesne, S. (eds.) MICCAI 2017. LNCS, vol. 10435, pp. 382–390. Springer, Cham (2017). https://doi.org/10.1007/978-3-319-66179-7_44
13. Mahapatra, D., et al.: Image super-resolution using progressive generative adversarial networks for medical image analysis. Comput. Med. Imaging Graph. **71**, 30–39 (2019)
14. Mao, Y., et al.: DisC-Diff: disentangled conditional diffusion model for multi-contrast MRI super-resolution. arXiv preprint arXiv:2303.13933 abs/2303.13933 (2023)
15. Martin, J., et al.: SigPy. RF: comprehensive open-source RF pulse design tools for reproducible research. In: Proceedings of the International Society for Magnetic Resonance in Medicine. ISMRM Annual Meeting, vol. 1045 (2020)
16. Peled, S., Yeshurun, Y.: Suprresolution in MRI: application to human white matter fiber tract visualization by diffusion tensor imaging. Mag. Reson. Med. **45**, 29–35 (2001)
17. Ravuri, S., Vinyals, O.: Classification accuracy score for conditional generative models. In: Advances in Neural Information Processing Systems, vol. 32 (2019)
18. Remedios, S.W., et al.: Deep filter bank regression for super-resolution of anisotropic MR brain images. In: Wang, L., Dou, Q., Fletcher, P.T., Speidel, S., Li, S. (eds.) 25th International Conference on Medical Image Computing and Computer Assisted Intervention (MICCAI 2022). LNCS, vol. 13436, pp. 613–622. Springer, Cham (2022). https://doi.org/10.1007/978-3-031-16446-0_58
19. Remedios, S.W., et al.: Joint image and label self-super-resolution. In: Svoboda, D., Burgos, N., Wolterink, J.M., Zhao, C. (eds.) Simulation and Synthesis in Medical Imaging: 6th International Workshop, SASHIMI 2021, Held in Conjunction with MICCAI 2021, Strasbourg, France, 27 September 2021, Proceedings 6, pp. 14–23. Springer, Cham (2021). https://doi.org/10.1007/978-3-030-87592-3_2
20. Remedios, S.W., et al.: Self-supervised super-resolution for anisotropic MR images with and without slice gap. In: Wang, L., Dou, Q., Fletcher, P.T., Speidel, S., Li, S. (eds.) Medical Image Computing and Computer Assisted Intervention – MICCAI 2022. MICCAI 2022. LNCS, vol. 13436, pp. 613–622. Springer, Cham (2022). https://doi.org/10.1007/978-3-031-16446-0_58
21. Saharia, C., et al.: Image super-resolution via iterative refinement. IEEE Trans. Pattern Anal. Mach. Intell. **45**(4), 4713–4726 (2022)
22. Sohl-Dickstein, J., et al.: Deep unsupervised learning using nonequilibrium thermodynamics. In: International Conference on Machine Learning, pp. 2256–2265. PMLR (2015)
23. Song, J., et al.: Denoising diffusion implicit models. arXiv preprint arXiv:2010.02502 (2020)
24. Van Essen, D.C., et al.: The WU-Minn human connectome project: an overview. Neuroimage **80**, 62–79 (2013)

25. Wang, X., et al.: ESRGAN: enhanced super-resolution generative adversarial networks. In: Leal-Taixé, L., Roth, S. (eds.) ECCV 2018. LNCS, vol. 11133, pp. 63–79. Springer, Cham (2019). https://doi.org/10.1007/978-3-030-11021-5_5
26. Wu, Z., et al.: AniRes2D: anisotropic residual-enhanced diffusion for 2D MR super-resolution. In: Medical Imaging 2024: Clinical and Biomedical Imaging, vol. 12930, pp. 567–574. SPIE (2024)
27. Wu, Z., et al.: Super-resolution of brain MRI images based on denoising diffusion probabilistic model. Biomed. Signal Process. Control **85**, 104901 (2023)
28. Zhao, C., et al.: SMORE: a self-supervised anti-aliasing and super-resolution algorithm for MRI using deep learning. IEEE Trans. Med. Imaging **40**(3), 805–817 (2020)
29. Zheng, H., et al.: Truncated diffusion probabilistic models and diffusion-based adversarial auto-encoders. arXiv preprint arXiv:2202.09671 (2022)

PDM: A Plug-and-Play Perturbed Multi-path Diffusion Module for Simultaneous Medical Image Segmentation Improvement and Uncertainty Estimation

Bo Zhou[1,2(✉)], Tianqi Chen[3,4], Jun Hou[3,4], Yinchi Zhou[2], Huidong Xie[2], Chi Liu[2,3], and James S. Duncan[2,3]

[1] Department of Radiology, Northwestern University, Chicago, USA
bo.zhou@northwestern.edu
[2] Department of Biomedical Engineering, Yale University, New Haven, USA
[3] Department of Radiology and Biomedical Imaging, Yale University, New Haven, USA
[4] Department of Computer Science, University of California Irvine, Irvine, USA

Abstract. Segmentation is a key step in medical image analysis. However, previous state-of-the-art segmentation deep models are largely deterministic, i.e. outputting a class-wise binary segmentation without uncertainty estimation. Further improvement for these already deployed segmentation models with uncertainty estimation is therefore highly desirable. In this work, we proposed a simple and efficient plug-and-play module, named Perturbed Multi-path Diffusion Model (PDM), that can be directly concatenated to segmentation results from previous methods to simultaneously improve the segmentation and produce a segmentation uncertainty estimation. In this module, we first randomly perturb the segmentation input with varied morphological operations, generating multiple perturbed segmentation. Then, based on the conditional Denoising Diffusion Probabilistic Model (cDDPM), we proposed to use the noise-added perturbed segmentation as initial inputs to multiple cDDPM reverse paths. The final outputs of these paths are ensemble to produce an improved segmentation, where the uncertainty is also calculated via the pixel-wise standard deviation from these reverse results. Lastly, we design a cascade framework with densely connected averaging where PDM are embedded for further improving the performance. We collected three different segmentation datasets and demonstrated that our proposed method can consistently improve the previous methods' segmentation while producing reasonable uncertainty maps that are potentially useful in clinics.

Keywords: Plug-and-Play · Diffusion Model · Segmentation · Uncertainty

1 Introduction

Automatic segmentation in medical imaging is an important post-processing step for diagnosis and treatment planning and thus has been widely studied for different clinical applications, such as in oncology, cardiology, and neurology [1]. UNet and its variants have been extensively investigated and have shown state-of-the-art performance for these clinical applications [2–4]. However, most of these methods' segmentation processes are deterministic, meaning they directly generate the class-wise binary segmentation. A parallel estimation of segmentation uncertainty is highly desirable for these black-box models in clinics, especially for applications such as radiation therapy treatment planning. Even though previous uncertainty estimation methods can be applied, e.g. Bayesian [5,6] or Monte Carlo Dropout strategies [7,8], they would require architecture modifications and model retraining which could be infeasible for already deployed deterministic models [9]. On the other hand, the diffusion model has been emerging as a powerful alternative to CNN that has shown promising performance in image generation [10–12]. Previous works have attempted to adapt the diffusion model for medical imaging segmentation [13,14]. While an uncertainty map can be generated from the diffusion model with different random noise initialization, its segmentation performance is still non-ideal as compared to the previous state-of-the-art UNet segmentation method, such as nnUNet [4]. Moreover, previous UNet-based segmentation methods and diffusion-based methods were often employed individually rather than collectively, which may lead to better performance.

To overcome these challenges, we proposed a simple and efficient plug-and-play module, named Perturbed Multi-path Diffusion Model (PDM), that can be directly concatenated to segmentation results from previous methods to simultaneously improve the segmentation results while producing a segmentation uncertainty estimation. In each PDM block, we first randomly perturb the segmentation input with varied morphological operations, generating multiple perturbed segmentation. Then, we revised the reverse process in the conditional Denoising Diffusion Probabilistic Model [10,11] by using the perturbed segmentation with noise as the initial input. The reverse results obtained from different perturbed segmentation with different noises are then ensemble to produce the improved segmentation. Meanwhile, the uncertainty map can be also generated by computing the pixel-wise standard deviation from these reverse results. To further improve the performance, we also propose to apply the PDM module in a cascaded manner with densely connected averaging steps. Our experimental results on three different segmentation tasks demonstrated that the proposed method can consistently improve the segmentation results from previous methods and produce accurate uncertainty maps.

2 Methods

2.1 Perturbed Multi-path Diffusion Module

The design of our Perturbed Multi-path Diffusion Model (PDM) and how it is used as a plug-and-play module for the prior segmentation methods is illustrated in Fig. 1. The input to the PDMs is the segmentation output from prior methods, e.g. nnUNet [4]. Each PDM consists of multiple conditional denoising diffusion probabilistic models (DDPMs) to improve this prior segmentation based on a cascaded and multi-path fashion. The training and inference details are as follows.

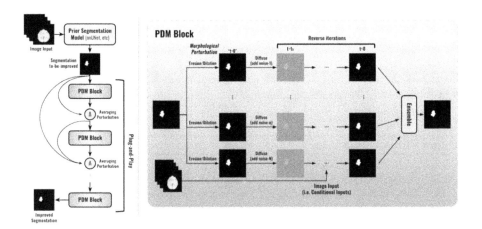

Fig. 1. The architecture of our plug-and-play perturbed multi-path diffusion module (PDM). The flow for improving prior segmentation is shown on the left, and the details of the PDM block are shown on the right.

Training: The building block of the PDM is simply the conditional DDPM with the image-to-be-segmented as conditional input. Denoting the input image as x and the segmentation ground truth as y_0. Like the conventional DDPM [10,11], the diffusion model consists of a forward diffusion process and a reverse denoising process. The forward diffusion process gradually adds Gaussian noise to the ground truth segmentation y_0 over T iterations, defined as:

$$q(y_{1:T}|y_0) = \prod_{t=1}^{T} q(y_t|y_{t-1}), \quad (1)$$

where $q(y_{t+1}|y_t) = \mathcal{N}(y_{t-1}; \sqrt{\alpha_t}y_{t-1}, (1-\alpha_t)I)$, and α_t are the noise schedule parameters. T is empirically set to 1000 here such that y_T is visually indistinguishable from Gaussian noise. Then, the forward process can be further marginalized at each step as:

$$q(y_t|y_0) = \mathcal{N}(y_t; \sqrt{\gamma_t}y_0, (1-\gamma_t)I), \quad (2)$$

where $\gamma_t = \prod_{s=0}^{t} \alpha_s$. Then, the posterior distribution of y_{t-1} given (y_0, y_t) can be derived as:

$$q(y_{t-1}|(y_0, y_t)) = \mathcal{N}(y_{t-1}|\mu, \sigma^2 I), \quad (3)$$

where $\mu = \frac{\sqrt{\gamma_{t-1}}(1-\alpha_t)}{1-\gamma_t}y_0 + \frac{\sqrt{\alpha_t}(1-\gamma_{t-1})}{1-\gamma_t}y_t$ and $\sigma^2 = \frac{(1-\gamma_{t-1})(1-\alpha_t)}{1-\gamma_t}$. With this, the noisy image during the forward process can thus be written as

$$\hat{y}_t = \sqrt{\gamma_t}y_0 + \sqrt{1-\gamma_t}\epsilon \tag{4}$$

where $\epsilon \sim \mathcal{N}(0, I)$. Here, the goal is to estimate the noise and thus gradually remove it during the reverse process to recover the target image y_0. In our conditional diffusion model, we utilized a network $f_{dm}(\cdot)$ to estimate the noise with a L2 loss function:

$$\mathcal{L}_{dm} = ||f_{dm}(x, \hat{y}_t, \gamma_t) - \epsilon||_2^2 \tag{5}$$

where x is the input image used as conditional input here. $\hat{y}_t$ is the noisy segmentation, and γ_t is the current noise level.

Algorithm 1: Plug-and-Play with PDM — Inference

Input: $x \in N^{d_1 \times d_2}$: segmentation prediction from previous network
Initialize #1: $t_s \in [0, T]$: the start timestep of denoising process
Initialize #2: N_c: # of cascade; N_p: # of path; K: morphological kernel
Initialize #3: $f_{dm}(\cdot)$: conditional diffusion network
for $c = 1, 2, 3, ..., N_c$ do
  if $c = 1$ then
    $y_0^c = x$; ▷ Initial input from prior segmentation
  else
    $y_0^c = \frac{1}{c}(y_0^{ens} + \sum_{n=1}^{c-1} y_0^{c-1})$; ▷ Averaging perturbation
  $y_0^c(i,j) = 1$ if $y_0^c(i,j) > 0.5$, else 0 ; ▷ Segmentation with Threshold
  for $p = 1, 2, 3, ..., N_p$ do
    if $\mathcal{N}(0, I) > 0.5$ then
      $y_0^c = y_0^c \oplus K$; ▷ Dilation perturbation
    else
      $y_0^c = y_0^c \ominus K$; ▷ Erosion perturbation
    $y_{t_s}^p = \sqrt{\gamma_{t_s}} y_0^c + \sqrt{1-\gamma_{t_s}} \epsilon_p$, $\epsilon_p \sim \mathcal{N}(0, I)$; ▷ Adding noise till t_s
    for $t = t_s, t_s - 1, t_s - 2, ..., 0$ do
      $\epsilon_t \sim \mathcal{N}(0, I)$ if $t > 0$, else $\epsilon_t = 0$
      $y_{t-1}^p = \frac{1}{\sqrt{\alpha_t}}(y_t^p - \frac{1-\alpha_t}{\sqrt{1-\gamma_t}} f_{dm}(x, y_t^p, \gamma_t)) + \sqrt{1-\alpha_t}\epsilon_t$
  $y_0^{ens} = \frac{1}{N_p} \sum_{p=1}^{N_p} y_0^p$; ▷ Ensemble multiple predictions

---

Inference: Once the network $f_{dm}(\cdot)$ in conditional DDPM are well trained, we can directly plug-and-play it for improving the prior segmentation. The inference pipeline is illustrated in Fig. 1 which consists of four key parts. First, given the prior segmentation y_{prior}, we apply random morphological perturbations, with:

$$y_{prior} = \begin{cases} y_{prior} \oplus K, & \text{if } \mathcal{N}(0, I) > 0.5. \\ y_{prior} \ominus K & \text{otherwise.} \end{cases} \tag{6}$$

where $\oplus$ is the dilation operator and $\ominus$ is the erosion operator. K is a circle kernel with a radius of 2. This process generates multiple y_{prior} for the diffusion model. Second, instead of starting the reverse process from a standard normal distribution $\mathcal{N}(y_T|0, I)$ at T for a single path, we start the reverse process at a pre-defined time point $t_s \in [0, T]$ with

$$\hat{y}_{t_s} = \sqrt{\gamma_{t_s}} y_{prior} + \sqrt{1 - \gamma_{t_s}} \epsilon_{prior} \tag{7}$$

where y_{prior} is the perturbed prior segmentation from Eq. 6, and $\epsilon_{prior} \sim \mathcal{N}(0, I)$. By rearranging Eq. 4, we can approximate the target segmentation y_0 as

$$y_0 = \frac{y_t - \sqrt{1 - \gamma_t} f_{dm}(x, \hat{y}_t, \gamma_t)}{\sqrt{\gamma_t}}. \tag{8}$$

Then, by substituting this estimation of y_0 into the posterior distribution of $q(y_{t-1}|(y_0, y_t))$ in Eq. 3. Each iteration of the reverse process can be formulated as

$$y_{t-1} = \frac{1}{\sqrt{\alpha_t}}(y_t - \frac{1-\alpha_t}{\sqrt{1-\gamma_t}} f_{dm}(x, y_t, \gamma_t)) + \sqrt{1-\alpha_t}\epsilon_t \tag{9}$$

where $\epsilon_t \sim \mathcal{N}(0, I)$. The starting reverse time point t_s is empirically set depending on the segmentation application. Thirdly, given multiple perturbed prior segmentation from Eq. 6, we perform multiple reverse paths starting at t_s, and ensemble these multi-path predictions with

$$y_0^{ens} = \frac{1}{N_p} \sum_{p=1}^{N_p} y_0^p, \tag{10}$$

where y_0^p is the prediction from a single reverse path and N_p is the number of paths. Lastly, to further improve the segmentation, we perform this operation in a cascaded style along with a densely connected averaging strategy. Specifically, each cascade consists of the same PDM structure. The prior segmentation for inputting into the next cascade is the averaged segmentation probabilities from the previous cascade's outputs. The algorithm is summarized in Algorithm 1.

2.2 Data Preparation and Evaluation Strategy

We collected three different datasets to validate our method. In the first dataset, we collected MRI brain tumor patient data from the BraTs challenge [15], consisting of 1,251 patient samples. Each patient sample consists of four different MRI sequences, i.e. T1, T1CE, T2, FLAIR, and we stacked them into a 4-channel volume. The provided ground truth labels contain four classes, which are background, GD-enhancing tumor, peritumoral edema, and necrotic and non-enhancing tumor core. We merged three different tumor classes into one class and therefore define the segmentation problem as a pixel-wise binary classification. The training set includes 73,065 axial images originating from 1,000 patients, while the test set includes the rest of 23,240 images originating from 251 patients. In the second dataset, we collected CT kidney tumor patient data from the KiTS challenge [16], consisting of 489 CT patient samples. Here, the provided ground truth labels contain four classes, which are background, normal kidney, cyst, and tumor. Similarly, we extracted the tumor-only label and merged the rest classes as background, thus defining the segmentation problem as a pixel-wise binary classification. The training set includes 18,202 axial images from 400 patients, and the test set includes the rest of 4,049 images from 89 patients. In the last dataset, we collected MRI prostate patient data from the Prostate158 challenge [17], consisting of 139 patient samples. We extracted the T2W MRI and one of the reader's prostate labels. Specifically, the provided ground truth labels contain three classes, which are background, central gland, and peripheral zone. We merged all the prostate region labels into one class and also defined the segmentation problem as a pixel-wise binary classification as above. The training set includes 2,304 axial images originating from 100 patients, while the test set includes the rest of 889 images originating from 39 patients. Dice score (DSC) and Jaccard Index (JACC) were used to evaluate the segmentation performance. Since our method is a plug-and-play module that aims to improve the segmentation from prior methods, we chose nnUNet [4], TransUNet [18], and UNETR [19] as the baseline methods to validate the improvement from PDM. In addition, we compared with the previous pure diffusion model-based segmentation method, i.e. EnsDDPM [13]. Ablative studies for different settings in PDM, i.e. number of paths, reverse starting time point, and number of cascades, were also conducted.

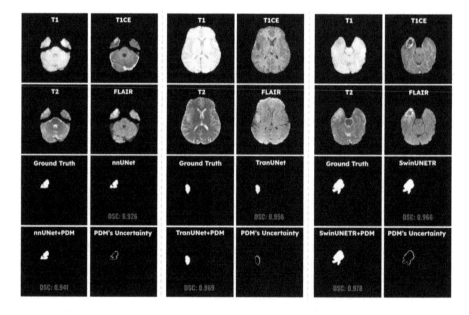

Fig. 2. Examples of segmentation improvement for different previous segmentation methods, including nnUNet (Left), TransUNet (Middle), and UNETR (Right). The PDM can not only improve the segmentation but also generate the uncertainty map of the segmentation improvements (bottom right of each patient sample).

2.3 Implementation Details

We implemented our method in Pytorch and performed experiments using an NVIDIA H100 GPU. We train the diffusion model in PDM with a batch size of 8 for 500k training steps. The Adam solver was used to optimize our models with $lr = 1 \times 10^{-4}$, $\beta_1 = 0.9$, and $\beta_2 = 0.99$. We used an EMA rate of 0.9999. A 10k linear learning rate warm-up schedule was implemented. We used a linear noise schedule with 1000 time steps.

Table 1. Quantitative comparison of segmentation results on three different segmentation datasets. $+PDM$ means adding our plug-and-play PDM to the corresponding prior segmentation.

Evaluation Segment	BraTs [15]		KiTs [16]		Prostate158 [17]	
	DSC	JACC	DSC	JACC	DSC	JACC
EnsDDPM [13]	0.902(0.189)	0.869(0.198)	0.818(0.312)	0.791(0.323)	0.912(0.262)	0.876(0.288)
nnUNet [4]	0.912(0.185)	0.887(0.203)	0.831(0.328)	0.802(0.330)	0.926(0.271)	0.897(0.297)
+ PDM	0.925(0.184)	0.898(0.201)	0.842(0.332)	0.810(0.333)	0.929(0.272)	0.899(0.298)
UNETR [19]	0.908(0.183)	0.877(0.212)	0.822(0.318)	0.797(0.329)	0.918(0.267)	0.886(0.293)
+ PDM	0.918(0.181)	0.896(0.220)	0.838(0.330)	0.808(0.332)	0.922(0.270)	0.898(0.298)
TransUNet [18]	0.884(0.192)	0.866(0.201)	0.802(0.298)	0.782(0.323)	0.902(0.260)	0.870(0.283)
+ PDM	0.905(0.189)	0.878(0.210)	0.821(0.308)	0.806(0.327)	0.918(0.267)	0.890(0.289)

3 Experimental Results

PDM can be easily applied to the previous methods' segmentation outputs in a post-processing manner. In Fig. 2, we show three visual examples of how our PDM helps improve MRI brain tumor segmentation from previous methods, including nnUNet, TransUNet, and UNETR. For example, we can see that the nnUNet slightly under-segmented the tumor, resulting in a DSC of 0.926. With our PDM plug-and-play, we can correct this under-segmentation with an improved DSC of 0.941. Similar improvements from PDM for TransUNet and UNETR can be found in the second and third examples. Notably, our PDM can also output an uncertainty map by computing the pixel-wise standard deviation from the multiple perturbed diffusion paths, as shown in the bottom right of each patient sample. The corresponding quantitative results on the whole test set are summarized in Table 1. For brain tumor segmentation with the BraTs dataset, PDM can improve the mean DSC from 0.912 to 0.925 for nnUNet, from 0.908 to 0.918 for UNETR, and from 0.884 to 0.905 for TransUNet, respectively. For kidney tumor segmentation with the KiTs dataset, PDM can improve the mean DSC from 0.831 to 0.842 for nnUNet, from 0.822 to 0.838 for UNETR, and from 0.802 to 0.821 for TransUNet, respectively. Similar enhancement can be found for the prostate segmentation task based on the Prostate158 dataset. Compared to the previous diffusion-based segmentation method, i.e. EnsDDPM [13], the enhanced CNN-based methods with PDM also consistently outperform it.

Table 2. Ablative studies on the perturbation operation in the PDM based on nnUNet prior segmentation.

Evaluation Segment	BraTs [15]		KiTs [16]	
	DSC	JACC	DSC	JACC
nnUNet [4]	0.912(0.185)	0.887(0.203)	0.831(0.328)	0.802(0.330)
+ PDM w/o Perturb	0.918(0.185)	0.890(0.203)	0.834(0.331)	0.803(0.331)
+ PDM w Perturb	0.925(0.184)	0.898(0.201)	0.842(0.332)	0.810(0.333)

We conducted ablative studies to investigate different components of PDM. Based on the segmentation from nnUNet, we firstly investigated the impact of the morphological perturbation in PDM (Table 2). We can see that adding PDM without the perturbation leads to subtle segmentation improvement, given that applying diffusion multiple times in a cascade fashion may overfit the diffusion prediction. On the other hand, PDM with perturbation leads to more robust segmentation improvement. Second, since PDM is a plug-and-play module with adjustable 1) number of paths, 2) number of cascades, and 3) reverse starting time, we studied the impact of those components (Fig. 3). Based on the BraTs dataset, we found that near-optimal performance was reached when the reverse starting time t_s was around 500 (Left Plot). The performance started to converge when the number of paths reached 10 (Middle Plot). The performance reached its peak when 3 cascades of PDM were used, and the model started to overfit after that (Right Plot).

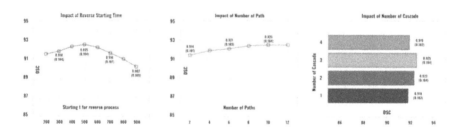

Fig. 3. Ablative studies on using different settings of PDM for improving segmentation. The impact of the reverse starting time point (Left), the impact of the number of paths (Middle), and the number of cascades (Right) are included.

4 Discussion and Conclusion

In this work, we proposed a simple and efficient plug-and-play module, called Perturbed Multi-path Diffusion Module (PDM), for simultaneous segmentation improvement of prior segmentation and estimation of pixel-wise uncertainty. We demonstrated successful applications on MRI brain tumors, CT kidney tumors, and prostate segmentation, with improvement added to three different previous segmentation methods. The current framework is implemented in 2D due to memory and computation constraints. However, we believe our method can be easily extended to 3D for further performance improvement, with either memory-efficient diffusion model strategies [20] or orthogonal dual-view 2D/2.5D diffusion model strategies [21–23]. Notably, because each path in PDM starts at a shortcut time point t_s, the inference process is significantly accelerated as compared to conventional DDPM, i.e. 1000 steps v.s. 500 steps. Since the conditional diffusion model in PDM is theoretically versatile, we believe replacing the diffusion model with other types of accelerated diffusion models, e.g. DDIM [12] and Resshift [24], could further improve the inference efficiency which is an important topic of our future studies.

References

1. Hesamian, M.H., Jia, W., He, X., Kennedy, P.: Deep learning techniques for medical image segmentation: achievements and challenges. J. Digit. Imaging **32**, 582–596 (2019)
2. Du, G., Cao, X., Liang, J., Chen, X., Zhan, Y.: Medical image segmentation based on u-net: a review. J. Imaging Sci. Technol. (2020)
3. Ronneberger, O., Fischer, P., Brox, T.: U-net: convolutional networks for biomedical image segmentation. In: International Conference on Medical Image Computing and Computer-assisted Intervention, pp. 234–241. Springer (2015)
4. Isensee, F., Jaeger, P.F., Kohl, S.A., Petersen, J., Maier-Hein, K.H.: nnu-net: a self-configuring method for deep learning-based biomedical image segmentation. Nat. Methods **18**(2), 203–211 (2021)
5. Sagar, A.: Uncertainty quantification using variational inference for biomedical image segmentation. In: Proceedings of the IEEE/CVF Winter Conference on Applications of Computer Vision, pp. 44–51 (2022)
6. Kwon, Y., Won, J.H., Kim, B.J., Paik, M.C.: Uncertainty quantification using Bayesian neural networks in classification: application to biomedical image segmentation. Comput. Statist. Data Anal. **142**, 106816 (2020)
7. Nair, T., Precup, D., Arnold, D.L., Arbel, T.: Exploring uncertainty measures in deep networks for multiple sclerosis lesion detection and segmentation. Med. Image Anal. **59**, 101557 (2020)
8. Mehrtash, A., Wells, W.M., Tempany, C.M., Abolmaesumi, P., Kapur, T.: Confidence calibration and predictive uncertainty estimation for deep medical image segmentation. IEEE Trans. Med. Imaging **39**(12), 3868–3878 (2020)
9. Abdar, M., et al.: A review of uncertainty quantification in deep learning: techniques, applications and challenges. Inf. Fusion **76**, 243–297 (2021)
10. Ho, J., Jain, A., Abbeel, P.: Denoising diffusion probabilistic models. Adv. Neural. Inf. Process. Syst. **33**, 6840–6851 (2020)
11. Saharia, C., et al.: Palette: image-to-image diffusion models. In: ACM SIGGRAPH 2022 Conference Proceedings, pp. 1–10 (2022)
12. Song, J., Meng, C., Ermon, S.: Denoising diffusion implicit models. arXiv preprint arXiv:2010.02502 (2020)

13. Wolleb, J., Sandkühler, R., Bieder, F., Valmaggia, P., Cattin, P.C.: Diffusion models for implicit image segmentation ensembles. In: International Conference on Medical Imaging with Deep Learning, pp. 1336–1348. PMLR (2022)
14. Wu, J., et al.: Medsegdiff: medical image segmentation with diffusion probabilistic model. In: Medical Imaging with Deep Learning, pp. 1623–1639. PMLR (2024)
15. Menze, B.H., et al.: The multimodal brain tumor image segmentation benchmark (brats). IEEE Trans. Med. Imaging **34**(10), 1993–2024 (2014)
16. Heller, N., et al.: The state of the art in kidney and kidney tumor segmentation in contrast-enhanced CT imaging: results of the kits19 challenge. Med. Image Anal. **67**, 101821 (2021)
17. Adams, L.C., et al.: Dataset of prostate MRI annotated for anatomical zones and cancer. Data Brief **45**, 108739 (2022)
18. Chen, J., et al.: Transunet: transformers make strong encoders for medical image segmentation. arXiv preprint arXiv:2102.04306 (2021)
19. Hatamizadeh, A., Nath, V., Tang, Y., Yang, D., Roth, H.R., Xu, D.: Swin unetr: swin transformers for semantic segmentation of brain tumors in MRI images. In: International MICCAI Brainlesion Workshop, pp. 272–284. Springer (2021)
20. Bieder, F., Wolleb, J., Durrer, A., Sandkuehler, R., Cattin, P.C.: Denoising diffusion models for memory-efficient processing of 3d medical images. In: Medical Imaging with Deep Learning, pp. 552–567. PMLR (2024)
21. Lee, S., Chung, H., Park, M., Park, J., Ryu, W.S., Ye, J.C.: Improving 3d imaging with pre-trained perpendicular 2d diffusion models. arXiv preprint arXiv:2303.08440 (2023)
22. Xie, H., et al.: Dose-aware diffusion model for 3d ultra low-dose pet imaging. arXiv preprint arXiv:2311.04248 (2023)
23. Chen, T., et al.: 2.5 d multi-view averaging diffusion model for 3d medical image translation: application to low-count pet reconstruction with CT-less attenuation correction. arXiv preprint arXiv:2406.08374 (2024)
24. Yue, Z., Wang, J., Loy, C.C.: Resshift: efficient diffusion model for image super-resolution by residual shifting. Adv. Neural Inf. Process. Syst. **36** (2024)

DyNo: Dynamic Normalization based Test-Time Adaptation for 2D Medical Image Segmentation

Yihang Fu[1], Ziyang Chen[1], Yiwen Ye[1], and Yong Xia[1,2,3(✉)]

[1] National Engineering Laboratory for Integrated Aero-Space-Ground-Ocean Big Data Application Technology, School of Computer Science and Engineering, Northwestern Polytechnical University, Xi'an 710072, China
yhfu.nwpu@gmail.com, {zychen,ywye}@mail.nwpu.edu.cn
[2] Ningbo Institute of Northwestern Polytechnical University, Ningbo 315048, China
[3] Research & Development Institute of Northwestern Polytechnical University in Shenzhen, Shenzhen 518057, China
yxia@nwpu.edu.cn

Abstract. Medical images often exhibit domain shifts owing to varying imaging protocols and scanners across different medical centres. To address this issue, Test-Time Adaptation (TTA) enables pre-trained models to adapt to test samples during inference. In this paper, we propose a novel method, termed **Dy**namic **No**rmalization (DyNo), for medical image segmentation. Composed of two components, DyNo successfully alleviates domain shifts by adaptively mixing the statistics of multiple domains. We first demonstrated the feasibility of statistics-based methods which merge source and test statistics simply through a supervised toy experiment. Then, we introduce a synthetic domain that synthesizes the distribution information from both the source and target domains using moving average, thereby gradually bridging large domain shifts through the statistics of our synthetic domain. Next, we propose an adaptive fusion strategy, enabling our model to adapt to dynamically changing test data by estimating domain shifts in a fully hyperparameter-free manner. Our DyNo outperforms six competing TTA methods on two benchmark medical image segmentation tasks with multiple scenarios. Extensive ablation studies also demonstrate the effectiveness of synthetic statistics and our adaptive fusion strategy. The code and weights of pre-trained source models are available at https://github.com/Yihang-Fu/DyNo.

Keywords: Test-time adaptation · Dynamic normalization · Medical image segmentation

---
Y. Fu and Z. Chen—Contributed equally.

Supplementary Information The online version contains supplementary material available at https://doi.org/10.1007/978-3-031-73284-3_27.

1 Introduction

Medical image segmentation is a critical component of computer-aided diagnosis. Despite considerable success in this field in recent years [1], segmentation models pre-trained on labeled datasets (source domain) often experience unforeseen performance decline when deployed across various medical centres (target domain). This performance degradation is primarily due to domain shifts [17], commonly resulting from the differences in scanners and imaging protocols.

To address this issue, extensive research has been dedicated to domain adaptation [10,25]. However, most of these methods necessitate a large and representative target dataset for distribution alignment, a requirement that is not always feasible. Consequently, test-time adaptation (TTA) has emerged as a promising approach, adapting the pre-trained model to the target domain using only test data during inference.

Existing TTA methods can be broadly divided into two categories. The first category is backward-based, which updates model parameters through self-training during inference. For instance, TTT [20] designs an auxiliary branch to predict the rotation angle of test data for adaptation, while others [15,21,27] fine-tune pre-trained models by minimizing the entropy of predictions to alleviate the domain shift. These methods, while effective, may encounter error accumulation and even model collapse due to the absence of reliable supervised information. To tackle this challenge, other methods conduct test-time adaptation without updating model parameters in a backward-free manner, which mainly focus on tuning the Batch Normalization (BN) [9] statistics to alleviate the domain shift. These statistics-based methods [12,23,26] typically fuse the

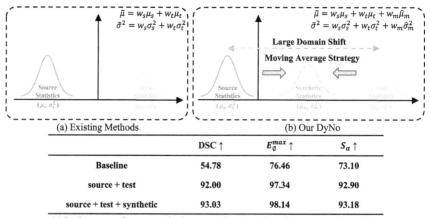

(c) Performance of our supervised toy experiment on the polyp segmentation task

Fig. 1. (a) Existing methods fuse the source and test statistics during inference. (b) We fuse the source and test statistics using a moving average strategy to generate the synthetic domain to alleviate large domain shifts for better adaptation. (c) We construct a TTA scenario using CVC-ClinicDB as the source domain and ETIS-LaribPolypDB as the target domain and train the learnable fusion weights (w_s, w_t, w_m) with 10 iterations for each test sample. We report the mean values of the Dice similarity coefficient (DSC), enhanced alignment (E_ϕ^{max}), and structural similarity (S_α) metrics for evaluation.

source and test statistics for adaptation to alleviate domain shifts based on the facts that source statistics contain discriminative information of corresponding tasks and test statistics contain distribution information of test domain data (see Fig. 1(a)).

However, these statistics-based methods still suffer from limitations when deploying in clinical applications, including: (1) these methods only demonstrate feasibility through experimental results but lack more detailed evidence; (2) it is challenging to address large domain shifts using only source and test statistics for inference; (3) these methods typically pre-set the fusion weights as fixed hyperparameter, making it difficult for the model to adapt to dynamically changing test data. They also require complex hyperparameter searching for different datasets, resulting in weak model adaptability.

In this study, we first conducted a supervised toy experiment by training the learnable weights assigned with the source and test statistics for each test data using the typical supervised loss in one iteration. As shown in Fig. 1(c), significant performance improvements have been achieved after training, indicating that optimizing weights to fuse the source and test statistics enhances the model's ability to fit the test data distribution. This can be served as an evidence of the effectiveness of statistics-based methods [12,23,26] in improving performance. To further alleviate domain shifts, we introduced a synthetic domain by mixing the source and test statistics using moving average strategy (see Fig. 1(b)), which can provide relevant information on the domain shift and improve adaptation performance, as shown in Fig. 1(c). Based on these observations, we propose a novel backward-free method for TTA, called **Dy**namic **No**rmalization (DyNo). DyNo dynamically assigns weights to various statistics to alleviate domain shifts. We first construct the synthetic domain by generating synthetic statistics using the moving average strategy. Then, we consider the distance between distinct statistics to produce weights to fuse source, test, and synthetic statistics. Since our DyNo is hyperparameter-free, it can be deployed across diverse scenarios without manual tuning. We evaluated DyNo, along with other state-of-the-art TTA methods, on two medical segmentation tasks: (1) polyp segmentation on endoscopic images from four domains, and (2) joint segmentation of the optic disc and optic cup using fundus images from five domains.

In summary, our contributions are three-fold: (1) We validate the feasibility of statistics-based methods and the effectiveness of introducing a synthetic domain to alleviate large domain shifts through a toy experiment. (2) We propose an adaptive fusion strategy, enabling our model to adapt to dynamically changing test data in a fully hyperparameter-free manner. (3) Extensive experiment results demonstrate that our approach outperforms other state-of-the-art TTA methods on two benchmark medical image segmentation tasks.

2 Methodology

2.1 Problem Definition

Let x and y denote the image and its corresponding label. Given a model $f_\theta : x \to y$ pre-trained on the labeled source data $\{(x_j^s, y_j^s)\}_{j=1}^{N_s}$. Test-time adaptation

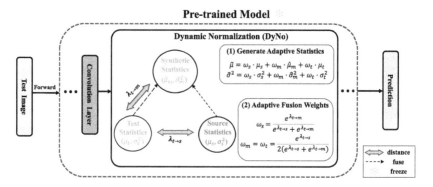

Fig. 2. Pipeline of our proposed DyNo. We first sustain a global synthetic domain by mixing the source and test statistics and employing the moving average strategy. Then the adaptive statistics are produced by utilizing the adaptive fusion strategy. We consider the distance between distinct domains ($\lambda_{t \to m}$ and $\lambda_{t \to s}$) to assign the fusion weights. Finally, the pre-trained model outputs the adapted prediction using the adaptive statistics for the current test image.

aims at adapting the model f_θ to the unlabeled target dataset $\{x_j^t\}_{j=1}^{N_t}$ during inference, where the marginal distributions are different, $p_s(x) \neq p_t(x)$, but the conditional label distributions are the same, $p_s(y|x) = p_t(y|x)$. To achieve this goal, we focus on tuning the statistics (mean μ and variance σ^2) within BN layers of the model f_θ. An overview of our method is illustrated in Fig. 2. Since we perform the same operation on all BN layers, the layer index is omitted for simplicity. We now delve into the details.

2.2 Synthetic Statistics

To overcome large domain shifts during inference, we generate the synthetic statistics for the ith test image by averaging the source and test statistics, defined as:

$$\mu_{m,i} = \frac{\mu_s + \mu_{t,i}}{2}, \quad \sigma_{m,i}^2 = \frac{\sigma_s^2 + \sigma_{t,i}^2}{2}, \tag{1}$$

where the subscripts m, s, and t denote the synthetic, source, and test statistics respectively, and the index i starts from 1. Inspired by [29], we employ the moving average strategy to sustain a global synthetic domain with statistics $\hat{\mu}_m$ and $\hat{\sigma}_m^2$ as follows:

$$\hat{\mu}_m = \frac{\hat{\mu}_m \cdot (i-1) + \mu_{m,i}}{i}, \quad \hat{\sigma}_m^2 = \frac{\hat{\sigma}_m^2 \cdot (i-1) + \sigma_{m,i}^2}{i}. \tag{2}$$

The synthetic statistics can gradually bridge the domain shifts between the source and target domains during inference, enhancing the adaptation process.

Algorithm 1: Algorithm of DyNo

Initialize: Frozen source model f_θ with source statistics μ_s and σ_s^2, initial $\hat{\mu}_m$ and $\hat{\sigma}_m^2$ with zeros
Input: For ith test image x_i^t
1: Calculate the test statistics $\mu_{t,i}$ and $\sigma_{t,i}^2$ on the current test image x_i^t
2: Obtain the synthetic statistics $\mu_{m,i}$ and $\sigma_{m,i}^2$ by Eq. (1)
3: Produce the fusion weights $\omega_{m,i}$, $\omega_{s,i}$ and $\omega_{t,i}$ by Eq. (5)
4: Update the global synthetic statistics $\hat{\mu}_m$ and $\hat{\sigma}_m^2$ by Eq. (2)
5: Generate the adaptive statistics $\tilde{\mu}_i$ and $\tilde{\sigma}_i^2$ by Eq. (4)
Output: The adaptive statistics $\tilde{\mu}_i$, $\tilde{\sigma}_i^2$ for inference

2.3 Adaptive Fusion Strategy

To measure the domain shift between distinct domains, we calculate the distance λ by:

$$\lambda_{t\to m,i} = |\mu_{t,i} - \hat{\mu}_m| + |\sigma_{t,i}^2 - \hat{\sigma}_m^2|, \quad \lambda_{t\to s,i} = |\mu_{t,i} - \mu_s| + |\sigma_{t,i}^2 - \sigma_s^2|. \quad (3)$$

Then we produce the fusion weights ω based on λ to generate the adaptive statistics ($\tilde{\mu}_i$ and $\tilde{\sigma}_i^2$) for inference by fusing the source, test, and synthetic statistics as follows:

$$\begin{aligned}\tilde{\mu}_i &= \omega_{s,i} \cdot \mu_s + \omega_{m,i} \cdot \hat{\mu}_m + \omega_{t,i} \cdot \mu_{t,i}, \\ \tilde{\sigma}_i^2 &= \omega_{s,i} \cdot \sigma_s^2 + \omega_{m,i} \cdot \hat{\sigma}_m^2 + \omega_{t,i} \cdot \sigma_{t,i}^2,\end{aligned} \quad (4)$$

where

$$\omega_{s,i} = \frac{e^{\lambda_{t\to m,i}}}{e^{\lambda_{t\to s,i}} + e^{\lambda_{t\to m,i}}}, \quad \omega_{m,i} = \omega_{t,i} = \frac{e^{\lambda_{t\to s,i}}}{2\left(e^{\lambda_{t\to s,i}} + e^{\lambda_{t\to m,i}}\right)}. \quad (5)$$

We argue that both test and synthetic statistics hold equal significance, thus assigning the same weights. Equation (5) reveals that when the test sample is closer to the synthetic/source domain, we advocate for fusing more synthetic/source statistics to facilitate adaptation, and conversely. The overall process of DyNo is summarized in Algorithm 1.

3 Experiments

3.1 Datasets and Metrics

Datasets. In this study, we conducted two medical image segmentation tasks: joint Optic Disc (OD) and Optic Cup (OC) segmentation task and polyp segmentation task. For the OD/OC segmentation task, we utilized five public datasets collected from various medical centres, denoted as domain A (RIM-ONE-r3 [7]), B (REFUGE-Training [16]), C (ORIGA [28]), D (REFUGE-Validation/Test [16]), and E (Drishti-GS [19]). Each dataset contains 159, 400,

650, 800, and 101 images, respectively. Following [3], we cropped the 800 × 800 region of interest (ROI) centered on the OD and resized it to 512 × 512, followed by the min-max normalization. For the polyp segmentation task, we collected four public datasets derived from different medical centres, denoted as domain A (BKAI-IGH-NeoPolyp [14]), B (CVC-ClinicDB [2]), C (ETIS-LaribPolypDB [18]), and D (Kvasir-Seg [11]). Each dataset contains 1000, 612, 196, and 1000 images, respectively. Following [6], we resized the input images to 352 × 352 and normalized them according to the statistics computed on the ImageNet dataset.

Metrics. For the OD/OC segmentation task, we utilized the mean Dice Similarity Coefficient (DSC) of OD and OC as our evaluation metric. For the polyp segmentation task, we reported DSC, enhanced-alignment metric (E_ϕ^{max}) [5], and structural similarity metric (S_α) [4]. We adopt each domain as the source domain and each of remaining domains as the target domain to conduct diverse TTA scenarios. The average performance across all remaining domains is displayed for evaluation.

3.2 Implementation Details

We pre-trained the source models following [3]. For the OD/OC segmentation task, we pre-trained the ResUNet-34 [8] model using the SGD optimizer with a momentum of 0.99 and a weight decay of 5E-4 for 200 epochs. The initial learning rate η_0 was set to 1E-3, and the learning rate for each epoch was computed according to $\eta_t = \eta_0 \times (1 - \frac{t}{T})^{0.9}$, where t is the current epoch and T is the total number of epochs. The batch size was set to 8. For the polyp segmentation task, we pre-trained the PraNet [6] model using the Adam optimizer with an initial learning rate of 1E-4 for 20 epochs. The batch size was set to 16. During the test-time adaptation process, we set the batch size to 1 for our DyNo and other competing methods. All the backward-based competing methods conduct fine-tuning with one iteration on each test sample. To deploy our DyNo, we introduced our adaptive fusion strategy to all the BN layers for the OD/OC segmentation task. For the polyp segmentation task, we only applied DyNo to the encoder due to the complexity of the decoder.

3.3 Results

We compared our DyNo with Baseline (test on the target domain without adaptation) and six competing TTA methods, including: (1) Two methods based on entropy minimization (TENT [21] and TIPI [15]); (2) A method based on the student-teacher framework (CoTTA [22]); (3) A method based on dynamic learning rate (DLTTA [24]); (4) A backward-free method based on mixing source statistics and test statistics (DUA [12]); (5) A method combining statistics fusion and entropy minimization (DomainAdaptor [27]).

Comparison on the Polyp Segmentation Task. As shown in Table 1, all backward-based methods exhibit worse performance on the polyp segmentation

Table 1. Performance of our DyNo, baseline, and six competing methods on the polyp segmentation task. The best and second-best results in each column are highlighted in **bold** and underline, respectively.

Methods	Domain A			Domain B			Domain C			Domain D			Average		
	DSC	E_ϕ^{max}	S_α	DSC	E_ϕ^{max}	S_α	DSC	E_ϕ^{max}	S_α	DSC	E_ϕ^{max}	S_α	DSC↑	E_ϕ^{max}↑	S_α↑
Baseline	75.32	84.96	82.60	63.23	77.96	75.75	**73.85**	**84.58**	**81.22**	79.45	88.25	86.17	72.96	83.94	81.43
TENT [21]	74.38	85.27	80.95	60.71	72.99	75.31	70.24	82.26	78.61	66.56	77.18	78.85	67.97	79.43	78.43
CoTTA [22]	71.00	82.72	79.03	59.31	71.90	74.30	70.33	82.45	78.58	64.77	75.66	77.56	66.35	78.18	77.37
DLTTA [24]	74.27	85.19	80.85	60.72	73.00	75.34	69.89	81.98	78.41	66.77	77.39	79.09	67.92	79.39	78.42
DUA [12]	74.74	85.56	81.59	72.36	85.54	81.77	70.57	81.76	79.30	82.26	90.72	87.67	74.98	85.90	82.58
DomainAdaptor [27]	75.22	86.08	81.37	61.03	73.32	75.51	71.34	83.20	79.14	67.08	77.84	77.92	68.67	80.11	78.48
TIPI [15]	76.30	86.42	82.64	59.85	72.81	72.28	68.75	80.40	78.62	80.99	89.89	85.97	71.47	82.38	79.88
DyNo (Ours)	**77.98**	**87.48**	**83.51**	**74.06**	**86.32**	**82.79**	73.46	83.94	81.13	**85.80**	**93.38**	**89.37**	**77.82**	**87.78**	**84.20**

Table 2. Performance of our DyNo, baseline, and six competing methods on the OD/OC segmentation task. The best and second-best results in each column are highlighted in **bold** and underline, respectively.

Methods	Domain A	Domain B	Domain C	Domain D	Domain E	Average
	DSC	DSC	DSC	DSC	DSC	DSC↑
Baseline	66.80	80.14	75.94	56.60	70.19	69.93
TENT [21]	72.68	**82.16**	76.25	61.03	71.74	72.77
CoTTA [22]	72.69	77.82	69.69	68.13	67.66	71.20
DLTTA [24]	72.35	81.19	**76.68**	57.87	71.49	71.92
DUA [12]	69.02	80.03	76.47	64.11	72.88	72.50
DomainAdaptor [27]	73.92	79.50	73.92	**69.96**	69.07	73.28
TIPI [15]	69.08	82.09	76.49	64.62	73.84	73.23
DyNo (Ours)	**74.25**	81.49	76.43	65.22	**74.21**	**74.32**

task, even failing to surpass the baseline. This phenomenon can be attributed to the fact that polyps are hidden objectives, posing challenges in detection amidst domain shifts, thus potentially leading to unreliable supervision and error accumulation. However, DUA and our DyNo still exhibit stable performance gains, demonstrating the stability and universality of backward-free methods. In contrast to DUA, our DyNo obtains an average performance improvement of 2.84%, 1.88%, and 1.62% on DSC, E_ϕ^{max}, and S_α respectively, demonstrating the superiority of our approach. Furthermore, we observed that all methods failed in Domain C, and our DyNo neither. Since Domain C possesses the least amount of data (only 196 cases), the model pre-trained on this domain is susceptible to overfitting, thus posing challenges to generalize effectively across other domains. Despite this, our DyNo achieved the second-best results in this domain with marginal deviation from the best result, underscoring that our approach exhibits broad applicability and can maintain the performance of pre-trained models even in challenging scenarios.

Table 3. Performance of various statistics combinations on the polyp segmentation task. The best and second-best results in each column are highlighted in **bold** and underline. 'Src': Abbreviation of 'Source'. 'Syn': Abbreviation of 'Synthetic'.

Combinations			Methods	Domain A			Domain B			Domain C			Domain D			Average		
Src	Test	Syn		DSC	E_ϕ^{max}	S_α	DSC	E_ϕ^{max}	S_α	DSC	E_ϕ^{max}	S_α	DSC	E_ϕ^{max}	S_α	DSC↑	E_ϕ^{max}↑	S_α↑
✓			Baseline	<u>75.32</u>	84.96	<u>82.60</u>	63.23	77.96	75.75	**73.85**	**84.58**	**81.22**	79.45	88.25	86.17	72.96	83.94	81.43
	✓		BN Stat [13]	75.17	<u>86.10</u>	81.58	72.18	85.23	<u>81.96</u>	71.05	82.47	79.34	80.56	88.95	87.14	74.74	85.69	82.50
✓	✓		DUA [12]	74.74	85.56	81.59	<u>72.36</u>	<u>85.54</u>	81.77	70.57	81.76	79.30	<u>82.26</u>	<u>90.72</u>	<u>87.67</u>	<u>74.98</u>	<u>85.90</u>	<u>82.58</u>
✓	✓		DomainAdaptor [27]	75.22	86.08	81.37	61.03	73.32	75.51	71.34	83.20	79.14	67.08	77.84	77.92	68.67	80.11	78.48
✓	✓	✓	DyNo (Ours)	**77.98**	**87.48**	**83.51**	**74.06**	**86.32**	**82.79**	<u>73.46</u>	<u>83.94</u>	<u>81.13</u>	**85.80**	**93.38**	**89.37**	**77.82**	**87.78**	**84.20**

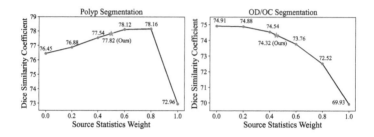

Fig. 3. Performance of our DyNo and utilizing various fixed weights on both tasks.

Comparison on the OD/OC Segmentation Task. As shown in Table 2, the backward-based methods exhibit more effective performance in this task. Although our DyNo cannot achieve the best performance on all domains, it obtains stable gains on each domain and attains the best average performance, surpassing the second-best method (DomainAdaptor) by 1.04%. We demonstrate that our DyNo outperforms even in the task where backward-based methods exhibit advantages.

Comparison on Various Combinations of Statistics. We conducted ablation studies to validate the effectiveness of our method on the polyp segmentation task, as shown in Table 3. We compared our DyNo with other methods utilizing various combinations of statistics, including: (1) Source statistics only (Baseline); (2) Test statistics only (BN Stat [13]); (3) Mixing source and test statistics (DUA [12] and DomainAdaptor [27]). The results reveal that (1) introducing test statistics during inference can better capture the distribution of test data; (2) compared to DUA, although DomainAdaptor utilizes a similar statistics-fusion strategy, the incorporation of unreliable supervision via entropy minimization degrades the performance significantly; (3) the best performance is achieved by our DyNo, demonstrating the effectiveness of introducing the synthetic statistics.

Compared to the Fixed Fusion Weights. To explore the best fusion weights for both tasks, We repeated the experiments by utilizing a series of fixed weights for the source statistics, while allocating the remaining weights to the synthetic and test statistics equally, ensuring the sum of weights is 1. As shown in Fig. 3, we observed that the performance varies differently under various fusion weights across two tasks, which means the patterns explored in one task cannot directly apply to another one. By comparison, our DyNo generates the adaptive weights in a hyperparameter-free manner, which can achieve stable performance on both tasks, demonstrating the stability and effectiveness of DyNo on varying domain shifts. Moreover, the performance we achieved is close to the best performance achieved by searching weights on both tasks, which implies that our DyNo approximates the upper-bound performance of searching weights.

4 Conclusion

In this paper, we propose DyNo, a hyperparameter-free TTA method based on dynamic normalization designed to alleviate domain shifts during inference. We first bridge the domain gap between the source domain and test data by constructing a global synthetic domain. Then we present an adaptive fusion strategy which fuses the source, test, and synthetic statistics to dynamically generate the adaptive ones considering the divergence between distinct domains. Extensive experiment results on two medical image segmentation benchmarks verify the superiority of our DyNo over other state-of-the-art TTA methods.

Acknowledgment. This work was supported in part by the National Natural Science Foundation of China under Grant 62171377, in part by the Ningbo Clinical Research Center for Medical Imaging under Grant 2021L003 (Open Project 2022LYKFZD06), in part by Shenzhen Science and Technology Program under Grants JCYJ20220530161616036, in part by the China Postdoctoral Science Foundation 2021M703340/BX2021333, and in part by the Innovation Foundation for Doctor Dissertation of Northwestern Polytechnical University under Grant CX2024016.

References

1. Azad, R., et al.: Medical image segmentation review: the success of u-net. arXiv preprint arXiv:2211.14830 (2022)
2. Bernal, J., Sánchez, F.J., Fernández-Esparrach, G., Gil, D., Rodríguez, C., Vilariño, F.: WM-DOVA maps for accurate polyp highlighting in colonoscopy: validation vs. saliency maps from physicians. Computeriz. Med. Imag. Graph. **43**, 99–111 (2015)
3. Chen, Z., Pan, Y., Ye, Y., Lu, M., Xia, Y.: Each test image deserves a specific prompt: continual test-time adaptation for 2d medical image segmentation. In: Proceedings of the IEEE/CVF Conference on Computer Vision and Pattern Recognition, pp. 11184–11193 (2024)
4. Fan, D.P., Cheng, M.M., Liu, Y., Li, T., Borji, A.: Structure-measure: a new way to evaluate foreground maps. In: Proceedings of the IEEE International Conference on Computer Vision, pp. 4548–4557 (2017)
5. Fan, D.P., Gong, C., Cao, Y., Ren, B., Cheng, M.M., Borji, A.: Enhanced-alignment measure for binary foreground map evaluation. arXiv preprint arXiv:1805.10421 (2018)
6. Fan, D.-P., et al.: PraNet: parallel reverse attention network for polyp segmentation. In: Martel, A.L., et al. (eds.) MICCAI 2020. LNCS, vol. 12266, pp. 263–273. Springer, Cham (2020). https://doi.org/10.1007/978-3-030-59725-2_26

7. Fumero, F., Alayón, S., Sanchez, J.L., Sigut, J., Gonzalez-Hernandez, M.: Rimone: an open retinal image database for optic nerve evaluation. In: 2011 24th International Symposium on Computer-Based Medical Systems (CBMS), pp. 1–6. IEEE (2011)
8. He, K., Zhang, X., Ren, S., Sun, J.: Deep residual learning for image recognition. In: Proceedings of the IEEE Conference on Computer Vision and Pattern Recognition, pp. 770–778 (2016)
9. Ioffe, S., Szegedy, C.: Batch normalization: accelerating deep network training by reducing internal covariate shift. In: International Conference on Machine Learning, pp. 448–456. PMLR (2015)
10. Javanmardi, M., Tasdizen, T.: Domain adaptation for biomedical image segmentation using adversarial training. In: 2018 IEEE 15th International Symposium on Biomedical Imaging (ISBI 2018), pp. 554–558. IEEE (2018)
11. Jha, D., et al.: Kvasir-SEG: a segmented polyp dataset. In: Ro, Y.M., et al. (eds.) MMM 2020. LNCS, vol. 11962, pp. 451–462. Springer, Cham (2020). https://doi.org/10.1007/978-3-030-37734-2_37
12. Mirza, M.J., Micorek, J., Possegger, H., Bischof, H.: The norm must go on: dynamic unsupervised domain adaptation by normalization. In: Proceedings of the IEEE/CVF Conference on Computer Vision and Pattern Recognition, pp. 14765–14775 (2022)
13. Nado, Z., Padhy, S., Sculley, D., D'Amour, A., Lakshminarayanan, B., Snoek, J.: Evaluating prediction-time batch normalization for robustness under covariate shift. arXiv preprint arXiv:2006.10963 (2020)
14. Ngoc Lan, P., et al.: Neounet: towards accurate colon polyp segmentation and neoplasm detection. In: Proceedings of the Advances in Visual Computing: 16th International Symposium, ISVC 2021, Virtual Event, 4–6 October 2021, Part II, pp. 15–28. Springer (2021)
15. Nguyen, A.T., Nguyen-Tang, T., Lim, S.N., Torr, P.H.: Tipi: test time adaptation with transformation invariance. In: Proceedings of the IEEE/CVF Conference on Computer Vision and Pattern Recognition, pp. 24162–24171 (2023)
16. Orlando, J.I., et al.: Refuge challenge: a unified framework for evaluating automated methods for glaucoma assessment from fundus photographs. Med. Image Anal. 59, 101570 (2020)
17. Quinonero-Candela, J., Sugiyama, M., Schwaighofer, A., Lawrence, N.D.: Dataset Shift in Machine Learning. MIT Press (2008)
18. Silva, J., Histace, A., Romain, O., Dray, X., Granado, B.: Toward embedded detection of polyps in wce images for early diagnosis of colorectal cancer. Int. J. Comput. Assist. Radiol. Surg. 9, 283–293 (2014)
19. Sivaswamy, J., Krishnadas, S., Joshi, G.D., Jain, M., Tabish, A.U.S.: Drishti-GS: retinal image dataset for optic nerve head (ONH) segmentation. In: 2014 IEEE 11th International Symposium on Biomedical Imaging (ISBI), pp. 53–56. IEEE (2014)
20. Sun, Y., Wang, X., Liu, Z., Miller, J., Efros, A., Hardt, M.: Test-time training with self-supervision for generalization under distribution shifts. In: International Conference on Machine Learning, pp. 9229–9248. PMLR (2020)
21. Wang, D., Shelhamer, E., Liu, S., Olshausen, B., Darrell, T.: Tent: Fully test-time adaptation by entropy minimization. arXiv preprint arXiv:2006.10726 (2020)
22. Wang, Q., Fink, O., Van Gool, L., Dai, D.: Continual test-time domain adaptation. In: Proceedings of the IEEE/CVF Conference on Computer Vision and Pattern Recognition, pp. 7201–7211 (2022)

23. Wang, W., et al.: Dynamically instance-guided adaptation: a backward-free approach for test-time domain adaptive semantic segmentation. In: Proceedings of the IEEE/CVF Conference on Computer Vision and Pattern Recognition, pp. 24090–24099 (2023)
24. Yang, H., et al.: Dltta: dynamic learning rate for test-time adaptation on cross-domain medical images. IEEE Trans. Med. Imaging **41**(12), 3575–3586 (2022)
25. Yang, J., Dvornek, N.C., Zhang, F., Chapiro, J., Lin, M.D., Duncan, J.S.: Unsupervised domain adaptation via disentangled representations: application to cross-modality liver segmentation. In: Shen, D., et al. (eds.) MICCAI 2019. LNCS, vol. 11765, pp. 255–263. Springer, Cham (2019). https://doi.org/10.1007/978-3-030-32245-8_29
26. You, F., Li, J., Zhao, Z.: Test-time batch statistics calibration for covariate shift. arXiv preprint arXiv:2110.04065 (2021)
27. Zhang, J., Qi, L., Shi, Y., Gao, Y.: Domainadaptor: a novel approach to test-time adaptation. In: Proceedings of the IEEE/CVF International Conference on Computer Vision, pp. 18971–18981 (2023)
28. Zhang, Z., et al.: Origa-light: an online retinal fundus image database for glaucoma analysis and research. In: 2010 Annual International Conference of the IEEE Engineering in Medicine and Biology, pp. 3065–3068. IEEE (2010)
29. Zhao, B., Chen, C., Xia, S.T.: Delta: degradation-free fully test-time adaptation. arXiv preprint arXiv:2301.13018 (2023)

Accurate Delineation of Cerebrovascular Structures from TOF-MRA with Connectivity-Reinforced Deep Learning

Shoujun Yu[1,2], Cheng Li[2], Yousuf Babiker M. Osman[1,2], Shanshan Wang[1,2(✉)], and Hairong Zheng[1,2(✉)]

[1] University of Chinese Academy of Sciences, Beijing, China
[2] Paul C. Lauterbur Research Center for Biomedical Imaging, Shenzhen Institute of Advanced Technology, Chinese Academy of Sciences, Shenzhen, China
{ss.wang,hr.zheng}@siat.ac.cn

Abstract. Automatic and accurate delineation of cerebrovascular structures from Time-of-Flight Magnetic Resonance Angiography (TOF-MRA) images is crucial for diagnosing and treating cerebrovascular diseases. While most deep learning approaches have presented encouraging capabilities for vessel delineation, their results still suffer from disconnected and incomplete segments due to the insufficient exploration of cerebrovascular structures and topologies. To tackle this issue, this paper proposes a connectivity-reinforced deep learning approach to protect the topological information of the vessels. In detail, the cerebral vessels are tracked and highlighted with emphasis on both the central and edge voxels by the specially designed CONnectivity Attention Module (CONAM). Furthermore, a specially designed adaptive connectivity loss ($\mathcal{L}_{AC}$) is introduced to reinforce the network training by balancing the global penalty and the auto-adjusted regional penalty, subsequently optimizing the overall connectivity of the vessel predictions. Extensive experiments have been conducted on the well-known IXI dataset, and the proposed method has been compared to seven state-of-the-art approaches. The results demonstrate that our method outperforms these approaches both quantitatively and qualitatively. Code will be available at https://github.com/Yusjlalala/CONA-Net.

Keywords: Deep learning · Cerebrovascular · Connectivity reinforcement · 3D delineation

1 Introduction

Cerebrovascular diseases, including stroke, are responsible for a significant number of deaths worldwide, accounting for approximately 11% of total deaths in 2019 [2]. These diseases often originate from vascular malformations [12]. The connectivity of cerebrovascular structures plays a crucial role in the screening and detection of vascular malformations. Time-of-flight magnetic resonance angiography (TOF-MRA) is a widely utilized imaging technique for the visualization, detection, and quantitative analysis of cerebrovascular structures [21]. Accurately delineating the cerebrovascular structures while preserving their connectivity from TOF-MRA images holds great importance for the diagnosis and quantitative analysis of cerebrovascular diseases, such as stroke and aneurysms [5]. However, manual delineation is time-consuming and cumbersome due to the inherent slender and elongated nature of cerebrovascular structures.

Over the past decade, significant progress has been made in developing deep learning-based methods for automatic medical image analysis [11,17,22,23,27]. Especially, various approaches have been proposed to tackle the cerebrovascular structures delineation task. For example, Sanchesa et al. proposed the Uception network, which enhances the 3D U-Net with inception modules to address the issue of inter-patient variability in cerebrovascular structure delineation [19]. Mou et al. introduced CS2-Net, which can classify the curvilinear structure more effectively by extracting both channel- and spatial-attention feature representations [16]. Building upon the concept of reverse attention [7], Xia et al. developed ER-Net, which incorporates an edge-reinforced module and an edge-constrained loss to prioritize the network's focus on the edge voxels, thus improving the delineation performance [24]. Chen et al. leveraged the consistency of the topological connection between semantic predictions and ground truth to guide the model towards learning topological structures [4]. Despite the promising performance of existing deep learning-based methods, challenges remain in achieving fully connected and complete vessel delineation. While some methods focus on enhancing edge [25] or centerline features [6], issues with disconnected and incomplete delineation persist.

In this paper, we propose a novel connectivity-reinforced deep learning method for the delineation of cerebrovascular structures, aiming at improving the accuracy and completeness of vessel delineation results. The main contributions of our work can be summarized as follows:

(1) We develop an effective model, the CONnectivity Attention Net (CONA-Net), for cerebrovascular structure delineation. The model tracks and highlights cerebral vessels with emphases on both central and edge voxels through its CONnectivity Attention Module (CONAM).

(2) We propose an innovative Adaptive Connectivity Loss function ($\mathcal{L}_{AC}$) to reinforce fully connected and accurate vessel delineation. $\mathcal{L}_{AC}$ supervises the overall delineation process and uniquely incorporates a region-specific loss. This allows for adaptively adjustment of penalties for regional false positive (FP) and false negative (FN) errors targeting both the centerline and edge predictions during the training process.

Through extensive experiments on an open-source IXI dataset, we demonstrate the effectiveness of our proposed method for cerebrovascular structure delineation.

2 Methods

The architecture of our proposed CONA-Net is illustrated in Fig. 1. Overall, it comprises three main modules: a central voxel tracking module for central voxel extraction, a reverse edge attention module for edge voxel extraction, and the CONAM that integrates the forces of the first two modules for more complete vessel delineation. Additionally, an innovative loss function is introduced to train the network and reinforce vessel connectivity. This section provides detailed explanations of the three modules and the novel loss function.

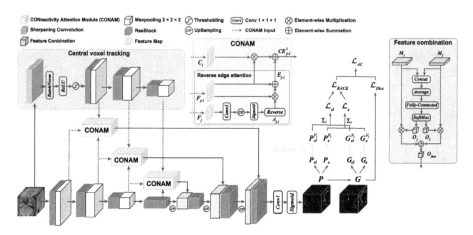

Fig. 1. The architecture of CONA-Net, which mainly consists of three modules, central voxel tracking module, reverse edge attention module, and CONAM. A novel loss function, $\mathcal{L}_{AC}$, is designed to promote vessel connectivity.

2.1 Connectivity Attention Net

The proposed CONA-Net follows a classic encoder-decoder structure, which has proven effective in various image delineation tasks. The encoder contains four stages with the building unit of ResBlock [13]. In the decoder, each of the three stages consists of an upsampling operation followed by two convolutions with ReLU activations. To enhance the capability of CONA-Net in capturing the complete vessel structures, we introduce several novel network modules.

Previous studies have indicated that integrating prior anatomical knowledge into network architectures significantly enhances the accuracy of delineation [14]. Both the centerline and edge are crucial topological features for maintaining the connectivity of cerebrovascular structures. In response, we introduce CONAM, which enhances cerebrovascular structure connectivity by utilizing centerline and edge features extracted by central voxel tracking and reverse edge attention modules, respectively. (see Fig. 1).

In the designed central voxel tracking module, we track the central voxels of cerebral vessels, which encompass the centerline and its small vicinity. Initially, we apply a sharpening convolution with the kernel $S(*) \in \mathbb{R}^{1\times 3\times 3\times 3}$, followed by batch normalization and ReLU function. Then, a central voxel thresholding operation with a threshold value CV_{thr} is appended to track the central voxels. After thresholding, features are binarized with values greater than CV_{thr} set to 1 and smaller than CV_{thr} set to 0. These features can serve as masks to track the central voxels. A series of ResBlocks and max pooling operations then generate multi-scale central voxel masks. The masks generated at the ith stage are denoted as $C_i \in \mathbb{R}^{\frac{C}{2^i}\times\frac{H}{2^i}\times\frac{W}{2^i}\times\frac{D}{2^i}}$ ($1 \le i \le 3$), where H, W, and D represent the height, width, and depth of the input image, respectively, and C represents the number of channels of the mask.

Inspired by [26], we introduce a reverse edge attention module between every adjacent layer of the encoder to capture the edge information of cerebral vessels. Let $F_j \in \mathbb{R}^{\frac{C}{2^j}\times\frac{H}{2^j}\times\frac{W}{2^j}\times\frac{D}{2^j}}$ ($2 \le j \le 4$) denote the feature maps generated at the jth stage of the encoder. In this module, we first apply a $1 \times 1 \times 1$ convolution to compress the feature F_j in channel dimension, generating the compressed feature map $X_j \in \mathbb{R}^{1\times H\times W\times D}$. This compressed feature is then upsampled to match the resolution of F_{j-1}. The weight map of the background region can be obtained as $A_{j-1} = 1 - sigmoid(Up(X_j))$. By combining F_{j-1} and A_{j-1}, we obtain the edge-reinforced features of the vessels, expressed as $E_{j-1} = F_{j-1} \circ A_{j-1} + F_{j-1}$, where $\circ$ is element-wise product.

Using the central voxel masks C_i generated from the central voxel tracking module and the edge-reinforced features E_{j-1} generated from the reverse edge attention module, the output features of CONAM that emphasize both central voxels and edge voxels are obtained as $CE_{j-1}^i = F_{j-1} \circ C_i + E_{j-1}$.

To incorporate the topological-related features from CONAM to the decoder, we employ a feature combination module (see Fig. 1). Let $M_1 \in \mathbb{R}^{\frac{C}{2^k}\times\frac{H}{2^k}\times\frac{W}{2^k}\times\frac{D}{2^k}}$ be the feature from the kth stage of decoder and $M_2 \in \mathbb{R}^{\frac{C}{2^k}\times\frac{H}{2^k}\times\frac{W}{2^k}\times\frac{D}{2^k}}$ be the corresponding feature from CONAM. In this module, M_1 and M_2 are first fused by concatenation. Then, a channel-wise attention weight map $S \in \mathbb{R}^{2C\times 1\times 1\times 1}$ is generated by average pooling: $S = Avgpool(Concat(M_1, M_2))$. After a fully-connected layer and a SoftMax function, we obtain the attention weight map $O \in \mathbb{R}^{2C\times 1\times 1\times 1}$ as $O = Sigmoid(FC(S))$, encoding the channel-wise interdependencies. Finally, O is divided into $O_1 \in \mathbb{R}^{C\times 1\times 1\times 1}$ and $O_2 \subset \mathbb{R}^{C\times 1\times 1\times 1}$, which are treated as the feature-oriented weights for M_1 and M_2, respectively. The fused feature maps $O_{fuse} \in \mathbb{R}^{\frac{C}{2^k}\times\frac{H}{2^k}\times\frac{W}{2^k}\times\frac{D}{2^k}}$ are obtained by: $O_{fuse} = O_1 \circ M_1 + O_2 \circ M_2$.

2.2 Adaptive Connectivity Loss

To train neural networks for medical image delineation, the Dice loss ($\mathcal{L}_{Dice}$) is commonly used [15], which is defined as:

$$\mathcal{L}_{Dice}(P, G) = 1 - \frac{2\sum_{n-1}^{N} p_n g_n + \varepsilon}{\sum_{n-1}^{N} p_n^2 + \sum_{n-1}^{N} g_n^2 + \varepsilon} \quad (1)$$

where N denotes the number of voxels in the image. P and G are the predictions and ground truth masks, respectively. $p_n \in [0, 1]$ and $g_n \in [0, 1]$ are the nth voxel of the predictions and ground truth masks. $\varepsilon = 1e^{-8}$ is the smoothing factor applied to ensure the numerical stability. Although $\mathcal{L}_{Dice}$ has demonstrated effectiveness in a wide range of image delineation tasks, it reveals limitations in delineation of cerebrovascular structures in MRA images, especially when delineating small vessel and cross-vessel connections [20].

Inspired by the adaptive loss scheme in [3], we propose a novel loss function $\mathcal{L}_{AC}$, which penalizes both global and regional errors. It is defined as:

$$\mathcal{L}_{AC} = \begin{cases} \mathcal{L}_{Dice}, & \text{DSC} \leq \theta \\ \mathcal{L}_{Dice} + \mathcal{L}_{RACE}, & \text{DSC} > \theta \end{cases} \quad (2)$$

where, $\mathcal{L}_{Dice}$ regularizes the global delineation, whereas the Regional Adaptive Centerline and Edge Loss ($\mathcal{L}_{RACE}$) focuses on the regional delineation. The Dice Similarity Coefficient (DSC) is calculated in every iteration and set a hyperparameter θ as the switching condition of the two options. The θ is empirically set to 0.8.

$\mathcal{L}_{RACE}$ consists of three terms, a centerline loss ($\mathcal{L}_{cl}$), an edge loss ($\mathcal{L}_e$), and a regularization:

$$\mathcal{L}_{RACE} = \frac{\zeta}{\kappa^2} \mathcal{L}_{cl}(P_{cl}, G_{cl}) + \frac{1}{\tau^2} \mathcal{L}_e(P_e, G_e) + log(1 + \kappa\tau) \quad (3)$$

where κ and τ are trainable parameters and $\zeta = 1$ is empirically determined. P_{cl} and G_{cl} represent the centerline maps extracted from the predictions and ground truth masks following the method outlined in [20]. P_e and G_e denote the edge maps extracted from the predictions and ground truth masks using a three-dimensional Laplacian convolutional kernel [24]. $\mathcal{L}_{cl}$ and $\mathcal{L}_e$ are the losses calculated from centerline maps and edge maps, respectively.

The network is more prone to false negative (FN) than false positive (FP) errors when segmenting small targets against a large background, affecting the accuracy of results [9]. The Tversky loss allows for a flexible trade-off between FN and FP errors, but is sensitive to hyperparameters [18]. Therefore, we introduce an adaptive regional penalty to fine-tune this trade-off for accurate vessel delineation. Specifically, $\mathcal{L}_{cl}$ and $\mathcal{L}_e$ are calculated as follows:

$$\mathcal{L}_{cl}(P_{cl}, G_{cl}) = \sum_l (1 - \frac{TP(P_{cl}^m, G_{cl}^m) + \varepsilon}{P(P_{cl}^m, G_{cl}^m) + \alpha_{cl}^m FP(P_{cl}^m, G_{cl}^m) + \beta_{cl}^m FN(P_{cl}^m, G_{cl}^m) + \varepsilon}), m \in V_l \quad (4)$$

$$\mathcal{L}_e(P_e, G_e) = \sum_l (1 - \frac{TP(P_e^m, G_e^m) + \varepsilon}{P(P_e^m, G_e^m) + \alpha_e^m FP(P_e^m, G_e^m) + \beta_e^m FN(P_e^m, G_e^m) + \varepsilon}), m \in V_l \quad (5)$$

where we divide a volume V into l non-overlapping sub-volumes $V_1, V_2, \ldots, V_l$. m denotes the mth sub-volume of V. The parameters α_{cl}^m, β_{cl}^m, α_e^m, and β_e^m are fine-tuned during the training process, calculated as following equations:

$$\begin{cases} \alpha_{cl}^m = A + B \times \frac{FP(P_{cl}^m, G_{cl}^m)}{FP(P_{cl}^m, G_{cl}^m) + FN(P_{cl}^m, G_{cl}^m)} \\ \beta_{cl}^m = A + B \times \frac{FN(P_{cl}^m, G_{cl}^m)}{FP(P_{cl}^m, G_{cl}^m) + FN(P_{cl}^m, G_{cl}^m)} \end{cases} \quad (6)$$

$$\begin{cases} \alpha_e^m = A + B \times \frac{FP(P_e^m, G_e^m)}{FP(P_e^m, G_e^m) + FN(P_e^m, G_e^m)} \\ \beta_e^m = A + B \times \frac{FN(P_e^m, G_e^m)}{FP(P_e^m, G_e^m) + FN(P_e^m, G_e^m)} \end{cases} \quad (7)$$

where A and B are two constant coefficients, both empirically set to 0.5 in this study. The values of these four parameters range from A to A+B. Both α_{cl}^m and α_e^m linearly increase with the proportion of FP errors. β_{cl}^m and β_e^m linearly increase with the proportion of FN errors.

3 Experiments

3.1 Setup

Dataset. All our experiments were conducted on the open-source dataset provided by [8], originating from the IXI dataset. This dataset includes 45 manually annotated TOF-MRA volumes, all with a uniform spatial resolution of 0.264 × 0.264 × 0.8 mm³. Each volume was center-cropped to a consistent matrix size of 400 × 400 × 64 voxels.

Evaluation Metrics. Two metrics were calculated to quantitatively evaluate the delineation performance: Centerline Dice (clDice) and Dice Similarity Coefficient (DSC). Notably, clDice is a connectivity-preserving metric to evaluate tubular and linear structure delineation based on the intersection of the centerline skeleton of the predictions and corresponding ground truth [20].

Implementation Details. CONA-Net was implemented using the PyTorch framework with two GPUs (NVIDIA RTX A6000). Adaptive moment estimation (Adam) was employed for network optimization. The initial learning rate was set to 0.001, with a weight decay of 0.0005. A ReduceLROnPlateau scheme was used [1], with the mode set to 'min', a factor of 0.5, and patience of 30. The batch size was set to 2 during training. The network was trained for 50 epochs.

3.2 Comparison with Seven State-of-the-Art Methods

Figure 2 visualizes the delineation results of different methods for one case. Overall, all methods demonstrate good delineation performance for large vessels. However, the proposed CONA-Net surpasses all existing models in terms of accurately delineating small vessels and preserving their connectivity, as highlighted by the yellow arrows with enlarged views within the yellow rectangles.

The quantitative results of our proposed CONA-Net and state-of-the-art deep learning-based methods are presented in Table 1. Compared to the seven existing methods, our CONA-Net generated better results with larger clDice and DSC values.

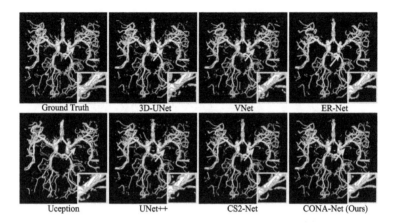

Fig. 2. Comparison of cerebrovascular structure delineation using different methods on the same MRA volume.

Table 1. Quantitative delineation results of different methods on the IXI dataset.(mean ± std, best results in **bold**). (* did not report their clDice result)

Method	clDice (%)	DSC (%)
3D-UNet [10]	82.40 ± 1.46	84.73 ± 0.87
V-Net [15]	84.07 ± 1.55	85.69 ± 1.00
Uception [19]	84.87 ± 1.35	88.08 ± 0.93
A-SegAN* [8]	\	86.38+3.06
U-Net++ [28]	85.88 ± 1.37	87.37 ± 1.01
ER-Net [24]	87.88 ± 2.61	88.95 ± 1.11
CS2-Net [16]	89.01 ± 1.47	88.61 ± 1.27
CONA-Net(Ours)	**89.52 ± 1.59**	**90.25 ± 1.13**

3.3 Ablation Study

Ablation studies were conducted to thoroughly evaluate the effectiveness of each introduced module in CONA-Net (see Fig. 3). The baseline is a 3D U Net with the residual block [13]. Net-1 and Net-2 are our CONA-Net without the central voxel tracking module and reverse edge attention module, respectively. Net-3 replaces the CONAMs in CONA-Net with direct skip connections. Net-4 utilizes feature summation to replace the feature combination module in CONA-Net. Net-5 employs $\mathcal{L}_{Dice}$ instead of $\mathcal{L}_{AC}$ in CONA-Net.

The results demonstrate that our final model, CONA-Net, achieves the best performance with a clDice value of 89.52%, and a DSC value of 90.25%. Compared to CONA-Net, Net-1 and Net-2 show reductions in clDice by 2.71% and 2.07%, respectively. Replacing the CONAMs in Net-3 leads to a significant drop in delineation performance, with a 7.02% reduction in clDice. The absence

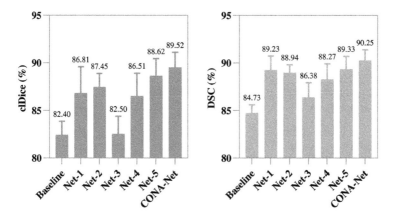

Fig. 3. Results of ablation studies on the different modules of CONA-Net and adaptive connectivity loss function.

of feature combination module in CONA-Net (Net-4) results in 3.01% clDice reduction.

Notably, the results from Net-3, Net-4 and CONA-Net indicate that using the feature combination module alone does not improve the model performance, but using it together with CONAM can effectively enhance the performance, suggesting that CONAM is the critical role in boosting effectiveness.

Compared to Net-5, the results from CONA-Net validate the effectiveness of the proposed loss function $\mathcal{L}_{AC}$, which improves the clDice by 0.9% and the DCS by 0.92%. This confirms that $\mathcal{L}_{AC}$ reinforces the connectivity of cerebrovascular structures and improves the overall delineation accuracy.

4 Conclusion

In this paper, we proposed a connectivity-reinforced network called CONA-Net for delineating cerebrovascular structures from TOF-MRA images. The proposed approach comprises a novel module CONAM and a specially designed adaptive connectivity loss function $\mathcal{L}_{AC}$. CONAM integrates information from both the central voxel tracking module and the reverse edge attention module to reinforce the network's attention on central and edge voxels, improving the delineation performance and cerebrovascular connectivity. $\mathcal{L}_{AC}$ not only constrains the global and regional predictions but also provides region-specific reinforced and adaptive error penalties in topological feature maps, thereby improving the overall accuracy of cerebrovascular delineation. Experimental results on the open-source IXI dataset demonstrate that our method can achieve better delineation results with improved accuracy and vessel connectivity compared to existing state-of-the-art methods.

Acknowledgments. This research was partly supported by the National Key Research and Development Program of China (2022YFA1004200), National Natural Science Foundation of China (52293425, 62222118, U22A2040), Guangdong Provincial Key Laboratory of Artifcial Intelligence in Medical Image Analysis and Application (2022B1212010011), Shenzhen Science and Technology Program (RCYX20210706092104034, JCYJ20220531100213029), and Youth Innovation Promotion Association CAS.

References

1. Torch.optim.lr_scheduler–PyTorch 2.2 documentation. https://pytorch.org/docs/stable/_modules/torch/optim/lr_scheduler.html#ReduceLROnPlateau
2. The top 10 causes of death. Technical Report, WHO (2019). https://www.who.int/news-room/fact-sheets/detail/the-top-10-causes-of-death
3. Cao, et al.: Learning crisp boundaries using deep refinement network and adaptive weighting loss. IEEE Trans. Multimedia **23**, 761–771 (2020)
4. Chen, C., et al.: Cerebrovascular segmentation in phase-contrast magnetic resonance angiography by multi-feature fusion and vessel completion. Comput. Med. Imaging Graph. **98**, 102070 (2022)
5. Chen, C., et al.: All answers are in the images: a review of deep learning for cerebrovascular segmentation. Comput. Med. Imaging Graph. **107**, 102229 (2023)
6. Chen, C., et al.: Cerebrovascular segmentation in TOF-MRA with topology regularization adversarial model. In: Proceedings of the 31st ACM International Conference on Multimedia, pp. 4250–4259. ACM, Ottawa ON Canada (2023)
7. Chen, S., et al.: Reverse attention for salient object detection. In: Proceedings of the European Conference on Computer Vision (ECCV), pp. 234–250 (2018)
8. Chen, Y., et al.: Attention-assisted adversarial model for cerebrovascular segmentation in 3d TOF-MRA volumes. IEEE Trans. Med. Imaging **41**(12), 3520–3532 (2022)
9. Chen, Y., et al.: Adaptive region-specific loss for improved medical image segmentation. IEEE Trans. Pattern Anal. Mach. Intell. **45**(11), 13408–13421 (2023)
10. Çiçek, Ö., Abdulkadir, A., Lienkamp, S.S., Brox, T., Ronneberger, O.: 3D U-Net: Learning Dense Volumetric Segmentation from Sparse Annotation. In: Ourselin, S., Joskowicz, L., Sabuncu, M.R., Unal, G., Wells, W. (eds.) MICCAI 2016. LNCS, vol. 9901, pp. 424–432. Springer, Cham (2016). https://doi.org/10.1007/978-3-319-46723-8_49
11. Diakite, A., et al.: Lesen: label-efficient deep learning for multi-parametric MRI-based visual pathway segmentation. In: 21st International Symposium on Biomedical Imaging (ISBI), IEEE (2024)
12. Flemming, K.D., et al.: Population-based prevalence of cerebral cavernous malformations in older adults: mayo clinic study of aging. JAMA Neurol. **74**(7), 801–805 (2017)
13. He, K., et al.: Deep residual learning for image recognition. In: 2016 IEEE Conference on Computer Vision and Pattern Recognition (CVPR), pp. 770–778. IEEE, Las Vegas, NV, USA (2016)
14. Huang, H., et al.: Medical image segmentation with deep atlas prior. IEEE Trans. Med. Imaging **40**(12), 3519–3530 (2021)
15. Milletari, F., et al.: V-net: Fully convolutional neural networks for volumetric medical image segmentation. In: 2016 Fourth International Conference on 3D Vision (3DV), pp. 565–571. IEE (2016)

16. Mou, L., et al.: Cs2-net: deep learning segmentation of curvilinear structures in medical imaging. Med. Image Anal. **67**, 101874 (2021)
17. Qi, K., Yang, H., Li, C., Liu, Z., Wang, M., Liu, Q., Wang, S.: X-Net: Brain Stroke Lesion Segmentation Based on Depthwise Separable Convolution and Long-Range Dependencies. In: Shen, D., Liu, T., Peters, T.M., Staib, L.H., Essert, C., Zhou, S., Yap, P.-T., Khan, A. (eds.) MICCAI 2019. LNCS, vol. 11766, pp. 247–255. Springer, Cham (2019). https://doi.org/10.1007/978-3-030-32248-9_28
18. Salehi, S.S.M., et al.: Tversky loss function for image segmentation using 3d fully convolutional deep networks (2017)
19. Sanchesa, P., et al.: Cerebrovascular network segmentation of MRA images with deep learning. In: 2019 IEEE 16th International Symposium on Biomedical Imaging (ISBI 2019), pp. 768–771. IEEE (2019)
20. Shit, S., et al.: clDice-a novel topology-preserving loss function for tubular structure segmentation. In: Proceedings of the IEEE/CVF Conference on Computer Vision and Pattern Recognition, pp. 16560–16569 (2021)
21. Subramaniam, P., et al.: Generating 3D TOF-MRA volumes and segmentation labels using generative adversarial networks. Med. Image Anal. **78**, 102396 (2022)
22. Sun, H., et al.: AUNet: attention-guided dense-upsampling networks for breast mass segmentation in whole mammograms (2020)
23. Wang, S., et al.: Annotation-efficient deep learning for automatic medical image segmentation, p. 5915 (2021)
24. Xia, L., et al.: 3D vessel-like structure segmentation in medical images by an edge-reinforced network. Med. Image Anal. **82**, 102581 (2022)
25. Yang, C., et al.: Contour attention network for cerebrovascular segmentation from TOF-MRA volumetric images. Med. Phys. **51**(3), 2020–2031 (2023)
26. Zhang, H., et al.: Cerebrovascular segmentation in MRA via reverse edge attention network. In: Martel, A.L., et al. Medical Image Computing and Computer Assisted Intervention - MICCAI 2020, MICCAI 2020, LNCS, vol. 12266, pp. 66–75. Springer, Cham (2020). https://doi.org/10.1007/978-3-030-59725-2_7
27. Zhou, et al.: D-UNet: a dimension-fusion u shape network for chronic stroke lesion segmentation, pp. 940–950 (2021)
28. Zhou, Z., et al.: UNet++: redesigning skip connections to exploit multiscale features in image segmentation. IEEE Trans. Med. Imaging **39**(6), 1856–1867 (2020)

Learning Instance-Discriminative Pixel Embeddings Using Pixel Triplets

Long Chen[1](✉) and Dorit Merhof[2]

[1] Institute of Imaging and Computer Vision, RWTH Aachen University, Aachen, Germany
long.chen@lfb.rwth-aachen.de
[2] Faculty of Informatics and Data Science, University of Regensburg, Regensburg, Germany

Abstract. Clustering pixels based on learned instance-discriminative pixel embeddings is a promising approach for the instance segmentation task, particularly with highly cluttered objects. The pixel embedding space is typically trained with proxy-based losses due to the large number of pixel samples. However, we have found that training guided by randomly sampled pixel triplets is not only feasible but also consistently yields better results. With our proposed loss, a basic convolution-based model achieves state-of-the-art results, with minimal pre- and post-processing, on a variety of biomedical datasets: FluoDSB (fluorescence cell), CVPPP (Arabidopsis leaf), Celegans (*C. elegans* nematode), and Neuroblastoma (cultured neuroblastoma cell).

Keywords: Instance segmentation · Instance discriminative · Pixel embedding · Pixel triplet

1 Introduction

A wide range of vision-based applications rely on instance segmentation, which aims to associate each foreground pixel to an object in the image plane. Recent advances in deep learning techniques have made neural network-based approaches predominant in this field. Beyond the performance improvements of the interchangeable backbone network, which benefits all deep learning-based approaches, the output design has a more direct influence on the performance for a specific task. An accurate, robust, and easily learnable representation of the targets is desirable.

State-of-the-art instance segmentation follows two main strategies. The popular top-down approach [1–4] identifies objects using a manageable representation, such as bounding boxes [1,2] or polygons [3,4]. Bounding boxes are further refined through a segmentation task, whereas polygons are typically used

Supplementary Information The online version contains supplementary material available at https://doi.org/10.1007/978-3-031-73284-3_29.

© The Author(s), under exclusive license to Springer Nature Switzerland AG 2025
X. Xu et al. (Eds.): MLMI 2024, LNCS 15241, pp. 290–299, 2025.
https://doi.org/10.1007/978-3-031-73284-3_29

to directly reconstruct pixel masks. However, the limitations of these representations become evident when objects become denser and shapes more complex. Bounding box-based approaches, which rely on non-maximum suppression (NMS) to remove duplicates, suffer from false suppression. This is due to denser objects leading to a significant bounding box overlap. The situation worsens for inherently overlapping objects and when object crossovers make the bounding boxes nearly identical, as seen in the *C. elegans* (*Caenorhabditis elegans*) data. The performance of the polygon-based approach degrades when the object shape is irregular, which is often the case in the biomedical domain.

On the other hand, the alternative bottom-up approach converts pixel values into representations that can be readily used for grouping pixels into objects. The explicit prediction of object boundaries [5] is often used to separate objects in biomedical images, partially due to its straightforward implementation. However, this method is sensitive to blurred and broken boundaries. Other grouping techniques include graph partitioning based on pixel pair affinity [6], applying the watershed method on a predicted distance transform map [7], and tracking pixels with a vector field [8]. All these methods follow the same strategy: they use deep models to map all pixels into an intermediate presentation, which ideally can be grouped efficiently using low-level processing techniques. The performance bottleneck of bottom-up approaches is the robustness and universality of the corresponding post-processing technique.

Recent research [9–11] proposes a more concise paradigm: mapping pixels into an embedding space, in which pixels of the same object are proximate, while those from different, particularly adjacent, objects are separated. Grouping can be accomplished using mature clustering algorithms such as MeanShift [12] and DBSCAN [13]. The pixel embedding approach demonstrates greater robustness among bottom-up methods and offers advantages in handling densely distributed objects and irregular shapes over top-down methods.

All previous work uses a proxy-based loss, which indirectly shifts a group of pixels via a proxy, typically the mean embedding of objects (Sect. 2.1). Even though triplet loss [14,17] and other proxy-free losses [15,16] are common in image-level contrastive learning, they have not been applied to pixel embedding learning. This is partly due to the study conclusion on image-level contrastive learning, which states that the sampling strategy, such as batch-hard, is crucial for training [17]. The large pixel space, combined with the need to consider all pixels, makes the application of many sampling strategies infeasible.

In this study, we propose a novel, carefully constructed triplet loss (Sect. 2.2). It allows randomly sampled pixel triplets to achieve superior results compared to the proxy-based loss. Furthermore, our research shows that the hard-case mining contributes minimally to the pixel-level contrastive learning, which deviates from its effects in image-level scenarios. With a basic convolutional backbone model and minimal pre- and postprocessing, our approach delivers state-of-the-art results on four datasets, each featuring different image modalities, object shapes, distributions, and densities: FluoDSB (fluorescence cell) [20], CVPPP (arabidopsis leaf) [19], Celegans (*C. elegans* nematode) [21], and Neuroblastoma (cultured neuroblastoma cell) [22].

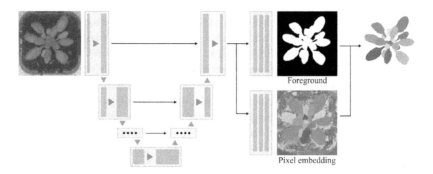

Fig. 1. Overview of the experimental model: the pixel embedding and an auxillary foreground are predicted by two separate light-weighted modules, sharing the same feature map from the backbone network.

2 Approach

In this study, we employ a concise convolution-based encoder-decoder model, depicted in Fig. 1. The foreground and the instance-discriminative embedding are generated by two distinct modules sharing the same feature map as input. While any architecture capable of dense predictions can be used in principle, we chose a standard U-Net [25] implementation due to our primary focus on the learning process.

The foreground branch excludes background pixels from further processing. All foreground pixels are grouped into objects using the Mean-Shift [12] clustering approach, based on the predicted pixel embedding. Correspondingly, pixel embedding learning only considers the foreground pixels.

2.1 Proxy-Based Loss

The core concept of instance-discriminative pixel embedding is to make the distance between different objects larger than the intra-object distance. A practical approach to consider all pixels involves using a centralized form: pushing representative centers (proxies) apart and pulling all pixels towards their respective centers. Typically, the average embedding of all pixels in an object is used as the proxy.

All previous works train the model using proxy-based losses, generally in the following form:

$$\mathcal{L} = \mathcal{L}_{attr}(\mu_\mathbf{c}, \mathbf{x}_p) + \mathcal{L}_{rep}(\mu_\mathbf{c}), \tag{1}$$

where the repelling term $\mathcal{L}_{rep}$ is a function of proxies $\mu_\mathbf{c}$, and the attracting term $\mathcal{L}_{attr}$ penalizes the difference between pixel embeddings $\mathbf{x}_p$ and their corresponding proxy $\mu_\mathbf{c}$. Specific losses [9,10] can utilize various distance metrics, such as Euclidean distance or cosine distance, and follow different repelling strategies, such as pushing apart all objects or just those spatially close.

2.2 Training with Pixel Triplets

An alternative method of constructing the supervision involves using pixel triplets, composed of an anchor pixel, a positive pixel, and a negative pixel. The positive pixel belongs to the same object as the anchor, while the negative pixel belongs to a different object. Unlike the method of separating pixels from different objects through a proxy, the triplet training aims to maximize the similarity between the negative-anchor and minimize the similarity between the positive-anchor across a set of triplets.

In this work, we use the cosine similarity, denoted as $D_{cos}(\mathbf{x},\mathbf{y}) = \frac{\mathbf{x}^\top \mathbf{y}}{\|\mathbf{x}\|\|\mathbf{y}\|}$, to construct the loss function:

$$\mathcal{L}_{triplet} = \sum_{i=1}^{P} D_{cos}^2(\mathbf{x}_i^{anchor}, \mathbf{x}_i^{neg}) - D_{cos}(\mathbf{x}_i^{anchor}, \mathbf{x}_i^{pos}), \tag{2}$$

where the index i traverses all triplets $\mathbf{t}_i = (\mathbf{x}_i^{anchor}, \mathbf{x}_i^{pos}, \mathbf{x}_i^{neg})$. The term $D_{cos}^2(\mathbf{x}_i^{anchor}, \mathbf{x}_i^{neg})$ reaches its minimum when the embeddings of two pixels are orthogonal, while $D_{cos}(\mathbf{x}_i^{anchor}, \mathbf{x}_i^{pos})$ attains its minimal value when $\mathbf{x}_i^{anchor}$ and $\mathbf{x}_i^{pos}$ are identical.

Triplet Sampling. For each pixel, the positive sample is randomly selected from pixels belonging to the same object. Negative sampling is performed among pixels of adjacent objects. Our object separation focuses only on spatially close objects, as there is no risk of distant objects being falsely merged. Furthermore, we believe that long-distance connections might act as noise to the training of convolution-based models due to their shift-equivalence property [18].

Additionally, an object on the outskirts can sometimes have no adjacent objects. In these instances, a constant dummy negative sample is used. This dummy negative sample is a zero embedding vector, which ensures a zero cosine similarity with any other vector. Two objects are considered adjacent if the distance between any two of their pixels is less than 20 pixels.

Gradient Attenuation. In designing the loss function, we found that reducing the gradients from objects that are already well predicted can significantly enhance the model's performance. The gradient passed through the cosine similarity operation is as follows:

$$\frac{\partial D_{cos}(\mathbf{x},\mathbf{y})}{\partial \mathbf{y}} = \frac{1}{\|\mathbf{y}\|}(\hat{\mathbf{x}} - \hat{\mathbf{y}}\hat{\mathbf{y}}^\top \hat{\mathbf{x}})^\top, \tag{3}$$

where the hat symbol denotes a normalized vector. From a geometric perspective, the gradient signal is equal to the projection of $\hat{\mathbf{x}}$ onto the direction perpendicular to $\hat{\mathbf{y}}$. In other words, the angle between $\hat{\mathbf{x}}$ and $\hat{\mathbf{y}}$ modulates the gradient amplitude. If the angle is zero, the gradient diminishes, while it is maximized when the two vectors are perpendicular.

The anchor-positive term inherently has a decreasing gradient magnitude as the two embeddings gradually become identical. In contrast, the gradient of the anchor-negative term maximizes as it approaches convergence (when the two vectors are orthogonal). Overwhelming gradients from the "easy cases" can hinder the learning progress of objects not yet separated. Therefore, instead of the absolute value, we use a squared version of the first term in Eq. 2. Our experiments show a significantly improved performance (Table 2).

3 Experiments

3.1 Data and Evaluation

We use four diverse datasets to show various challenges in biomedical data.
FluoDSB contains a variety of fluorescent nuclei relevant to various cell biology studies. The data was first collected for the 2018 Data Science Bowl [20]. To ensure error-free annotation, Schmidt et al. [4] selected and revised a subset of 497 fluorescence microscopy images from the original collection.
CVPPP records the growth process of multiple tobacco and Arabidopsis plants, with each leaf manually annotated [19]. The leaves form a dense clump in each image. Our experiments utilize the publicly available training set, excluding the 28 tobacco images (subset A3). A total of 783 images are used.
Neuroblastoma comprises 100 images of cultured neuroblastoma cells [22]. Each image captures both cytoplasmic and nuclear stains. In this work, we focus on the more irregular and extended cytoplasm.
Celegans collects 100 bright-field images, each containing multiple alive/dead *C. elegans* nematodes [21]. These slender worms can adopt straight, curved, or ring-shaped postures and may come into contact or cross over each other.

We use the sortedAP score [23] for evaluation. This metric is parameter-free and more sensitive to typical instance segmentation errors compared to others, such as mean Average Precision (mAP) and Panoptic Quality (PQ).

3.2 Implementation

Focusing on training supervision, we use a pretty basic fully convolutional encoder-decoder model, with one extra encoder and decoder stage compared to U-Net [25]. Filter numbers are halved for a lightweight model. All output modules consist of two 3×3 convolutional layers, two post-activation batch normalization layers and a 1×1 convolutional layer for output. The foreground output is Sigmoid-activated, and the embedding output uses linear activation.

The training loss is the equal-weighted sum of all output module losses. The foreground module is supervised by a standard binary cross-entropy loss. The model is optimized using the Adam [24] optimizer, with an initial learning rate of $1e^{-3}$, decaying to 0.9 every 10k steps. The training follows the train-validate-test practice and lasts for 300k steps. Validation is conducted after each epoch. The model snapshot that has the highest validation score is used for the final evaluation. For details on dataset splits, see the Supplementary Material.

We apply minimal pre- and post-processing without any data-specific operations. Images are resampled to a uniform size of 512×512 pixels. Per-image standardization is applied by subtracting the image mean and dividing by the standard deviation. During training, we augment the data with geometric transformations, including random flipping, shifting, rotation, and elastic deformation. A trivial threshold of 0.5 identifies the foreground pixels, which are then clustered into objects based on the pixel embedding using Mean-Shift [12]. For Mean-Shift, a binning size of 0.2 and a bandwidth of 0.5 are used. In the end, we remove noisy objects smaller than 100 pixels, with their pixels merged into the nearest object.

3.3 Results and Discussion

(A) Triplet Number and Hard Cases. As shown in Table 2, increasing the number of triples does not improves the score as intuitively expected. Even when we only use triplets of 0.1-fold foreground pixels, the performance does

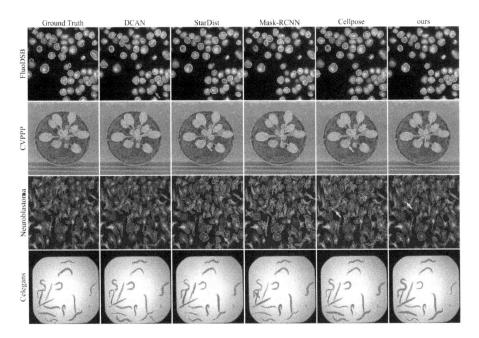

Fig. 2. Qualitative results of different instance segmentation approaches. A few typical errors are marked with arrows: missed objects (blue), false merging (red), inaccurate outlines (green) and over-segmentation (yellow). (Color figure online)

not deteriorate. We believe that due to the high correlation of pixels in the vicinity, very dense sampling is not necessary. Furthermore, the random process conducted during each training step reduces the need for intensive sampling.

Unlike the decisive impact in image-level contrastive learning, the effect of hard cases in pixel embedding learning is marginal. Training with 50th-percentile hard cases only improves the sortedAP score by an average of 1.07% across four datasets. A comparable outcome is observed with the 75th-percentile hard cases, as shown in Table 2.

(B) Gradient Attenuation. Gradient attenuation is a crucial component, without which the triplet approach cannot surpass the proxy-based loss. It enhances the sortedAP score by 11.94% (see Table 2, 1st section). A comparable improvement 13.50% is also observed in the proxy-based loss (see the Supplementary Material).

(C) Augmentation. Geometric augmentation significantly improves the performance by an average of 68.32% across four datasets, while appearance augmentation has a negligible impact (Table 2, 2nd section). Unless the dataset is extremely small, such a performance gain from data augmentation is unexpected. In instance-discriminative learning, the CNN model heavily relies on postures, such as object orientation, and the positional arrangement of close objects. Geometric augmentation naturally enriches this information. Limited by article length, we will explain the mechanisms of shift-equivalence models assigning different embedding to pixels in the same image in a separate publication.

(D) Comparison with Other Approaches. Our triplet method consistently outperforms the proxy-based loss by 1.83% on average. Compared with all methods, it tops two out of four datasets and closely matches the best scores on the rest (Table 1). This is largely due to its robustness against various error types. While other approaches perform well on specific datasets, like StarDist on roundish cells (FluoDSB) and DCAN on Neuroblastoma, they are prone to specific error types in certain situations. For an insightful understanding, we discuss the results from the perspective of error types.

Missed objects (blue arrows, Fig. 2). For high object density, false suppression is a rarely avoidable problem for NMS, in which almost all bounding box based approaches involve. Denser objects lead to higher overlap ratio, making a perfect suppression threshold hard to determine. A larger NMS threshold can alleviate the problem (also increase the risk of duplicated prediction), it cannot be completely avoided, even using large thresholds over 0.8. Certain arrangements, such as two parallel leaves and two cross-over *C. elegans* (column 4 in Fig. 2) are inherently indistinguishable by the bounding boxes.

False merging (red arrows, Fig. 2). Although all approaches take the separation of touching objects into consideration, some bottom-up approaches are less robust in terms of the false merging error. For instance, DCAN relies on continual boundaries to segregate objects. However, boundaris are very vulnerable structure. Even when the accuracy of predicting boundary pixels is high, a few miss-classified pixels can directly lead to false merge of objects.

Inaccurate outline (green arrows, Fig. 2). This is a major issue of approaches that represent object outline with polygons. This approximation is good enough for many daily objects, but inferior to more irregular shapes that appear more often in biomedical data. For instance, StarDist has difficulties in reconstructing fine boundaries of neuroblastoma cells and leaves.

Over-segmentation (yellow arrows, Fig. 2). Another strenuosity of polygon-based approaches is reconstructing slender shapes, such as the C. elegans body, resulting in an over-segmentation. Over-segmentation is more common in bottom-up approaches, observed both in our method and Cellpose, which is basically a less grouped situation. As stated, for most bottom-up approaches, the robustness of the grouping process is the crucial point.

Table 1. Comparison to state-of-the-art. The approaches are evaluated using the sortedAP score. The best and second best scores are in bold.

Dataset	FluoDSB	CVPPP	Neuroblastoma	Celegans
DCAN [5]	.5647	.5552	**.6860**	.6860
StarDist [4]	.6997	.6163	.5744	.2600
CellPose [8]	**.7163**	.6345	.6774	.7465
MRCNN [1]	.6682	**.6682**	.6212	.6643
Proxy-based [10]	.6942	.6866	.6577	**.7849**
Triplet	**.7132**	**.6881**	.6776	**.7955**

Table 2. Ablation study of learning with triplets. The reference (underlined) experiment by default applies gradient attenuation, geometric data augmentation and no hard-case mining. The other experiments consider the effects by controlling variables. The sortedAP score is used here.

gradAtt.	aug.	hard-case	#triplet	FluoDSB	CVPPP	Neuroblastoma	Celegans
✓	geo.	–	1-fold	.7132	.6881	.6776	.7955
	geo.	–	1-fold	.6840	.5049	.6531	.7688
✓	none	–	1-fold	.5946	.4312	.3354	.4149
✓	app.	–	1-fold	.6067	.4275	.3335	.4095
✓	geo.	0.5	1-fold	.7198	.6891	.6889	.8078
✓	geo.	0.75	1-fold	.7181	.6855	.6984	.8102
✓	geo.	–	0.1-fold	.7034	.6874	.6756	.8015
✓	geo.	–	0.5-fold	.7109	.6839	.6921	.7887
✓	geo.	–	2-fold	.7016	.6835	.6834	.8148
✓	geo.	–	4-fold	.7042	.6776	.6932	.7914

4 Conclusion and Outlook

We propose a pixel triplet-based approach that consistently outperforms proxy-based losses in training an instance-discriminative pixel embedding. Our method, using a basic convolution-based backbone with minimal pre-/post-processing, achieves superior performance among other competing methods. This demonstrates the effectiveness of pixel embedding paradigms in instance segmentation tasks. We've explored how shift-equivalence convolution-based models achieve instance discrimination on the same image plane. These findings will be published in a separate publication. Based on this research, our future work will focus on exploring the optimal learning architecture, particularly long-distance pixel connections, the concept popularized by transformers.

References

1. He, K., Gkioxari, G., Dollár, P., Girshick, R.: Mask R-CNN. In: Proceedings of the 2017 IEEE International Conference on Computer Vision, pp. 2980–2988 (2017)
2. Chen, X., Girshick, R., He, K., Dollár, P.: TensorMask: a foundation for dense object segmentation. In: Proceedings of the 2019 IEEE/CVF International Conference on Computer Vision, pp. 2061–2069 (2019)
3. Xie, E., et al.: PolarMask: single shot instance segmentation with polar representation. In: Proceedings of the 2020 IEEE/CVF Conference on Computer Vision and Pattern Recognition, pp. 12193–12202 (2020)
4. Schmidt, U., Weigert, M., Broaddus, C., Myers, G.: Cell detection with star-convex polygons. In: Frangi, A.F., Schnabel, J.A., Davatzikos, C., Alberola-López, C., Fichtinger, G. (eds.) Medical Image Computing and Computer Assisted Intervention – MICCAI 2018: 21st International Conference, Granada, Spain, September 16-20, 2018, Proceedings, Part II, pp. 265–273. Springer International Publishing, Cham (2018). https://doi.org/10.1007/978-3-030-00934-2_30
5. Chen, H., Qi, X., Yu L., Dou, Q., Qin, J., Heng, P.A.: DCAN: deep contour-aware networks for accurate gland segmentation. In: Proceedings of the 2016 IEEE Conference on Computer Vision and Pattern Recognition, pp. 2487–2496 (2016)
6. Gao, N., et al.: SSAP: single-shot instance segmentation with affinity pyramid. In: Proceedings of the 2019 IEEE/CVF International Conference on Computer Vision, pp. 642–651 (2019)
7. Bai, M., Urtasun, R.: Deep watershed transform for instance segmentation. In: Proceedings of the 2017 IEEE Conference on Computer Vision and Pattern Recognition, pp. 5221–5229 (2017)
8. Stringer, C., Wang, T., Michaelos, M., Pachitariu, M.: Cellpose: a generalist algorithm for cellular segmentation. Nat. Methods **18**(1), 100–106 (2021)
9. De Brabandere, B., Neven, D., Van Gool, L.: Semantic instance segmentation for autonomous driving. In: Proceedings of the 2017 IEEE Conference on Computer Vision and Pattern Recognition Workshops, pp. 7–9 (2017)
10. Chen, L., Strauch, M., Merhof, D.: Instance segmentation of biomedical images with an object-aware embedding learned with local constraints. In: Proceedings of 2019 International Conference on Medical Image Computing and Computer-Assisted Intervention, pp. 451–459 (2019)

11. Payer, C., Štern, D., Feiner, M., Bischof, H., Urschler, M.: Segmenting and tracking cell instances with cosine embeddings and recurrent hourglass networks. Med. Image Anal. **57**, 106–119 (2019)
12. Comaniciu, D., Meer, P.: Mean shift: a robust approach toward feature space analysis. IEEE Trans. Pattern Anal. Mach. Intell. **24**(5), 603–619 (2002)
13. Campello, R.J., Moulavi, D., Sander, J.: Density-based clustering based on hierarchical density estimates. In: Proceedings of 2013 Pacific-Asia Conference on Knowledge Discovery and Data Mining, pp. 160–172 (2013)
14. Weinberger, K. Q., Saul, L. K.: Distance metric learning for large margin nearest neighbor classification. J. Mach. Learn. Res. **10**(2), 207–244 (2009)
15. Sohn, K.: Improved deep metric learning with multi-class N-pair loss objective. In: Proceedings of the 30th Conference on Neural Information Processing Systems
16. Oh Song, H., Xiang, Y., Jegelka, S., Savarese, S.: Deep metric learning via lifted structured feature embedding. In: Proceedings of the 2016 IEEE Conference on Computer Vision and Pattern Recognition, pp. 4004–4012 (2016)
17. Hermans, A., Beyer, L., Leibe, B.: In defense of the triplet loss for person re-identification (2017). arXiv preprint arXiv:1703.07737
18. Zhang, R.: Making convolutional networks shift-invariant again. In: Proceedings of 2019 International Conference on Machine Learning, pp. 7324–7334 (2019)
19. Scharr, H., Pridmore, T. P., Tsaftaris, S. A.: Computer vision problems in plant phenotyping, CVPPP 2017–Introduction to the CVPPP 2017 workshop papers. In: Proceedings of the 2017 IEEE International Conference on Computer Vision Workshops, pp. 2020–2021 (2017)
20. Caicedo, J.C., et al.: Nucleus segmentation across imaging experiments: the 2018 data science Bowl. Nat. Methods. **16**(12), 1247–1253 (2019)
21. Ljosa, V., Sokolnicki, K.L., Carpenter, A.E.: Annotated high-throughput microscopy image sets for validation. Nat. Methods **9**(7), 637–637 (2012)
22. Yu, W., Lee, H.K., Hariharan, S., Bu, W.Y., Ahmed, S.: CCDB: 6843, mus musculus. Neuroblastoma. CIL, Dataset (2019)
23. Chen, L., Wu, Y., Stegmaier, J., Merhof, D.: SortedAP: rethinking evaluation metrics for instance segmentation. In: Proceedings of the 2023 IEEE/CVF International Conference on Computer Vision, pp. 3923–3929 (2023D)
24. Kingma, D. P., Ba, J.: Adam: a method for stochastic optimization (2014). arXiv preprint arXiv:1412.6980
25. Ronneberger, O., Fischer, P., Brox, T.: U-Net: convolutional networks for biomedical image segmentation. In: Proceedings of 2015 International Conference on Medical Image Computing and Computer Assisted Intervention, pp. 234–241 (2015)

Geo-UNet: A Geometrically Constrained Neural Framework for Clinical-Grade Lumen Segmentation in Intravascular Ultrasound

Yiming Chen[1], Niharika S. D'Souza[2(✉)], Akshith Mandepally[3],
Patrick Henninger[3], Satyananda Kashyap[2], Neerav Karani[1], Neel Dey[1],
Marcos Zachary[3], Raed Rizq[3], Paul Chouinard[3], Polina Golland[1],
and Tanveer F. Syeda-Mahmood[2]

[1] Massachusetts Institute of Technology, Boston, MA, USA
[2] IBM Research, Almaden, San Jose, CA, USA
Niharika.Dsouza@ibm.com
[3] Boston Scientific Corporation, Maple Grove, MN, USA

Abstract. Precisely estimating lumen boundaries in intravascular ultrasound (IVUS) is needed for sizing interventional stents to treat deep vein thrombosis (DVT). Unfortunately, current segmentation networks like the UNet lack the precision needed for clinical adoption in IVUS workflows. This arises due to the difficulty of automatically learning accurate lumen contour from limited training data while accounting for the radial geometry of IVUS imaging. We propose the Geo-UNet framework to address these issues via a design informed by the geometry of the lumen contour segmentation task. We first convert the input data and segmentation targets from Cartesian to polar coordinates. Starting from a convUNet feature extractor, we propose a two-task setup, one for conventional pixel-wise labeling and the other for *single boundary* lumen-contour localization. We directly combine the two predictions by passing the predicted lumen contour through a new activation (named CDFeLU) to filter out spurious pixel-wise predictions. Our unified loss function carefully balances area-based, distance-based, and contour-based penalties to provide near clinical-grade generalization in unseen patient data. We also introduce a lightweight, inference-time technique to enhance segmentation smoothness. The efficacy of our framework on a venous IVUS dataset is shown against state-of-the-art models.

Keywords: Lumen Segmentation · Intravascular Ultrasound · Geometric Contour Modeling · CDF Error Linear Units

---

Y. Chen and N. S. D'Souza—Joint first-authorship.

Supplementary Information The online version contains supplementary material available at https://doi.org/10.1007/978-3-031-73284-3_30.

1 Introduction

Deep Vein Thrombosis (DVT) is a serious condition that can cause significant short-term discomfort and lead to irreversible venous system damage that may be limb or life-threatening [14]. It is a precursor to pulmonary embolism, a critical condition where a clot travels to the lungs, impeding blood oxygenation. To manage DVT, clinicians often utilize Intravascular Ultrasound (IVUS) [15] to guide endovascular treatments, where a catheter equipped with an ultrasound transducer is inserted to visualize internal structures and pinpoint anatomical landmarks. IVUS samples are organized into pullbacks, where consecutive frames of images are captured as the catheter travels through the blood vessel, emitting sound waves that are reflected by/pass through structures based on their densities [15]. The physician may remove the thrombus and insert balloons or stents in its place to keep the vessel open. These devices are sized based on nearby healthy regions, where accurate measurement of the vessel's lumen is crucial for avoiding complications like pain from improper device sizes or fatal stent migration [16]. Automatic segmentation of venous IVUS (v-IVUS) images is challenging owing to variability/irregularity in tissue/vessel appearance across subjects due to thin vessel walls, noise, stents, artifacts, and the manual nature of the pullback (i.e. variable longitudinal frame rate across pullbacks due to manual control of the catheter by the physician).

Deep Neural Networks (DNNs) for vascular segmentation [1] have soared in popularity due to their ability to provide improved performance without manual intervention during deployment. Variants of the UNet [13] have been successful for plaque/calcification detection and vessel segmentation [1,2,6,20] as well as stent [18] and lesion detection/classification [11] for coronary artery disease. These use either 2D images [2] or 3D image blocks [6,9] as inputs and produce a pixel-wise map of the segmentation target as the output. The IVUS segmentation literature focuses on arterial acquisitions which provide a different field of view (FoV) and use a motorized pullback providing a fixed longitudinal frame rate. However, venous acquisitions are not well-studied, and most existing techniques do not generalize well to v-IVUS data due to under/over-segmentation of lumen regions in the presence of imaging artifacts and their predilection to output spurious, fragmented predictions when there are nearby vessels or tissue structures. We posit that this is due to their inability to reflect the radial geometry of the imaging modality and constrain the output to be a single contiguous lumen region, as dictated by the anatomy under consideration.

In this paper, we alleviate the issues above by designing a new neural framework, named Geo-UNet—a fully convolutional architecture for lumen segmentation from venous IVUS images that satisfies radial contour-geometry constraints directly through the imaging-representation, architecture, and loss functions (as opposed to imposing anatomical constraints via regularization alone [12]). Our method features 3 main components: **1) Input representation:** we operate on 2D-image inputs converted from Cartesian to polar coordinates which better reflect inherent IVUS imaging physics [2,17]. **2) Anatomically Constrained Self-informing Network**: We propose a two-task setup with a shared UNet

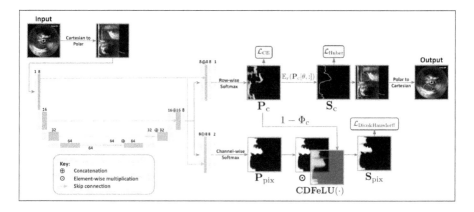

Fig. 1. Geo-UNet Architecture for Lumen Segmentation: The feature extractor is a fully convolutional UNet module with inputs of polar 2D IVUS frames. The top branch produces a probability map for the lumen contour ($\mathbf{P}_c$) via a row-wise softmax, which is converted to a single contour segmentation ($\mathbf{S}_c$) via a row-wise expectation function. The bottom branch produces a per-pixel probability map ($\mathbf{P}_{\text{pix}}$) via a channel-wise softmax. CDFeLU(·) allows the top branch to inform the bottom, refining the pixel-wise probabilities to give the segmentation ($\mathbf{S}_{\text{pix}}$) that is compared against the (polar) ground-truth lumen mask. The loss functions are highlighted in grey.

feature extraction module. In polar space, the lumen boundary is a single contour. While the natural prediction target is a standard pixel-level segmentation, we design a second objective to predict a single lumen boundary contour. Using this prediction as a guide, we refine the pixel-level segmentation via a new activation function—*CDFeLU*, based on the cumulative distribution function. This regularization mitigates spurious predictions from pixel-level segmentation without the need for additional post-processing, a known shortcoming of prior approaches. During training, our unified loss function combines area-based, distance-based, and contour-based penalties to provide improved generalization.
3) Inference-time Continuity Enhancement: Based on the radial geometry in imaging and properties of the convolutional UNet, we propose a continuity enhancement technique, coined Geo-UNet++, which is a lightweight, inference-time procedure to address wrap-around discontinuities at $0/2\pi$ angles in the segmentation estimation. Our framework compares favorably against state-of-the-art segmentation baselines with consistent improvements in segmentation Dice scores and derived lumen diameter estimation for stent sizing.

2 Geo-UNet for Lumen Segmentation from v-IVUS

Figure 1 illustrates the Geo-UNet framework. The shared convUNet feature extractor connects to two prediction branches as detailed below.

Lumen Contour Estimation Branch: In polar space, the horizontal and vertical axes of an IVUS image correspond to radii (r) and angles (θ), respectively.

$\mathbf{Y}_{\text{pix}}$ denotes the ground-truth binary lumen mask of size $R \times R$ ($R = 256$ pixels). Summing along the r coordinate for each θ gives the contour lumen map $\mathbf{Y}_c[\cdot]$ of size $R \times 1$. $\mathbf{Y}_c[\theta] = \sum_r \mathbf{Y}_{\text{pix}}[\theta, r]$ captures the lumen depth at each θ, a distinct value in $\{0, \ldots, R\}$. The lumen boundary is a single, smooth contour with no self-intersection (i.e. has a distinct depth $r \in \{0, \ldots, R-1\}$ for each $\theta \in \{0, \ldots, R-1\}$, after discretizing the range $[0, 2\pi]$ into R intervals). The top network branch captures the lumen contour by computing a softmax across each row of the single-channel output to obtain a row-sparse probability map $\mathbf{P}_c$ of size $R \times R$. The entries $\mathbf{P}_c[\theta, r] \in [0, 1]$ denote the probability that the contour depth at θ is r and is ideally high along the lumen contour and near 0 elsewhere. We convert $\mathbf{P}_c$ into a segmentation contour $\mathbf{S}_c$ with an expectation across radii values: $\mathbf{S}_c[\theta] = \mathbb{E}_r(\mathbf{P}_c[\theta, :]) = \sum_{r=0}^{R-1} r * \mathbf{P}_c[\theta, r]$, accounting for the uncertainty along boundary pixels in a differentiable operation [4]. We use two training losses for $\mathbf{P}_c$ and $\mathbf{S}_c$. First, we compute the cross entropy between $\mathbf{P}_c$ and $\mathbf{Y}_c$:

$$\mathcal{L}_{\text{CE}} = \frac{-1}{R^2} \sum_{\theta, r=0,0}^{R-1} \mathbb{1}[\mathbf{Y}_c[\theta] = r]\log(\mathbf{P}_c[\theta, r]) + \mathbb{1}[\mathbf{Y}_c[\theta] \neq r]\log(1 - \mathbf{P}_c[\theta, r]) \quad (1)$$

We then encourage $\mathbf{S}_c[\theta]$ to match $\mathbf{Y}_c[\theta]$ using a Huber loss [7]:

$$\mathcal{L}_{\text{Huber}}(\cdot) = \sum_{\theta=0}^{R-1} \frac{d_\theta^2}{2} \mathbb{1}(|\mathbf{d}_\theta| < 1) + (|\mathbf{d}_\theta| - 0.5)\mathbb{1}(|\mathbf{d}_\theta| \geq 1), \quad (2)$$

where $\mathbf{d}_\theta = \mathbf{Y}_c[\theta] - \mathbf{S}_c[\theta]$. To get a polar binary segmentation that guarantees a single lumen in Cartesian space, for each θ, we have 1s for pixels to the left of/along $\mathbf{S}_c[\theta]$ and 0 elsewhere. This serves as the final prediction output.

Pixel-wise Segmentation Branch with Probabilistic Contour Maps: Applying a conventional channel-wise softmax operation [1], the bottom branch outputs a pixel-wise probability map $\mathbf{P}_{\text{pix}}$ of size $R \times R$, where $\mathbf{P}_{\text{pix}}[\theta, r]$ denotes the probability that pixel $[\theta, r]$ is in or on the lumen boundary. To reconcile this with the lumen contour estimate, we compute a dense probability map from $\mathbf{P}_c$ via a novel activation function based on the cumulative distribution function (CDF). Let $\Phi_c[\theta, r] = \text{CDF}(\mathbf{P}_c[\theta, r])$, the transformation $(1 - \Phi_c[\theta, r])$ models the confidence that the pixel $[\theta, r]$ is contained within the lumen that is larger at smaller radii, serving as a probabilistic mask for $\mathbf{P}_{\text{pix}}$. We compute the refined pixel-wise segmentation $\mathbf{S}_{\text{pix}}$ of size $R \times R$ via the activation CDFeLU($\mathbf{P}_{\text{pix}}, \mathbf{P}_c$):

$$\mathbf{S}_{\text{pix}}[\theta, r] = \mathbf{P}_{\text{pix}}[\theta, r] * (1 - \Phi_c[\theta, r]) = \mathbf{P}_{\text{pix}}[\theta, r] * \left[1 - \sum_{j=0}^{r} \mathbf{P}_c[\theta, j]\right]$$

CDF error Linear Units (CDFeLU) is analogous to Gaussian Error Linear Units (GELU) [5], where the CDF error is estimated based on the geometry of the lumen boundary as opposed to a normal distribution. Finally, we impose a combination of area-based (Dice) and distance-based (Hausdorff [3]) losses on $\mathbf{S}_{\text{pix}}$ to match the ground-truth pixel-wise lumen mask $\mathbf{Y}_{\text{pix}}$:

$$\mathcal{L}_{\text{Dice\&Hausdorff}}(\cdot) = \lambda * \mathcal{L}_{\text{Dice}}(\mathbf{S}_{\text{pix}}, \mathbf{Y}_{\text{pix}}) + (1 - \lambda) * \mathcal{L}_{\text{Haus.}}(\mathbf{S}_{\text{pix}}, \mathbf{Y}_{\text{pix}}) \quad (3)$$

with the trade-off $\lambda \in (0,1)$ determined experimentally to be 0.9. Note that by design, CDFeLU($\mathbf{P}_{\text{pix}}, \mathbf{P}_{\text{c}}$) de-emphasises regions outside the lumen (right of $\mathbf{Y}_{\text{c}}[\theta]$), filtering out potentially spurious predictions in $\mathbf{P}_{\text{pix}}$, a task usually reserved for manual/semi-automated post-processing. At the same time, it reinforces overlaps between $\mathbf{P}_{\text{c}}$ and $\mathbf{P}_{\text{pix}}$, effectively encouraging Geo-UNet to focus on estimates that align well across the two branches during training.

Geo-UNet++ to Alleviate Wrap-Around Artifacts During Inference: Recall that we map pixel intensities from Cartesian space to r-θ space to generate polar images, where $\theta \in \{0,\ldots,2\pi\}$. A consequence is that the intensities of the model predictions are not constrained to align at $\theta = 0$ and $\theta = 2\pi$, as they lie at the top and bottom borders of the polar image. This often results in a wrap-around discontinuity when converting back to Cartesian coordinates that consistently induces errors in the diameter estimation. To alleviate this, we introduce an inference-time technique based on the radial nature of the Cartesian v-IVUS images and properties of convolution. We apply vertical wrap-padding to yield a rectangular, continuous input ranging $\theta = \{-\pi/2,\ldots,2\pi\}$ by copying over the additional $\pi/2$ context. With frozen weights, the trained Geo-UNet model can be applied as is to the padded input. We slice the output across the middle section $\theta = \{-\pi/3,\ldots,-\pi/3 + 2\pi\})$ to avoid edge effects in the padded input predictions, before finally presenting the result on the rotated Cartesian input. We observe improved prediction alignment along the re-sliced output for the padded input. See Fig. 3 (supplementary) for a walk-through. This increases deployment time marginally (0.3–0.4 ms/frame) to enhance accuracy.

3 Data, Experimental Evaluation, and Results

Data: Our images are acquired using the Boston Scientific OC35 peripheral imaging catheter, which uses a rotating transducer to generate cross-sectional views. The catheter has a 70 mm imaging diameter and a 15 MHz operating frequency. It is typically used in the detection and treatment of venous disease (e.g. DVT, non-thrombotic iliac venous lesions, chronic post-thrombotic syndrome, and more). No registration is needed to align the IVUS frames by the nature of the acquisition. We obtained data for 79 patients with 166 pullbacks of varying durations. The data is labeled per frame and partitioned into two groups: diseased and normal. The former refers to regions with acute/subacute clots and chronic Post Thrombotic Syndrome (PTS). The latter contains labels N1 (frames with typical geometry despite variability in appearance shown in Fig. 4 (supplementary)) and N2 (frames with irregular geometry due to compression from nearby vessels but no thrombus present). Since stent-sizing is performed on healthy frames, all N1/N2 frames were labeled by expert annotators, for a total of 77,917 annotated image frames. Given the increased variability in appearance and subjectivity in annotation, the lumen in N2 frames is qualitatively harder to segment as compared to N1 frames.

Implementation Details: We train all models on healthy images (frames marked N1 and N2) and adopt a three-fold cross-validation which stratifies pullbacks across patients (53/21/5 train/test/validation). Input IVUS frames and model outputs are of size 256 × 256 ($R = 256$). Hyperparameters across all experiments are determined using the validation set. We use a batch size of 3 with 16 gradient accumulation steps. The Adam optimizer is used with a scheduler that linearly decreases the learning rate from 10^{-4} to 10^{-7} over 50,000 training iterations. We apply stacked augmentations including rotation, translation, shear, contrast enhancement, Gaussian blur, intensity scaling, and speckle noise on Cartesian inputs for better generalization [23]. The training loss sums Eqs. (1)–(3). To save on compute time, we only retain $\mathcal{L}_{\text{Huber}}(\cdot)$ (Eq. 2) at each validation step to guide model optimization for Geo-UNet/ablations and $\mathcal{L}_{\text{Dice}}(\cdot)$ for the vanilla U-Net baselines. Our machine has 50 CPU cores and 2 A-100 NVIDIA GPUs with 32 GB RAM, resulting in an average training time of 3.5–4 h per cross-validation fold. To estimate the lumen diameter from a segmentation mask, we pass lines through the center of mass (COM) of the largest component at 5° increments. The longest and shortest lengths of intersection with the mask border are the major and minor diameters, respectively.

Finally, to further encourage spatial contiguity in the predicted masks, we tried introducing implicit smoothness constraints via 1D average pooling on the polar representation across the θ axis and as a separate post-processing mechanism [2] on the output. However, both strategies provided negligible performance improvements when weighed against the additional training/inference times.

Evaluation/Clinical Targets: In addition to the test-Dice, we evaluate the measurement error in the diameter of the major/minor axes of the predicted lumen against that of the ground-truth lumen [16]. Commercial stents are sized on N1 frames, are available in 0.5 mm increments, and are sized against the average of the major and minor diameter [16]. Per a clinician, the models need to achieve a major and minor axis diameter error within 0.25/0.5/0.75 mm for 50/90/95% of all N1 frames. N2 frames are mainly used for vessel compression detection and not for stent-sizing. Thus, they have less stringent clinical targets of 50/70% of frames within errors of 0.5/0.75 mm.

3.1 Baselines Comparisons and Ablations

We curate our baselines to reflect the state-of-the-art in the fields of medical image segmentation and automated processing of IVUS images.

MedSAM: Medical Segment Anything Model [10] is a general-purpose, promptable 2D-segmentation model with a ViT backbone [19], trained on multiple modalities (CT, MRI, ultrasound, etc.). The inputs are 2D medical images and a user-specified bounding box to produce a binary pixel-wise segmentation without fine-tuning. We input the Cartesian v-IVUS images and a fixed bounding box based on the FoV to accommodate lumen regions with the largest diameters.

BoundaryReg: BoundaryReg is a recent approach [4] based on convolutional UNets that was designed to produce layer surface segmentation for retinal Optical Computed Tomography (OCT). Like GeoUNet, this model estimates both dense pixel and sparse contour predictions using a shared UNet followed by two output convolutional layers without distinct skip connections. It also has additional topology modules to separate retinal layers. As A-scan OCT images and retinal layer segmentation have analogous geometric properties to polar v-IVUS representations and lumen boundary estimation, respectively, we implement this baseline for our application according to the details in [4].

Cartesian Dice & Hausdorff: Convolutional UNets are commonly used for lumen segmentation from 2D (arterial) IVUS images [1]. To adopt these baselines to v-IVUS, we use the architecture from Fig. 1 with only the bottom branch where inputs are Cartesian v-IVUS images and outputs are Cartesian masks. We train using $\mathcal{L}_{\text{Dice\&Hausdorff}}(\cdot)$ between predictions and ground truths [3,8].

Polar Dice & Hausdorff: In line with prior work [2,17], we adopt a similar architecture and loss function as the previous baseline, but convert the inputs and targets to polar representations. This baseline also serves as an ablation for Geo-UNet where the contribution of the contour estimation branch is omitted. We obtain a single lumen region from the potentially fragmented pixel-wise predictions by post-processing the outputs to retain the largest connected component [1], both in this approach and the previous baseline.

Ablation Studies: To evaluate Geo-UNet, we perform two ablations that systematically remove its key constituent components. These comparisons are (1) Geo-UNet excluding the CDFeLU re-weighting and (2) Geo-UNet without the pixel-wise prediction branch. The former uses the same loss function as Geo-UNet while the latter trains the model on a combination of $\mathcal{L}_{\text{CE}}(\cdot)$ and $\mathcal{L}_{\text{Huber}}(\cdot)$.

3.2 Lumen Segmentation Performance Analysis

To quantify the generalization performance, we report the test-Dice and percentage of frames with major and minor diameter error within 0.25/0.5/0.75 mm for N1 and N2 frames in the test subjects in Table 1 for all models. We observe that MedSAM [10] severely under-performs all the conv-UNet frameworks trained on v-IVUS, due to an inability to meaningfully discern the lumen region without a more carefully curated manual prompt and generalization limitations.

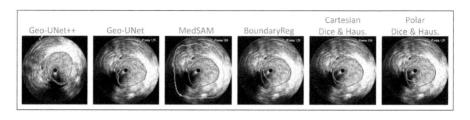

Fig. 2. Example of lumen segmentation performance. (Green-Predicted, Blue-Truth) (Color figure online)

Table 1. Model Performance. Best performance is in bold, second best is underlined.

Methodology	Test Dice (avg/std)	% Frames w. Maj. Dia. err. within 0.25/0.50/0.75 mm	% Frames w. Min. Dia. err. within 0.25/0.50/0.75 mm
Against Baselines (N1 frames)			
Geo-UNet++	0.95/0.045	66/**84**/**90**	**73**/**89**/**94**
Geo-UNet	**0.95**/**0.034**	**69**/**84**/**90**	69/85/<u>91</u>
MedSAM [10]	0.31/0.087	0/0/0	0/0/0
BoundaryReg [4]	0.94/0.043	60/78/86	<u>70</u>/<u>86</u>/<u>91</u>
Cart. Dice & Haus.	0.93/0.051	61/77/83	62/79/87
Polar Dice & Haus.	0.94/0.038	<u>66</u>/<u>80</u>/<u>87</u>	67/84/90
Against Baselines (N2 frames)			
Geo-UNet++	**0.88**/**0.094**	<u>41</u>/<u>59</u>/<u>69</u>	**60**/**80**/**87**
Geo-UNet	<u>0.87</u>/0.10	**47**/**64**/**73**	<u>57</u>/<u>76</u>/<u>85</u>
MedSAM [10]	0.23/0.085	0/0/0	0/0/0
BoundaryReg [4]	<u>0.87</u>/0.093	36/54/65	55/74/84
Cart. Dice & Haus.	0.83/0.12	32/44/52	44/63/74
Polar Dice & Haus.	0.86/0.12	40/58/<u>69</u>	55/74/83
Against Ablations (N1 frames)			
Geo-UNet	0.95/**0.034**	**69**/**84**/**90**	**69**/<u>85</u>/**91**
w/o CDFeLU	0.94/0.035	**69**/<u>82</u>/<u>88</u>	65/83/<u>90</u>
w/o pixel-wise pred.	0.95/0.039	67/81/87	**69**/**85**/**91**
Against Ablations (N2 frames)			
Geo-UNet	0.87/0.10	**47**/**64**/**73**	**57**/**76**/**85**
w/o CDFeLU	0.86/0.10	45/<u>63</u>/<u>72</u>	<u>53</u>/<u>71</u>/<u>81</u>
w/o pixel-wise pred.	**0.88**/**0.092**	<u>46</u>/62/71	**57**/**76**/**85**

BoundaryReg underperforms Geo-UNet due to architectural differences and the lack of IVUS anatomy-rooted design decisions. The model trained in Cartesian space uniformly performs worse than all polar models, reinforcing our choice to use polar representations. The polar UNet trained on only pixel-wise segmentation performs worse than the Geo-UNet on several comparisons. Upon a qualitative examination (see Fig. 2, 5 (supplementary)), the last two baselines can result in fragmented predictions with multiple components, as they are not constrained to predict a single lumen contour. This problem is not resolved by post-processing to choose the largest component given the heterogeneity across pullbacks and anatomical locations. Taking the output from the contour prediction branch inherently ensures a single prediction region. The combination of the two branches is effective as seen by comparing Geo-UNet and its ablated version without the pixel-wise prediction (Table 1). Removing the re-weighting (CDFeLU) worsens performance on both N1 and N2 frames. Finally, Geo-UNet++, featuring continuity enhancement during inference, provides improvements in the estimates of the minor diameter, while maintaining the quality of the major diameter estimates for the N1 frames[1]. Overall, these observations make a strong case for adopting geometry-informed principles into the design of neural frameworks for lumen segmentation from v-IVUS imaging.

---

[1] Errors on N2 major diams, remain above clinical precision despite slightly worsening.

Future Directions: Our framework can be easily extended beyond v-IVUS to applications with radial acquisitions/geometry such as multiple vessel boundary segmentation in arterial IVUS, retinal/airway OCT, laparoscopy, etc. Another natural extension of our framework is to 3D models that incorporate contextual information from adjacent frames [21]. This is not entirely straightforward due to 1) Large shape variances among normal v-IVUS frames 2) Normal training frames constituting non-contiguous segments within a pullback between interspersed and anatomically distinct diseased frames and 3) Variable frame rates due to manual pullbacks. We envision that 3D rendering techniques such as Neural Implicit Functions [22] could potentially circumvent these issues.

4 Conclusion

We develop a novel geometry-informed neural model, Geo-UNet, for precise lumen segmentation on venous IVUS imaging for automated stent-sizing. The two-task design, i.e. lumen contour estimation and dense pixel prediction, ensures appropriate constraints per data geometry. The CDFeLU re-weighting allows us to unify the distinct prediction targets probabilistically and effectively mitigate spurious predictions. The inclusion of complementary losses provides sufficient regularization to ensure reliable and robust generalization across unseen pullbacks (patients) despite the modest dataset size. Finally, the inference time enhancement improves performance with negligible cost. Overall, Geo-UNet/Geo-UNet++ achieves a majority of clinical targets, with only a narrow gap in others, making it an attractive assistive tool for interventional specialists.

References

1. Arora, P., Singh, P., Girdhar, A., Vijayvergiya, R.: A state-of-the-art review on coronary artery border segmentation algorithms for intravascular ultrasound (IVUS) images. Cardiovasc. Eng. Technol. **14**(2), 264–295 (2023)
2. Blanco, P.J., et al.: Fully automated lumen and vessel contour segmentation in intravascular ultrasound datasets. Med. Image Anal. **75**, 102262 (2022)
3. Cardoso, M., Li, W., Brown, R., et al.: Monai: an open-source framework for deep learning in healthcare. arXiv preprint arXiv:2211.02701 (2022)
4. He, Y., et al.: Fully convolutional boundary regression for retina OCT segmentation. In: Shen, D., et al. (eds.) MICCAI 2019. LNCS, vol. 11764, pp. 120–128. Springer, Cham (2019). https://doi.org/10.1007/978-3-030-32239-7_14
5. Hendrycks, D., Gimpel, K.: Gaussian error linear units (GELUs). arXiv preprint arXiv:1606.08415 (2016)
6. Huang, X., et al.: Post-IVUS: a perceptual organisation-aware selective transformer framework for intravascular ultrasound segmentation. Med. Image Anal. **89**, 102922 (2023)
7. Huber, P.J.: Robust estimation of a location parameter. Annal. Statist. **53**(1), 73–101 (1964). https://doi.org/10.1214/aoms/1177703732
8. Karimi, D., Salcudean, S.E.: Reducing the Hausdorff distance in medical image segmentation with convolutional neural networks. IEEE Trans. Med. Imaging **39**(2), 499–513 (2019)

9. Kashyap, S., et al.: Feature selection for malapposition detection in intravascular ultrasound - a comparative study. In: Wu, S., Shabestari, B., Xing, L. (eds.) MICCAI 2023, pp. 165–175. Springer, Cham (2024). https://doi.org/10.1007/978-3-031-47076-9_17
10. Ma, J., He, Y., Li, F., Han, L., You, C., Wang, B.: Segment anything in medical images. Nat. Commun. **15**(1) (2024). https://doi.org/10.1038/s41467-024-44824-z
11. Meng, L., Jiang, M., Zhang, C., Zhang, J.: Deep learning segmentation, classification, and risk prediction of complex vascular lesions on intravascular ultrasound images. Biomed. Signal Process. Control **82**, 104584 (2023)
12. Oktay, O., et al.: Anatomically constrained neural networks (ACNNs): application to cardiac image enhancement and segmentation. IEEE Trans. Med. Imaging **37**(2), 384–395 (2018). https://doi.org/10.1109/TMI.2017.2743464
13. Ronneberger, O., Fischer, P., Brox, T.: U-net: convolutional networks for biomedical image segmentation. In: Navab, N., Hornegger, J., Wells, W.M., Frangi, A.F. (eds.) MICCAI 2015. LNCS, vol. 9351, pp. 234–241. Springer, Cham (2015). https://doi.org/10.1007/978-3-319-24574-4_28
14. Scarvelis, D., Wells, P.S.: Diagnosis and treatment of deep-vein thrombosis. CMAJ **175**(9), 1087–1092 (2006)
15. Secemsky, E.A., et al.: Intravascular ultrasound guidance for lower extremity arterial and venous interventions. EuroIntervention **18**(7), 598 (2022)
16. Stähr, P., et al.: Importance of calibration for diameter and area determination by intravascular ultrasound. Int. J. Cardiac Imaging **12**, 221–229 (1996)
17. Szarski, M., Chauhan, S.: Improved real-time segmentation of intravascular ultrasound images using coordinate-aware fully convolutional networks. Comput. Med. Imaging Graph. **91**, 101955 (2021)
18. Wissel, T., et al.: Cascaded learning in intravascular ultrasound: coronary stent delineation in manual pullbacks. J. Med. Imaging **9**(2), 025001 (2022)
19. Xiao, H., Li, L., Liu, Q., Zhu, X., Zhang, Q.: Transformers in medical image segmentation: a review. Biomed. Signal Process. Control **84**, 104791 (2023)
20. Xie, M., et al.: Two-stage and dual-decoder convolutional u-net ensembles for reliable vessel and plaque segmentation in carotid ultrasound images. In: 2020 19th IEEE International Conference on Machine Learning and Applications (ICMLA), pp. 1376–1381. IEEE (2020)
21. Xu, X., Sanford, T., Turkbey, B., Xu, S., Wood, B.J., Yan, P.: Polar transform network for prostate ultrasound segmentation with uncertainty estimation. Med. Image Anal. **78**, 102418 (2022)
22. Yariv, L., Gu, J., Kasten, Y., Lipman, Y.: Volume rendering of neural implicit surfaces. Adv. Neural. Inf. Process. Syst. **34**, 4805–4815 (2021)
23. Zhang, L., et al.: Generalizing deep learning for medical image segmentation to unseen domains via deep stacked transformation. IEEE Trans. Med. Imaging **39**(7), 2531–2540 (2020). https://doi.org/10.1109/TMI.2020.2973595

Domain Influence in MRI Medical Image Segmentation: Spatial Versus k-Space Inputs

Erik Gösche[1](✉)[iD], Reza Eghbali[2,3][iD], Florian Knoll[1][iD], and Andreas M. Rauschecker[3][iD]

[1] Department of Artificial Intelligence in Biomedical Engineering, University of Erlangen-Nuremberg, Erlangen, Germany
{erik.goesche,florian.knoll}@fau.de

[2] Berkeley Institute for Data Science (BIDS), University of California, Berkeley, Berkeley 94720, USA
eghbali@berkeley.edu

[3] Department of Radiology and Biomedical Imaging, University of California, San Francisco, San Francisco 94158, USA
andreas.rauschecker@ucsf.edu

Abstract. Transformer-based networks applied to image patches have achieved cutting-edge performance in many vision tasks. However, lacking the built-in bias of convolutional neural networks (CNN) for local image statistics, they require large datasets and modifications to capture relationships between patches, especially in segmentation tasks. Images in the frequency domain might be more suitable for the attention mechanism, as local features are represented globally. By transforming images into the frequency domain, local features are represented globally. Due to MRI data acquisition properties, these images are particularly suitable. This work investigates how the image domain (spatial or k-space) affects segmentation results of deep learning (DL) models, focusing on attention-based networks and other non-convolutional models based on MLPs. We also examine the necessity of additional positional encoding for Transformer-based networks when input images are in the frequency domain. For evaluation, we pose a skull stripping task and a brain tissue segmentation task. The attention-based models used are PerceiverIO and a vanilla Transformer encoder. To compare with non-attention-based models, an MLP and ResMLP are also trained and tested. Results are compared with the Swin-Unet, the state-of-the-art medical image segmentation model. Experimental results indicate that using k-space for the input domain can significantly improve segmentation results. Also, additional positional encoding does not seem beneficial for attention-based networks if the input is in the frequency domain. Although none of the models matched the Swin-Unet's performance, the less complex models showed promising improvements with a different domain choice.

Supplementary Information The online version contains supplementary material available at https://doi.org/10.1007/978-3-031-73284-3_31.

Keywords: Medical Image Segmentation · Frequency Domain Analysis · Attention-based Networks

1 Introduction

The success of the Transformer model [22] and later the Vision Transformer [5] has led to various attention-based models achieving state-of-the-art performance in medical vision tasks [6,8,15]. The attention mechanism enables an excellent way to capture long-range dependencies within the input, which has been shown to be an advantage in vision tasks [25]. However, Transformer-based networks struggle with large-scale inputs like medical images due to their quadratic complexity with respect to input size [22]. To address this issue, input images are typically subdivided into patches. Training of Transformer-based models requires large amounts of data to capture relationships among these image patches. The reason for this comes from the lack of inherent biases in Transformer models, such as local receptive fields or shared weights, which facilitate feature learning in CNNs. Particularly for pixel data, the explicit correlation of adjacent pixels, often strongly correlated, and translation in-variance are desirable characteristics [24]. An additional consequence of the absent receptive field is that every input value contributes to attention computation, resulting in interrelated values. While this aids the extraction of global features, it makes it difficult to capture local features [3].

Enhancing attention-based models can be achieved by using images in the frequency domain. Converting images to the frequency domain allows local features to be represented globally, leveraging the attention mechanism's ability to capture long-range dependencies. This approach is particularly relevant for medical imaging tasks, such as magnetic resonance imaging (MRI), where the properties of the data acquisition process make frequency domain representations particularly suitable. The convolution theorem, stating that convolution in image space is equivalent to element-wise product in Fourier space, inspired this method. This means that the convolutional layer in the image space can be replaced by a simple linear projection layer where each neuron applies a weight to one of the Fourier coefficients in the frequency space. The attention layer is a special case of a fully connected layer. However, the existence of non-linearities prevents us from further extending this line of thought. Furthermore, for natural language processing (NLP) tasks, it has been observed that in the frequency domain, additional position coding for input data, which is typically required for attention operations, may be unnecessary [14]. This research investigates how the choice of image domain (spatial or k-space) affects segmentation results of DL models, focusing on simple models, including attention-based networks and other non-convolutional networks based on multilayer perceptrons (MLPs). Additionally, it examines the necessity of additional positional encoding for Transformer-based networks when input images are in the frequency domain.

To address these research questions, two segmentation tasks are posed: skull stripping and brain tissue segmentation. The attention-based models evaluated in this study include the PerceiverIO [11] and a vanilla Transformer encoder. To provide a comprehensive comparison, non-attention-based models such as an MLP and the ResMLP [21] are also trained and tested. The performance of these models is compared with that of the Swin-Unet [2], a state-of-the-art medical image segmentation model.

1.1 Related Work

Several authors have proposed DL models for image classification and reconstruction tasks that utilize the representation of images in the frequency domain [12,17,19,27]. Recently, Wang et al. proposed a model for lesion segmentation in brain MRIs that uses masked image modeling in the frequency domain as a self-supervised pre-training stage [23]. These models generally include sub-blocks that consist of a 2D Fast Fourier Transform (FFT) layer, a specific processing operation in the frequency domain, and a 2D inverse FFT to transform the data back to the spatial domain. Since our goal is to compare the performance of Transformer and MLP based architectures across different domains, we do not include any FFT or inverse FFT layer inside our models and keep the model architecture consistent across different domains.

Another related line of research is machine learning-based undersampled MRI reconstruction (see [7,18] and references therein) and simultaneous reconstruction and segmentation [9,20]. This is inherently a cross-domain problem where the network has to predict spatial space values from k-space samples. These methods most often use convolutional layers, variational networks, and recently transformers [26]. However, segmentation is done on the reconstructed image in the spatial domain.

2 Method

2.1 Model Architectures

In this work, we employ the PerceiverIO and a vanilla Transformer encoder as exemplars of attention-based models. Additionally, we incorporate an MLP and the ResMLP for comparative analysis. Convolution-based models are not selected due to their incompatibility with the frequency domain, where each point represents information across the entire spatial domain, making these models misleading. This selection involves relatively simple models in terms of their complexity. The following experiments can therefore be used to determine how the lack of model capacity can be compensated by selecting suitable domains. To enable further evaluation of the results, widely used models such as nnU-Net [10] and Swin-Unet, the latter of which is also recognized as state of the art, are also included. The MLP is the simplest model among the four, serving as a proof-of-concept. It consists of linear input and output embeddings, N hidden fully connected layers, each followed by a tanh activation layer. Input reshaping reduces complexity, and both the dimensionality (M) of the latent space and the number of hidden layers (N) are hyperparameters. Similarly, the Transformer encoder employs linear embeddings for input and output, incorporating N encoder components from the Transformer model. The architecture resembles BERT [4], with the number of encoder blocks and latent space dimensionality being hyperparameters. For the implementation of the Transformer encoder, the corresponding PyTorch class was used. Fourier position encoding may be concatenated with the input. The PerceiverIO, implemented using Krasser and Stumpf's Python module [13], utilizes cross-attention to map input to a smaller latent space, reducing attention complexity from quadratic to linear. The ResMLP model, following documentation, resembles the vision transformer but lacks attention layers. Linear layers replace attention layers, and traditional normalizations are omitted in favor of affine transformations. For this work, the ResMLP was implemented so that the input is not divided into patches and embedded using a linear projection. Instead, the sagittal slices of the MRI brain data serve as channels and the remaining dimensions are flattened.

3 Experiments and Results

3.1 Datasets

We present two segmentation tasks, each accompanied by its own dataset. For the initial task of skull stripping, we employ the UPENN-GBM dataset [1]. As a follow-up task, focusing on brain tissue segmentation, we aim to highlight variations in complexity levels across segmentation tasks, utilizing the OASIS-1 dataset [16]. Both datasets are freely available to ensure the reproducibility of this work. Follow-up scans in both datasets are excluded from this work. For the brain tissue segmentation task, only the OASIS FreeSurfer output (brain mask as input and tissue segmentation for labels) is used. The number of tissue segmentation classes have been simplified to six classes (cortical gray matter, white matter, CSF, deep gray matter, brain stem and cerebellum). The exact mapping can be seen in the published source code. This way, 611 subject from the UPENN-GBM dataset and 407 subjects from the OASIS-1 dataset are used. All samples are converted to NIfTI format and sampled to an isotropic voxel size of 3 mm to reduce complexity. Furthermore, the samples are cropped to a size of $64 \times 64 \times 64$ and z-normalized. We apply the subsequent augmentation transformations: random affine transformation, random contrast adjustment, random Gaussian noise addition, random MRI motion artifact introduction, and random MRI bias field artifact inclusion. For both segmentation tasks, we partitioned the dataset into three subsets: training (80%), validation (10%), and testing (10%).

3.2 Implementation Details

This study uses Python (version 3.10.12), PyTorch (version 2.0.1), TorchIO (version 0.18.92), Ray (version 2.5.1), and PyTorch Lightning (version 2.0.5) for all implementations and analyses[1] To propagate the slices of each sample through

---
[1] The source code is publicly available at https://www.github.com/rauschecker-sugrue-labs/kspace-segmentation.

the network in the frequency domain, we employ the 2D real FFT implementation in PyTorch. This method operates under the assumption that the outcome of the FFT is Hermitian symmetric, which holds true in our case since we are using already reconstructed MRI data. By exploiting this symmetry, the input is halved. As the real part and imaginary part of the resulting complex numbers are saved separately in two vectors, the size of the input remains the same even after the Fourier transformation. The training routine for all non-baseline models is designed to allow independent specification of the target domain for inputs and labels at startup. This flexibility enables the definition of various combinations of input and label domains. When labels are in the spatial domain, the task becomes a classification problem and the model is trained using a cross-entropy loss; when labels are transformed into the frequency domain by using the 2D FFT, it becomes a regression problem and the model is trained using mean squared error loss. We use three domain configurations. In the spatial domain, both input and output are in the spatial domain. In the k-space domain, the input is transformed into the frequency domain using FFT, and the output is predicted in the frequency domain. In the k-space-to-spatial domain, the input is in the frequency domain, and the output is predicted in the spatial domain.

3.3 Segmentation Performance in Different Domains

This section presents quantitative segmentation results from all experiments. Public source code records hyperparameter configurations used for obtaining these results. The results of the skull stripping task among the models are displayed in Table 1. Notably, in the spatial domain, Dice scores are similar across models, with MLP slightly underperforming in recall and specificity. In k-space, MLP performs worse than in the spatial domain, while other models perform similarly. MLP remains weakest in the k-space domain.

Table 1. Dice similarity coefficient (DSC), sensitivity (sens) and specificity (spec) for different models on skull-stripping in spatial domain, k-space and k-space-to-spatial domain.

	Spatial			K-Space			K-Space to Spatial		
	DSC	Sens	Spec	DSC	Sens	Spec	DSC	Sens	Spec
MLP	0.964	0.960	0.992	0.898	0.888	0.976	0.966	0.968	0.990
ResMLP	**0.978**	0.973	0.995	**0.978**	0.980	0.993	**0.976**	0.977	0.993
PerceiverIO	0.930	0.923	0.984	0.927	0.919	0.984	0.929	0.919	0.985
Trans. Encoder	0.971	0.971	0.992	0.971	0.969	0.993	0.972	0.973	0.993
nnU-Net	0.986	0.987	0.996						
Swin-UNet	0.994	0.995	0.974						

For brain tissue segmentation, Fig. 1 displays results for spatial, k-space, and k-space-to-spatial domains, respectively. Detailed evaluations can be found in the supplementary material. There are notable differences in model performance in contrast to skull stripping. Starting in spatial domain, ResMLP outperforms others non-baseline models significantly with a Dice score of 0.876, followed by the Transformer encoder and the MLP. This means, the Transformer encoder surpasses MLP by approximately 12% in spatial domain. PerceiverIO performs poorly with a Dice score of only 0.418. In k-space, ResMLP's performance declines to a Dice score of 0.815, while other models improve. The ResMLP is no longer able to detect the fine structures of the cortical gray matter and therefore classifies these areas as too large (see Fig. 2). The Transformer encoder notably improves by about 13% to a Dice score of 0.790. PerceiverIO's performance for this class slightly improves with the domain change to a score of 0.461. The MLP achieves an improved Dice score of 0.690. In the k-space-to-spatial domain, ResMLP excels, achieving a Dice score of 0.883. All models benefit from this domain regarding cortical gray matter segmentation except PerceiverIO. MLP's performance slightly worsens compared to k-space but improves compared to the spatial domain. The Transformer encoder achieves its best performance in the k-space-to-spatial domain with a Dice score of 0.861. As it can be seen in the displayed segmentation mask, the Transformer encoder was able to delineate the individual classes more sharply.

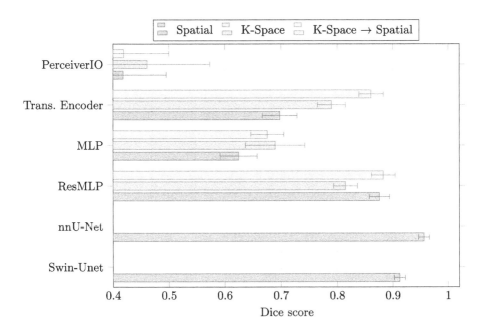

Fig. 1. Mean and standard deviations of the Dice score among all models on brain tissue segmentation in spatial, k-space, and k-space-to-spatial domain. The mean values are represented by the bars and the standard deviations are indicated by the error bars.

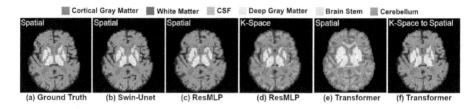

Fig. 2. Segmentation masks of the different models for brain tissue segmentation among varies domains.

The following Table 2 presents the results regarding the necessity of additional position coding for attention-based networks when working with k-space input. The table outlines the comparison between PerceiverIO and Transformer encoding, with and without additional positional encoding, across three distinct domains for the skull stripping task and the brain tissue segmentation task. Notably, no substantial differences are discernible within each model, across all three metrics.

Table 2. Dice scores for attention-based models with and without positional encoding (PE) on skull stripping and brain tissue segmentation in k-space. Average Dice score is reported for brain tissue segmentation task.

Architecture	Skull Stripping		Brain Tissue Segmentation	
	PE	no PE	PE	no PE
PerceiverIO	0.927	0.930	0.461	0.463
Trans. Encoder	0.971	0.970	0.790	0.790

3.4 Complexity Comparison with State-of-the-Art Models

The training and testing of the nnU-Net and Swin-Unet in spatial domain shows that both models outperform the other models in any domain constellation. However, in skull stripping the differences are relatively small. The reason for this is most likely because this task is not complex enough to see differences between domains. Looking at the results of the brain tissue segmentation in Fig. 1, the differences become more obvious. In this analysis, it becomes clear that nnU-Net and the Swin-Unet exhibits the most effective performance in brain tissue segmentation. However, both the Transformer encoder in the k-space-to-spatial domain and ResMLP offer competitive segmentation outcomes. This becomes particularly clear when comparing the complexity of these models. Table 3 shows the floating point operations (FLOPs) required for the forward and backward pass of the models used. The number of model parameters is also documented. The results clearly show that although the Swin-Unet outperforms the other non-baseline models, it is also much more complex. Comparing the Transformer encoder with the Swin-Unet, it can be seen that more than twice as many FLOPs are required for the forward and backward pass. In addition, the ResMLP requires less then 600×10^9 FLOPs for both passes. Taking this into consideration, the performance of the Transformer encoder and ResMLP is impressive. In summary, the choice of domain significantly impacts the results for brain tissue segmentation. The correct domain can compensate for a lack of model capacity and thus good segmentation results can be achieved with simple models.

Table 3. Comparison of FLOPs and parameters in different architectures (FFT/iFFT not included)

Architecture	FLOPs (Forward)	FLOPs (Backward)	Parameters (M)
MLP	128.85G	231.93G	83.91
ResMLP	208.57G	391.38G	134.43
Trans. Encoder	361.38G	696.99G	234.97
PerceiverIO	943.30G	1,886.60G	47.54
Swin-Unet	746.88G	1,491.20G	234.97

4 Conclusion and Limitations

This study examined DL model performance, especially attention-based and non-convolutional types, for brain segmentation tasks across domains, emphasizing both spatial and frequency domains. The study demonstrated how brain segmentation outcomes vary when input and label data are independently presented in either the spatial or frequency domain during supervised learning. Four models were implemented: PerceiverIO, a Transformer encoder, an MLP, and ResMLP, focusing on three domain configurations: spatial-to-spatial, k-space-to-k-space, and k-space-to-spatial. Skull stripping and brain tissue segmentation tasks were selected. Results indicated that domain configuration significantly impacts segmentation performance for sufficient complex tasks. For example, using the Transformer encoder, brain tissue segmentation performance improved by over 23% when data was transformed into the frequency domain, measured by the Dice score. This supports the idea that Fourier-transformed input data is better suited for attention-based networks like the Transformer encoder. However, this was not observed for the skull stripping task, likely due to its simplicity. The Transformer encoder and ResMLP performed best in the k-space-to-spatial domain configuration, possibly due to easier prediction of segmentation masks in the spatial domain, which exhibit an imbalance in frequency components. Additionally, additional positional encoding is unnecessary when input data is in the frequency domain, extending findings from NLP tasks to computer vision.

Results were compared with Swin-Unet, the baseline model. Despite Swin-Unet outperforming the implemented models, ResMLP showed competitive performance in the k-space-to-spatial domain, considering its relative simplicity.

Some aspects of this work limit the scope and applicability of the results obtained. One aspect is that only the binary cross-entropy loss, cross-entropy loss and mean squared error loss were considered as loss functions in this work. Other loss functions such as Dice Loss or Focal Loss would also be worth considering. The OASIS dataset is freely accessible and offers isotropic resolution, which simplifies our pre-processing pipeline. However, it should also be noted that this dataset contains presumably healthy subjects, which makes the segmentation task easier. Also, the approach used in this work cannot be directly translated into the clinical setting. By using the real 2D FFT it was assumed that the input is Hermitian symmetric. However, this assumption cannot be made for raw MRI k-space data.

Acknowledgements. We gratefully acknowledge the scientific support and HPC resources provided by the Erlangen National High Performance Computing Center (NHR @FAU) of the Friedrich-Alexander-Universität Erlangen-Nürnberg (FAU). The hardware is funded by the German Research Foundation (DFG).

Disclosure of Interests. We have no competing interests.

References

1. Bakas, S., et al.: The University of Pennsylvania Glioblastoma (UPenn-GBM) cohort: advanced MRI, clinical, genomics, & radiomics. Sci. Data **9**(1), 453 (2022). https://doi.org/10.1038/s41597-022-01560-7
2. Cao, H., et al.: Swin-Unet: Unet-like pure transformer for medical image segmentation. In: Karlinsky, L., Michaeli, T., Nishino, K. (eds.) ECCV 2022, vol. 13803, pp. 205–218. Springer, Cham (2023). https://doi.org/10.1007/978-3-031-25066-8_9
3. Chen, J., et al.: TransUNet: transformers make strong encoders for medical image segmentation. arXiv preprint arXiv:2102.04306v1 (2021)
4. Devlin, J., Chang, M.W., Lee, K., Toutanova, K.: BERT: pre-training of deep bidirectional transformers for language understanding. arXiv preprint arXiv:1810.04805 (2019)
5. Dosovitskiy, A., et al.: An image is worth 16x16 words: transformers for image recognition at scale. In: International Conference on Learning Representations (2021). https://openreview.net/forum?id=YicbFdNTTy
6. Gutsche, R., et al.: Automated brain tumor detection and segmentation for treatment response assessment using amino acid PET. J. Nucl. Med.: Off. Publ. Soc. Nucl. Med. **64**(10), 1594–1602 (2023). https://doi.org/10.2967/jnumed.123.265725
7. Hammernik, K., Schlemper, J., Qin, C., Duan, J., Summers, R.M., Rueckert, D.: Systematic evaluation of iterative deep neural networks for fast parallel mri reconstruction with sensitivity-weighted coil combination. Magn. Reson. Med. **86**(4), 1859–1872 (2021)
8. Hatamizadeh, A., et al.: UNETR: transformers for 3D medical image segmentation. In: 2022 IEEE/CVF Winter Conference on Applications of Computer Vision (WACV), pp. 1748–1758 (2022). https://doi.org/10.1109/WACV51458.2022.00181

9. Huang, Q., Chen, X., Metaxas, D., Nadar, M.S.: Brain segmentation from k-space with end-to-end recurrent attention network. In: Shen, D., et al. (eds.) MICCAI 2019. LNCS, vol. 11766, pp. 275–283. Springer, Cham (2019). https://doi.org/10.1007/978-3-030-32248-9_31
10. Isensee, F., Jaeger, P.F., Kohl, S.A.A., Petersen, J., Maier-Hein, K.H.: nnU-Net: a self-configuring method for deep learning-based biomedical image segmentation. Nat. Methods **18**(2), 203–211 (2021). https://doi.org/10.1038/s41592-020-01008-z
11. Jaegle, A., et al.: Perceiver IO: a general architecture for structured inputs and outputs. arXiv preprint arXiv:2107.14795 (2022)
12. Jiang, L., Dai, B., Wu, W., Loy, C.C.: Focal frequency loss for image reconstruction and synthesis. In: Proceedings of the IEEE/CVF International Conference on Computer Vision, pp. 13919–13929 (2021)
13. Krasser, M., Stumpf, C.: A PyTorch implementation of perceiver, perceiver IO and perceiver AR with PyTorch lightning scripts for distributed training (2023). https://github.com/krasserm/perceiver-io
14. Lee-Thorp, J., Ainslie, J., Eckstein, I., Ontanon, S.: FNet: mixing tokens with Fourier transforms. arXiv peprint arXiv:2105.03824 (2022)
15. Lin, Y., Liu, L., Ma, K., Zheng, Y.: Seg4Reg+: consistency learning between spine segmentation and cobb angle regression. arXiv preprint arXiv:2208.12462 (2022)
16. Marcus, D.S., Wang, T.H., Parker, J., Csernansky, J.G., Morris, J.C., Buckner, R.L.: Open access series of imaging studies (OASIS): cross-sectional MRI data in young, middle aged, nondemented, and demented older adults. J. Cogn. Neurosci. **19**(9), 1498–1507 (2007). https://doi.org/10.1162/jocn.2007.19.9.1498
17. Rao, Y., Zhao, W., Zhu, Z., Lu, J., Zhou, J.: Global filter networks for image classification. Adv. Neural. Inf. Process. Syst. **34**, 980–993 (2021)
18. Singh, D., Monga, A., de Moura, H.L., Zhang, X., Zibetti, M.V., Regatte, R.R.: Emerging trends in fast MRI using deep-learning reconstruction on undersampled k-space data: a systematic review. Bioengineering **10**(9), 1012 (2023)
19. Stuchi, J.A., Boccato, L., Attux, R.: Frequency learning for image classification. arXiv preprint arXiv:2006.15476 (2020)
20. Tolpadi, A.A., et al.: K2s challenge: from undersampled k-space to automatic segmentation. Bioengineering **10**(2), 267 (2023)
21. Touvron, H., et al.: ResMLP: feedforward networks for image classification with data-efficient training. arXiv preprint arXiv:2105.03404 (2021)
22. Vaswani, A., et al.: Attention Is All You Need. arXiv preprint arXiv:1706.03762v7 (2017)
23. Wang, W., et al.: Fremim: Fourier transform meets masked image modeling for medical image segmentation. In: Proceedings of the IEEE/CVF Winter Conference on Applications of Computer Vision, pp. 7860–7870 (2024)
24. Wu, H., et al.: CVT: introducing convolutions to vision transformers. arXiv preprint arXiv:2103.15808 (2021)
25. Zhang, D., Tang, J., Cheng, K.T.: Graph reasoning transformer for image parsing. arXiv preprint arXiv:2209.09545 [cs] (2022)
26. Zhao, Z., Zhang, T., Xie, W., Wang, Y.F., Zhang, Y.: K-space transformer for undersampled MRI reconstruction. In: BMVC, p. 473 (2022)
27. Zhou, M., et al.: Deep Fourier up-sampling. arXiv preprint arXiv:2210.05171 (2022)

Enhanced Small Liver Lesion Detection and Segmentation Using a Size-Focused Multi-model Approach in CT Scans

Abdullah F. Al-Battal[1,2](✉), Van Ha Tang[3], Steven Q. H. Truong[3], Truong Q. Nguyen[1], and Cheolhong An[1]

[1] Electrical and Computer Engineering Department, UC San Diego, California, La Jolla, USA
[2] Electrical Engineering Department, King Fahd University of Petroleum and Minerals, Dhahran, Saudi Arabia
aalbatta@ucsd.edu
[3] Vinbrain JSC, Hanoi, Vietnam

Abstract. This paper presents a novel approach to enhance the detection and segmentation of small liver lesions in computed tomography (CT) scans using a size-focused multi-model framework. Current state-of-the-art segmentation models, primarily based on the UNet architecture, often exhibit inferior performance on small lesions due to severe class and size imbalances. We introduce a model architecture incorporating a configurable attention mechanism within the model's skip connections and a lesion selection algorithm that compares predictions from multiple models, including a general lesion segmentation model and a small lesion-focused model, selecting the most suitable prediction. The approach was evaluated on a clinical 3-phase CT dataset and the public LiTS dataset. Results show improvements in overall lesion segmentation performance by 1.5% and 1.9% for the clinical and LiTS datasets, respectively. Additionally, the detection of small lesions improved by 4.4% and 1.8% for both datasets, respectively.

Keywords: Medical image segmentation · Liver lesion segmentation · Small lesion segmentation · Convolutional neural networks

1 Introduction

Liver cancer is the 6th most common cancer and the 3rd most common cause of cancer death [22,28]. Detecting and segmenting lesions within the liver accurately is an important step in the diagnosis, treatment planning, and management of cancer. Computed tomography (CT) scans are the standard of care for this task [8,19]. Nevertheless, detecting liver lesions within a CT scan is a complex and challenging task for radiologists and oncologists where the detection rate can be as low as 72% for 10–20 mm lesions, and 16% for smaller ones [9,29].

This work is supported by a grant from Vinbrain, JSC.

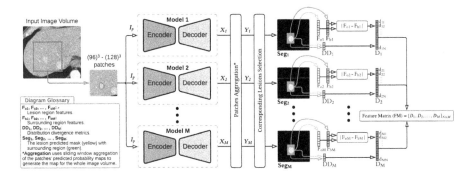

Fig. 1. The proposed lesion mask prediction selection approach. The 3D image patch is fed to all models. Corresponding lesions are selected based on the overlap Dice score.

Segmenting lesions is a highly imbalanced task as they are severely underrepresented. To overcome class imbalance, a common practice [27] assigns loss weights inversely proportional to the class distribution [26,27], but it is ineffective for highly imbalanced datasets [2]. It also does not address the lesion size imbalance (The size of liver lesions can differ by up to four orders of magnitude). State-of-the-art (SOTA) segmentation models, such as the SwinUNetR [10], Model Genesis [30], nnUNet [11], MedNext [21], and FF&A UNet [1] are based on the UNet architecture [20]. These models use convolutional layers [12,14,24], except for the SwinUNetR, which uses the shifted windows transformer architecture [7,16]. Nevertheless, these models still struggle with detecting and segmenting small lesions. Selecting from the predictions of two models (SE 2) has demonstrated its ability to improve the detection and segmentation of liver lesions, including small ones [1].

To improve the overall liver lesion segmentation and enhance the detection and segmentation of small lesions, we propose a lesion selection algorithm that selects predictions from multiple models. We test it on two datasets: A clinical 3-phase CT dataset and the Liver tumor segmentation challenge (LiTS) dataset [4]. We compare the proposed approach to the current SOTA segmentation models as well as soft voting ensembling (SVE), hard voting ensembling (HVE), and the two model selection approach (SE 2) [1]. Our main contributions are:

- A segmentation model architecture with a configurable attention mechanism in the skip connection between the encoder and decoder.
- A lesion selection algorithm that can compare the predictions of multiple models and select the most suitable prediction.
- A size-based loss function weighting approach to improve small lesion detection and segmentation with lesion border weight decay (LBWD) to reduce the impact of annotation and registration errors.

2 Methodology

Our approach combines the predictions of multiple models through a lesion mask selection algorithm as described in Fig. 1, and Sect. 2.4. The multiple models are acquired through multiple checkpoints of two model architectures, which are the main segmentation model (MSM) and the small lesion focused model (SLF).

2.1 The Segmentation Model

In our approach, both segmentation models (MSM and SFL) are based on the UNet architecture as shown in Fig. 2. We use the ConvNext [17] convolutional block in the encoder and decoder. To improve the model's segmentation performance, we designed a configurable attention (CA) module in the skip connection. This attention module is configured differently by changing the kernel size and stride for each of the depth stages of the model, as well as whether the model is designed for general segmentation purposes (MSM) or focused on small lesions (SFL). The kernel and the stride decrease with the model's depth, corresponding to the reduced spatial resolution at these stages. For the SFL model, we reduce the stride in stages 1 and 2 to allow the attention module to extract spatial features on a finer scale. The model architecture is outlined in Fig. 2.

Both models are trained using a compound loss of the weighted sum of the Dice loss and the weighted binary cross-entropy (WBCE) loss [2]. When training both models, we implemented an adaptive weighting of the WBCE loss function to ensure a recall value that is at least 5% higher than precision to recover lesions effectively due to the large imbalance, especially for small ones. This adaptive algorithm [1] is an extension of the adaptive recall and precision balancing algorithm we proposed in [2] for 2D medical image segmentation.

2.2 The Intensity-Based Features Used for Prediction Comparison

These features are the mean, standard deviation, median, 5th and 95th percentiles, skewness, kurtosis, and Shannon entropy [23] of the intensity values distribution. In addition to these features, we use three symmetric distribution divergence metrics, which are: The absolute Standardized Mean Difference (ASMD) [5], the Kolmogorov-Smirnov statistic (KS) [3,25], and the Jensen-Shannon divergence (JSD) [15]. These metrics are defined as:

$$\text{ASMD} = \frac{|\mu_L - \mu_{ST}|}{\sqrt{\frac{\sigma_L^2 + \sigma_{ST}^2}{2}}}, \tag{1}$$

$$\text{KS} = \sup_i |F_L(i) - F_{ST}(i)|, \tag{2}$$

$$\text{JSD}(L, ST) = \frac{1}{2} D_{KL}(p_L \| p_M) + \frac{1}{2} D_{KL}(p_{ST} \| p_M), \tag{3}$$

where μ and σ are the mean and standard deviation of intensity values (i). The L and ST subscripts refer to the lesion (L) and surrounding tissue (ST). F is the cumulative distribution functions while p is probability distribution function (PDF). $D_{KL}(p\|q)$ is the Kullback-Leibler divergence between two distributions p and q [6,13], and $p_M = \frac{1}{2}(p_L + p_{ST})$ is the average of the two PDFs.

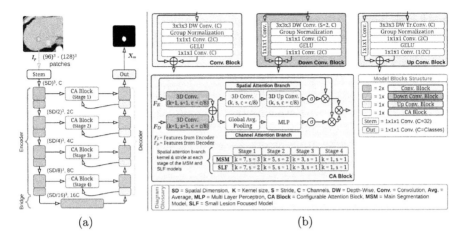

Fig. 2. The structure of the proposed segmentation model in (a) with the model building blocks in (b).

2.3 Multi-model Corresponding Lesions Selection Algorithm

The most significant challenge when identifying corresponding lesions from multiple models is that a lesion can belong to multiple correspondences. To overcome this, the algorithm initiates an empty collection of sets C to store the final lesion correspondence sets. The algorithm iterates over the segmentation masks of each model (denoted as $Seg_1, Seg_2, \ldots, Seg_M$). For each model Seg_m, it counts the number of lesions, denoted as I_m. Then, for each lesion l_i^m where $i \in \{1, \ldots, I_m\}$, it initializes a set $C[m, i]$ with the tuple (m, i), representing the model index m and lesion index i. The algorithm then proceeds to find corresponding lesions in other models by comparing the current lesion l_i^m against each lesion l_k^j in other segmentation maps Seg_j (where $j \in \{1, \ldots, M\}$ and $j \neq m$). The number of lesions in each model is counted as K_j. The Dice score (DSC) [31] between lesions l_i^m and l_k^j is evaluated and if $DSC(l_i^m, l_k^j) \geq 0.5$, they are considered corresponding and (j, k) is added to the set $C[m, i]$. After all the correspondence sets are identified, duplicate correspondences are resolved by keeping only the largest correspondence set for each lesion belonging to more than one set.

2.4 Segmentation Mask Prediction

The prediction selection approach outlined in Fig. 1 uses the separation features vectors $(D_1, D_2, \ldots, D_M)$ to identify the best prediction mask for each of the lesions in correspondence sets. $D_1, D_2, \ldots, D_M$ are generated by concatenating the lesion versus surrounding tissue separation features $|F_{oi} - F_{bi}|$ and the distribution divergence metrics DD_i for $i = 1, 2, \ldots, M$ such that $D_i = [|F_{oi} - F_{bi}|, DD_i] : D_i = [d_{11}, d_{12}, \ldots, d_{1N}]$. The feature matrix (FM) containing these vectors is then created as follows: $FM = [D_1, D_2, \ldots, D_M]$, where N is the number of features and M is the number of models. The predicted mask from the model with the largest count of maximum separation values across all the features is then selected for each lesion.

2.5 Enhancing Small Lesions Recall Via Size-Based Loss Weighting

The weighting approach spatially weighs both the WBCE and Dice losses for each lesion individually, with weights that increase as the lesion size decreases. We also incorporate a weight decay at the lesions' borders to minimize the impact of annotation and registration errors, as borders would be the most impacted by such errors. This decay, which we name the Lesion Border Weight Decay (LBWD), is defined as:

$$\text{LBWD} = \begin{cases} 1 - e^{-c_i d_i} & \text{if Seg}_{gt} \geq 0.5 \\ 1 - e^{-c_o d_o} & \text{otherwise,} \end{cases} \quad (4)$$

where c_i and c_o are coefficients controlling the decay rate inside and outside lesions, respectively. d_i and d_o are the normalized distances from the lesion border inside and outside lesions, respectively. Seg_{gt} refers to the ground truth segmentation map. If c_i and c_o are chosen universally across lesions, the loss weights within the lesion will vary significantly based on the lesion shape and size as shown in Fig. 3 (c). Therefore, we choose the coefficient (c_i) for each lesion individually by iteratively increasing c_i until 80% of the lesion volume is higher than 0.8. Higher thresholds can cause excessively steep decays for small lesions. To allow the LBWD to recover in regions between lesions, we identify the boundaries in the background that separate each of the lesions as shown in Fig. 3 (b). The coefficient c_o is then iteratively chosen so that the LBWD reaches a value of 1 at the point on the background boundary closest to the lesion. The final weight map is generated by multiplying the LBWD map by the lesion size-based weight map. For the lesion size-based weight map, we found that a weighting factor that linearly decays from 20 for lesions with a span of 4 mm to 1 for lesions with a span of 45 mm performs well in its ability to enhance lesion detection while maintaining a reasonable balance between precision and recall.

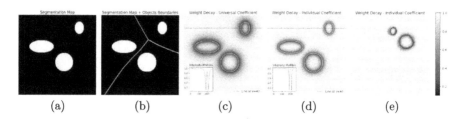

Fig. 3. Synthetically generated segmentation map with three lesions (a). The boundaries that separate the lesions' regions (b). The LBWD map for universally (c) and individually (d) chosen coefficients. The LBWD map for an example from the clinical dataset (e). The lesions' borders are represented by white contours in (c) and (d).

3 Experiments and Results

3.1 Dataset and Implementation

We evaluated the proposed approach's segmentation and detection performance on both the public LiTS dataset [4] and the internal 3-phase clinical dataset. The clinical dataset contains scans from 354 subjects annotated with liver lesion segmentation labels by two radiologists experienced in liver oncology. Each subject has 3 contrast-enhanced scans at three different phases: arterial, delayed, and venous. The arterial and delayed phases are registered on the venous phase for each subject. The dataset's average axial spacing is 0.66 mm, while the average slice thickness is 0.9 mm. The average number of lesions per scan is 2.2, with lesions as large as 129 mm and as small as 2.7 mm in diameter. The LiTS dataset contains CT scans of the liver from 131 subjects. For both datasets, we trained both models (the MSM and SLF models) in the proposed approach and used five model weights checkpoints to generate five predictions for each subject to apply

Table 1. Lesion segmentation performance by subject. Best and 2nd best results are boldfaced. All metrics are in the range 0 to 100. Values in parentheses represent the standard deviation across subjects.

Dataset	Approach	Model	Dice	IoU	Recall	Precision
3-Phase Clinical	Single Model	SwinUNetR	68.2 (23.2)	55.8 (23.5)	68.1 (25.0)	78.0 (23.6)
		Model Genesis	70.7 (22.4)	58.7 (22.8)	73.4 (23.5)	76.6 (23.2)
		nnUNet	73.4 (22.4)	62.0 (23.0)	74.8 (24.2)	79.4 (22.0)
		MedNext	75.1 (20.1)	63.6 (21.1)	76.8 (22.6)	**80.3** (19.1)
		FF&A UNet	75.5 (19.8)	**63.9** (20.8)	76.8 (22.2)	**80.9** (18.9)
		Ours (MSM)	**76.2** (17.5)	**64.2** (19.8)	**80.1** (18.8)	78.9 (19.5)
		Ours (SLF)	75.9 (17.8)	63.9 (19.9)	**81.5** (19.6)	76.6 (20.1)
	Ensemble	HVE	76.0 (18.1)	64.1 (20.0)	78.2 (20.6)	**80.1** (19.3)
		SVE	**76.5** (17.9)	**64.7** (19.8)	**80.9** (19.1)	78.2 19.9)
		SE 2	75.8 (19.8)	64.4 (20.7)	78.4 (21.7)	**80.1** (18.9)
		Ours-SE 2–5	**76.6** (17.3)	**64.7** (19.5)	**81.1** (18.1)	78.1 (20.6)
LiTS	Single Model	SwinUNetR	68.4 (28.3)	58.1 (29.5)	60.0 (28.9)	**81.4** (18.7)
		Model Genesis	73.1 (28.0)	63.4 (27.4)	71.4 (28.6)	75.8 (16.3)
		nnUNet	74.3 (27.6)	63.8 (28.2)	71.4 (30.2)	**78.1** (15.7)
		MedNext	73.5 (27.2)	63.8 (27.3)	74.4 (30.8)	72.2 (18.7)
		FF&A UNet	74.4 (27.1)	64.3 (27.0)	77.2 (27.2)	74.9 (19.2)
		Ours (MSM)	**75.3** (25.1)	**64.6** (26.0)	**78.7** (25.3)	75.6 (22.4)
		Ours (SLF)	**75.8** (25.0)	**64.9** (25.9)	**78.9** (24.8)	76.1 (18.8)
	Ensemble	HVE	73.3 (27.6)	63.5 (27.9)	78.9 (25.3)	73.1 (18.5)
		SVE	**75.8** (25.0)	**65.2** (25.9)	**79.2** (25.3)	**76.0** (16.5)
		SE 2	75.6 (24.8)	64.9 (25.6)	79.1 (27.2)	73.8 (14.4)
		Ours-SE 2–5	**75.8** (24.9)	**65.3** (25.6)	**79.6** (24.7)	**75.7** (14.5)

Table 2. Lesion segmentation and detection performance by lesion. The Dice score (DSC) for lesions overall and detected lesions at detection DSC thresholds of 0.2, 0.5, and 0.75 are outlined, as well as the percentage of lesions detected at these thresholds. The best results are boldfaced. All metrics are in the range 0 to 100.

Dataset	Approach	DSC				Percentage Detected		
		DSC	0.2	0.5	0.75	0.2	0.5	0.75
3-Phase Clinical (All Lesions)	SwinUNetR	56.0	71.8	77.7	84.4	77.2	66.4	43.3
	Model Genesis	55.5	72.3	77.9	83.7	76.1	65.7	45.1
	nnUNet	60.4	76.7	79.8	85.5	78.4	72.4	52.6
	MedNext	60.1	76.4	79.8	85.4	78.4	72.0	53.0
	FF&A UNet	60.6	76.7	79.8	85.4	78.7	73.1	53.7
	Ours-SE 2-5	**65.0**	**76.8**	**80.8**	**86.1**	**84.4**	**76.3**	**56.9**
LiTS (All Lesions)	SwinUNetR	50.2	73.9	78.3	84.9	67.4	60.9	40.6
	Model Genesis	55.9	75.3	78.9	84.5	73.6	67.0	48.3
	nnUNet	56.1	75.2	79.1	84.8	73.9	66.7	47.1
	MedNext	59.6	74.3	76.9	83.7	79.7	74.3	46.7
	FF&A UNet	62.7	77.7	79.6	85.3	80.1	76.2	53.3
	Ours-SE 2-5	**64.2**	**78.4**	**80.1**	**85.6**	**81.3**	**77.6**	**55.6**
3-Phase Clinical (Small Lesions)	SwinUNetR	26.6	57.9	69.1	80.6	44.8	29.9	9.2
	Model Genesis	27.6	63.0	70.3	79.0	42.5	33.3	13.8
	nnUNet	32.7	65.1	72.7	82.7	49.4	39.1	18.4
	MedNext	31.9	62.6	71.9	81.0	50.6	37.9	18.4
	FF&A UNet	33.3	64.0	72.0	81.1	51.7	40.2	19.5
	Ours-SE 2-5	**38.5**	**65.5**	**73.1**	**83.1**	**58.3**	**45.2**	**21.4**
LiTS (Small Lesions)	SwinUNetR	29.4	64.7	70.6	82.1	44.9	38.1	14.4
	Model Genesis	36.5	67.2	72.3	81.9	54.2	45.8	21.2
	nnUNet	36.2	66.2	72.7	82.6	54.2	44.1	20.3
	MedNext	48.9	68.4	72.0	82.8	71.2	63.6	24.6
	FF&A UNet	49.9	71.1	73.8	**84.3**	69.5	63.6	27.1
	Ours-SE 2-5	**50.2**	**71.3**	**74.2**	82.6	**70.9**	**64.1**	**30.1**

the selection approach (SE 2-5) outlined in Sects. 2.3 and 2.4. For the clinical dataset, the scans were split into 200 for training and 154 for testing, while for the LiTS dataset, the split was 100 for training and 31 for testing. For all the datasets, we used a spatial resolution of $1\,\mathrm{mm}^3$, and patches of size 128^3 voxels for the clinical dataset and 96^3 voxels for the LiTS dataset. The models were trained for 800 to 1200 epochs depending on the learning rate that is suitable for the model, which ranged from $1e^{-4}$ to $1e^{-2}$. All the models were trained using the AdamW [18] optimizer and the compound loss function of the Dice and BCE loss, except for the nnUNet and Model Genesis models that are trained using their recommended optimizer, which is the stochastic gradient descent method.

3.2 Evaluation and Results

The segmentation performance of the proposed approach and the MSM and SFL models is summarized in Table 1, and is compared to the benchmark models. We also compare the proposed selection approach (SE 2–5) to the two model selection approach (SE 2), soft voting ensembling (SVE) and hard voting ensembling (HVE) for both datasets. The proposed models and approach constantly outperformed the benchmark models, improving the overall relative segmentation performance by 1.5% and 1.9% for the clinical and LiTS datasets, respectively. This performance improvement results in better segmentation maps of lesions as shown in Fig. 4, where different lesions from both datasets are shown together with the ground truth and predicted segmentation boundaries. Reducing variability and improving consistency across subjects is essential to promote equity in patient outcomes. Our approach reduced the variability in segmentation performance across subjects by 12.6% and 3.9% for the clinical and LiTS datasets, respectively. For the LiTS dataset, the SLF model outperforms the MSM model in our approach as the LiTS dataset contains a large number of small lesions compared to the clinical dataset.

The detection performance of the proposed approach is summarized in Table 2. We computed the Dice score by lesion after identifying lesion mask matches between the predicted and ground truth masks for each lesion individually. We also evaluated the Dice score of detected lesions using three different lesion detection thresholds based on the Dice scores of 0.2, 0.5, and 0.75, as well as the percentage of lesions detected at these thresholds. Overall, the proposed approach improves the detection and segmentation of lesions as demonstrated in Table 2. It is worth noting that although the proposed approach improved liver lesion segmentation and detection, the performance was not improved for all lesions. Therefore, a promising direction for future research would be to investigate the impact of other features to identify the best mask for the selection

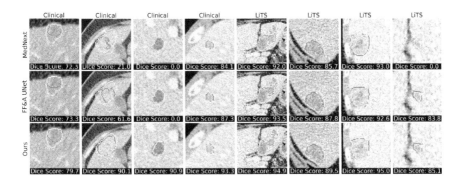

Fig. 4. Qualitative comparison of the proposed approach to the best two benchmark models for the clinical (columns 1 to 4) and LiTS (columns 5 to 8) dataset. The ground truth mask boundary is in yellow, while the predicted mask boundary is in red. (Color figure online)

algorithm as well as other approaches to using these features to compare the prediction generated by different models.

4 Conclusion

In this paper, we proposed a prediction selection approach for liver lesion segmentation and detection in CT scans that uses the predictions from multiple weight checkpoints of two models. The first model is a general lesion segmentation model (MSM), while the second is a small lesion focused model (SLF), which is designed to improve small lesion segmentation and detection while maintaining comparable performance on large ones. For both models, we proposed a configurable attention (CA) module that is modified based on the model depth and purpose. We also proposed a size-based loss function weighting approach with lesion border weight decay (LBWD). The selection approach extracts intensity-based features from the lesion region and surrounding tissue and compares the predictions from the different models, selecting the prediction that has the largest count of the highest feature separations. We tested the proposed approach a 3-phase clinical dataset and the LiTS dataset. Our approach improved the segmentation performance by 1.5% for the clinical and 1.9% for the LiTS datasets, while improving the percentage of detected small lesions by 4.4% and 1.8% for the two datasets, respectively. Our results demonstrate that, compared to the current state-of-the-art segmentation models, our approach can improve the detection and segmentation of lesions, including small ones.

Acknowledgements. This work is supported by a grant from Vinbrain, JSC.

References

1. Al-Battal, A.F., et al.: Efficient in-training adaptive compound loss function contribution control for medical image segmentation. In: 2024 46th Annual International Conference of the IEEE Engineering in Medicine & Biology Society (EMBC). IEEE (2024)
2. Al-Battal, A.F., et al.: Enhancing lesion detection and segmentation via lesion mask selection from multi-specialized model predictions in ct scans of the liver and kidney. arXiv preprint (2024)
3. An, K.: Sulla determinazione empirica di una legge didistribuzione. Giorn Dell'inst Ital Degli Att **4**, 89–91 (1933)
4. Bilic, P., et al.: The liver tumor segmentation benchmark (lits). Med. Image Anal. **84**, 102680 (2023)
5. Cohen, J.: Statistical power analysis for the behavioral sciences. Routledge (2013)
6. Csiszár, I.: I-divergence geometry of probability distributions and minimization problems. The annals of probability, pp. 146–158 (1975)
7. Dosovitskiy, A., et al.: An image is worth 16x16 words: Transformers for image recognition at scale. arXiv preprint arXiv:2010.11929 (2020)
8. Elbanna, K.Y., Kielar, A.Z.: Computed tomography versus magnetic resonance imaging for hepatic lesion characterization/diagnosis. Clin. Liver Disease **17**(3), 159–164 (2021)

9. Freitas, P.S., Janicas, C., Veiga, J., Matos, A.P., Herédia, V., Ramalho, M.: Imaging evaluation of the liver in oncology patients: a comparison of techniques. World J. Hepatol. **13**(12), 1936 (2021)
10. Hatamizadeh, A., Nath, V., Tang, Y., Yang, D., Roth, H.R., Xu, D.: Swin unetr: swin transformers for semantic segmentation of brain tumors in mri images. In: International MICCAI Brainlesion Workshop. pp. 272–284. Springer (2021). https://doi.org/10.1007/978-3-031-08999-2_22
11. Isensee, F., Jaeger, P.F., Kohl, S.A., Petersen, J., Maier-Hein, K.H.: nnu-net: a self-configuring method for deep learning-based biomedical image segmentation. Nat. Methods **18**(2), 203–211 (2021)
12. Krizhevsky, A., Sutskever, I., Hinton, G.E.: Imagenet classification with deep convolutional neural networks. Commun. ACM **60**(6), 84–90 (2017)
13. Kullback, S., Leibler, R.A.: On information and sufficiency. Ann. Math. Stat. **22**(1), 79–86 (1951)
14. LeCun, Y., Bottou, L., Bengio, Y., Haffner, P.: Gradient-based learning applied to document recognition. Proc. IEEE **86**(11), 2278–2324 (1998)
15. Lin, J.: Divergence measures based on the shannon entropy. IEEE Trans. Inf. Theory **37**(1), 145–151 (1991)
16. Liu, Z., et al.: Swin transformer: hierarchical vision transformer using shifted windows. In: Proceedings of the IEEE/CVF International Conference on Computer Vision, pp. 10012–10022 (2021)
17. Liu, Z., Mao, H., Wu, C.Y., Feichtenhofer, C., Darrell, T., Xie, S.: A convnet for the 2020s. In: Proceedings of the IEEE/CVF Conference on Computer Vision and Pattern Recognition, pp. 11976–11986 (2022)
18. Loshchilov, I., Hutter, F.: Decoupled weight decay regularization. arXiv preprint arXiv:1711.05101 (2017)
19. van Oostenbrugge, T.J., Fütterer, J.J., Mulders, P.F.: Diagnostic imaging for solid renal tumors: a pictorial review. Kidney Cancer **2**(2), 79–93 (2018)
20. Ronneberger, O., Fischer, P., Brox, T.: U-Net: convolutional networks for biomedical image segmentation. In: Navab, N., Hornegger, J., Wells, W.M., Frangi, A.F. (eds.) MICCAI 2015. LNCS, vol. 9351, pp. 234–241. Springer, Cham (2015). https://doi.org/10.1007/978-3-319-24574-4_28
21. Roy, S., et al.: Mednext: transformer-driven scaling of convnets for medical image segmentation. In: International Conference on Medical Image Computing and Computer-Assisted Intervention, pp. 405–415. Springer (2023). https://doi.org/10.1007/978-3-031-43901-8_39
22. Rumgay, H., et al.: Global burden of primary liver cancer in 2020 and predictions to 2040. J. Hepatol. **77**(6), 1598–1606 (2022)
23. Shannon, C.E.: A mathematical theory of communication. Bell Syst. Tech. J. **27**(3), 379–423 (1948)
24. Simonyan, K., Zisserman, A.: Very deep convolutional networks for large-scale image recognition. arXiv preprint arXiv:1409.1556 (2014)
25. Smirnov, N.: Table for estimating the goodness of fit of empirical distributions. Ann. Math. Stat. **19**(2), 279–281 (1948)
26. Sudre, C.H., Li, W., Vercauteren, T., Ourselin, S., Jorge Cardoso, M.: Generalised dice overlap as a deep learning loss function for highly unbalanced segmentations. In: Cardoso, M., et al. (eds.) DLMIA/ML-CDS -2017. LNCS, vol. 10553, pp. 240–248. Springer, Cham (2017). https://doi.org/10.1007/978-3-319-67558-9_28
27. Sugino, T., Kawase, T., Onogi, S., et al.: Loss weightings for improving imbalanced brain structure segmentation using fully convolutional networks. In: Healthcare. vol. 9, p. 938. MDPI (2021)

28. Sung, H., et al.: Global cancer statistics 2020: Globocan estimates of incidence and mortality worldwide for 36 cancers in 185 countries. CA: Cancer J. Clinicians **71**(3), 209–249 (2021)
29. Wiering, B., Ruers, T.J., Krabbe, P.F., Dekker, H.M., Oyen, W.J.: Comparison of multiphase ct, fdg-pet and intra-operative ultrasound in patients with colorectal liver metastases selected for surgery. Ann. Surg. Oncol. **14**, 818–826 (2007)
30. Zhou, Z., Sodha, V., Pang, J., Gotway, M.B., Liang, J.: Models genesis. Med. Image Anal. **67**, 101840 (2021)
31. Zijdenbos, A.P., Dawant, B.M., Margolin, R.A., Palmer, A.C.: Morphometric analysis of white matter lesions in mr images: method and validation. IEEE Trans. Med. Imaging **13**(4), 716–724 (1994)

Generation and Segmentation of Simulated Total-Body PET Images

Arnau Farré-Melero[1], Pablo Aguiar-Fernández[2,3,4], and Aida Niñerola-Baizán[1,5,6](✉)

[1] Universitat de Barcelona, Barcelona, Spain
afarre.m@ub.edu
[2] CIMUS - Universidade de Santiago de Compostela (USC), Santiago de Compostela, Spain
pablo.aguiar@usc.es
[3] Nuclear Medicine Department, Instituto Investigación Sanitaria IDIS, Hospital Clínico Universitario de Santiago de Compostela, Santiago de Compostela, Spain
[4] Neurodegenerative Diseases Research Area of Biomedical Research Networking Center (CIBERNED), ISCIII, Santiago de Compostela, Spain
[5] Nuclear Medicine Department, Hospital Clínic Barcelona - IDIBAPS, Barcelona, Spain
[6] Biomedical Research Networking Center of Bioengineering, Biomaterials and Nanomedicine (CIBER-BBN), ISCIII, Barcelona, Spain
ninerola@clinic.cat

Abstract. Positron Emission Tomography (PET) is a Nuclear Medicine technique with a wide range of applications, particularly on Oncology. The imminent clinical introduction of Long Axial Field-of-View (LAFOV) scanners has raised the interest of the community, due to its enhanced sensitivity and lower doses required. In this project, we aim to develop a novel framework on Monte Carlo techniques, in which a set of Total-Body simulated images is used to train networks for organ and lesion segmentation. Our methodology can be divided into phantom generation, simulation and segmentation. We established a distribution of parameters to obtain map cohorts with anatomical variability, which were simulated using the open-source platform SimPET and finally segmented at organ and lesion levels using U-Net architectures. Usually, PET images are segmented together with Computerized Tomography (CT), integrated on PET/CT scans for attenuation correction. Direct PET segmentation might be useful in the cases of low-quality CT, in novel attenuation correction approaches without using anatomical imaging and specifically for non-anatomical lesions. Our model gives excellent overlapping results on most of the organs among simulated test images. The model also presents good performance on lesion segmentation, if they are large and well-contrasted enough. Furthermore, our method overcomes the sparsity of data limitation and automatically annotates the image database by using the activity maps information.

Keywords: Positron Emission Tomography · Monte Carlo Simulation · Total-Body Imaging · Image Segmentation

© The Author(s), under exclusive license to Springer Nature Switzerland AG 2025
X. Xu et al. (Eds.): MLMI 2024, LNCS 15241, pp. 331–339, 2025.
https://doi.org/10.1007/978-3-031-73284-3_33

1 Introduction

Positron Emission Tomography (PET) is a Nuclear Medicine imaging modality whose functional information has converted it the choice diagnostic technique in many disciplines such as Neurology, Cardiology and Oncology. PET presents a low signal-to-noise ratio and spatial resolution, and many degrading effects, particularly the attenuation effect, that can be corrected by adding anatomical information, e.g., from Computerized Tomography (CT), during reconstruction. Thus, PET scanners are frequently coupled together with a CT scanner.

Each scanner is characterized by its Field-of-View (FOV), defined as the spatial length, in the longitudinal axis, that it is capable to image in a single scan. Traditional scanners present FOV values that range 15–25 cm approximately. Hence, Total-Body imaging requires sequential acquisitions to cover the full subject, and to join them by overlapping their ends. In recent years appeared a new generation of scanners, named Long Axial FOV (LAFOV) scans, with FOVs up to 2 m. These not only allow faster scanning, but also present an enhanced sensitivity, leading to lower doses required. Total-Body PET images offer a wide range of applications, including the detection of metastasis or pharmacokinetic studies. Furthermore, they could improve the diagnostic in respect of traditional scans, as they can find smaller lesions, leading to an earlier and more accurate detection.

PET image analysis is based on the differential radiotracer uptake of the tissues, resulting on intensity variations in the final images. Thus, regions that present a higher uptake are named hypermetabolic, being hypometabolic in the opposite case. Consequently, it is common in Nuclear Medicine to apply some quantification methods apart from the traditional qualitative analysis, as a support for the clinician decision-making. Here we can see an important application of segmentation on PET images. We can find many models with great performances on organ segmentation [7], which is usually performed on a CT [3,4], but direct segmentation might be advantageous in cases like low-quality CT, attenuation correction approaches without using anatomical information [2] and particularly for non-anatomical lesions. Hence, in the literature there are some examples of models with good but not excellent direct segmentations [1,10], claiming in most cases the sparsity of data as a limitation [6].

Nevertheless, these quantification methods demand reference values, which can be obtained with phantoms. These can be physical objects, but digital phantom simulation may offer a more flexible and accurate modelling. Hence, Monte Carlo simulations are a strongly implemented technique in Nuclear Medicine to efficiently sample complex system parameters but with a high computational cost. In this work, we aim to explore another potential application of simulation, concerning the capacity of building a database for training Deep Learning segmentation networks. Thus, the main objective is to develop a framework that allows to train these networks from our custom generated database.

2 Material and Methods

2.1 Phantom Generation

We generated our phantom database using the Extended Cardiac-Torso Phantom Program, version 2 (XCAT2) from Duke's University [5] which is focused on building anthropomorphic Total-Body phantoms and lesions. XCAT2 outputs an activity map (i.e., the theoretical dose distribution of the radiotracer), and an attenuation map (i.e., the attenuation coefficient for each tissue). We modified some anatomical parameters to ensure enough variability, but we limited the complexity by discarding all the individual organ modifications, leaving the main axes of the body, torso, extremities and head, the size of the musculature and the breast, and the bone and skin thickness. For all the parameters, we assumed a Gaussian distribution.

For the activity map, we took as a first approximation the uptake values from [11], and started a trial-and-error characterization process of the activities of the subjects. Note that they are took as fixed values for each generated phantom, as a good first-instance approximation. We worked analogously for the lesions, which are modelled as ideal, spherical lesions and whose position is limited to the liver and lungs. Patients that ended up with extremely small lesions were discarded. We added some variability in the activity as well, with values substantially higher than the background. Note that the attenuation map remains unaltered, as we are assuming functional lesions that do not significantly modify the anatomy of the patient.

The annotation could be simply retrieved by using the information from the generated activity maps. We could annotate the lesions, main organs and even at regional level.

2.2 Subject Data

In order to train the different Networks, we propose five different cohorts of simulated subjects, that were obtained under the same conditions, with a total of 350 simulations. Each cohort had an independent phantom generation. The two first cohorts include 100 male and female subjects with a single small lesion (i.e., a mean diameter of 2 cm, equivalent to a tumour stage between T1 and T2) in the liver and the lungs, respectively. Cohorts III and IV are analogous, but with half the subjects and larger lesions (i.e., a mean diameter of 5 cm, equivalent to a stage between T2 and T3). Cohort V includes both 25 males and females with up to 6 lesions, emulating a metastatic scenario.

2.3 Simulation and Reconstruction

We simulated fluorodeoxyglucose PET images using SimPET, an open-source platform for the realistic Monte Carlo simulations. This program accepts as input a pair of activity and attenuation maps, which are the ones from our generated database. The attenuation map emulates the CT for further corrections. SimPET makes use of two basic modules: Simulation System for Emission Tomography (SimSET), in charge of the photon tracking and detection, and Software for Tomographic Image Reconstruction (STIR), that manages the reconstruction of the images.

For the reconstruction, we used the tridimensional Ordered Subset Expectation Maximization (OSEM3D) algorithm as it is well-established in the clinical practice [9]. The resulting images present a voxel size of $3.46 \times 3.46 \times 3.27$ mm, with each bed having a dimension of $256 \times 256 \times 47$ voxels. Note that in this work we are still using the traditional bed-to-bed scheme, as the LAFOV scans were not yet implemented.

2.4 Image Processing

Firstly, we removed some characteristic STIR bright artifacts that appear at the edges of the simulations. Next, we removed the leftover slices above the head acquired in the last bed and aligned it with the previously generated Ground Truths. Following we applied a median filter to mitigate the effect of the typical Poisson noise in Nuclear Medicine imaging, and we standardized the images to the z-scores. Finally, images were resized to the same suitable input dimensions for the segmentation network (i.e., $256 \times 256 \times 512$).

To improve the final results, we applied some post-processing to the obtained masks. For the organs, we applied a closing and an opening filter, while for lesions we applied a dilation operation and a median filter. In both cases, we removed non-plausible spatial information (e.g., organ confusion) for each organ, and applied a final median filter.

2.5 Segmentation Methods

Our segmentation approach is based on the use of U-Nets. For the organ segmentation, we tested two architectures: a classical 2D U-Net (feeding with slices on the coronal plane) and a simplification of the Triplanar Ensemble U-Net model (TrUE-Net), which is an ensemble of three 2D U-Nets described in the work of [8]. This last net trains three 2D U-Nets, one with the slices on each plane (i.e., sagittal, coronal and axial) and averages the three obtained probability maps to extract the final label map. This architecture is schematised on Fig. 1.

For the organ segmentation, we used an Adam optimizer (with a learning rate of 10^{-4}) and a cross entropy loss. We trained the nets with cohorts I (for males) and II (for females), splitting the dataset into a 70% training, 15% validation and 15% test sets. For the lesion segmentation, we switched the loss to a focal binary cross entropy, in a try to mitigate the accentuated imbalance between classes. As we are dealing with a smaller objective, we also proposed an alternative, which consisted on a first ensemble segmentation targeting the liver or the lung (depending on the location of the lesion), which helps us defining a region of interest (ROI). On this newly delimited ROI, after applying a median filter with 3x3 kernel, we targeted the lesion by two approaches. On the one hand, we inputted these ROIs to a second U-Net. On the other hand, we applied a

Multiotsu thresholding with 3 classes, and removed the voxels below the lowest threshold (as a conservative criteria). For the metastatic scenario we tried again these approaches, but targeting the liver and the lungs at the same time.

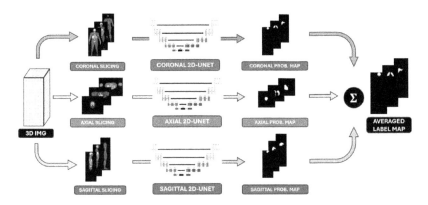

Fig. 1. Scheme of the TrUE-Net ensemble, with each branch corresponding to one plane training and a final map averaging. The ensemble is also used to patch the organ of interest in lesion segmentation.

To evaluate the performance of the segmentation models, we took as a first instance metric the Dice Similarity Coefficient as the overlapping with the Ground Truths. We also computed the Hausdorff Distance and, particularly for the lesion masks, the Volumetric Similarity. For a deeper understanding, we also computed the sensitivity and the precision of the masks voxel-to-voxel. In all cases we presented the average on the test sets.

3 Results and Discussion

3.1 Phantom Anatomical Model and Simulations

After setting all the chosen parameters to obtain plausible human anatomies, we applied them for each of the 350 generated phantoms. An example of an obtained pair for a male and female subject is show in Fig. 2, together with an approximation of the height, weight and waist parameters of all the considered subjects. These show that our database presents an acceptable anatomical variability for the scope of this work.

In Fig. 2 we also present their corresponding simulated images, whose noise has been analyzed as well, having an acceptable level of realism. The simulations present a mean correlation coefficient of 0.907 ± 0.002 after pre-processing with the reference activity map.

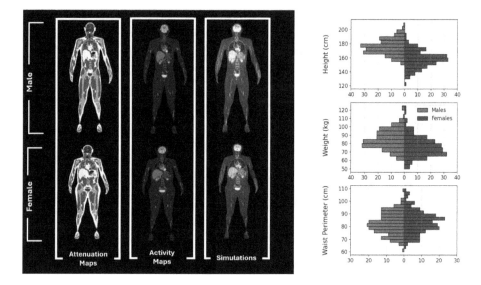

Fig. 2. Example of an attenuation and activity map for a male and female and its corresponding simulation. The approximation of the height, weight and waist parameters of all the 350 subjects is shown at the right.

3.2 Organ Segmentation

In Fig. 3 we compare the overlappings for male and female subjects of the organs considered. In general, performance is reasonably good, especially for the large, well-contrasted organs (e.g., brain and lungs), with DICE scores exceeding 0.8 in almost all the cases. TrUE-Net also shows a better performance compared to the simple U-Net, not only by increasing the overlapping, but also by decreasing the Hausdorff Distance. The metrics of the best characterized organs is shown in Table 1, where sensitivity and specificity generally present comparable values.

Furthermore, we also evaluated the performance at regional level. In Fig. 3 we also show the performance for some regions in the brain, kidneys and large intestine. We find comparable results for large, well-contrasted regions (e.g., colon portions) but in general there is a slight decrease in the performance, accentuated on small regions (e.g., brain ventricles or adrenal gland). Thus, the model seems to be effective in segmenting not only organs, but also some regions, but appears to fail in capturing small details.

3.3 Lesion Segmentation and Detection

In terms of lesion segmentation, all four methods present a very poor performance for the small lesions of cohorts I and II and should not be considered. The best of the four, the single U-Net, presents an overlapping of 0.62 ± 0.10 and 0.41 ± 0.10 for the liver and lungs, respectively. Thus, in this scenario the use of

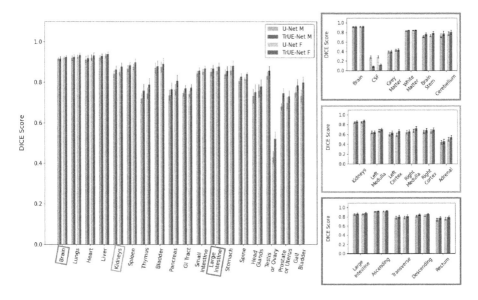

Fig. 3. Comparison of the organ DICE scores for U-Net and TrUE-Net architectures on cohorts I and II. The comparison with some regions of the brain, kidneys and large intestine is displayed at the right as well.

Table 1. Average metrics of cohorts I and II obtained for the segmentation of the six best characterised organs of our phantoms, using TrUE-Net.

Tissue	DICE	HD	VS	Sensitivity	Precision
Brain	0.923 ± 0.010	13 ± 8	0.972 ± 0.004	0.934 ± 0.016	0.915 ± 0.004
Lungs	0.933 ± 0.012	3.5 ± 0.3	0.996 ± 0.002	0.935 ± 0.013	0.931 ± 0.013
Heart	0.927 ± 0.017	3.7 ± 0.8	0.991 ± 0.002	0.934 ± 0.017	0.920 ± 0.017
Liver	0.936 ± 0.014	3.6 ± 0.4	0.992 ± 0.001	0.942 ± 0.013	0.929 ± 0.014
Kidneys	0.88 ± 0.03	7.7 ± 0.5	0.983 ± 0.004	0.86 ± 0.03	0.89 ± 0.03
Bladder	0.89 ± 0.03	15 ± 10	0.981 ± 0.003	0.89 ± 0.03	0.88 ± 0.03

the ensemble seems to even worsen the results, probably due to the inconsistency among planes.

In case of larger lesions, however, the whole performance clearly improves, as shown in Fig. 4. Particularly, the thresholding approach is the one with the worst results, while the other methods offer similar results, although the use of the U-Net on the organ of interest is clearly faster on the training. In Table 2 we present thus the metrics for the lesion segmentation on this approach. We also observe that in metastasis cases (cohort V) there is a slight decrease on the performance compared to the single-lesion, probably due to the interaction of both organs rather than focusing on only one. In general, we see a high Volumetric Similarity and a higher precision than sensitivity, which suggests that the masks present

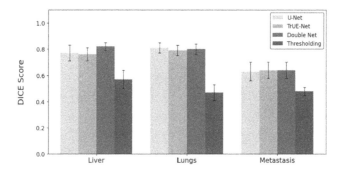

Fig. 4. Comparison of the four methods described in Sect. 2.5 for the large lesions on the liver and lungs and the metastatic scenario.

more false negatives than positives. The misalignment of the lesion zone with the ground truth due to a slight dilation of the bright zones during reconstruction might contribute to this fact.

Table 2. Average metrics obtained for the segmentation of the lesions on cohorts III (liver), IV (lungs) and V (metastasis), using the U-Net on the organ of interest method.

Cohort	DICE	HD	VS	Sensitivity	Precision
Liver	0.82 ± 0.03	2.6 ± 0.4	0.93 ± 0.02	0.82 ± 0.04	0.84 ± 0.05
Lung	0.80 ± 0.04	3.0 ± 0.4	0.95 ± 0.03	0.77 ± 0.05	0.84 ± 0.05
Metastasis	0.64 ± 0.06	25 ± 11	0.96 ± 0.01	0.64 ± 0.06	0.65 ± 0.07

We also studied the lesion detection capability of the models. If we set a detection above a 0.50 DICE score, the U-Net and TrUE-Net detected all the lung lesions and a 86% of liver lesions on the test set. Over a 0.80 DICE score, these values change to a 83% for the lung and a 71% for the liver lesions with the U-Net, and a 83% for the lung and 57% for the liver lesions with the TrUE-Net. For the double-net approach, all the lesions were detected over a 0.50 DICE, and we got a 83% and 57% over a 0.80 DICE.

4 Conclusions

In this work we satisfactorily developed a framework from the generation of a Total-Body phantom database to the segmentation of their simulations. Furthermore, our database presents enough anatomical variability and the simulations show an acceptable level of noise, which let us train models that show good performance in segmenting organs and lesions In this simulated domain.

Our approach is promising as it overcomes the sparsity of data limitation, added to the automatic, almost perfect annotation that can be obtained by using the information from the activity maps. In the near future, we aim to improve this work by validating our models on real PET images, by increasing our cohorts and improving the activity characterization of our simulations.

Disclosure of Interests. The authors have no competing interests to declare that are relevant to the content of this article.

References

1. Dirks, I., Keyaerts, M., Neyns, B., Vandemeylebroucke, J.: Computer-aided detection and segmentation of malignant melanoma lesions on whole-body 18F-FDG PET/CT using an interpretable deep learning approach. Comput. Methods Programs Biomed. **221**, 106902 (2022). https://doi.org/10.1016/j.cmpb.2022.106902
2. Dong, X., et al.: Deep learning-based attenuation correction in the absence of structural information for whole-body positron emission tomography imaging. Phys. Med. Biol. **65**, 055011 (2020). https://doi.org/10.1088/1361-6560/ab652c
3. Okada, T., Linguraru, M., Hori, M., Summers, R., Tomiyama, N., Sato, Y.: Abdominal multi-organ segmentation from CT images using conditional shape-location and unsupervised intensity priors. Med. Image Anal. **26**, 1–18 (2015). https://doi.org/10.1016/j.media.2015.06.009
4. Pan, S., et al.: Abdomen CT multi-organ segmentation using token-based MLP-mixer. Med. Phys. **50**, 3027–3038 (2023). https://doi.org/10.1002/mp.16135
5. Segars, W., Sturgeon, G., Mendoca, S., Grimes, J., Tsui, B.: 4D XCAT phantom for multimodality imaging research. Med. Phys. **37**, 4902–4915 (2010). https://doi.org/10.1118/1.3480985
6. Shiyam-Sundar, L., Beyer, T.: Is automatic tumor segmentation on whole-body 18F-FDG pet images a clinical reality? J. Nucl. Med. **1978**(65), 1–3 (2024). https://doi.org/10.2967/jnumed.123.267183
7. Shiyam-Sundar, L., et al.: Fully automated, semantic segmentation of whole-body 18F-FDG PET/CT images based on data-centric artificial intelligence. J. Nucl. Med. **1978**(63), 1941–1948 (2022). https://doi.org/10.2967/jnumed.122.264063
8. Sundaresan, V., Zamboni, G., Rothwell, P., Jenkinson, M., Griffanti, L.: Triplanar ensemble U-net model for white matter hyperintensities segmentation on MR images. Med. Image Anal. **73**, 102184–102184 (2021). https://doi.org/10.1016/j.media.2021.102184
9. Tsuda, K., Suzuki, T., Toya, K., Sato, E., Fujii, H.: 3D-OSEM versus FORE + OSEM: optimal reconstruction algorithm for FDG PET with a short acquisition time. World Jo. Nucl. Med. **22**, 234–243 (2023). https://doi.org/10.1055/s-0043-1774418
10. Zhang, J., Huang, Y., Zhang, Z., Shi, Y.: Whole-body lesion segmentation in 18F-FDG PET/CT. arXiv (Cornell University) (2022). https://doi.org/10.48550/arxiv.2209.07851
11. Zincirkeser, S., Sahin, E., Halac, M., Sager, S.: Standardized uptake values of normal organs on 18f-fluorodeoxyglucose positron emission tomography and computed tomography imaging. J. Int. Med. Res. **35**, 231–236 (2007). https://doi.org/10.1177/147323000703500207

Integrating Convolutional Neural Network and Transformer for Lumen Prediction Along the Aorta Sections

Yichen Yang[1,2,4], Pengbo Jiang[2], Xiran Cai[4], Zhong Xue[2], and Dinggang Shen[1,2,3(✉)]

[1] School of Biomedical Engineering and State Key Laboratory of Advanced Medical Materials and Devices, ShanghaiTech University, Shanghai, China
[2] Department of Research and Development, United Imaging Intelligence, Shanghai, China
[3] Shanghai Clinical Research and Trial Center, Shanghai, China
dinggang.shen@gmail.com
[4] School of Information Science and Technology, ShanghaiTech University, Shanghai, China

Abstract. Aortic dissection is defined as a separation of layers of the aortic wall leading high mortality. Accurately segmenting both the true and false lumens is critical for revealing geometrical characteristics for diagnosis and evaluation of the dissection. Existing aortic dissection segmentation methods are mainly convolutional neural network (CNN)-based and are limited in precisely distinguishing these lumens due to lack of long-range dependencies along the aorta. To address this issue, we propose an integrated CNN and transformer prediction network (ICTP-Net) to capture both low-level spatial details and long-range global dependencies. Rather than a simple concatenation and fusion, an attention fusion (AF) block is employed to merge features from two branches. Additionally, due to the vascular anatomy, the proposed network is trained and applied in a sliding-context-dependent manner where we use partial previous segmentation for the prediction of the next section, further enhancing the spatial continuity along the aorta. 726 data samples were used in the experiments, and comparative and ablation studies show that the proposed ICTP-Net achieves the best aortic dissection true lumen segmentation compared with other state-of-the-art methods, demonstrating the effectiveness of the model integration, AF module and the sliding-context-dependent design.

Keywords: Aortic Dissection Segmentation · CT Angiography · Transformer · Convolutional Neural Network

1 Introduction

Aortic Dissection (AD) is a severe disease of the main artery of human body caused by intimal tear, leading blood to flow into a new cavity, forming a true lumen (TL) and a false lumen (FL). Thoracic endovascular aortic repair (TEVAR) is a common treatment of AD where true and false lumen segmentation is needed to observe detailed pathology for surgical planning and prognosis [17]. Manual segmentation often requires slice-by-slice or semi-automatic segmentation, and hence automated segmentations are critical for efficient and effective processing.

With the development of deep learning, more automated segmentation methods have been proposed [1,2,6,7,19,20]. In the literature, Cao et al. [1] proposed a cascaded convolutional neural network (CNN) model for multi-task segmentation and labelling. After segmenting the whole aorta, the entire aorta mask as additional input is further classified into TL/FL regions. To address the tortuous and complex morphological structure of the aorta, Chen et al. [2] first straightened the aorta after aorta segmentation, then performed lumen segmentation and labelling refinement. Hahn et al. [7] followed the straightening scheme but employed only 2D models that naturally ignores spatial context. Moreover, Wobben et al. [19] explored cascaded models using 3D residual symmetric U-Net for TL, FL and thrombosis segmentation. Feng et al. [6] designed a skip connection attention refinement module enabling CNN to use low-level features efficiently.

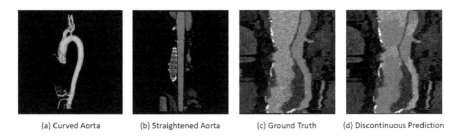

(a) Curved Aorta   (b) Straightened Aorta   (c) Ground Truth   (d) Discontinuous Prediction

Fig. 1. Illustration of aorta anatomy and discontinuous prediction. (a) The curved aorta structure; (b) a slice view of the straightened aorta image; (c) a slice view of ground truth which is continuous in space; (d) the discontinuous prediction issue.

However, relying solely on CNNs, existing methods may overlook spatial continuity, causing issues like discontinuous lumen prediction (see Fig. 1 (d)). Transformers [18], initially used in natural language processing and later in vision tasks (e.g., Vision Transformer [4]), excel in global information modelling but require extensive pre-training on large datasets. Integrating Transformers and CNNs could mitigate these challenges, leveraging both for improved long-range dependency modelling while benefiting from CNN's spatial feature bias. Several attempts have been made on benefiting Transformer from inductive bias of convolution, either by combining CNN with Transformer [21,22], or integrating convlolution operations into Transformer structures [10,11], showing promising performance in medical image segmentation. However, none of existing methods have studied capturing spatial changes of the aorta with combined structures. Xiang et al. [20] integrated Transformer layers within a CNN structure to enhance flap features, but did not fully exploit global aortic information. Flap information aids boundary identification but does not distinguish true lumen from false lumen, limiting their model's ability to model long-range dependencies effectively. To deal with vascular structure of the aorta and various input size, overlapped sliding window segmentation can be employed. But it ignores interrelations between windows, causing loss of aorta spatial context, and potentially leads to discontinuous prediction.

To address the above issues, an Integrated CNN and Transformer Prediction Network (ICTP-Net) is proposed to accurately segment aortic dissection true lumen by focusing on spatial continuity. The major contributions of ICTP-Net are as follows:

- Dual-branch CNN and Transformer encoders are employed, capable of modelling long-range dependencies along the aorta while preserving local details.
- An attention fusion (AF) module is presented for feature merging to leverage global information from the Transformer branch and enhance feature fusion across semantic spaces.
- A sliding-context-dependent scheme is used for training and application. We employ an mask encoder to overcome the shortcomings of the sliding window method. This approach better integrates segmentations within each window during application, utilizing segmentation context information and ensuring continuity in prediction results.

2 Method

The objective is aortic true lumen segmentation and ensuring correctness and spatial continuity of TL prediction. As shown in Fig. 2, ICTP-Net has three components: 1) dual parallel encoders, one built with CNN and another with Transformer, to generate hierarchical complementary representations; 2) attention fusion (AF) modules to merge the dual-branch features and connect to the decoder with skip connections to produce the final result; 3) a sliding-context-dependent manner for training and testing the prediction procedure.

2.1 Dual-Branch Encoding

CNN has insufficient long-range dependency modelling ability due to the natural inherent bias of convolution, which, on the other hand, gives CNN its extraordinary generalization and local detail modelling ability compared to Transformer. Capturing morphological changes of the lumen along the aorta is critical for aortic dissection segmentation. However, the aorta has a length usually five times its diameter [5], making it challenging for CNNs to capture continuous morphological changes of aortic dissection. To address this issue, we employ ViT as an independent branch in the proposed encoder, in addition to a hierarchical CNN encoder, to capture long-range dependencies along the aorta.

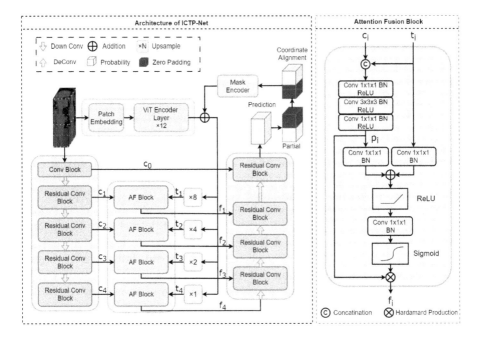

Fig. 2. Architecture of proposed ICTP-Net.

CNN Encoder. To obtain contextual features and maintain certain spatial details by convolutional neural networks, we build CNN branch with 4 encoding blocks of V-Net [15]. CNN encoder overall produces 5 hierarchical features c_i, where $i = 0, 1, 2, 3, 4$, each of size $\left(\frac{D}{2^i}, \frac{H}{2^i}, \frac{W}{2^i}\right)$. Specifically, an input image first goes through a simple convolution block and generates a low-level local feature c_0, followed by hierarchical feature generation through four encoding blocks. Each encoding block composes certain number of convolution operations with a residual path connecting block input and output. Four residual blocks consist of 1, 2, 3 and 3 convolution operations, respectively.

Transformer Encoder. For the Transformer branch, input images are first divided into patches of equal size $(16, 16, 16)$, then flattened and linear projected by point-wise convolution. Image embeddings are then fed into a series of 12 Transformer blocks. Unlike [8,9] which incorporate features from internal Transformer blocks, we exclusively use the last output feature of the ViT encoder, as it passes through all attention blocks and is expected to have the strongest representation. This feature map then passes through a neck consisting of convolution and layer normalization to reduce channel dimensions, before being reshaped back to $\left(\frac{H}{16}, \frac{D}{16}, \frac{W}{16}\right)$. Finally it undergoes four distinct upsample modules to generate hierarchical feature maps t_i where $i = 1, 2, 3, 4$, each of size $\left(\frac{D}{2^i}, \frac{H}{2^i}, \frac{W}{2^i}\right)$.

2.2 Attention Fusion Module

In aortic dissection segmentation, we aim to enhance local features c_i captured by CNN with Transformer features t_i that capture long-range dependencies, enabling the model to better understand continuous spatial changes in the lumen while preserving local details. c_i and t_i pair with same resolution, where $i = 1, 2, 3, 4$, are fed into AF module respectively to generate fused feature f_i, before fed into hierarchical CNN decoder.

Specifically, as depicted in Fig. 2, c_i and t_i are first concatenated and convolved to obtain the pre-fused feature p_i. Subsequently, p_i and t_i are further fused with the guidance of global information. In this step, we employ an attention gate structure similar to that first used in pancreas segmentation [16]. However, in AF module, this structure processes mixed local-global features p_i and global features t_i, aiming to leverage information from both branches and emphasize long-range dependencies in fused features. Specifically, p_i and t_i are initially projected by point-wise convolution separately, then added and passed through ReLU, another point-wise convolution and finally a Sigmoid activation to generate a global information-aware weighting matrix w_i. The process can be formulated as:

$$w_i = \sigma(\delta(\gamma(\phi \cdot p_i + \psi \cdot t_i))), \tag{1}$$

where σ and γ represents Sigmoid and ReLU activation, and δ, ϕ, ψ are kernel weights of point-wise convolutions. The pre-fused feature p_i is then gated by w_i through pixel-wise multiplication to produce the ultimately fused feature f_i, which can be formulated as:

$$f_i = w_i \circ p_i, \tag{2}$$

where $\circ$ represents the Hadamard product. Through AF module, features carrying spatial details and long-range dependencies are effectively fused and the long range dependencies of morphological changes along aorta are emphasized in fused features.

2.3 Sliding-Context-Dependent Training and Application

Sliding window approach divides a single segmentation task into independent submissions, disregarding their relationships, which can lead to issues like discontinuous lumen predictions. Resampling to a uniform size could address this but risks losing original image information. Inspired by interactive segmentation approaches [12,14], where each iteration refines predictions based on previous outputs, we introduce a mask encoder that takes partial predictions from the previous iteration as input. The mask embedding is directly added to ViT features before upsampling, as illustrated in Fig. 2.

Unlike interactive segmentation tasks that refine predictions on the same input image across iterations with backpropagation, our approach processes different inputs in each iteration due to the sliding window method as shown in Fig. 3. Therefore, before embedding the mask in each iteration, we align the region of the previous prediction overlapping with the current input block to synchronize their coordinates. In the first iteration without a previous prediction, we employ a learnable token instead of the mask embedding added to ViT features. With this design, each submission within a window benefits from insights gained from the prediction of previous iteration, ensuring better continuity across the entire prediction process.

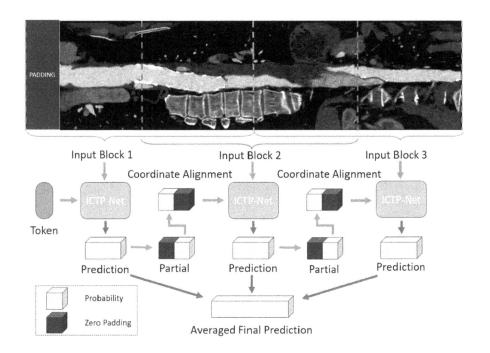

Fig. 3. The sliding-context-dependent scheme.

3 Experiments and Results

Straightening of Aorta. To simplify complex torsion and curved morphology of aorta, straightening is performed on original Computed Tomography Angiography (CTA) images. Specifically, aorta mask is first generated by a pre-trained V-Net for aorta segmentation, skeleton of aorta is then extracted using a traditional method [13] and the longest path of skeleton is chosen as the centerline. Finally, 2D multi-planar reformations orthogonal to centerline are generated and stacked to form the 3D straightened aorta image.

Dataset and Evaluation Metrics. The dataset comprises 726 straightened Aortic Dissection CTA images, each with an in-plane size of 200×200 and varying slice numbers from 206 to 770, all having isotropic spacing of 0.5 mm. Manual annotations were conducted by an experienced expert using medical processing software mimics (Materialise, Leuven, Belgium). The dataset is divided randomly into a training set (517 cases), a validation set (57 cases) and a testing set (152 cases). For quantitative evaluation, Dice Similarity Coefficient (Dice), False Positive Rate (FPR), Intersection over Union (IoU) and 95th percentile Hausdorff Distance (HD95) serve as evaluation metrics.

Implementation Details. All our experiments are conducted with PyTorch on a single A100-SXM4-40GB GPU. We train our model with Adam optimizer, using a batch size of 2 for 300 epochs. The initial learning rate is set to 2e-3, and we employ a milestone learning rate decay strategy with a milestone at 150 epochs and a decay rate of 0.1. During training, we first resample the in-plane size to [100, 100], and then randomly sample a crop center from the ground truth and extract a sub-volume of shape [96, 96, 256]. The cropped image is randomly rotated along the Z-axis. To effectively train our model with a mask encoder, we crop a sub-volume of shape [96, 96, 512] (padding if needed) and divide it into 3 consecutive blocks of shape [96, 96, 256] with a moving stride 128 along Z-axis. Each set of 3 consecutive blocks is treated as a unit, and the predictions of one block are partially passed to the next iteration. Images are normalized using a fixed window level of 300 HU and window width of 800 HU, followed by intensity truncation to the range $[-1, 1]$.

Table 1. Quantitative comparison with state-of-the-arts.

Method	Dice (%) ↑	FPR (%) ↓	IoU (%) ↑	HD95 (mm) ↓
3D U-Net [3]	92.58 ± 5.70	0.32 ± 0.29	86.65 ± 8.75	8.48 ± 9.85
UNETR [9]	86.67 ± 9.70	0.66 ± 0.72	77.60 ± 13.31	10.96 ± 8.19
SwinUNETR [8]	91.37 ± 7.16	0.39 ± 0.43	84.80 ± 10.49	8.36 ± 8.96
Cao et al. [1]	92.55 ± 5.82	0.33 ± 0.32	86.62 ± 9.00	5.27 ± 7.07
Chen et al. [2]	92.42 ± 6.87	0.38 ± 0.53	86.53 ± 9.99	6.07 ± 8.87
Feng et al. [6]	92.70 ± 5.22	0.32 ± 0.27	86.80 ± 8.28	5.55 ± 7.76
Xiang et al. [20]	91.37 ± 5.36	0.30 ± 0.27	84.51 ± 8.43	7.35 ± 7.96
Ours	**93.57 ± 4.27**	**0.28 ± 0.22**	**88.19 ± 6.88**	**4.61 ± 5.85**

Comparison with State-of-the-Art Methods. To evaluate our proposed method, we perform quantitative comparisons with state-of-the-art 3D medical image segmentation models including CNN based 3D U-Net [3] and Transformer based models such as SwinUNETR [8] and UNETR [9]. We also compare our method with state-of-the-art Aortic Dissection segmentation methods [1,2,6,20]. Quantitative and qualitative results are demonstrated in Table 1 and Fig. 4, respectively.

Table 2. Quantitative results of ablation study. Features are fused using convolution operations in setting E.3 and E.5.

Setting	Method	Dice (%) ↑	FPR (%) ↓	IoU (%) ↑	HD95 (mm) ↓
E.1	Pure CNN encoder	92.52 ± 6.10	0.31 ± 0.27	86.61 ± 9.24	5.01 ± 6.17
E.2	Pure Trans encoder	84.48 ± 8.82	0.80 ± 0.53	74.05 ± 12.06	12.90 ± 8.96
E.3	CNN + Trans encoder	92.87 ± 5.57	0.32 ± 0.37	87.13 ± 8.62	5.53 ± 8.17
E.4	CNN + Trans + AF	93.19 ± 4.92	0.31 ± 0.27	87.60 ± 7.80	4.88 ± 7.31
E.5	CNN + Trans + Mask Encoder	93.41 ± 4.48	**0.28 ± 0.25**	87.93 ± 7.19	5.41 ± 8.10
E.6	**Ours**	**93.57 ± 4.27**	**0.28 ± 0.22**	**88.19 ± 6.88**	**4.61 ± 5.85**

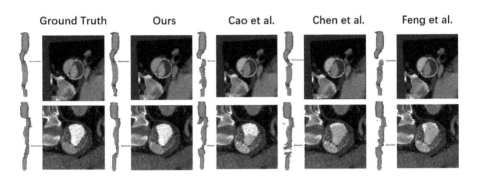

Fig. 4. Qualitative segmentation results. An axial slice view is attached to each 3D segmentation result to show the details.

As shown in Table 1, our method achieves the highest mean Dice of 93.57% compared to competing methods. It also achieves the best FPR of 0.28%, IoU of 88.19% and HD95 of 4.61 mm, respectively. Our method also exhibits the lowest standard deviation across all metrics, indicating superior stability. Quantitative results show that our method consistently outperforms the competing methods in aortic dissection TL segmentation task. Qualitatively, our proposed method excels in maintaining spatial continuity of the lumen. In Fig. 4, it is evident that other methods sometimes misclassify false lumen as true lumen when the false lumen becomes brighter and narrower, a challenge our method effectively addresses by ensuring spatial continuity in lumen prediction.

Ablation Study. To validate the effectiveness of parallel encoder, AF module and the sliding-context-dependent design, we conduct ablation studies by incrementally combining these components. Quantitative results are detailed in Table 2. First by comparing E.3 to E.1 and E.2, we can find that combining CNN and Transformer leads to a slightly better performance than single encoder models. Effectiveness of AF module and sliding-context-dependent design can be concluded by comparing E.4 to E.3 and E.5 to E.3, respectively. E.6 shows that with all components, the ICTP-Net has remarkable improvements over single encoder models across all metrics.

4 Conclusion

In this paper, we have proposed a novel method for Aortic Dissection true lumen segmentation. Specifically, the proposed ICTP-Net leverages the long-range dependency modelling ability of Transformers and the inductive bias of CNNs. An attention Fusion module is proposed to effectively fuse CNN and Transformer features and utilize global information from the Transformer stream. Furthermore, we design the model to operate in an sliding-context-dependent scheme with a mask encoder, which further ensures the spatial continuity of predictions. Experimental results show that ICTP-Net achieves state-of-the-art performance on the task of aortic dissection true lumen segmentation compared to other competing methods. In the future, we plan to further explore ability of ICTP-Net on multi-task segmentation, including distinguishing true/false lumen and identifying thrombosis in aortic dissections.

Disclosure of Interests. The authors have no competing interests to declare that are relevant to the content of this article.

References

1. Cao, L., et al.: Fully automatic segmentation of type B aortic dissection from CTA images enabled by deep learning. Eur. J. Radiol. **121**, 108713 (2019). https://doi.org/10.1016/j.ejrad.2019.108713
2. Chen, D., et al.: Multi-stage learning for segmentation of aortic dissections using a prior aortic anatomy simplification. Med. Image Anal. **69**, 101931 (2021). https://doi.org/10.1016/j.media.2020.101931
3. Çiçek, Ö., Abdulkadir, A., Lienkamp, S.S., Brox, T., Ronneberger, O.: 3D U-Net: learning dense volumetric segmentation from sparse annotation. In: Ourselin, S., Joskowicz, L., Sabuncu, M.R., Unal, G., Wells, W. (eds.) MICCAI 2016. LNCS, vol. 9901, pp. 424–432. Springer, Cham (2016). https://doi.org/10.1007/978-3-319-46723-8_49
4. Dosovitskiy, A., et al.: An Image is Worth 16 × 16 Words: transformers for image recognition at scale (2020). arXiv preprint arXiv:2010.11929
5. Dotter, C.T., et al.: Aortic length: angiocardiographic measurements. In: Circulation, pp. 915–920. (1950). https://doi.org/10.1161/01.CIR.2.6.915
6. Feng, H., et al.: Automatic segmentation of thrombosed aortic dissection in post-operative CT-angiography images. Med. Phys. **50**(6), 3538–3548 (2023). https://doi.org/10.1002/mp.16169
7. Hahn, L. D., et al.: CT-based true-and false-lumen segmentation in type B aortic dissection using machine learning. Radiol. Cardiothorac. Imag. D **2**(3), e190179 (2020). https://doi.org/10.1148/ryct.2020190179
8. Tang, Y., et al.: Self-supervised pre-training of swin transformers for 3D medical image analysis. In: Proceedings of the IEEE/CVF Conference on Computer Vision and Pattern Recognition, pp. 20730–20740 (2022). https://doi.org/10.48550/arXiv.2111.14791
9. Hatamizadeh, A., et al.: UNETR: transformers for 3D medical image segmentation. In: Proceedings of the IEEE/CVF Winter Conference on Applications of Computer Vision, pp. 574–584 (2022). https://doi.org/10.48550/arXiv.2103.10504

10. He, A., et al.: H2Former: an efficient hierarchical hybrid transformer for medical image segmentation. IEEE Trans. Med. Imag. **42**(9), 2763–2775 (2023). https://doi.org/10.1109/TMI.2023.3264513
11. He, Y., et al. SwinUNETR-V2: stronger swin transformers with stagewise convolutions for 3D medical image segmentation. In: Greenspan, H., et al. MICCAI 2023. LNCS, vol. 14223, pp. 416–426. Springer, Cham (2023). https://doi.org/10.1007/978-3-031-43901-8_40
12. Kirillov, A., et al.: Segment anything. In: Proceedings of the IEEE/CVF International Conference on Computer Vision, pp. 4015–4026 (2023). https://doi.org/10.48550/arXiv.2304.02643
13. Lee, T., et al.: Building skeleton models via 3-D medial surface axis thinning algorithms. CVGIP: graphical models and image processing **56**(6), 462–478 (1994). https://doi.org/10.1006/cgip.1994.1042
14. Liu, Q., et al.: SimpleClick: interactive image segmentation with simple vision transformers. In: Proceedings of the IEEE/CVF International Conference on Computer Vision, pp. 22290–22300 (2023). https://doi.org/10.48550/arXiv.2210.11006
15. Milletari, F., et al.: V-Net: fully convolutional neural networks for volumetric medical image segmentation. In: 2016 Fourth International Conference on 3D Vision, pp. 565–571 (2016). https://doi.org/10.1109/3DV.2016.79
16. Oktay, O., et al.: Attention U-Net: learning where to look for the pancreas (2018). arXiv preprint arXiv:1804.03999
17. Pepe, A., et al.: Detection, segmentation, simulation and visualization of aortic dissections: a review. Med. Image Anal. **65**, 101773 (2020). https://doi.org/10.1016/j.media.2020.101773
18. Vaswani, A., et al.: Attention is all you need. In: Advances in Neural Information Processing Systems, vol. 30 (2017)
19. Wobben, L. D., et al.: Deep learning-based 3D segmentation of true lumen, false lumen, and false lumen thrombosis in type-B aortic dissection. In: 2021 43rd Annual International Conference of the IEEE Engineering in Medicine & Biology Society (EMBC), pp. 3912–3915 (2021). https://doi.org/10.1109/EMBC46164.2021.9631067
20. Xiang, D., et al.: ADSeg: a flap-attention-based deep learning approach for aortic dissection segmentation. Patterns **4**(5), 100727 (2023). https://doi.org/10.1016/j.patter.2023.100727
21. Yuan, F., et al.: An effective CNN and transformer complementary network for medical image segmentation. Pattern Recogn. **136**, 109228 (2023). https://doi.org/10.1016/j.patcog.2022.109228
22. Zhang, Y., et al. TransFuse: fusing transformers and CNNs for medical image segmentation. In: de Bruijne, M., et al. MICCAI 2021. LNCS, vol. 12901, pp. 14–24. Springer, Cham (2021). https://doi.org/10.1007/978-3-030-87193-2_2

CSSD: Cross-Supervision and Self-denoising for Hybrid-Supervised Hepatic Vessel Segmentation

Qiuting Hu[1], Li Lin[1,2], Pujin Cheng[1,2], and Xiaoying Tang[1,3(✉)]

[1] Department of Electronic and Electrical Engineering, Southern University of Science and Technology, Shenzhen, China
tangxy@sustech.edu.cn
[2] Department of Electrical and Electronic Engineering, The University of Hong Kong, Hong Kong SAR, China
[3] Jiaxing Research Institute, Southern University of Science and Technology, Jiaxing, China

Abstract. Hepatic vessel segmentation from Computer Tomography (CT) plays a crucial role in the diagnosis and treatment of various diseases. However, manually delineating hepatic vessels is a time-consuming, arduous task that demands expertise, rendering the procurement of substantial, high-fidelity annotated data from experts a challenge. Public available datasets often comprise unlabelled data or labels afflicted by noise. Recent research efforts have focused on utilizing these unlabelled data or noisy labels to extract additional information. However, these methods typically involve single-level supervision, lacking the joint use of unlabelled data and noisy labels. Therefore, developing a model training paradigm that can effectively combine multiple hierarchical supervision levels is desirable. To address this issue, we propose a novel framework that robustly learns segmentation from a small amount of relatively high-quality labelled data, a large amount of noisy labelled data and unlabelled data through Hybrid-Supervision Learning (HSL). Specifically, we employ two parallel segmentation networks to learn from unlabelled data using perturbation consistency of pseudo labels, and introduce the Cross-Sample Mutual Attention (CMA) module to transfer prior knowledge of relatively high-quality labelled data to unlabelled data, and employ a self-denoising approach to address potential issues with model generalization stemming from inaccuracies within noisy labels. Extensive experiments on public datasets (3DIRCADb, MSD3, MSD8) demonstrate the superiority of the proposed framework.

Keywords: Cross-supervision · Self-denoising · Hybrid-supervision learning · Cross-sample mutual attention

1 Introduction

Hepatic vessel segmentation is crucial for the diagnosis and treatment of liver diseases. However, both computer-aided and manual segmentation of hepatic vessel faces significant challenges due to factors such as noise, insufficient contrast, the complexity of vessel structures in CT images, and the imbalance between vessels and background. In recent years, deep learning (DL) method have developed into the state of the art (SOTA) for hepatic vessel segementation. However, due to the scarcity of accurately labelled hepatic vessel data, collecting sufficient high-quality annotated data for the effective training of DL segmentation models is practically challenging. In addition, the publicly available hepatic vascular datasets often lack precise labeling. For example, although the annotation of the 3DIRCADb dataset is not very accurate, it is already among the highest quality of public datasets. Therefore, in our method, the annotations in the 3DIRCADb dataset are considered to be of high quality, as shown in Fig. 1(a). Most datasets, such as MSD8, as shown in Fig. 1(b), have low-quality annotations with numerous unlabelled or mislabelled pixels, referred to as "noise". Moreover, images with no annotations are more prevalent.

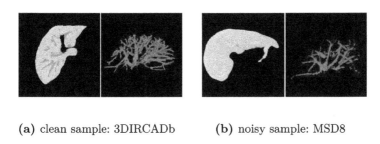

(a) clean sample: 3DIRCADb (b) noisy sample: MSD8

Fig. 1. The example cases from (a) the 3DIRCADb dataset [1] and (b) the Medical Segmentation Decathlon (MSD) dataset [9]. Red color indicates the vessels. Additionally, in (b) yellow arrows denote some of the unlabelled pixels. (Color figure online)

To address the scarcity of labelled data, a machine learning paradigm known as semi-supervised learning (SSL) has been proposed, and achieved notable advancements in image segmentation. SSL method can be primarily categorized into two popular methods, i.e., pseudo labeling [2,21,23] and smoothness assumption based consistency regularization [3,10,12,18]. Pseudo labeling methods involve training a model with limited labelled data to generate pseudo labels for unlabelled images. Subsequently, these newly generated pseudo labelled data are combined with the original labelled set for further optimization. Consistency regularization methods are based on the smoothness assumption, encouraging predictions to remain consistent under different perturbations.

Furthermore, to leverage low-quality labels, a machine learning paradigm called noise label learning (NLL) has been introduced. However, training networks with noisy labels presents significant challenges as networks are prone to memorizing inaccurate labels, thereby diminishing their generalization capability. Thus, various endeavors have been undertaken to mitigate the effects of noisy labels. A common strategy is to perform sample selection and train the network on the selected samples [6]. Xu et al. [14] also uses confident learning (CL) to turn noisy labels into valuable labels. It is noteworthy that although their method is also applied to hepatic vessel segmentation, it relies solely on labelled datasets. In addition, during the training process, noisy labels are updated, which could lead to error accumulation if the previous noise label correction is not effective. Tu et al. [11] introduced a label purifier based on meta-learning, decoupling the dual-level optimization problem into representation and label distribution learning. However, these paradigms typically only utilize two types of data, and do not effectively integrate high-quality annotated samples, low-quality annotated samples, and unlabelled samples. Considering the challenges in obtaining accurate hepatic vessel annotations, research into Hybrid-Supervision Learning (HSL) is imperative. Nevertheless, it remains relatively scarce, underscoring the urgency of developing a training paradigm that can judiciously leverage multi-level supervision models.

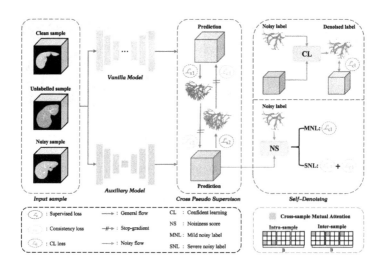

Fig. 2. The framework of our proposed method.

Inspired by the cross pseudo supervision-like (CPS-like) paradigm [3], we further propose a novel **C**ross-**S**upervision and **S**elf-**D**enoising based (CSSD) hybrid-supervision framework for hepatic vessel segmentation. It is a challenging task to capture the intricate structures of vessels under limited supervision. Therefore, within our HSL framework, we integrate not only clean labelled samples but also noisy labelled samples and unlabelled samples, thereby fully utilizing the images and diverse supervisory information. Specifically, CSSD uses two parallel networks with the same structure but different initializations, namely the vanilla network and the auxiliary network. By encouraging consistent segmentation of the same input across networks, the networks can further exploit the image information of the noisy samples and unlabelled samples. Then through the Cross-Sample Mutual Attention (CMA) module in the auxiliary network, transferring knowledge from labelled to unlabelled data. Subsequently, we incorporate a CL strategy to refine noisy labels and maximize the exploitation of information within noisy labels. Extensive testing demonstrates the excellent performance of our proposed approach in hepatic vessel segmentation.

2 Methods

In this section, we firstly give an overview of our proposed CSSD in Sect. 2.1. Then we introduce the CMA module about facilitating the transfer of information between clean labelled and unlabelled samples in Sect. 2.2. Finally, we describe how to effectively utilize noisy labels by scoring the noisy labels with the help of pesudo labels and employ confident learning to rectify labels for mildly noisy samples in Sect. 2.3.

2.1 Overview

We denote the clean samples set as $D_c = \{x_i^c, y_i^c\}_{i=1}^{N_c}$, the noisy samples set as $D_n = \{x_i^n, y_i^n\}_{i=1}^{N_n}$, and the unlabelled samples set as $D_u = \{x_i^u\}_{i=1}^{N_u}$, where x_i^c, x_i^n and $x_i^u \in R^{H \times W \times D}$ represent the volumes with height H, width W and depth D. Figure 2 gives an overview of CSSD, which follows the typical SSL framework with two models, named vanilla model and auxiliary model respectively. These two models have the same backbone V-Net [8], but differ in their bottleneck layer and weight initialization methods. At each iteration, we equally sample $\mathcal{M}_c$ clean samples, $\mathcal{M}_u$ unlabelled samples and $\mathcal{M}_n$ noisy samples. We first feed them into two model to get segmentation confidence maps p_v and p_a. Then, we apply the argmax operation on these confidence maps to derive the predicted one-hot label maps $\hat{y}_v$ and $\hat{y}_a$. For clean samples, our goal is to minimize the supervised loss $\mathcal{L}_s$ which is consisted of Dice loss and cross-entropy (CE) loss:

$$\mathcal{L}_s = \frac{1}{|\mathcal{M}_c|} \sum_{X \in \mathcal{M}_c} (\ell_{dice}(p_{vi}; y^c) + \ell_{dice}(p_{ai}; y^c) + \ell_{ce}(p_{vi}; y^c) + \ell_{ce}(p_{ai}; y^c)), \quad (1)$$

where ℓ_{dice} is the Dice loss and ℓ_{ce} is the CE loss and y^c is the ground truth. As for unlabelled sample, Our goal is to minimize the consistency loss $\mathcal{L}_c$:

$$\mathcal{L}_c = \frac{1}{|\mathcal{M}_u|} \sum_{X \in \mathcal{M}_u} (\ell_{ce}(p_{vi}; \hat{y}_{ai}^u) + \ell_{ce}(p_{ai}; \hat{y}_{vi}^u)). \quad (2)$$

Finally, we use the noisy label quality grading module and the confident learning module to denoise the noisy label, which will be introduced in Sect. 2.3. And we also use the CE loss to compute the CL loss $\mathcal{L}_{cl}$ between the denoised label $\hat{y}^{dn}$ and segmentation confidence map p_v^n. In the first 2,000 iterations, noisy samples are treated as unlabelled samples, thus the overall loss goes as follows:

$$\mathcal{L} = \begin{cases} \lambda_s(\mathcal{L}_{s1} + \mathcal{L}_{s2}) + \lambda_c(\mathcal{L}_{c1} + \mathcal{L}_{c2} + \mathcal{L}_{c3} + \mathcal{L}_{c4}), & \text{iteration} \leq 2000, \\ \lambda_s(\mathcal{L}_{s1} + \mathcal{L}_{s2}) + \lambda_c(\mathcal{L}_{c1} + \mathcal{L}_{c2}) + \lambda_{cl}\mathcal{L}_{cl}, & \text{otherwise,} \end{cases} \quad (3)$$

where λ_s is the trade-off weight of the supervision loss while λ_c and λ_{cl} are the trade-off weights of $\mathcal{L}_c$ and $\mathcal{L}_{cl}$. $\mathcal{L}_{c1}$ and $\mathcal{L}_{c2}$ are the consistency loss of unlabelled sample, while $\mathcal{L}_{c3}$ and $\mathcal{L}_{c4}$ are the consistency loss of noisy sample. λ_{cl} is set as 0 in the first 2,000 iterations to evalute the quality of noisy labels.

2.2 Cross-Sample Mutual Attention Module

To facilitate the transfer of prior knowledge from labelled data to unlabelled data and enhance the representation of common significant features between them, we introduce the CMA module, as shown in bottom right panel of Fig. 2. The proposed CMA module is incorporated into the bottleneck layer of the auxiliary model. We employ two self-attention modules arranged in different dimensional orders to achieve efficient mutual attention computation among all pixels. These are sample-wise intra-sample self-attention along spatial dimensions and inter-sample self-attention along batch dimensions. Firstly, intra-sample self-attention is applied along the spatial dimensions of each sample to model the information propagation paths between each pixel position within the sample, constructing intra-sample relationships. Subsequently, inter-sample self-attention is implemented along the batch dimension, where pixels from the same spatial position enter the self-attention module, thereby constructing inter-sample relationships and facilitating information propagation between different samples.

2.3 Noisy Label Self-Denoising

Definition of Noisy Labels. We call relatively precise and accurate annotations clean labels, and the corresponding images clean samples. In addition, there are annotations that are less reliable and may contain labelling errors. And these annotations are referred to as noise labels, and the corresponding images are referred to as noise samples.

Noisy Label Quality Grading. The quality of noise labels varies, and different degrees of noise can have various effects on the training process. Therefore, we devise a noise scoring system to grade the noisy label, which includes both mild and severe noise labels. For severe noise labels, we discard the original labels and treat the samples as unlabelled samples, and implement a perturbation consistency method to mitigate its influence. For mild noise labels, we refine the labels, thus robustly exploiting noise labels to mitigate the potential misguidance introduced by noise labels. Given that labels for clean pixels should have semantic similarity to the ground truth, while noise pixels may not, we measure the noisiness score by assessing the CE between predicted outputs and noise labels. A larger CE indicates a severe noise annotation, vice versa. We define the noisiness score for noisy label y^n as

$$I(y^n) = -[y^n \log(p_a^n) + (1-y^n)\log(1-p_a^n)], \quad (4)$$

where $I(\cdot)$ means the noisiness score, p_a^n denotes the prediction of the noisy sample in the auxiliary network. We employ a threshold value τ to split noisy

labels into severe and mild noisy labels. Considering that the noisiness score is subject to change during the training process, it is more appropriate to utilize a dynamic threshold values that adapts to the statistics of I. Therefore, we define τ as the h-th percentile of I over a specific number of iterations (e.g., 2000).
Label Noise Estimation Module. We employ confident learning to evaluate the accuracy of labels, expressed as confidence degree η. The quantification of confidence degree η is determined by leveraging results obtained from both noisy labels and an auxiliary model. For binary labels, higher logit values in the output indicate greater reliability. η mumerically equals to the average of the logits in the vessel region. We caculate the η as

$$\eta = \frac{\sum_h \sum_w \sum_d p_a^n(h,w,d) y^n(h,w,d)}{\sum_h \sum_w \sum_d p_a^n(h,w,d)}, \quad (5)$$

where y^n denotes the noisy labels, and p_a^n means the outputs of noisy samples in the auxiliary model, $\eta \in [0,1]$.
Self-denoising. Self-denoising aims to obtain more precise labels for guiding the training process. Correction of noisy labels is exclusively implemented when dealing with labels exhibiting mild noise. Through the utilization of label noise detection, a confidence degree η is derived, which is subsequently employed to refine labels with mild noise in the following way

$$\hat{y}^{dn} = \eta y^n + (1-\eta) p_a^n, \quad (6)$$

where $\hat{y}^{dn}$ means the denoised label.

3 Experiments and Results

3.1 Dataset and Preprocessing

Three publicly datasets are utilized in our study: (1) **3DIRCADb** is denoted as clean sample and serves as the training subset and test set with a ratio of 1:1. It consists of 20 volumes of CT images, each annotated with vessel and liver masks. (2) **MSD8** is referred to as noisy sample and is exclusively employed as a training subset. It provides 303 volumes of CT scans with vessel masks only. However, compared to 3DIRCADb, the vessel in MSD8 are coarser. (3) **MSD3** is designated as unlabelled sample, solely utilized for training. MSD3 provides 131 volumes of CT scans with liver masks only.

In the pre-processing stage, we follow a standardised procedure: (1) cropped the region of interest (ROI). As for MDS8 dataset, we use the trained Uniseg model [17] to get live masks. Subsequently, all cropped images are resized to $D \times 320 \times 320$. (2) All images are resampled to $1.5 \times 0.8 \times 0.8$ mm^3. (3) The HU values are truncated to the range $[-100, 250]$ and then normalized using the Min-Max method. For data augmentations, we utilize random rotations and random crops, performed random 90-degree rotations on the data along three axes, and performed random flips. We use a patch size of [64, 128, 128].

Table 1. Quantitative comparison on the hepatic vessel dataset. The best results are highlighted in bold, and the second-best results are underlined. FS denotes fully-supervised.

Paradigm	Methods	DSC↑	RAVD↓	ASD↓
FS	V-Net [8]	62.98 ± 4.47	0.39 ± 0.11	1.75 ± 0.48
SSL	MT [10]	66.25 ± 2.18	0.34 ± 0.15	1.36 ± 0.26
	UMAT [18]	66.63 ± 3.18	0.30 ± 0.12	1.71 ± 0.34
	CPS [3]	65.72 ± 3.66	0.38 ± 0.16	1.55 ± 0.35
	CAML [5]	68.13 ± 3.25	<u>0.28 ± 0.11</u>	1.23 ± 0.38
	GenericSSL [13]	63.44 ± 2.29	0.75 ± 0.26	1.88 ± 0.45
	AC-MT [15]	<u>68.30 ± 2.45</u>	0.29 ± 0.09	<u>1.21 ± 0.37</u>
NLL	GCE [22]	44.12 ± 0.29	0.68 ± 0.10	1.27 ± 0.18
	COT [6]	48.47 ± 1.52	0.56 ± 0.13	1.36 ± 0.26
	TriNet [20]	48.65 ± 1.57	0.54 ± 0.25	1.34 ± 0.25
	CLSLS [19]	40.14 ± 1.26	0.72 ± 0.12	<u>1.13 ± 0.35</u>
	DAST [16]	64.86 ± 3.12	0.35 ± 0.12	1.24 ± 0.48
HSL	Ours	**68.42 ± 2.75**	**0.28 ± 0.10**	**1.08 ± 0.24**

3.2 Implementation and Evaluation Metrics

Implementation Details. The framework is based on the Pytorch implementation using an NVIDIA TITAN RTX GPU. During training, a batch of 2 clean, 2 unlabelled and 2 noisy samples was fed into the network. However, in the SSL methods, the batch size is set to 4, including 2 clean and 2 unlabelled samples. And for NLL methods, the batch size is set to 4, including 2 clean and 2 noisy samples. The initial learning rate is set to 0.01 with linear decay. Using the SGD optimizer, there is a weight decay of 0.0001 and a momentum of 0.9. In terms of SSL parameters, the initial consistency weight is set at 0.1 and the EMA decay rate is set at 0.99.

Evaluation Metrics. We utilize three metrics to comprehensively evaluate the segmentation results, including Dice Similarity Coefficient (DSC), Relative Absolute Volume Difference (RAVD), and Average Surface Distance (ASD).

3.3 Comparison with State-of-the-Art Methods

We conducte a comprehensive comparison of CSSD with two categories of methods: (1) SSL method and (2) NNL method. All methods utilize the same backbone V-Net for a fair comparison.

Comparison with SSL Methods. We compare CSSD with 6 SSL methods, as shown in Table 1. CSSD exhibits superior performance across all three metrics when compared to other comparison with SSL methods methods. As illustrated

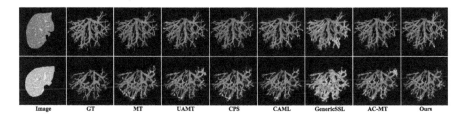

Fig. 3. Visualization of the segmentation results of different SSL methods. Red color voxels denote the ground truth, while green color voxels denote the false positive and the bule color voxels denote the false negative. (Color figure online)

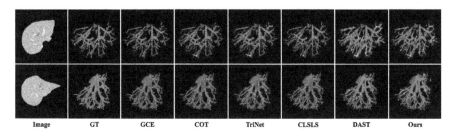

Fig. 4. Visualization of the segmentation results of different NLL methods. (Color figure online)

in Fig. 3, the visualisation results also show the superiority of CSSD in local detail and topological coherence.

Comparison with NLL Methods. Additionally, we conduct a comparative analysis with 5 NLL methods, and the results are shown in Table 1. And the representative visualisation results are shown in Fig. 4. We find that the majority of NLL methods exhibit inferior performance compared to fully supervised models (Only 3DIRCADb in Table 1) trained using only the clean samples. Suggesting that the additional noisy labelled samples adversely affects model performance, particularly under noisy conditions.

Table 2. Ablation study on the hepatic vessel dataset. CMA denotes the cross-sample mutual attention module, and CL denotes the confident learning.

CMA	CL	DSC ↑	RAVD ↓	ASD ↓
		65.72 ± 3.15	0.38 ± 0.15	1.55 ± 0.21
✓		67.96 ± 3.41	0.31 ± 0.14	1.46 ± 0.37
	✓	67.59 ± 2.15	0.31 ± 0.13	1.63 ± 0.34
✓	✓	**68.42 ± 2.75**	**0.28 ± 0.10**	**1.08 ± 0.24**

3.4 Ablation Study

In this section, we analyze the effectiveness of the proposed CMA module and CL module. The results of our ablation study, as presented in Table 2, reveal the substantial impact of both the CMA and CL modules in enhancing the performance of the baseline model. Notably, the combination of these two modules yields a remarkable absolute improvement of 2.7% in the Dice coefficient.

4 Conclusion

In this paper, we propose a novel HSL framework named CSSD, which addresses the challenge of hepatic vessel segmentation task by leveraging a limited number of clean samples, a substantial number of noisy samples, and unlabelled samples. Comprehensive experiments are conducted on publicly available datasets, revealing that the integration of CMA module and CL module significantly enhances network performance, surpassing the performance of alternative methods.

Acknowledgments. This study was supported by the National Natural Science Foundation of China (62071210); the National Key Research and Development Program of China (2023YFC2415400); the Shenzhen Science and Technology Program (RCYX2021 0609103056042); the Shenzhen Science and Technology Innovation Committee (KCXFZ2020122117340001); the Guangdong Basic and Applied Basic Research (2021A1515220131); the high level of special funds (G030230001).

References

1. 3DIRCADb Dataset. https://www.ircad.fr/research/3d-ircadb-01/
2. Bai, W., Oktay, O., et al.: Semi-supervised learning for network-based cardiac MR image segmentation. In: Descoteaux, M., et al. (eds.) MICCAI 2017. LNCS, vol. 10434, pp 253-260. Springer, Cham (2017). https://doi.org/10.1007/978-3-319-66185-8_29
3. Chen, X., Yuan, Y., Zeng, G., Wang, J.: Semi-supervised semantic segmentation with cross pseudo supervision. In: Proceedings of the IEEE Conference on Computer Vision and Pattern Recognition, pp. 2613–2622 (2021)
4. Cui, W., et al.: Semi-supervised brain lesion segmentation with an adapted mean teacher model. In: Chung, A., et al. (eds.) MICCAI 2021. LNCS, vol. 11492, pp. 554–565. Springer, Cham (2021). https://doi.org/10.1007/978-3-030-20351-1_43
5. Gao, S., Zhang, Z., Ma, J., Li, Z., Zhang, S.: Correlation-aware mutual learning for semi-supervised medical image segmentation. In: Greenspan, H., et al. (eds.) MICCAI 2023. LNCS, vol. 14220, pp. 98–108. Springer, Cham (2023). https://doi.org/10.1007/978-3-031-43907-0_10
6. Han, B., et al.: Co-teaching: robust training of deep neural networks with extremely noisy labels. In: Advances in Neural Information Processing Systems, pp. 8527–8537 (2018)
7. Liu, L., Tian, J., Zhong, C., Shi, Z., Xu, F.: Robust hepatic vessels segmentation model based on noisy dataset. In: Medical Imaging 2020: Computer-Aided Diagnosis, vol. 11314, pp. 122–128. SPIE (2020)

8. Milletari, F., Navab, N., Ahmadi, S.A.: V-net: fully convolutional neural networks for volumetric medical image segmentation. In: 2016 Fourth International Conference on 3D Vision, pp. 565–571 (2016)
9. Simpson, A.L., Antonelli, M., et al.: A large annotated medical image dataset for the development and evaluation of segmentation algorithms. arXiv preprint arXiv:1902.09063 (2019)
10. Tarvainen, A., Valpola, H.: Mean teachers are better role models: Weight-averaged consistency targets improve semi-supervised deep learning results. In: Advances in Neural Information Processing Systems, pp. 1195–1204 (2017)
11. Tu, Y., et al.: Learning from noisy labels with decoupled meta label purifier. In: Proceedings of the IEEE Conference on Computer Vision and Pattern Recognition, pp. 19934–19943 (2023)
12. Vu, T.-H., Jain, H., Bucher, M., Cord, M., Pérez, P.: Advent: adversarial entropy minimization for domain adaptation in semantic segmentation. In: Proceedings of the IEEE Conference on Computer Vision and Pattern Recognition, pp. 2517–2526 (2019)
13. Wang, H., Li, X.: Towards generic semi-supervised framework for volumetric medical image segmentation. In: Advances in Neural Information Processing Systems, pp. 1833–1848 (2023)
14. Xu, Z., Lu, D., et al.: Noisy labels are treasure: mean-teacher-assisted confident learning for hepatic vessel segmentation. In: de Bruijne, M., et al. (eds.) MICCAI 2021. LNCS, vol. 12901, pp. 3–13. Springer, Cham (2021). https://doi.org/10.1007/978-3-030-87193-2_1
15. Xu, Z., Wang, Y., Lu, D., Luo, X., Yan, J., Zheng, Y., Tong, R.K.: Ambiguity-selective consistency regularization for mean-teacher semi-supervised medical image segmentation. Med. Image Anal. **88**, 102880 (2023)
16. Yang, S., et al.: Learning COVID-19 pneumonia lesion segmentation from imperfect annotations via divergence-aware selective training. IEEE J. Biomed. Health Inf. **26**(8), 3673–3684 (2022). https://doi.org/10.1109/JBHI.2022.3172978
17. Ye, Y., Xie, Y., Zhang, J., Chen, Z., Xia, Y.: UniSeg: a prompt-driven universal segmentation model as well as a strong representation learner. In: Greenspan, H., et al. (eds.) MICCAI 2023. LNCS, vol. 14222, pp. 508–518. Springer, Cham (2023). https://doi.org/10.1007/978-3-031-43898-1_49
18. Yu, L., Wang, S., Li, X., Fu, C.-W., Heng, P.-A.: Uncertainty-aware self-ensembling model for semi-supervised 3D left atrium segmentation. In: Proceedings of the IEEE Conference on Computer Vision and Pattern Recognition, pp. 605–613 (2019)
19. Zhang, M., et al.: Characterizing label errors: confident learning for noisy-labeled image segmentation. In: Martel, A.L., et al. (eds.) MICCAI 2020. LNCS, vol. 12261, pp. 721–730. Springer, Cham (2020). https://doi.org/10.1007/978-3-030-59710-8_70
20. Zhang, T., Yu, L., Hu, N., Lv, S., Gu, S.: Robust medical image segmentation from non-expert annotations with tri-network. In: Martel, A.L., et al. (eds.) MICCAI 2020. LNCS, vol. 12264, pp. 249–259. Springer, Cham (2020). https://doi.org/10.1007/978-3-030-59719-1_25
21. Zhang, W., et al.: Boostmis: boosting medical image semi-supervised learning with adaptive pseudo labeling and informative active annotation. In: Proceedings of the IEEE Conference on Computer Vision and Pattern Recognition, pp. 20666–20676 (2022)

22. Zhang, Z., Sabuncu, M.: Generalized cross entropy loss for training deep neural networks with noisy labels. In: Advances in Neural Information Processing Systems, pp. 8778–8788 (2018)
23. Zheng, H., Motch Perrine, S. M., et al.: Cartilage segmentation in high-resolution 3D micro-CT images via uncertainty-guided self-training with very sparse annotation. In: Martel, A.L., et al. (eds.) MICCAI 2020. LNCS, vol. 12261, pp. 802–812. Springer, Cham (2020). https://doi.org/10.1007/978-3-030-59710-8_78

Calibrated Diverse Ensemble Entropy Minimization for Robust Test-Time Adaptation in Prostate Cancer Detection

Mahdi Gilany[1](✉), Mohamed Harmanani[1], Paul Wilson[1], Minh Nguyen Nhat To[2], Amoon Jamzad[1], Fahimeh Fooladgar[2], Brian Wodlinger[3], Purang Abolmaesumi[2], and Parvin Mousavi[1]

[1] School of Computing, Queen's University, Kingston, Canada
mahdi.gilany@queensu.ca
[2] Department of Electrical and Computer Engineering, University of British Columbia, Vancouver, Canada
[3] Exact Imaging, Markham, Canada

Abstract. High resolution micro-ultrasound has demonstrated promise in real-time prostate cancer detection, with deep learning becoming a prominent tool for learning complex tissue properties reflected on ultrasound. However, a significant roadblock to real-world deployment remains, which prior works often overlook: model performance suffers when applied to data from different clinical centers due to variations in data distribution. This distribution shift significantly impacts the model's robustness, posing major challenge to clinical deployment. Domain adaptation and specifically its test-time adaption (TTA) variant offer a promising solution to address this challenge. In a setting designed to reflect real-world conditions, we compare existing methods to state-of-the-art TTA approaches adopted for cancer detection, demonstrating the lack of robustness to distribution shifts in the former. We then propose Diverse Ensemble Entropy Minimization (DEnEM), questioning the effectiveness of current TTA methods on ultrasound data. We show that these methods, although outperforming baselines, are suboptimal due to relying on neural networks output probabilities, which could be uncalibrated, or relying on data augmentation, which is not straightforward to define on ultrasound data. Our results show a significant improvement of 5% to 7% in AUROC over the existing methods and 3% to 5% over TTA methods, demonstrating the advantage of DEnEM in addressing distribution shift.

Keywords: Ultrasound Imaging · Prostate Cancer · Computer-aided Diagnosis · Distribution Shift Robustness · Test-time Adaptation

---

Supplementary Information The online version contains supplementary material available at https://doi.org/10.1007/978-3-031-73284-3_36.

© The Author(s), under exclusive license to Springer Nature Switzerland AG 2025
X. Xu et al. (Eds.): MLMI 2024, LNCS 15241, pp. 361–371, 2025.
https://doi.org/10.1007/978-3-031-73284-3_36

1 Introduction

Prostate cancer (PCa) diagnosis remains a pivotal challenge, where early and accurate detection can significantly influence treatment outcomes. The standard diagnostic approach is transrectal ultrasound-guided biopsy (TRUS), which *systematically* samples biopsy cores from uniform locations across the prostate. However, this method cannot specifically target suspicious areas due to similar acoustic tissue properties and low resolution, often resulting in missed diagnoses or unnecessary interventions. Alternatively, advanced ultrasound modalities aim to improve tissue characterization and offer targeted diagnostic strategies. In particular, high-frequency micro-ultrasound (micro-US) has recently emerged for visualization of prostate tissue at significantly higher resolutions than conventional imaging [4], facilitating the identification of potential cancer lesions. Using a qualitative approach, micro-US has demonstrated comparable PCa detection to multi-parametric MRI targeted biopsy [1,3], with the distinct advantage of operating in real-time. However, quantitative approaches interpreting this modality is in early-stage research with few studies [5,22] proposed for user-independent and objective PCa detection. Recent studies have utilized deep learning to analyze micro-US images for identifying PCa [6,27]. While deep learning can effectively extract complex tissue features from noisy ultrasound images, its adaptability to any data distribution can undermine model *robustness* during deployment when data distributions shift. Such shifts significantly impact model performance, as the model may learn biases in the training distribution and form spurious correlations between input images and tissue types [14,23]. This study argues that, despite recent successes, the performance and effectiveness of existing micro-US PCa detection studies are overestimated. These studies typically assume similar data distributions between training and deployment scenarios, which is unrealistic in clinical settings where shifts in data distribution are likely to occur due to differences in patient populations, operator techniques, and imaging devices.

To improve PCa detection robustness and mitigate distribution shifts, we explore the promising field of domain adaptation, where techniques are developed to adjust trained models to perform well on a target domain with a shift in distribution [15]. Specifically, we investigate a more challenging variant, *test-time adaptation* (TTA), which employs unsupervised objectives to fine-tune trained models on a single test sample during inference [18]. This approach eliminates the need to access the entire test dataset and accounts for variability in data distributions of test patients or biopsy cores. In practice, several challenges hinder the direct application of state-of-the-art (SOTA) TTA methods in PCa detection. Recent studies indicate that prior methods may fail in real-world scenarios with multiple types of distribution shifts [20], and their effectiveness varies depending on the shift type [31]. Moreover, the most common and promising TTA methods utilize either self-supervised learning (SSL) or entropy loss in their core [25,26]. However, choice of SSL data augmentation for ultrasound is not straightforward, and current entropy-based methods rely on neural networks generated probabilities (thus, entropy), known to be biased and uncalibrated [9]. This issue is exacerbated by distribution shifts and ultrasound artifacts and noisy labels [21].

An ideal TTA method should be tailored for micro-US shifts, produce calibrated entropy, and not rely on augmentations. To this end, we propose DEnEM, a novel TTA method for addressing clinical center distribution shifts in micro-US PCa detection. DEnEM utilizes an ensemble of neural networks diversified through mutual information minimization to produce calibrated marginal entropy without the need for data augmentations. Our key contributions are:

1. We investigate the existing SOTA PCa detection methods in a real-world setting, revealing the impact of distribution shift on their performance.
2. We propose leveraging SOTA TTA methods, for the first time, to address clinical center distribution shifts in PCa detection. We demonstrate their effectiveness in significantly improving over existing methods while also examining their limitations on micro-US data.
3. Inspired by TTA methods, we propose DEnEM, a novel approach tailored for micro-US that addresses entropy calibration and the reliance on data augmentations. We demonstrate that DEnEM significantly outperforms both baselines and SOTA TTA methods across several evaluation metrics.

2 Related Works

Micro-US PCa Detection: Recent studies have employed quantitative approaches for micro-US PCa detection [5,22,24,28]. Rohrbach et al. [22] introduced the first quantitative method using SVMs on manually extracted ultrasound spectral features, which is less effective than deep learning and requires a reference phantom, limiting its usability. Gilany et al. [5] demonstrated the success of CNNs in automatically extracting features from RF images to detect PCa. More recent methods [6,27] addressed labeled data scarcity and weak labeling, improving detection performance. However, these studies often neglect clinical integration and robust deployment across various clinical centers, leading to overestimated and unrealistic performance due to shifting data distributions.

Test-Time Adaptation: TTA is a challenging form of unsupervised domain adaptation, requiring the adaptation of a model trained on a source domain to a shifted target domain during inference, without accessing the *source domain* to preserve data privacy, and without the *entire target domain*, enabling immediate application post-deployment [18]. The core of TTA lies in defining a proxy objective to adapt the model to test samples unsupervised. These objectives include entropy minimization [26,30], self-supervised learning (SSL) [2,25], and feature alignment [31]. "TENT" [26] fine-tunes the model, at inference, by minimizing the entropy on test prediction probabilities p: $H(p) = -\sum_c p^c \log p^c$. "MEMO" [30] extends TENT by minimizing the entropy of *marginal* probability distribution calculated across a set of augmentations. While entropy minimization is a theoretically well established TTA objective [7], neural networks often produce biased and uncalibrated entropy [9]. Moreover, "TTT" [25] and "MT3" [2] use SSL proxy objectives. These methods modify the neural architecture by attaching a self-supervision head to the feature extractor network.

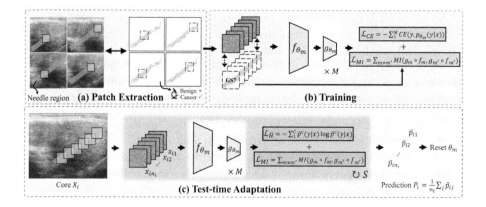

Fig. 1. Overview of DEnEM method. (a) RF patches extraction from needle region. (b) Deep ensemble training with cross entropy and mutual information losses. (c) Model adaptation at inference to each core with marginal entropy loss before the prediction.

During inference, this head and SSL objective adapt the feature extractor to the test distribution. However, these approaches have two drawbacks: they alter the neural architecture, potentially degrading performance in some tasks, and they only fine-tune the feature extractor, leaving the classification head unadapted. The latter is particularly impactful on distribution shift [13].

3 Materials and Method

3.1 Data

Our study employs a dataset collected from five clinical centers, with consented data from 693 patients who underwent systematic TRUS-guided prostate biopsy as part of a clinical trial (NCT02079025 clinicaltrials.gov). The procedure leverages high-frequency micro-US technology, typically capturing 10-12 biopsy cores per patient. Prior to firing the biopsy gun, a single RF image of the prostate with 28 mm depth and 46 mm width is captured. These images, paired with corresponding histopathology reports of the cores constitute our dataset. The reports include a binary label indicating the presence of cancer, an estimate of the cancer length or "involvement of cancer", and the aggressiveness of cancer as determined by the Gleason Score [19]. Overall, our dataset includes 6607 biopsy cores with 5727 of those being benign. More details of the data distribution among centres are provided in the supplementary material.

The needle trace region in the RF image, where histopathology labels are assigned, is first identified [5,6,24]. From this region, we perform patch extraction by capturing overlapping regions of interest (ROIs) that correspond to a tissue area of 5 mm by 5 mm, using 1 mm by 1 mm strides (see Fig. 1 (a)). An RF patch is considered to be within the needle region if there is at least a 60% overlap, a threshold set to compensate for potential inaccuracies in identifying the needle region [27]. RF patches are resized from their original dimensions of 1780 × 55 to 256 × 256 to align with our deep network architectures.

3.2 Method

Figure 1 shows an overview of our proposed approach. This approach trains a deep ensemble [16] of neural networks to minimize both cross-entropy loss and mutual information loss. During inference, marginal entropy is utilized to adapt the deep ensemble to the distribution of patches from a biopsy core in test set.

Deep Ensemble: Our model borrows ResNet10 [10] image encoder from prior PCa detection literature [5,27]. Deep ensemble is a collection of M encoders and classifiers with different random weight initialization. Unlike single network models, deep ensemble not only provides a highly calibrated entropy, but also this calibration remains less affected by distribution shift [21].

Mutual Information Loss: Randomly initialized neural networks in deep ensemble may still learn similar encoders and classifiers, providing biased and uncalibrated entropy. To overcome this, we propose ensuring statistical independence of predictions across the networks. Inspired by "DivDis" [17], we minimize the mutual information between the predictions of each network pair. The mutual information loss ($\mathcal{L}_{MI}$) and the total training loss ($\mathcal{L}$) are defined as follows:

$$\mathcal{L}_{MI}(g_m \circ f_m, g_{m'} \circ f_{m'}) = \mathcal{D}_{KL}(p_{\theta_m, \theta_{m'}}(y_m, y_{m'}) || p_{\theta_m}(y_m) \otimes p_{\theta_{m'}}(y_{m'})) \quad (1)$$

$$\mathcal{L} = \mathcal{L}_{CE} + \lambda \sum_{m \neq m'} \mathcal{L}_{MI}(g_m \circ f_m, g_{m'} \circ f_{m'}) \quad (2)$$

where $p_{\theta_m}(y|x)$ is the output probability distribution of a data sample from the m'th model with parameters θ_m. $\mathcal{D}_{KL}$ refers to $\mathcal{KL}$ divergence between joint probability distribution of models m and m', and is calculated empirically [17]. Note that unlike "DivDis", our mutual info. loss is calculated on the same training/adaptation data. $\mathcal{L}_{CE}$ represents the sum of cross-entropy loss for all networks, and $\lambda = 10$ controls the importance of mutual info. loss.

Marginal Entropy Loss: Entropy minimization [26] and marginal entropy minimization [30] are promising proxy objectives for test-time robustness and adaptation to distribution shift with theoretically well established literature [7]. However, neural networks may not provide a well calibrated entropy, significantly undermining the effectiveness of these methods. Additionally, data augmentations may not be feasible or properly studied for a specific data type like ultrasound RF images, proposing new challenges to the problem. Therefore, we propose to find marginal entropy across deep ensemble networks instead of across various augmentations [30]. We first find marginal distribution by averaging M output probabilities: $\bar{p}(y|x) = \mathbb{E}_M[p_{\theta_m}(y|x)] = \frac{1}{M}\sum_{m=1}^{M} p_{\theta_m}(y|x)$. Then, the marginal entropy loss is calculated as: $\mathcal{L}_{\overline{H}} = -\sum_{c=1}^{C} \bar{p}^c(y|x) \times \log \bar{p}^c(y|x)$.

Test-Time Adaptation: To adapt to test distribution, we fine-tune the ensemble networks by minimizing both unsupervised objectives of marginal entropy

($\mathcal{L}_{\overline{H}}$) and mutual info. ($\mathcal{L}_{MI}$) losses on RF patches of each test biopsy core $X_i = \{x_{i1}, x_{i2}, ..., x_{in_i}\}$, separately. After S iteration of adaptation, with learning rate lr_{adapt}, the parameters of ensemble networks, i.e. θ_m are recovered to the values after initial training. This is equivalent to episodic adaptation on each biopsy core, and we leave online adaptation for future work due to the challenging hyperparameter tuning [31]. Lastly, we follow SOTA TTA methods [29] and substitute all batch norm layers [11] with batch-agnostic group norm layers [29]. More discussions on this effective approach are presented in the next sections.

4 Experiments

We organize our experiments in following parts: (i) *SOTA baseline comparison*. We compare our model's performance with prior SOTA methods for PCa detection in micro-US. This includes patch classification with single or with an ensemble of ResNet10 [5], patch classification with SSL pre-training on unlabeled data [27], and core classification with TRUSFormer [6] which collects patch extracted features with a transformer aggregator. (ii) *SOTA TTA comparison*. Additionally, we compare our model against SOTA TTA methods, including TENT [26] with ResNet10 backbone, MEMO [30] with marginal entropy calculated across various augmentations proposed in [27], TTT [25] with BYOL [8] as the SSL objective, and its extended meta learning version MT3 [2]. We also include "SAR" which filters unreliable samples based on entropy and minimizes a smoothed entropy loss [20]. (iii) *Ablation studies*. We perform two ablation studies and a qualitative analysis. First, we evaluate the contribution of each of our model's component to the performance. Second, we follow TTA methods [20] and study the effect of batch-agnostic norm layers. In particular, we replace ResNet10 batch norm [11] layers with group norm [29] and compare.

Evaluation Strategy: We evaluate all methods using a leave-one-center-out (LOCO) strategy, holding all data from one clinical center for testing in each experiment. This approach reflects real-world deployment. The remaining four centers are divided into training and validation with patient-wise stratification. To ensure balanced representation, the dataset from each center is first divided into training and validation sets and then combined. The performance is measured at core-level by averaging output probabilities of RF patches inside a core. Following previous works [5,22], cores with cancer involvement $\leq$ 40% are removed from test for more stable evaluation and AUROC and balanced-accuracy (sensitivity and specificity average) are reported. We additionally include AUROC-All where all test set cores, regardless of cancer involvement, are included.

Implementation Details: We re-implement all baseline models to evaluate them in a realistic LOCO setting with data distribution shift. We use PyTorch 2.1, with training and validation of our method taking approximately 4.6 h on a single NVIDIA A40 GPU. Inference on a biopsy core, including adaptation, takes around 250 ms. We manually tune hyperparameters based on validation. Adam optimizer [12] with learning rate of $1e-4$ and a scheduler with cosine annealing and a linear warm-up were used. For adaptation during inference, we employed SGD with the best learning rate selected from $lr_{adapt} = \{1e-1, 1e-2, 1e-3\}$ and number of iterations from $S = \{1, 5\}$. Additional details are provided in the supplementary document.

5 Results and Discussion

Table 1. PCa detection performance for baselines, TTAs, and our proposed method.

Method	AUROC	AUROC-All	Balanced-Acc.
Baselines			
ResNet10	75.2 ± 7.0	68.3 ± 6.0	68.0 ± 4.6
Ensem. [16]	75.8 ± 6.4	68.1 ± 6.3	68.0 ± 6.2
SSL + ResNet [27]	75.1 ± 4.0	67.7 ± 5.3	68.0 ± 4.3
TRUSFormer [6]	75.3 ± 4.9	70.4 ± 2.6	68.8 ± 4.0
Test-time adaptations			
TENT [26]	77.3 ± 4.2	71.0 ± 3.6	69.7 ± 5.4
MEMO [30]	77.4 ± 4.2	70.9 ± 3.6	69.4 ± 6.4
TTT [25]	77.8 ± 5.1	71.1 ± 3.4	66.5 ± 8.8
MT3 [2]	77.7 ± 5.7	70.3 ± 4.4	63.3 ± 4.8
SAR [20]	77.6 ± 4.4	71.2 ± 3.7	70.7 ± 5.0
DEnEM (ours)	$\mathbf{80.9 \pm 4.5}$	$\mathbf{75.0 \pm 2.9}$	$\mathbf{75.6 \pm 3.4}$

Table 2. Ablation study on different components of the proposed method.

Method	Group-Norm	Ensem.	Mutual-Info.	Test-Adapt	AUROC-All
ResNet10	✓	–	–	–	71.7 ± 3.7
Ensem.	✓	✓	–	–	72.9 ± 4.1
Ensem.+$\mathcal{L}_{MI}$	✓	✓	✓	–	73.8 ± 3.5
Ensem.+$\mathcal{L}_{\overline{H}}$	✓	✓	-	✓	73.1 ± 4.0
Ensem.+$\mathcal{L}_{MI}$+$\mathcal{L}_{\overline{H}}$ (ours)	✓	✓	✓	✓	75.0 ± 2.9

SOTA Baseline Comparison: Table 1 summarizes the results of SOTA comparison. Overall, our proposed method (DEnEM) outperforms all baselines by a significant margin in core AUROC, AUROC-All, and Balanced-ACC by around 5% to 7%. In particular, our patch classification approach outperforms SOTA core classification TRUSFormer by ∼5% indicating the effectiveness of DEnEM. Future works may leverage the orthogonal benefits of these two methods. Moreover, comparing the baselines shows that the leading method in PCa detection may fall behind relatively simpler approaches in LOCO setting, highlighting the importance of this realistic evaluation. Lastly, note that the high standard deviations are due to averaging the results of *different* test datasets with high variability in performance. In ablation study, we will show the standard deviation for each individual test datasets.

SOTA TTA Comparison: Similar to baseline comparisons, our proposed method substantially improves over SOTA TTA methods in all metrics by 3% to 5%. Comparing TTA methods show that TENT and MEMO produce similar results, indicating that marginal entropy across various augmentations is not effective. Considering this in conjunction with SSL-pretraining results, we hypothesize that the proposed augmentations for RF images [27] might be ineffective. In conclusion, the performance of DEnEM compared with marginal differences between TTA results justifies the proposed marginal entropy in DEnEM.

Ablation Studies: Table 2 shows the effect of different components on the performance of our method. Overall, each component increases the performance individually, with $\mathcal{L}_{MI}$ improving by 0.9% and $\mathcal{L}_{\overline{H}}$ improving by 0.2%. However, when combined together, the gain is compounded with improvement of 2.1%. Unlike the baselines, deep ensemble improves AUROC-All over ResNet10 when group norm is adopted. This expected outcome demonstrates the harmful effect of batch norm on baselines with LOCO setting. Recent TTA methods [20] have also highlighted the potential drawbacks of batch norm under distribution shift as the mean and variance estimation in batch norm layers will be biased. In Fig. 2 (b), we compare the AUROC-All of ResNet10 using batch norm versus group norm (used in our experiment) with separate bar plots for each center left out for test. This figure further confirms the substantial improvement of replacing batch norm with group norm. Additionally, it reveals a high variability in performance, ranging from 68% to 77% across different centers. We hypothesize that the quantity and quality of data in both training and test sets contribute to this variability, though further research is needed in future works.

Qualitative analysis: We qualitatively analyze our model's prediction by using heatmaps, comparing the predictions when no TTA is adopted (e.g. ResNet10 baseline) with DEnEM, as detailed in Fig. 2 (a). These heatmaps are created by running the model across the entire RF image using sliding patches. They showcase two examples of benign (top row) and cancerous (bottom row) cores with red indicating cancer predictions and blue for benign predictions. These maps, overlaid on corresponding B-mode images, not only demonstrate the failure of baseline compared to DEnEM, but also qualitatively show the capability of our model in localizing cancer and offering reliable guidance for targeted biopsy.

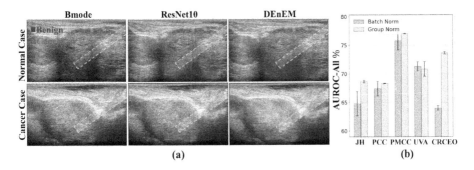

Fig. 2. (a) Heatmap comparison of ResNet10 and DEnEM with cancer (red) vs. benign (blue) areas. Top row: benign core; bottom row: cancerous core (Gleason score 3+4). (b) Baseline ResNet10 Batch norm vs. group norm comparison for different test center. (Color figure online)

6 Conclusion

This study examined the robustness of deep learning models to clinical center distribution shifts in micro-US PCa detection. We showed that existing methods are vulnerable to these shifts, with leading approaches sometimes outperformed by simpler ones. To address this, we adopted TTA, which improved detection but faced limitations due to ultrasound augmentations and biased entropy of neural networks. We proposed DEnEM, which calculates calibrated marginal entropy across a diverse ensemble of networks without requiring data augmentations. DEnEM significantly enhanced performance, outperforming previous TTA methods and demonstrating its potential to improve PCa detection robustness.

References

1. Abouassaly, R., Klein, E.A., El-Shefai, A., Stephenson, A.: Impact of using 29 mhz high-resolution micro-ultrasound in real-time targeting of transrectal prostate biopsies: initial experience. World J. Urol. **38**(5), 1201–1206 (2020)
2. Bartler, A., Bühler, A., Wiewel, F., Döbler, M., Yang, B.: Mt3: meta test-time training for self-supervised test-time adaption. In: International Conference on Artificial Intelligence and Statistics, pp. 3080–3090. PMLR (2022)
3. Cotter, F., Perera, S., Sathianathen, N., Lawrentschuk, N., Murphy, D., Bolton, D.: Comparing the diagnostic performance of micro-ultrasound-guided biopsy versus multiparametric magnetic resonance imaging-targeted biopsy in the detection of clinically significant prostate cancer: A systematic review and meta-analysis. Société Internationale d'Urologie Journal **4**(6), 465–479 (2023)
4. Ghai, S., et al.: Assessing cancer risk on novel 29 mhz micro-ultrasound images of the prostate: creation of the micro-ultrasound protocol for prostate risk identification. J. Urol. **196**(2), 562–569 (2016)
5. Gilany, M., Wilson, P., Jamzad, A., Fooladgar, F., To, M.N.N., Wodlinger, B., Abolmaesumi, P., Mousavi, P.: Towards confident detection of prostate cancer using high resolution micro-ultrasound. In: International Conference on Medical

Image Computing and Computer Assisted Intervention, pp. 411–420. Springer (2022). https://doi.org/10.1007/978-3-031-16440-8_40

6. Gilany, Met al.: Trusformer: improving prostate cancer detection from micro-ultrasound using attention and self-supervision. Inter. J. Comput. Assisted Radiol. Surgery, 1–8 (2023)
7. Goyal, S., Sun, M., Raghunathan, A., Kolter, J.Z.: Test time adaptation via conjugate pseudo-labels. Adv. Neural. Inf. Process. Syst. **35**, 6204–6218 (2022)
8. Grill, J.B., et al.: Bootstrap your own latent-a new approach to self-supervised learning. Adv. Neural. Inf. Process. Syst. **33**, 21271–21284 (2020)
9. Guo, C., Pleiss, G., Sun, Y., Weinberger, K.Q.: On calibration of modern neural networks. In: International Conference on Machine Learning, pp. 1321–1330 (2017)
10. He, K., Zhang, X., Ren, S., Sun, J.: Deep residual learning for image recognition. In: Proceedings of the IEEE Conference on Computer Vision and Pattern Recognition, pp. 770–778 (2016)
11. Ioffe, S., Szegedy, C.: Batch normalization: accelerating deep network training by reducing internal covariate shift. In: International Conference on Machine Learning, pp. 448–456. pmlr (2015)
12. Kingma, D.P., Ba, J.: Adam: A method for stochastic optimization. arXiv preprint arXiv:1412.6980 (2014)
13. Kirichenko, P., Izmailov, P., Wilson, A.G.: Last layer re-training is sufficient for robustness to spurious correlations. arXiv preprint arXiv:2204.02937 (2022)
14. Koh, P.W., et al.: Wilds: A benchmark of in-the-wild distribution shifts. In: International Conference on Machine Learning, pp. 5637–5664. PMLR (2021)
15. Kouw, W.M., Loog, M.: A review of domain adaptation without target labels. IEEE Trans. Pattern Anal. Mach. Intell. **43**(3), 766–785 (2019)
16. Lakshminarayanan, B., Pritzel, A., Blundell, C.: Simple and scalable predictive uncertainty estimation using deep ensembles. Adv. Neural Inform. Process. Syst. **30** (2017)
17. Lee, Y., Yao, H., Finn, C.: Diversify and disambiguate: out-of-distribution robustness via disagreement. In: The Eleventh International Conference on Learning Representations (2022)
18. Liang, J., He, R., Tan, T.: A comprehensive survey on test-time adaptation under distribution shifts. arXiv preprint arXiv:2303.15361 (2023)
19. Michalski, J.M., Pisansky, T.M., Lawton, C.A., Potters, L.: Chapter 53 - prostate cancer. In: Gunderson, L.L., Tepper, J.E. (eds.) Clinical Radiation Oncology (Fourth Edition), pp. 1038–1095.e18. Elsevier, Philadelphia, fourth edition edn. (2016)
20. Niu, S., et al.: Towards stable test-time adaptation in dynamic wild world. In: The Eleventh International Conference on Learning Representations (2023)
21. Ovadia, Y., et al.: Can you trust your model's uncertainty? evaluating predictive uncertainty under dataset shift. Adv. Neural Inform. Process. Syst. **32** (2019)
22. Rohrbach, D., Wodlinger, B., Wen, J., Mamou, J., Feleppa, E.: High-frequency quantitative ultrasound for imaging prostate cancer using a novel micro-ultrasound scanner. Ultrasound Med. Biol. **44**(7), 1341–1354 (2018)
23. Sagawa, S., Koh, P.W., Hashimoto, T.B., Liang, P.: Distributionally robust neural networks for group shifts: On the importance of regularization for worst-case generalization. arXiv preprint arXiv:1911.08731 (2019)
24. Shao, Y., Wang, J., Wodlinger, B., Salcudean, S.E.: Improving prostate cancer (pca) classification performance by using three-player minimax game to reduce data source heterogeneity. IEEE Trans. Med. Imaging **39**(10), 3148–3158 (2020)

25. Sun, Y., Wang, X., Liu, Z., Miller, J., Efros, A., Hardt, M.: Test-time training with self-supervision for generalization under distribution shifts. In: International Conference on Machine Learning, pp. 9229–9248. PMLR (2020)
26. Wang, D., Shelhamer, E., Liu, S., Olshausen, B., Darrell, T.: Tent: Fully test-time adaptation by entropy minimization. arXiv preprint arXiv:2006.10726 (2020)
27. Wilson, P.F., et al.: Self-supervised learning with limited labeled data for prostate cancer detection in high frequency ultrasound. IEEE Trans. Ultrasonics Ferroelectrics Frequency Control (2023)
28. Wilson, P., et al.: Toward confident prostate cancer detection using ultrasound: a multi-center study. Inter. J. Comput. Assisted Radiol. Surgery (2024)
29. Wu, Y., He, K.: Group normalization. In: Proceedings of the European conference on computer vision (ECCV), pp. 3–19 (2018)
30. Zhang, M., Levine, S., Finn, C.: Memo: test time robustness via adaptation and augmentation. Adv. Neural. Inf. Process. Syst. **35**, 38629–38642 (2022)
31. Zhao, H., Liu, Y., Alahi, A., Lin, T.: On pitfalls of test-time adaptation. arXiv preprint arXiv:2306.03536 (2023)

SpineStyle: Conceptualizing Style Transfer for Image-Guided Spine Surgery on Radiographs

R. Neeraja[1], S. Devadharshiniinst[1], N. Venkateswaran[2], Vivek Maik[1(✉)], Aparna Purayath[1], Manojkumar Lakshmanan[1], and Mohanasankar Sivaprakasam[1]

[1] Healthcare Technology Innovation Centre (HTIC), Indian Institute of Technology Madras (IITM), Chennai, India
{maik.vivek,aparna_p}@htic.iitm.ac.in
[2] Department of Biomedical Engineering, Sri Sivasubramaniya Nadar College of Engineering (SSN), Kalavakkam, India

Abstract. Intraoperative radiographs are used in image-guided spine surgery (IGSS) for vertebrae planning and other processes like 2D/3D registration, single image tomography, and 3D reconstruction. Since preoperative CT is also a prerequisite for the majority of these techniques, computationally connecting the intraoperative radiographs and preoperative CT aids in the extraction of 3D data and improving accuracy. However, since the CT and radiograph modalities have different imaging attributes, Digitally Reconstructed Radiographs (DRR) produced from the CTs must be domain-adjusted with radiographs before computationally linking the two modalities. The proposed research focuses on the domain-adjusting algorithm SpineStyle, which is a unique style transfer algorithm meant to transfer DRR style attributes to intraoperative radiographs. SpineStyle implements a customized VGG-19 architecture capable of dealing with spine anatomy and producing anatomically accurate style-transferred images from a small dataset. Because the preservation of anatomy is critical for IGSS, this work additionally provides a Semantic Content Loss Measure (SCLM) evaluation metric to measure anatomy preservation during style transfer. Experimental studies show that SpineStyle outperforms other standard CNNs while also preserving anatomical content. The implementation of the project can be found in this repository: **SpineStyle**.

Keywords: Image-guided surgery · Style Transfer · Digitally Reconstructed Radiographs (DRRs) · VGG-19 · Content and Style Loss · Semantic Content Loss Metric

1 Introduction

Image Guided Spine Surgery (IGSS) is a rapidly evolving technological arena, with new techniques being integrated into systems to increase surgical efficiency and patient outcomes. Two such critical advancements include 2D/3D registration where live intraoperative radiographs from C-Arm fluoroscopic images are aligned with preoperative Computed Tomography (CT) data [14], and 3D reconstruction from 2D radiographs [2,11,19]. Both of these procedures vastly aid surgeons in accurately identifying target locations, particularly in the case of minimally invasive spinal surgery (MIS) where they have to determine the position of the vertebral anatomy from the preoperative CTs and intraoperative radiographs. It becomes vital for the surgeons to interlink the preoperative 3D CT data with the intraoperative 2D radiographs so that they can utilize the 3D information from the CT during the surgery. Since CT and radiographs are separate modalities, the images produced by these two modalities will also differ. Hence, for any computational procedure between the two modalities, this imaging attribute disparity may be bridged by utilizing a technique called Style Transfer or Domain Adaptation. Since the CT and radiography domains are far too different to be subjected to any kind of style transfer, a DRR may be used as a synthetic intermediate [5].

Style Transfer using Generative Adversarial Networks (GANs) and Convolutional Neural Networks (CNNs) has already been used on medical images for segmentation [3,18], depth estimation [8] and multi-modal learning [4,10]. Little work has been implemented so far using style transfer for reconstruction and registration purposes [2,13,19], which mostly rely on GANs that require large amounts of training data samples. Since the dataset is more limited for spinal radiographs and CT scans, neural style transfer using CNNs proves more resourceful. Among the different CNNs available, style transfer using VGG-19 has been shown to provide the least total loss and greater output image quality [9]. Additionally, from the conception of style transfer [6] till now [12], there is a distinct lack of metrics to evaluate the effectiveness of this process especially when preserving the anatomical content of the X-ray images is important to keep up with the accuracy of the IGSS and the same should be evaluated with a metric as well. In this work we propose SpineStyle - a deep learning pipeline that uses a modified version of VGG-19 to transfer the style of a DRR onto C-Arm anteroposterior (AP) and lateral posterior (LP) view images of the lumbar (L1-L5) vertebrae with the following contributions:

1) By making changes to the normalization and image weights of the VGG-19, the model was customized as a feature extractor for style transfer specific to spinal radiographs.
2) The change in pooling layers from max to average pooling helps us to work better with the low contrast nature of the C-arm radiographs.
3) An additional regularization term added to the loss function aids in noise removal and produces a smooth output image.
4) A novel Semantic Content Loss Measure (SCLM) is proposed to measure the extent of the transfer of style and the degree of preservation of the spinal anatomy.

2 Methods

The proposed SpineStyle pipeline, as shown in Fig. 1, consists of an adapted form of VGG-19 that extracts features and performs style transfer based on two input images: the content image (C-Arm fluoroscopic image) and the style image (DRR image from CT). The output image appears in the style of the DRR with the contents of the C-Arm image.

2.1 Style Transfer Model SpineStyle

The VGG-19 architecture [16] has 16 convolutional layers that use 3×3 filters with a fixed stride of 1 and padding, followed by 2×2 max pooling layers with a stride of 2. This uniform architecture is beneficial for style transfer as it provides normalized hierarchical feature maps and allows easy scaling of the model by adding more layers. In SpineStyle, the max pooling layers are replaced with average pooling layers since max pooling retains only the most prominent features within each region whereas style transfer for spinal images requires capturing more general features such as overall texture and intensity distribution. Batch normalization layers, which are best suited for managing data in batches, are replaced by instance normalization layers, which normalize each feature map independently while neglecting batch information. As SpineStyle only processes one image at a time, instance normalization does not necessarily increase computational load or training time, making it more appropriate for IGSS.

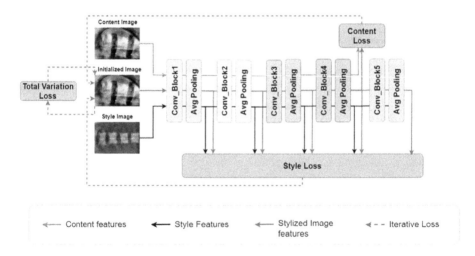

Fig. 1. Schematic of the SpineStyle pipeline - Style Transfer Model for input content (C-Arm) and style (DRR) images using content loss, style loss and total variation loss

2.2 Losses and Implementation Details

Feature maps from the 4th convolutional block are used to capture detailed structural information. This method allows for the extraction of more detailed shapes after multiple layers of nonlinear processing. All 5 convolutional blocks are used to extract features from the style image. This approach ensures effective style transfer since the lower layers are used to extract general textures and the upper layers for more abstract structures. The loss layers are added after each convolutional layer. Content Loss [6] is mathematically described as the Mean Square Error (MSE) between the two sets of feature maps in a particular layer.

$$C_{loss} = \frac{1}{2} \sum_{ij} \left(F_{ij}^l - P_{ij}^l \right)^2 \tag{1}$$

Here, F_{ij}^l and P_{ij}^l represent the feature maps of the content image and generated image at layer l respectively where i and j are the spatial dimensions of the feature maps. Style loss [6] is computed by finding the MSE between the gram matrix of the style image and the generated image. Equation 2 describes the gram matrix, where F_{ij}^l is a matrix containing all the vectorized feature maps of layer l and F_{ji}^l is the transpose of this matrix.

$$G_{ij}^l = \sum_k (F_{ij}^l F_{ji}^l) \tag{2}$$

Equation 3 describes style loss where G_{ij}^l and A_{ij}^l are the gram matrices of the style image and generated image at layer l respectively.

$$S_{loss} = \frac{1}{4N_l^2 M_l^2} \sum_{ij} \left(G_{ij}^l - A_{ij}^l \right)^2 \tag{3}$$

The gram matrix must be normalized to reduce the effect on gradient descent for feature maps with larger dimension sets. Using Total Variation Loss (TVL) [1], a form of L2 regularization using Euclidean losses, the network is encouraged to produce images where adjacent pixels have similar intensity values, producing a visually smooth output with less noise.

$$R_{loss} = \sum_{ij} \left((x_{i,j+1} - x_{i,j})^2 + (x_{i+1,j} - x_{ij})^2 \right)^{\frac{\beta}{2}} \tag{4}$$

In the above equation, x represents the style transferred output at every epoch for spatial locations $(i \times j)$. Thus, the total loss may be defined as:

$$Loss = w_c C_{loss} + w_s S_{loss} + w_r R_{loss} \tag{5}$$

where w_c, w_s and w_r are content weight, style weight and regularization weight respectively. Since the aim is to transfer the style of an image, $w_s \gg w_c$. Style weight ranges from 10^6 to 10^7 and content weight ranges from 1 to 100. The value of w_r ranges from 0 to 1. Increasing w_r results in smoother images, while decreasing it allows for more detail but may introduce artifacts.

3 Experimental Results and Discussion

DRRs were generated using Plastimatch [15] in the rotational mode where the source is assumed to rotate in a 360° circular orbit around the isocenter of the CT and a DRR is generated for every 1° of the circular orbit. The AP and LP radiographs were passed through a preprocessing pipeline of standard joint smoothing enhancement along with denoising using Gaussian and non-local means algorithms.

SpineStyle Transfer Results. SpineStyle was run up to 200 epochs with $w_c = 10$, $w_s = 1000000$ and $w_r = 0.0001$ to produce the output in Fig. 2. To get a better visual assessment of the SpineStyle transfer output, we compare the histograms of the content, style, and style transferred images as seen in Fig. 3. Ideally, the output image including the contents of the C-Arm fluoroscopic image should have a histogram comparable to that of the DRR, implying that the anatomy of the X-ray appears with the same intensity mapping as the DRR.

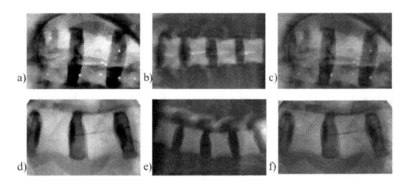

Fig. 2. Experimental Results of SpineStyle on AP(toprow) and LP(bottomrow) images where (a),(d) are the C-arm Images, (b),(e) are the DRRs and (c),(f) are the style transferred outputs

Semantic Content Loss Measure (SCLM). There are no proper metrics for assessing content retention in style-transferred images. Metrics like Structural Similarity Index Measure (SSIM) and Peak Signal-to-Noise Ratio (PSNR) require a reference ground truth image for comparison. These metrics rely on comparing pixel intensities of images, but in style transfer, the image undergoes significant intensity transformation, rendering SSIM and PSNR comparisons unsuitable. We propose SCLM, which compares the feature maps of the content image to those of the output image, focusing on edges and shapes that describe the overall anatomy rather than intensity. This comparison is carried out after every convolutional block of VGG-19 using the sum of absolute differences (SAD) as shown in Fig. 4. The mean obtained from the comparison of

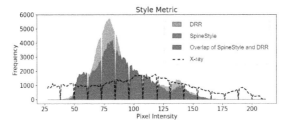

Fig. 3. Comparative image histograms for spinal AP radiograph (content), DRR (style) & style transferred output image from SpineStyle. The histogram for SpineStyle is comparable to the DRR's histogram.

Table 1. SCLM Values Across Convolutional Blocks of VGG-19 in AP and LP images with and without resizing of the images.

Metric	Block 1	Block 2	Block 3	Block 4	Block 5
No. of Feature Maps	64	128	256	512	512
SCLM AP	0.2930	0.2146	0.2049	**0.1471**	0.1471
SCLM AP (with resizing)	0.2985	0.2184	0.2068	**0.1443**	0.1993
SCLM LP	0.4157	0.4055	0.3366	**0.2280**	0.2546
SCLM LP (with resizing)	0.4574	0.3274	0.2812	**0.2022**	0.2370

feature maps from the 4th block using the SAD is the proposed SCLM value, which is described in Fig. 5.

A lower SCLM value indicates that the content of the output image is very similar to the original content image, suggesting better content preservation. This is crucial in style transfer for spinal radiographs where anatomy preservation is crucial. Table 1 shows that the SCLM value decreases till the 4th block and slightly increases in the 5th block. The decrease till the 4th block is because lower blocks capture general features and as we progress through the blocks the complexity of the feature maps increases, representing the content better and leading to lower SCLM. The slight increase in SCLM in the 5th block is attributed to the low spatial resolution of the filters (14×14) which causes many outliers that hamper the measure.

Pooling and Regularization Effects. Average pooling results in smoother and more consistent feature maps which lead to gradients that are more stable during back propagation. Additionally, using max pooling for AP images resulted in a high style loss (up to 700) compared to the content loss (7). Substituting the max pooling layers with average pooling layers reduced the initial style loss to 200, which facilitated faster loss convergence. With max pooling, the convergence was observed at 700 epochs whereas with average pooling, the convergence happened earlier at 200 epochs itself, implying faster training. The total variation loss minimizes the variation between neighboring pixels, whereas

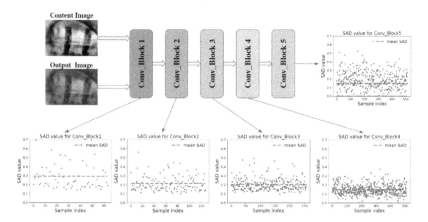

Fig. 4. SCLM plotted for feature maps of the convolutional blocks of VGG-19 which was used to quantify anatomical preservation. The 4th block, used to extract content features, shows the least mean SCLM values.

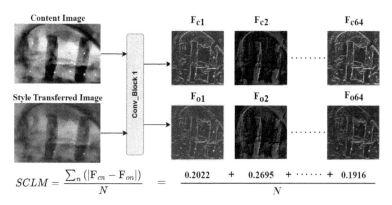

Fig. 5. SCLM for the first convolutional block computed across 64 filters from content and style transferred image. F_{cn} and F_{on} represent the nth feature map of the content and output image respectively. N is the number of feature maps in that block.

the regularization weight (w_r) determines the extent of minimization. Here, w_r = 0.1 gives an extremely smooth image to the point where style features are completely lost. Through trial and error, it was determined that w_r = 0.0001 gave an image with less noise but also preserved all the style features. Figure 6 shows the output images for different values of w_r.

Performance Comparison. The proposed SpineStyle is compared to various model architectures as shown in Fig. 7, including VGG-19, VGG-16, AlexNet [9], Inception v3 [17], and ResNet 50 [7]. Uniform architectures in the VGG family, such as VGG-19 and VGG-16, performed well but were not as effective as SpineStyle. Additionally, it was discovered that deeper models, such

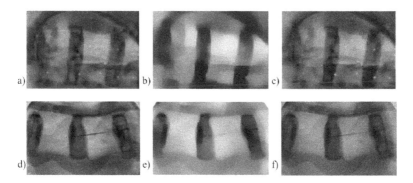

Fig. 6. Effect of Regularization on AP (top row) and LP (bottom row) images with: $w_r = 0$ in (a),(d), $w_r = 0.1$ in (b),(e), $w_r = 0.0001$ in (c),(f)

as ResNet, struggle with style transfer because residual connections interfere with the extraction of fine textural information by contributing outputs from a prior layer to a deeper layer. In contrast, AlexNet's shallower architecture creates fewer and less detailed feature maps than VGG, making it less suitable for detailed style transfer tasks. Inception v3 is also unsuited for style transfer due to its changing receptive fields, which make it difficult to record style information without mixing in content elements. The comparative histogram study of the VGG family from Fig. 8. shows that the standard VGG-19 achieves a style transfer that closely matches the DRR's style. However, the process requires more training time than SpineStyle and compromises on anatomical detail. In the case of VGG-16, the anatomy is better preserved but style is not effectively transferred. SpineStyle produces the most effective result in terms of style transfer and content preservation. The histogram of SpineStyle's output most closely resembles that of the DRR, and has the least SCLM, among all standard CNN models for style transfer as seen in Table 2.

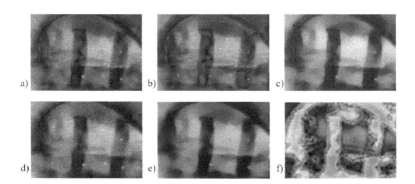

Fig. 7. Perfomance Comparison - VGG Family (top): SpineStyle(a), VGG19(b) and VGG16(c). Standard CNNs (bottom): AlexNet(d), Inception v3(e) and ResNet50(f)

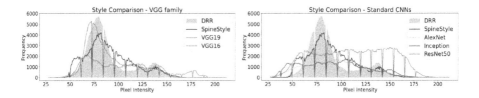

Fig. 8. Inter Model Comparison of Style for AP radiograph - VGG family (left) and Standard CNNs (right)

Table 2. Performance Comparison of the proposed algorithm using SCLM metric for AP and LP images. The anatomical preservation is achieved for values of SCLM closer to 0.

	Spinestyle	VGG19	VGG16	AlexNet	Inception v3	ResNet50
SCLM - AP	**0.1443**	0.3131	0.2698	0.3669	0.3657	0.4832
SCLM - LP	**0.2022**	0.3399	0.3535	0.2709	0.3563	0.4365

4 Conclusion

By the use of the proposed SpineStyle algorithm, critical anatomical details of the original radiographs are preserved while achieving the desired DRR-like appearance. The modifications made to the VGG-19 architecture proved crucial for optimizing the style transfer process, especially with a limited dataset. The results indicate that the modified VGG-19 model facilitates faster convergence of losses. The histogram studies and SCLM show the accuracy of the style transfer, ensuring that the output images maintain anatomical accuracy and stylistic features similar to DRRs. Future work involves integrating this pipeline into operations such as 2D/3D image registration and 3D reconstruction from biplanar radiographs.

References

1. Allard, W.: Total variation regularization for image denoising, i. geometric theory. SIAM J. Math. Anal. **39**, 1150–1190 (2007). https://doi.org/10.1137/060662617
2. Almeida, D.F., Astudillo, P., Vandermeulen, D.: Three-dimensional image volumes from two-dimensional digitally reconstructed radiographs: a deep learning approach in lower limb CT scans. Med. Phys. **48**(5), 2448–2457 (2021)
3. Cao, S., Konz, N., Duncan, J., Mazurowski, M.A.: Deep learning for breast mri style transfer with limited training data. J. Digit. Imaging **36**(2), 666–678 (2023). https://doi.org/10.1007/s10278-022-00755-z
4. Dou, Q., et al.: Pnp-adanet: plug-and-play adversarial domain adaptation network at unpaired cross-modality cardiac segmentation. IEEE Access (2019). https://doi.org/10.1109/ACCESS.2019.2929258

5. Galvin, J.M., Sims, C., Dominiak, G., Cooper, J.S.: The use of digitally reconstructed radiographs for three-dimensional treatment planning and CT-simulation. Int. J. Radiat. Oncol. Biol. Phys. **31**(4), 935–942 (1995)
6. Gatys, L., Ecker, A., Bethge, M.: A neural algorithm of artistic style. arXiv (2015). https://doi.org/10.1167/16.12.326
7. He, K., Zhang, X., Ren, S., Sun, J.: Deep residual learning for image recognition. In: 2016 IEEE Conference on Computer Vision and Pattern Recognition (CVPR), pp. 770–778 (2016). https://doi.org/10.1109/CVPR.2016.90
8. Karaoglu, M.A., et al.: Adversarial domain feature adaptation for bronchoscopic depth estimation. In: de Bruijne, M., et al. (eds.) MICCAI 2021. LNCS, vol. 12904, pp. 300–310. Springer, Cham (2021). https://doi.org/10.1007/978-3-030-87202-1_29
9. Kavitha, S., Dhanapriya, B., Vignesh, G.N., Baskaran, K.: Neural style transfer using vgg19 and alexnet. In: 2021 International Conference on Advancements in Electrical, Electronics, Communication, Computing and Automation (ICAECA), pp. 1–6 (2021). https://doi.org/10.1109/ICAECA52838.2021.9675723
10. Krishna, A., Yenneti, S., Wang, G., Mueller, K.: Image factory: A method for synthesizing novel ct images with anatomical guidance. Medical Physics **51** (2023). https://doi.org/10.1002/mp.16864
11. Kyung, D., Jo, K., Choo, J., Lee, J., Choi, E.: Perspective projection-based 3d ct reconstruction from biplanar x-rays. In: ICASSP 2023 - 2023 IEEE International Conference on Acoustics, Speech and Signal Processing (ICASSP), pp. 1–5 (2023). https://doi.org/10.1109/ICASSP49357.2023.10096296
12. Liao, J.: A study on neural style transfer methods for images. 2022 2nd International Conference on Big Data, Artificial Intelligence and Risk Management (ICBAR), pp. 60–64 (2022). https://api.semanticscholar.org/CorpusID:258379250
13. Liu, M., et al.: Alzheimer's disease neuroimaging initiative: style transfer generative adversarial networks to harmonize multisite mri to a single reference image to avoid overcorrection. Hum. Brain Mapp. **44**(14), 4875–4892 (2023)
14. Markelj, P., Tomaževič, D., Likar, B., Pernuš, F.: A review of 3d/2d registration methods for image-guided interventions. Med. Image Anal. **16**(3), 642–661 (2012). https://doi.org/10.1016/j.media.2010.03.005
15. Sharp, G.C.: Plastimatch drr documentation. https://plastimatch.org/drr.html, Accessed 22 June 2024
16. Simonyan, K., Zisserman, A.: Very deep convolutional networks for large-scale image recognition. arXiv: 1409.1556 (Sep 2014)
17. Szegedy, C., Vanhoucke, V., Ioffe, S., Shlens, J., Wojna, Z.: Rethinking the inception architecture for computer vision. In: 2016 IEEE Conference on Computer Vision and Pattern Recognition (CVPR), pp. 2818–2826 (2016). https://doi.org/10.1109/CVPR.2016.308
18. Xu, Y., Li, Y., Shin, B.S.: Medical image processing with contextual style transfer. HCIS **10**(1), 46 (2020). https://doi.org/10.1186/s13673-020-00251-9
19. Ying, X., Guo, H., Ma, K., Wu, J., Weng, Z., Zheng, Y.: X2ct-gan: reconstructing ct from biplanar x-rays with generative adversarial networks. In: 2019 IEEE/CVF Conference on Computer Vision and Pattern Recognition (CVPR), pp. 10611–10620 (2019). https://doi.org/10.1109/CVPR.2019.01087

SGSR: Structure-Guided Multi-contrast MRI Super-Resolution via Spatio-Frequency Co-Query Attention

Shaoming Zheng, Yinsong Wang, Siyi Du, and Chen Qin(✉)

Department of Electrical and Electronic Engineering and I-X,
Imperial College London, London, UK
{s.zheng22,y.wang23,s.du23,c.qin15}@imperial.ac.uk

Abstract. Magnetic Resonance Imaging (MRI) is a leading diagnostic modality for a wide range of exams, where multiple contrast images are often acquired for characterizing different tissues. However, acquiring high-resolution MRI typically extends scan time, which can introduce motion artifacts. Super-resolution of MRI therefore emerges as a promising approach to mitigate these challenges. Earlier studies have investigated the use of multiple contrasts for MRI super-resolution (MCSR), whereas majority of them did not fully exploit the rich contrast-invariant structural information. To fully utilize such crucial prior knowledge of multi-contrast MRI, in this work, we propose a novel structure-guided MCSR (SGSR) framework based on a new spatio-frequency co-query attention (CQA) mechanism. Specifically, CQA performs attention on features of multiple contrasts with a shared structural query, which is particularly designed to extract, fuse, and refine the common structures from different contrasts. We further propose a novel frequency-domain CQA module in addition to the spatial domain, to enable more fine-grained structural refinement. Extensive experiments on fastMRI knee data and low-field brain MRI show that SGSR outperforms state-of-the-art MCSR methods with statistical significance.

Keywords: Magnetic resonance imaging · Multi-contrast super-resolution · Co-Query Attention

1 Introduction

Magnetic Resonance Imaging (MRI) is one of the most useful and important imaging techniques in hospitals for the diagnosis of diseases, due to its safety without ionizing radiation and its superior soft tissue contrast. However, the acquisition of high-resolution (HR) MRI is often time-consuming, and thus acquiring multiple images of that with different contrasts becomes even more challenging. To mitigate the challenges, super-resolution (SR) of MRI has emerged as a promising approach to improve the spatial resolution of images and restore tissue contrast. Since multi-contrast images in MRI provide abundant and complementary information for characterizing tissues, they can be jointly exploited in multi-contrast SR (MCSR) for enhancing the SR performance. In principal, multi-contrast images of the same subject should share contrast-invariant information that corresponds to the underlying spatial structure such as object edges, whereas they also possess individual contrast-specific attribute that captures the rendering of structure determined by imaging physics, e.g., tissue contrast. Exploring such attributes in multi-contrast MRI is therefore assumed to be able to provide effective and complementary knowledge for the MCSR task.

Unlike most of the early SR models that mainly focus on single image SR (SISR) without leveraging complementary information within multiple correlated images, recent works have considered the multi-image SR via utilizing a reference (Ref) image as guidance to super-resolve low resolution (LR) images. Examples of those include TTSR [15] and MASA [10], which used the Ref image as a textural guidance for SR similar to style transfer tasks, without specifically considering the structural information across images, and thus could result in potential lack of structural coherence. Recent studies in MRI SR have also started to leverage the availability of multi-contrast images, where they similarly also utilize an available high-resolution contrast image as the Ref. Under this setup, Lyu et al. [13], MINet [2] and SANet [4] have been proposed and demonstrated the superiority of MCSR over SISR. They mainly used relatively simple contrast fusion mechanisms such as feature concatenation and separable attention for leveraging multi-contrast information. To improve upon that, DCAMSR [6] has introduced spatial-channel attention to better capture cross-contrast semantics. Besides, various works [3,5,11,14,18,19] have also proposed multi-contrast fusion mechanisms for MRI reconstruction. However, they are all designed to equally attend to each region without explicitly focusing on structural parts. Therefore they cannot distinguish structural and non-structural features, which can thus be inefficient in exploiting the cross-contrast information for MCSR.

To overcome this, there have been some efforts in investigating the explicit incorporation of structural information in MCSR. For instance, McMRSR++ [8] have used the high-frequency signals (commonly representing the structural information) of the HR Ref as an additional target for the SR, which enables the model to be structure-aware by enforcing it to recover the structural target from the latent space. WavTrans [7] similarly proposed to perform contrast fusion on the high-frequency wavelet components of the Ref to fully exploit the underlying structural knowledge. However, the structural information assumed in their works are extracted either from Ref or LR contrast, without exploiting the complementary structural information among them.

In this work, we propose SGSR, a novel super-resolution framework that leverages the prior knowledge in multi-contrast images for MCSR via exploiting contrast-invariant structural information. In particular, we propose a co-query attention (CQA) mechanism driven by the shared-structure characteristic in multi-contrast images, where contrast-invariant features (i.e., structures) are leveraged jointly from multiple contrasts, constituting the 'query', and

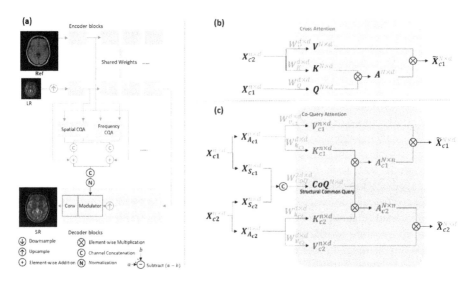

Fig. 1. (a) Overall architecture of our SGSR; (b) Standard cross-attention; (c) The proposed co-query attention (CQA) within the spatial module.

contrast-specific characteristics (i.e., appearances) are distilled and interacted with structural information in individual contrast respectively. We further propose to perform CQA in both spatial and frequency domains to exploit their complementary mechanisms in representing and processing signals. In particular, the spatial CQA can attend to local features while the frequency CQA enables more fine-grained structure-appearance interactions. We observe that CQA is more parameter-efficient compared to cross-attention methods. Our experiments on knee and brain MRI datasets showcase that SGSR outperforms the state-of-the-art MRI SISR and MCSR methods both quantitatively and qualitatively. To clarify our novelty, we are the first to explicitly exploit contrast-invariant structures with attention mechanisms in MCSR.

2 Methodology

Our proposed SGSR consists of three components as shown in Fig. 1(a). First, the spatial co-query attention (SCQA) module extracts and refines common structures from multiple contrasts via an efficient spatial attention between structures and appearances (Sect. 2.1). Second, the frequency co-query attention (FCQA) module performs CQA in parallel to the SCQA, formulating token as frequency components to further enable more fine-grained of structure-appearance interactions (Sect. 2.2). Finally, an encoder-decoder network serves as the backbone, enabling the attention learning on the feature pyramid of the image. (Sect. 2.3).

2.1 Spatial Co-Query Attention (SCQA)

Co-Query Attention (CQA): Conventionally, cross-attention can be used for dealing with information transfer between components, where the attention is performed between queries of one contrast and keys and values of another one, as shown in Fig. 1(b). However, in MSCR, as we discussed earlier, it is desirable to exploit common contrast-invariant information among multiple contrasts, whereas the cross attention mechanism is limited in this aspect due to its inherent design. To tackle this issue, we therefore propose the query-sharing design of CQA. Specifically, as shown in Fig. 1(c), CQA is composed of a shared 'query' representing the common structures among contrasts and individual 'keys' and 'values' corresponding to contrast-specific features. The attention between keys and query therefore reflects the structure-appearance interactions within images, exploring structure-related contrast information.

In detail, we assume that multi-contrast images can be extracted into structural features X_{S_c} of size $N \times d$ and appearance features X_{A_c} of size $n \times d$, which serve as the input of CQA. Here c represents different contrasts, N and n are the number of tokens, and d is the number of feature channels. As shown in Fig. 1(c), CQA first concatenates the input structural features X_{S_c} of each contrast along the channel dimension and map it from $2d$ to d-dimension to generate the co-query CoQ with size of $N \times d$. Then, for each contrast c, it further generates contrast-specific keys K_c and values V_c with size of $n \times d$ from the appearances features X_{A_c}. The CQA attention is then computed between the shared query and each of the contrast-specific keys and values, which can be formulated as:

$$\tilde{X}_c = \sigma \left(\frac{Q(K_c)^T}{\sqrt{d}} \right) V_c. \quad (1)$$

Here $\tilde{X}_c$ denote the output of CQA, which can be interpreted as structural features refined with the appearances of contrast c. Note that unlike the single attention output as in cross-attention, CQA produces output for each contrast.

CQA in Spatial Domain: We first propose to adapt CQA in the spatial domain to enable interactions among local image features. For extracting structural and appearance features as inputs to the SCQA, we propose to adopt a simple strategy of low pass filtering to approximate them based on the assumption that low frequency components contain most of the information of image contrast while high frequency signals correspond to information about the structures, e.g., edges and boundaries of the image [1]. In detail, the appearance features X_{A_c} in spatial domain can be obtained via a downsampling operation on the encoder features X_c, and the structural features X_{S_c} can be derived by the subtraction of the upsampled X_{A_c} from X_c. Given these, the computational complexity of the CQA in the spatial module would be $O(Nnd)$ (typically, $n \ll N$) instead of the quadratic $O(N^2 d)$ in vanilla cross-attention, which is therefore more efficient.

2.2 Frequency Co-Query Attention (FCQA)

To fully exploit the common structural knowledge and further refine it beyond local spatial features, we propose to extend CQA to the 2D Fourier frequency domain, named Frequency Co-Query Attention (FCQA). In contrast to SCQA,

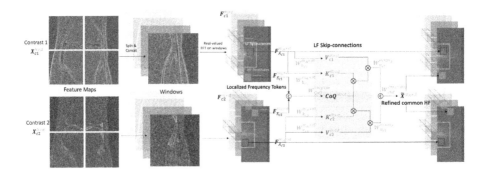

Fig. 2. Frequency co-query attention (FCQA) module. The above process is followed by Inverse FFT to transform back to spatial domain.

each token in FCQA represents a frequency component instead of a spatial position as shown in Fig. 2. This extends the bi-partition of high-frequency structures and low-frequency appearances in SCQA to more fine-grained frequency levels, which therefore allows the exploitation of finer structure-appearance relationships. In the FCQA module, it consists of a transformation from spatial features to localized frequency tokens, a CQA module in the frequency domain, and an inverse transformation back to the spatial domain.

Localized Frequency Tokens: To transform from spatial to the frequency domain, the FCQA module first performs 2D Fast Fourier Transform (FFT) within local windows of the encoder feature maps X_c. This aims to reduce the number of frequency components and thus the attention tokens compared with global FFT for efficient computation. We perform real-valued FFT on each window as it can halve the number of resulting frequency components (denoted as m) compared to complex FFT. Then for each 2D spatial frequency component, we concatenate its d-dimensional tokens in each window to form a final token of dimension $n_f d$, where n_f is the number of windows. Our final frequency domain F_c is therefore composed of m such localized frequency tokens. If we use these tokens for attention, the computational complexity would be $O(m^2 \cdot n_f d) = O((N/n_f)^2 \cdot n_f d) = O(N^2 d/n_f)$, which is also sub-quadratic as in SCQA. In essence, we found that concatenating windows along the channel significantly improve the efficiency.

CQA in Fourier Frequency Domain: Similar to the strategy in SCQA, we separate structures- and appearances-relevant features by grouping the localized frequency tokens into high-frequency structural tokens and low-frequency appearance tokens. Specifically, we take the central $1/f$ part of F_c as appearance features F_{A_c} with m_A tokens and the rest as structural features F_{S_c} with m_S tokens, as shown in Fig. 2, where f is the hyperparameter for frequency splitting. To achieve efficient computation and utilize the sparsity of frequency domain, we further reduce the number of both structural and appearance tokens to $\tilde{m}$ via linear projections. In this way, the computational complexity of attention can be improved to $O(\tilde{m}^2 d) < O(N^2 d/n_f)$. We then merge the attention result of each contrast with the original appearance tokens as shown in the "LF skip-connections" in Fig. 2 to restore the size of the original frequency domain F_c. Then we perform inverse FFT to transform the refined structural features back to spatial domain for downstream combination with spatial features.

2.3 Super-Resolution Backbone

We use an encoder-decoder network with 4 residual groups of convolutional layers as the base of our overall architecture with default configurations as in [6]. The encoder backbone encodes images of all contrasts into multi-scale spatial features X_c, which are provided to both SCQA and FCQA modules. In the decoder, we gradually recover the latent features of LR contrast to the pixel space by using the latent features to modulate the refined structural features from the CQA modules, similar to [6]. In this way, the upsampling process can obtain structural guidance at each scale. The overall architecture of the backbone is shown in Fig. 1(a). The model is then trained with an L1 loss between the SR and the HR.

3 Experiments

Experimental Setup. We adopted two datasets: (1) fastMRI knee [16], with PDFS contrast as LR and PD contrast as Ref. We used 227 train/validation and 24 test pairs. (2) M4Raw [12], a dataset of 0.3T low-field brain MRI from 183 participants, where each has T1, T2 and FLAIR contrasts. We use T1 images as Ref and T2 for SR. Following [6], we used 128 train/validation pairs and 30 testing pairs. To generate the $s\times$ downsampled LR images ($s \in \{2, 4\}$), we keep the $1/s^2$ values in the central square window of k-space of the images, followed by 2D inverse FFT to convert them back into image domain.

Implementation Details. We used Adam optimizer with a batch size of 4 on 4 NVidia A5000 GPUs. The initial learning rate for SANet [4] was set to 4×10^{-5} according to [6], and 2×10^{-4} for the other methods. The learning rate was decayed by a factor of 0.1 starting from the 40th epoch. The performance was evaluated for 2× and 4× SR in terms of PNSR and SSIM. In SCQA, we downsample the appearances features to 16×16. We set $f = 4$ in FCQA.

Quantitative Comparison Study. We compare our SGSR with representative methods in diverse categories: (1) SISR methods: SwinIR [9] and ELAN [17], (2) reference-guided SR methods: TTSR [15] and MASA [10], and (3) state-of-the-art methods specifically targeting MRI MCSR: SANet [4] and DCAMSR [6]. The results are summarized in Table 1. Our SGSR achieves the best performance across all metrics, scale factors, and datasets for both SISR and MCSR. For MCSR, SGSR achieves 0.1dB improvements in PSNR with 57% less model parameters than the SOTA (DCAMSR) in M4Raw 4× SR, validating the representation power and low parameter cost of designing inductive bias for structural

Table 1. Quantitative comparisons. * stands for values that our method significantly outperforms with p-value < 0.01. MASA-SR and TTSR are unable to perform 2× SR based on their official implementation.

Methods	fastMRI (knee)				M4Raw (brain)				
	2×		4×		2×		4×		
	PSNR↑	SSIM↑	PSNR↑	SSIM↑	PSNR↑	SSIM↑	PSNR↑	SSIM↑	#Params↓
SwinIR (SISR)	32.01*	0.715*	30.73*	0.628*	32.16*	0.777*	29.73*	0.709*	11.9M
ELAN (SISR)	32.03*	0.715*	30.42*	0.618*	31.71*	0.770*	28.72*	0.680*	8.3M
DCAMSR (SISR)	32.07*	0.717*	30.71*	0.627*	32.19*	0.777*	29.74*	0.709*	9.3M
SGSR (SISR)	**32.12**	**0.719**	**30.86**	**0.632**	**32.22**	**0.778**	**29.80**	**0.713**	**4.0M**
MASA-SR	-	-	30.77*	0.628*	-	-	29.48*	0.703*	4.03M
TTSR	-	-	30.64*	0.627*	-	-	29.86*	0.712*	6.42M
SANet	32.00*	0.716*	30.41*	0.622*	32.06*	0.775*	29.47*	0.704*	11M
DCAMSR	32.20*	0.721*	30.97*	0.637*	32.31*	0.779*	30.48*	0.728*	9.3M
DCAMSR (4M)	32.13*	0.720*	30.64*	0.626*	32.07*	0.775*	29.80*	0.712*	4.0M
SGSR (Ours)	**32.23**	**0.723**	**31.05**	**0.640**	**32.39**	**0.780**	**30.58**	**0.731**	**4.0M**

refinement in SGSR. For fairer comparison, when the number of parameters of DCAMSR [6] matches ours at 4.0M by reducing network complexity, ours achieves even much higher performance gain over it (+0.78dB for M4Raw 4×, +0.41dB for fastMRI 4×, +0.32dB for M4Raw 2×, +0.1dB for fastMRI 2×). As shown in Table 3, the attention of our modules are superior in all aspects (10× less FLOPs and 5–10× less memory), aligning with our complexity analysis in methodology. Compared with general reference-guided SR methods, SGSR shows even better improvements (+1.1dB in PSNR over MASA) as these methods do not explicitly attend to the structural context due to their limited receptive field and weak fusion mechanisms designed for local textures. We achieve similar gains on fastMRI 4× compared to how DCAMSR improves on MASA (second best) (+0.2dB). SGSR also generalizes well to SISR. In the SISR experiment of our SGSR, we use the copy the LR as Ref. In this way, SGSR tends to extract and refine the structures within the LR contrast and thus achieves better quality than both global attention (ELAN) and local attention (SwinIR) methods without any structure-exploiting designs.

Qualitative Results. Visual comparisons are shown in Fig. 3 for 4× SR. Our method produces overall the least errors and accurately recovers very small structures such as those indicated by arrows. We attribute this to our model's capacity to refine structures leveraging different contrasts compared to other methods.

Ablation Study. We conducted ablation studies on the M4Raw [12] 4× SR as shown in Table 2. Four variations of our SGSR are experimented: (1) *w/o Ref* (i.e. SISR): we replace Ref input by upsampled LR. (2) *CA* (cross-attention): i.e. DCAMSR [6]. (3) *FCQA*, and (4) *SCQA*. The gain of SGSR (*FCQA+SCQA*) from *CA* validates the superiority of exploiting common structural representa-

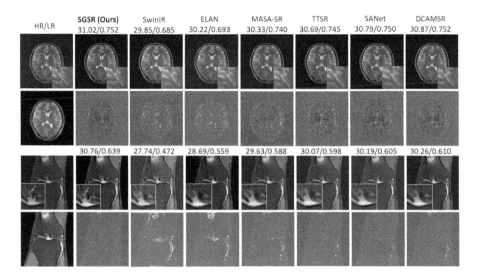

Fig. 3. Qualitative results on SR predictions and error maps of fastMRI [16] (top) and M4Raw [12] (bottom). Metrics are shown in PSNR/SSIM for each image. Input (bottom) and ground truth (top) are shown in the first column.

Table 2. Ablation study on M4Raw [12] 4× task. * for values that the full version significantly outperforms with p-value < 0.01.

Variants	Ref	CA	SCQA	FCQA	PSNR	SSIM
w/o Ref		✓	✓	✓	29.80*	0.713*
CA	✓	✓			30.48*	0.729*
FCQA	✓			✓	30.53*	0.729*
SCQA	✓		✓		30.55*	0.729*
FCQA+SCQA	✓		✓	✓	**30.58**	**0.731**

Table 3. Comparisons of FLOPs & GPU memory between attention of our modules and cross attention for 128×128 LR & Ref input.

Variants	FLOPs	Memory
SCQA attention	2.85×10^8	118 MB
FCQA attention	2.68×10^8	214 MB
Cross attention	2.28×10^9	1030 MB

tion which undergoes refinement with each contrast. In addition, the gains of both $SCQA$ and $FCQA$ from CA validate the effectiveness of integrating complementary mechanisms of structural fusion and refinement in both spatial and frequency domains. Their gain account for 70% and 50% of the gap between DCAMSR and our SGSR ($p \ll 0.01$). We attribute this to their respective advantages: SCQA attends to local features while FCQA enables fine-grained structure-appearance interactions. We visualize the structural queries in SCQA in the supplementary materials.

4 Conclusion

In this paper, we presented a novel and small structure-guided multi-contrast MRI super-resolution (SGSR) scheme. Exploiting the contrast-invariant structural information as an important prior, our method features common structural extraction and refinement from all the contrasts via a tailored co-query attention (CQA) mechanism. In addition to the CQA on spatial domain which processes local features, we further extend this attention paradigm to the frequency domain, adapting it to fine-grained structure-appearance interactions with a novel frequency co-query attention (FCQA) paradigm. Comparison study shows that our method outperforms other competing multi-contrast SR methods regarding both SR image quality and parameter cost. For future works, we will extend the method to multi-modal tasks such as super-resolution between MRI and CT.

Acknowledgements. This work was partially supported by the Engineering and Physical Sciences Research Council [grant number EP/X039277/1].

References

1. Du, T., et al.: Adaptive convolutional neural networks for accelerating magnetic resonance imaging via k-space data interpolation **72**, 102098 (2021). https://doi.org/10.1016/j.media.2021.102098
2. Feng, C.M., Fu, H., Yuan, S., Xu, Y.: Multi-contrast MRI super-resolution via a multi-stage integration network. In: MICCAI 2021, pp. 140–149 (2021). https://doi.org/10.1007/978-3-030-87231-1_14
3. Feng, C.M., Yan, Y., Chen, G., Xu, Y., Hu, Y., Shao, L., Fu, H.: Multimodal transformer for accelerated MR imaging **42**(10), 2804–2816 (2022). https://doi.org/10.1109/TMI.2022.3180228. https://ieeexplore.ieee.org/abstract/document/9796552, conference Name: IEEE Transactions on Medical Imaging
4. Feng, et al.: Exploring separable attention for multi-contrast MR image super-resolution. IEEE Trans. Neural Netw. Learn. Syst. 1–12. https://doi.org/10.1109/TNNLS.2023.3253557
5. 00000000 Guo, P., Mei, Y., Zhou, J., Jiang, S., Patel, V.M.: ReconFormer: accelerated MRI reconstruction using recurrent transformer **43**(1), 582–593 (2023). https://doi.org/10.1109/TMI.2023.3314747. https://ieeexplore.ieee.org/abstract/document/10251064, conference Name: IEEE Transactions on Medical Imaging
6. Huang, S., et al.: Accurate multi-contrast MRI super-resolution via a dual cross-attention transformer network. In: MICCAI 2023, pp. 313–322 (2023). https://doi.org/10.1007/978-3-031-43999-5_30
7. Li, G., Lyu, J., Wang, C., Dou, Q., Qin, J.: WavTrans: synergizing wavelet and cross-attention transformer for multi-contrast MRI super-resolution. In: MICCAI 2022, pp. 463–473 (2022). https://doi.org/10.1007/978-3-031-16446-0_44
8. Li, G., et al.: Rethinking multi-contrast MRI super-resolution: rectangle-window cross-attention transformer and arbitrary-scale upsampling, pp. 21230–21240 (2023)
9. Liang, J., Cao, J., Sun, G., Zhang, K., Van Gool, L., Timofte, R.: SwinIR: image restoration using swin transformer. In: Proceedings of the IEEE/CVF International Conference on Computer Vision, pp. 1833–1844 (2021)
10. Lu, L., Li, W., Tao, X., Lu, J., Jia, J.: MASA-SR: matching acceleration and spatial adaptation for reference-based image super-resolution. In: Proceedings of the IEEE/CVF Conference on Computer Vision and Pattern Recognition, pp. 6368–6377 (2021)

11. Lyu, J., Sui, B., Wang, C., Tian, Y., Dou, Q., Qin, J.: DuDoCAF: dual-domain cross-attention fusion with recurrent transformer for fast multi-contrast MR imaging. In: Medical Image Computing and Computer Assisted Intervention MICCAI 2022, pp. 474–484 (2022). https://doi.org/10.1007/978-3-031-16446-0_45
12. Lyu, M., et al.: M4raw: a multi-contrast, multi-repetition, multi-channel MRI k-space dataset for low-field MRI research **10**(1), 264 (264). https://doi.org/10.1038/s41597-023-02181-4, publisher: Nature Publishing Group
13. Lyu, Q., et al.: Multi-contrast super-resolution MRI through a progressive network. IEEE Trans. Med. Imag. **39**(9), 2738–2749 (2020)
14. Sun, L., Fan, Z., Fu, X., Huang, Y., Ding, X., Paisley, J.: A deep information sharing network for multi-contrast compressed sensing MRI reconstruction **28**(12), 6141–6153 (2019). https://doi.org/10.1109/TIP.2019.2925288. https://ieeexplore.ieee.org/abstract/document/8758456, conference Name: IEEE Transactions on Image Processing
15. Yang, F., Yang, H., Fu, J., Lu, H., Guo, B.: Learning texture transformer network for image super-resolution. In: Proceedings of the IEEE/CVF Conference on Computer Vision and Pattern Recognition, pp. 5791–5800 (2020)
16. Zbontar, J., et al.: fastMRI: an open dataset and benchmarks for accelerated MRI (2018). https://doi.org/10.48550/arXiv.1811.08839. http://arxiv.org/abs/1811.08839
17. Zhang, X., Zeng, H., Guo, S., Zhang, L.: Efficient long-range attention network for image super-resolution. In: ECCV 2022, pp. 649–667 (2022). https://doi.org/10.1007/978-3-031-19790-1_39
18. Zhao, L., et al.: JoJoNet: joint-contrast and joint-sampling-and-reconstruction network for multi-contrast MR (2022). https://doi.org/10.48550/arXiv.2210.12548. http://arxiv.org/abs/2210.12548
19. Zhou, B., Zhou, S.K.: DuDoRNet: learning a dual-domain recurrent network for fast MRI reconstruction with deep t1 prior, pp. 4273–4282 (2020)

Knowledge Distillation Based Dual-Branch Network for Whole Slide Image Analysis

Weiheng Fu[1], Meilan Xu[2(✉)], Jie Wu[1], Xiaoshuang Shi[1,4], Kang Li[3(✉)], and Xiaofeng Zhu[1]

[1] School of Computer Science and Engineering, University of Electronic Science and Technology of China, Chengdu 611731, China
[2] School of Electronic Information and Artificial Intelligence, Leshan Normal University, Leshan 614000, China
xu.meilian05@gmail.com
[3] West China Biomedical Big Data Center, West China Hospital, Sichuan University, Chengdu 610041, China
likang@wchscu.cn
[4] Sichuan Artificial Intelligence Research Institute, Yibin 644000, China

Abstract. Recently, a **M**ulti-scale **R**epresentation **A**ttention based deep multiple instance learning **N**etwork (**MRAN**) has been proposed to directly extract patch-level features from gigapixel whole slide images, and achieved promising performance on multiple popular datasets. However, it directly employs noisy bag-level labels for model training, thereby restricting the model performance. To overcome this issue, we propose a novel **K**nowledge **D**istillation based **D**ual-**B**ranch deep multiple instance learning framework, namely **KDDB**. Specifically, it integrates knowledge distillation in MRAN to generate soft targets for bag-level images, and designs a novel attention-semantic consistent loss to reduce the semantic inconsistency between the attention weights of bag-level images and their prediction class probabilities, so as to reduce the negative effect caused by bag-level wrong labels. Extensive experiments on multiple whole slide image datasets demonstrate the superior performance of the proposed framework over MRAN and other recent state-of-the-art methods, with better model interpretability. *All source codes will available online.*

Keywords: Whole slide images · deep multiple instance learning · knowledge distillation · attention-semantic consistent

1 Introduction

Automated analysis of histopathological images is pivotal in the realm of clinical cancer diagnosis, prognostic evaluation, and prediction of treatment responses.

Supplementary Information The online version contains supplementary material available at https://doi.org/10.1007/978-3-031-73284-3_39.

The emergence of digital whole slide images (WSIs) has significantly enhanced the application of deep learning techniques in the analysis of these images [9]. One of the major challenges posed by WSIs is their high resolution, often reaching a large size, like 50,000 × 50,000 pixels. This scale makes it impractical to directly feed WSIs into deep learning models. Typically, these WSIs are divided into smaller patches to facilitate processing and analysis.

However, this division into smaller patches introduces a substantial challenge of acquiring detailed manual annotations. Considering the vast number of patches in a single WSI, manual annotation becomes an exceedingly arduous task. As a result, obtaining patch-wise labels is usually expensive and time-consuming, thereby restricting the application of conventional supervised learning methods, which rely on accurate annotations. In light of these challenges, multiple instance learning (MIL) has been widely applied with deep learning for WSI analysis. In MIL-based methods [2,3,10,14,19,20], each WSI is often regarded as a bag, with containing patches as instances. Within the MIL paradigm, in a positive bag, there exists at least one positive instance, while for a negative bag, all instances should be negative.

Most of existing MIL-based methods utilize a pre-trained model on ImageNet-1K [12,16,20], unsupervised methods [4,15] and even downsampling [7,18] to extract patch-level features for WSI analysis. Unfortunately, these methods often fail to extract optimal patch-level feature representations [17], and even lose some significant semantic information [13]. Recently, MRAN [17] utilizes a two-branch deep neural network (in which the two branches employ bag-level and WSI-level labels for model training, respectively, but the bag-level labels are the ones from their corresponding WSIs, leading to some bag-level images with wrong labels.) and a multi-scale representation attention mechanism to directly extract patch-level features from gigapixel WSIs, and achieves state-of-the-art performance on some popular datasets. Although MRAN adopts the bag-level attention to reduce the effect of bag-level wrong labels, its performance is still restricte by the semantic inconsistency of attention weights [10], i.e., one significant semantic bag often has a very small attention weight, and vice versa.

To overcome aforementioned issues, this paper introduces a novel **K**nowledge **D**istillation based **D**ual-**B**ranch deep MIL framework, namely **KDDB**, which utilizes knowledge distillation and an attention-semantic consistent loss to reduce the effect of bag-level wrong labels in MRAN. For clarity, we show the structure of the proposed framework in Fig. 1. Specifically, the proposed framework employs MRAN as the backbone, but it utilizes the second branch as a teacher model to generate soft labels for bag-level images as the targets of the first branch, which is regarded as a student model. Additionally, it adopts a novel attention-semantic consistent loss to reduce the inconsistency between the bag-level attention weight and the bag-level image class prediction probability, so as to further reduce the effect of bag-level wrong labels. Extensive experiments on three WSI datasets demonstrate that our method can significantly boost the model performance of MRAN.

Compared to previous methods, our main contributions are two-fold: We introduce a knowledge distillation strategy based on MRAN to reduce the negative effect of bag-level wrong labels. Additionally, we design a novel attention-semantic consistent loss to reduce the semantic gap of the bag-level attention.

2 Methodology

2.1 Overall

In this section, we first briefly introduce the proposed framework. As shown in Fig. 1, the proposed framework utilizes MRAN as the backbone, which consists of the preprocessing stage and two branches. Specifically, in MRAN, (a) is the preprocessing stage to first divide each WSI into a set of bags and then divide each bag into multiple patches, Then, patch-level images are fed into the first branch (b), which first employs the convolutional layers of ResNet18 [5] to extract patch-level feature maps (each patch-level feature map can be viewed as a set of cell-level features), and then feed them into cell- and patch-level attention to obtain the attention weights (e.g., η_{nijk} and δ_{nij}) for cell- and patch-level images, respectively, so as to learn bag-level image feature representations by using the losses $loss_{pc}$ and $loss_b$. Next, bag-level image features are fed into the second branch (c), which contains two fully-connected layers (FC_d and FC_w), bag-level attention and their loss $loss_w$ for learning bag-level attention weights (e.g., γ_{ni}) and WSI classification. Additionally, different from MRAN directly using hard labels from the corresponding WSIs as the targets for bag-level images, the proposed framework adopts the knowledge distillation method ($loss_{kd}$) to learn soft targets for bag-level images to reduce the negative effect of wrong labels, where the second branch is employed as a teacher model and the first branch is used as a student one. Moreover, our framework adopts a novel attention-semantic consistent loss ($loss_{ac}$) to reduce the semantic inconsistency of bag-level attention. In the following, Sect. 2.2 introduces the notations and definitions used in this paper, Sect. 2.3 and 2.4 show the knowledge distillation method and the objective function with the consistent loss, respectively.

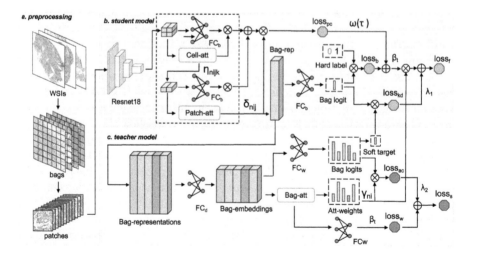

Fig. 1. The structure pass of the proposed framework, KDDB.

2.2 Notations and Definitions

Given a dataset consisting of N WSIs $\{\mathbf{X}_n\}_{n=1}^{N}$, and the WSI $\mathbf{X}_n$ has a corresponding label $Y_n \in \{0,1\}$, where $\mathbf{X}_n \in \mathbb{R}^{C \times H_w \times W_w}$ denotes the n-th WSI, C, H_w and W_w represent the number of channels, height and width of each WSI, respectively. In our work, each WSI is divided into a set of non-overlapping bag-level images $\{\mathbf{X}_{ni}\}_{i=1}^{N_b}$, where $\mathbf{X}_{ni} \in \mathbb{R}^{C \times H_b \times W_b}$ denotes the i-th bag in $\mathbf{X}_n$, H_b and W_b represent the height and width of the bag, respectively, $N_b = \frac{H_w W_w}{H_b W_b}$ is the number of bags in $\mathbf{X}_n$. Similarly, the bag-level images would be further divided into small patches $\{\mathbf{X}_{nij}\}_{j=1}^{N_p}$, where $\mathbf{X}_{nij} \in \mathbb{R}^{C \times H_p \times W_p}$ denotes the j-th patch of $\mathbf{X}_{ni}$, H_p and W_p represent the height and width of each patch, respectively, and $N_p = \frac{H_b W_b}{H_p W_p}$ is the number of patches in $\mathbf{X}_{ni}$. Moreover, in order to extract crucial cell-level information, the patch-level image is considered as consisting of a bunch of cell-level images $\{\mathbf{X}_{nijk}\}_{k=1}^{N_c}$, where $\mathbf{X}_{nijk} \in \mathbb{R}^{C \times H_c \times W_c}$ denotes the k-th cell within $\mathbf{X}_{nij}$, H_c and W_c represent the height and width of each cell-level image, respectively, and $N_c = \frac{H_p W_p}{H_c W_c}$ is the number of cells in $\mathbf{X}_{nij}$. In this paper, same as MRAN [17], we set $H_b = 2048$, $H_p = 128$ and $H_c = 32$.

2.3 Knowledge Distillation

The process of knowledge distillation involves training a teacher model and then distilling the knowledge from the teacher model to a student model. In the proposed framework, the teacher model is the MLP in the second branch, which consists of two fully connected layers and one bag-level attention, with using the features of bag-level images as the model input. In addition to the attention weights of bag-level images and WSI prediction, the teacher model can generate soft targets of bag-level images. For clarity, the soft targets can be obtained by:

$$s\left(\mathbf{z}_{ni}^{w}, T\right) = \frac{\exp\left(\mathbf{z}_{ni}^{w}/T\right)}{\sum_{i=1}^{K} \exp\left(\mathbf{z}_{ni}^{w}/T\right)}, \quad (1)$$

where $s\left(\mathbf{z}_{ni}^{w}, T\right) \in \mathbb{R}^K$ denotes the soft targets of the bag-level image $\mathbf{X}_{ni}$, $\mathbf{z}_{ni}^{w} \in \mathbb{R}^K$ is generated by the teacher model and denotes the logit vector of $\mathbf{X}_{ni}$, K is the number of classes and T is the distill temperature.

Additionally, the student model first utilizes hard targets of bag-level images for model training, where the hard targets of bag-level images are the labels of their corresponding WSIs, and then employs the soft targets for model training to reduce the negative effect of bag-level wrong labels. As stated in [6], soft targets contain the informative dark knowledge from the teacher model. To leverage the soft targets, we adopt the following loss function:

$$\text{loss}_{kd} = s\left(\mathbf{z}_{ni}^{w}, T\right) \log \frac{s\left(\mathbf{z}_{ni}^{w}, T\right)}{s\left(\mathbf{z}_{ni}^{b}, T\right)}, \quad (2)$$

where $\mathbf{z}_{ni}^{b} \in \mathbb{R}^K$ obtained by the student model is the logit vector of $\mathbf{X}_{ni}$.

2.4 Objective Function

As mentioned before, the first branch (student model) utilizes bag-level labels for model training, suppose that t is the true class of the WSI $\mathbf{X}_n$, and thus the loss for the bag-level image $\mathbf{X}_{ni}$ is:

$$\text{loss}_b = -\log\left(s(\mathbf{z}_{ni}^b)[t]\right). \tag{3}$$

Additionally, the first branch consists of cell- and patch-level attention, same as MRAN [17], we connect these two attentions with the loss function as follows:

$$\text{loss}_{pc} = -\sum_{j=1}^{N_p} \delta_{nij} \left\{ \log\left(s(\mathbf{z}_{nij})[t]\right) + \omega(\tau) \sum_{k=1}^{N_c} \eta_{nijk} \log\left(s(\mathbf{z}_{nijk})[t]\right) \right\}, \tag{4}$$

where δ_{nij} is obtained by the patch-level attention and denotes the attention weight of the patch-level image $\mathbf{X}_{nij}$, η_{nijk} is obtained by the cell-level attention and denotes the weight of the cell-level image $\mathbf{X}_{nijk}$, $\mathbf{z}_{nij} \in \mathbb{R}^K$ is the logit vector of the patch-level image $\mathbf{X}_{nij}$, $\mathbf{z}_{nijk} \in \mathbb{R}^K$ is the logit vector of the cell-level image $\mathbf{X}_{nijk}$, $\omega(\tau)$ is an unsupervised function to balance cell- and patch-level image classification, and τ denotes the number of current training epochs.

Moreover, to reduce the effect of bag-level wrong labels, we add the knowledge distillation loss in the first branch. Thus, the loss for the first branch is:

$$\text{loss}_f = \frac{1}{|\mathcal{S}|} \sum_{\mathbf{z}_{ni}^b \in \mathcal{S}} (\gamma_{ni}\beta_t(\text{loss}_b + \omega(\tau)\text{loss}_{pc}) + \lambda_1 \text{loss}_{kd}), \tag{5}$$

where $\mathcal{S}$ is a set containing bag-level images, $|\mathcal{S}|$ denotes the number of bag-level images, γ_{ni} is obtained by the bag-level attention in the second branch (teacher model) and denotes the attention weight for $\mathbf{X}_{ni}$, $\beta_t = \frac{\frac{N}{N_t}}{\sum_{r=1}^K \frac{N}{N_r}}$ is the weight of the t-th class, N_t and N_r are the numbers of WSIs in the t-th and r-th classes, respectively, $\omega(\tau)$ is to adjust the weight of loss_{pc}, and λ_1 is a constant to adjust the weight of loss_{kd} for model training.

In the second branch (teacher model), in addition to the loss for WSI classification, similar to cell- and patch-level attention, we also connect the bag-level attention with the loss function, and thus it can obtain the following loss:

$$\text{loss}_w = -\beta_t \left\{ \log\left(s(\mathbf{z}_n)[t]\right) + \omega(\tau) \sum_{i=1}^{N_b} \gamma_{ni} \log\left(s(\mathbf{z}_{ni}^w)[t]\right) \right\}, \tag{6}$$

where $\mathbf{z}_n \in \mathbb{R}^K$ is the logit vector of $\mathbf{X}_n$, and $\omega(\tau)$ is to balance the WSI- and bag-level image classification.

Moreover, in order to reduce the gap between the attention weight of bag-level image and its prediction class probability, we introduce a novel attention-semantic consistent loss as follows:

$$\text{loss}_{ac} = \frac{1}{|\mathcal{S}|} \sum_{\mathbf{z}_{ni}^w \in \mathcal{S}} \sum_{i=1}^{N_b} ||\gamma_{ni} - \frac{s(\mathbf{z}_{ni}^w)[t]}{\sum_{i=1}^n s(\mathbf{z}_{ni}^w)[t]}||_F^2, \tag{7}$$

where N_b denotes the number of bag-level images in the WSI $\mathbf{X}_n$. Equation (7) aims to obtain that the bag-level image $\mathbf{X}_{ni}$ with a larger γ_{ni} should also have a larger prediction class probability among all bag-level images in its corresponding WSI.

Then, we combine $loss_w$ with $loss_{ac}$ to obtain the loss for the second branch:

$$loss_s = \frac{1}{|\mathcal{B}|} \sum_{\mathbf{X}_n \in \mathcal{B}} (loss_w + \lambda_2 \, loss_{ac}), \tag{8}$$

where $\mathcal{B}$ denotes a set containing WSIs, $|\mathcal{B}|$ is the number of WSIs in $\mathcal{B}$, and λ_2 is a constant to adjust the weight of $loss_{ac}$.

Because the first branch and the second branch are alternatively trained and nested, based on Eq. (5) and Eq. (8), the objective function of our framework is:

$$loss = v \, loss_f + (1 - v) loss_s, \tag{9}$$

where $v \in \{0, 1\}$ is to control the parameters in which branch should be updated.

For clarity, we present the detailed training procedure in Algorithm 1, which is shown in the supplemental material.

3 Experiments

3.1 Experimental Settings

We conduct experiments on three datasets, including two public datasets: STAD and CAMELYON-16 (C-16) [1], and one private dataset HCC. **STAD** (Stomach Adenocarcinoma) is from The Cancer Genome Atlas (TCGA) and it contains 755 slides from 443 subjects, with 632 positive slides and 123 negative ones. We randomly select 60%, 20% and 20% subjects for model training, validation and testing, respectively. **C-16** consists of 399 WSIs of lymph nodes, with 160 positive slides and 239 negative ones. We randomly select 60%, 20%, 20% WSIs to construct training, validation and testing sets, respectively. **HCC** (Hepatocellular carcinoma) has 326 WSIs, with 156 positive slides and 170 negative ones, and each one has the average size of 102,952×90,369 pixels. We randomly utilize 80%, 10% and 10% WSIs for model training, validation and testing, respectively.

We compare the proposed method with eight popular methods, including CLAM_SB [11] and CLAM_MB [11], DSMIL [8], TransMIL [12], DTFD-MIL [19], MuRCL [20], IBMIL [10], and MRAN [17].

We adopt the same implementation and parameter settings as MRAN [17]. For the additional parameters in the proposed framework, we empirically set $T = 3$, $\lambda_1 = 3$ and $\lambda_2 = 5$. Additionally, for fairness, we employ the same batch size as the proposed framework for comparison methods, but we adopt their default augmentation to obtain the best performance of themselves.

To assess the proposed framework and comparison methods, we employ five popular metrics: accuracy (ACC), Area Under the Curve (AUC), F_1 score, Sensitivity (SE) and Specificity (SP). Additionally, we randomly split each dataset five times and report their average results obtained by each method.

Table 1. Classification results of different methods on the three datasets. We bold and underline the best and second-best results at each setting, respectively.

Dataset	Metrics	CLAM_SB	CLAM_MB	DSMIL	TransMIL	DTFD-MIL	MuRCL	IBMIL	MRAN	KDDB
STAD	ACC	90.2±2.2	91.5±1.5	62.5±7.2	92.7±2.0	89.9±3.7	91.9±1.8	77.7±11.5	<u>96.9</u>±1.2	**98.5**±1.1
	AUC	89.3±5.1	92.7±3.8	77.1±5.9	94.7±2.4	94.8±3.3	94.8±3.5	78.7±5.1	<u>99.0</u>±0.8	**99.5**±0.5
	F$_1$	94.1±1.3	94.9±0.9	68.6±7.7	85.7±2.6	93.7±2.2	95.1±1.2	84.9±9.1	<u>98.2</u>±0.7	**99.1**±0.6
	SE	94.7±1.2	96.3±3.3	59.8±11.4	83.4±2.8	90.4±4.6	94.3±1.8	79.1±1.5	**99.2**±0.1	<u>98.6</u>±1.5
	SP	67.2±8.6	67.2$_{13.1}$	<u>91.9</u>±3.4	84.1±1.1	88.0±9.9	78.7±10.1	70.8±10.9	85.0±9.2	**98.6**±1.9
C-16	ACC	74.0±5.5	75.0±8.9	59.2±3.4	74.8±2.0	64.4±7.8	73.0±11.9	67.6±9.0	<u>77.0</u>±4.8	**80.1**±9.4
	AUC	80.2±7.8	80.3±8.8	61.3±2.7	80.4±0.2	65.1±9.1	74.3±9.9	68.0±10.3	<u>83.8</u>±4.1	**88.9**±5.9
	F$_1$	66.0±5.9	63.1±20.3	57.8±6.3	<u>73.0</u>±1.0	53.4±5.2	57.9±15.0	59.7±6.3	72.6±6.9	**76.7**±8.0
	SE	64.3±14.7	61.2±27.8	<u>79.1</u>±14.5	72.7±1.0	57.6±15.2	50.2±12.2	66.6±16.5	69.6±4.4	**82.2**±9.2
	SP	80.4±14.8	<u>84.1</u>±15.6	48.4±9.9	74.9±1.5	67.6±17.9	**87.0**±11.7	66.6±19.2	81.7±5.1	75.4±13.2
HCC	ACC	51.5±2.1	55.1±1.3	60.5±5.3	49.4±3.8	<u>60.7</u>±5.6	50.9±9.4	60.5±5.7	58.2±7.0	**65.8**±4.4
	AUC	57.7±7.4	59.4±9.0	52.4±7.2	61.2±15.9	55.9±7.7	50.6±13.8	55.1±9.4	<u>65.0</u>±13.4	**69.3**±11.0
	F$_1$	62.5±3.9	<u>62.7</u>±6.5	59.2±9.1	42.0±7.1	58.7±11.8	58.6±8.6	57.2±8.9	62.6±8.3	**64.7**±6.7
	SE	**80.0**±15.3	<u>76.4</u>±22.4	57.4±15.9	49.5±1.7	59.9±19.4	61.4±3.8	53.0±13.6	74.6±19.4	66.5±13.9
	SP	21.2±19.0	32.5±25.1	62.7±17.2	50.5±1.3	61.2±14.4	37.3±16.8	**b69.1**±13.9	43.1±15.4	<u>66.1</u>±7.7

3.2 Results and Analysis

Table 1 shows the classification results of the proposed KDDB and eight comparison methods on three datasets. As we can see, KDDB achieves better performance than the other methods on all the three datasets in term of the three major metrics ACC, AUC and F$_1$ score. For example, the gain of KDDB is 4.0%, 6.1% and 5.6% over the best competitors on C-16 in term of ACC, AUC and F$_1$ score, respectively. Additionally, Table 1 suggests that KDDB can significantly boost the performance of MRAN. This might be caused by (i) the knowledge distillation to reduce the negative impact of bag-level wrong labels; (ii) the attention-semantic consistent loss to further reduce their negative effect.

3.3 Ablation Study

To evaluate the effect of the knowledge distillation method and the attention-semantic consistent loss in our framework, i.e., $loss_{kd}$ and $loss_{ac}$, we conduct experiments on MRAN, MRAN+KD, and KDDB on the dataset C-16, where MRAN+KD denotes only using $loss_{kd}$ for MRAN, and KDDB is to employ both $loss_{kd}$ and $loss_{ac}$ for MRAN. Table 2 shows the classification results of the three methods on three datasets. As we can see, on C-16, MRAN+KD can obtain better performance than MRAN in term of ACC, AUC, F$_1$ score and SE, and KDDB using $loss_{ac}$ can further boost the performance of MRAN+KD, especially on F$_1$ score and SE. Table S1 shows the results on STAD and HCC in the supplemental material, with obtaining similar findings. Thus, experimental results illustrate the effectiveness of the proposed $loss_{kd}$ and $loss_{ac}$.

Table 2. Classification results of ablation study on the dataset C-16. We bold and underline the best and second-best results at each setting, respectively.

Dataset	Method	ACC	AUC	F_1	SE	SP
C-16	MRAN	77.0 ± 4.8	83.8 ± 4.1	72.6 ± 6.9	69.6 ± 4.4	**81.7 ± 5.1**
	MRAN+KD	<u>79.5 ± 7.6</u>	<u>88.2 ± 5.9</u>	<u>73.9 ± 9.0</u>	<u>76.2 ± 14.7</u>	<u>81.2 ± 10.5</u>
	KDDB	**80.1 ± 9.4**	**88.9 ± 5.9**	**76.7 ± 8.0**	**82.2 ± 9.2**	75.4 ± 13.2

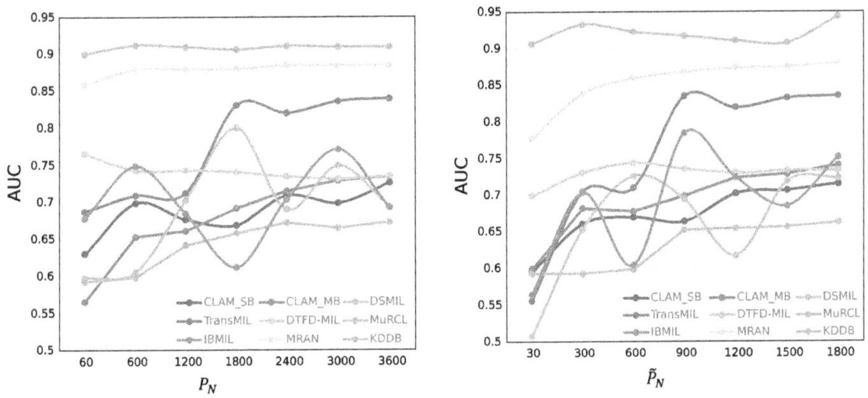

(a) AUC vs. P_N

(b) AUC vs. $\widetilde{P}_N$

Fig. 2. Classification results (AUC) of nine methods at different numbers of selected patches on C-16. Both P_N and $\widetilde{P}_N$ are the total number of selected patches, however, (a) MRAN and KDDB select the top 60 bags for each WSI, with variable top selected numbers of patches in each bag; (b) MRAN and KDDB select the top 30 patches for each bag, with the variable top selected numbers of bags in each WSI.

3.4 Interpretation Experiments

Apart from the classification experiments, we also perform interpretation experiments to assess the model interpretability of the proposed KDDB, by comparing it with the eight comparison methods. Similar to MRAN [17], we employ the popular metric AUC to assess their selected patch-level images, without using any additional prior knowledge. Specifically, we first utilize the selected patch-level images of each method from the original testing set to construct a new testing dataset for each method, and then assess their classification performance by using the new testing set. For fairness, all methods select the same number of patch-level images from each WSI. Figure 2 illustrates the AUC of different methods across varying numbers of selected patch-level images. It indicates that KDDB achieves the best AUC even using a very small number of selected patch-level images, which infers that our method has better model interpretability.

4 Conclusion

In this paper, we propose a novel deep MIL framework, which integrates knowledge distillation into MRAN to generate soft targets for bag-level images to reduce the negative effect of their wrong labels, and designs a novel attention-semantic consistent loss to reduce the semantic gap of the bag-level attention to further reduce the effect of the bag-level wrong labels. Experiments on popular WSI datasets demonstrate the superior classification and interpretation performance of our framework over recent state-of-the-art methods, and the effectiveness of our knowledge distillation method and the attention-semantic consistent loss. In the future, we will extend and apply the proposed framework to more difficult tasks, like multi-class classification, multi-modal data and image retrieval.

Acknowledgments. This work was supported by National Natural Science Foundation of China (No. 62276052) and Sichuan Science and Technology Program (No. 2024YFHZ0268).

References

1. Bejnordi, B.E., Veta, M., Van Diest, P.J., et al.: Diagnostic assessment of deep learning algorithms for detection of lymph node metastases in women with breast cancer. JAMA **318**(22), 2199–2210 (2017)
2. Bi, Q., et al.: Local-global dual perception based deep multiple instance learning for retinal disease classification. In: Medical Image Computing and Computer Assisted Intervention, pp. 55–64 (2021)
3. Chikontwe, P., Kim, M., Nam, S.J., Go, H., Park, S.H.: Multiple instance learning with center embeddings for histopathology classification. In: Medical Image Computing and Computer Assisted Intervention, pp. 519–528 (2020)
4. Fourkioti, O., Arampatzis, A., Jin, C., De Vries, M., Bakal, C.: CAMIL: context-aware multiple instance learning for whole slide image classification. arXiv preprint arXiv:2305.05314 (2023)
5. He, K., Zhang, X., Ren, S., Sun, J.: Deep residual learning for image recognition. In: Proceedings of the IEEE Conference on Computer Vision and Pattern Recognition, pp. 770–778 (2016)
6. Hinton, G., Vinyals, O., Dean, J.: Distilling the knowledge in a neural network. arXiv preprint arXiv:1503.02531 (2015)
7. Hou, L., Samaras, D., Kurc, T.M., Gao, Y., Davis, J.E., Saltz, J.H.: Patch-based convolutional neural network for whole slide tissue image classification. In: IEEE Conference on Computer Vision and Pattern Recognition, pp. 2424–2433 (2016)
8. Li, B., Li, Y., Eliceiri, K.W.: Dual-stream multiple instance learning network for whole slide image classification with self-supervised contrastive learning. In: IEEE Conference on Computer Vision and Pattern Recognition, pp. 14318–14328 (2021)
9. Li, X., Li, C., Rahaman, M.M., et al.: A comprehensive review of computer-aided whole-slide image analysis: from datasets to feature extraction, segmentation, classification and detection approaches. Artif. Intell. Rev. **55**(6), 4809–4878 (2022)
10. Lin, T., Yu, Z., Hu, H., Xu, Y., Chen, C.W.: Interventional bag multi-instance learning on whole-slide pathological images. In: IEEE Conference on Computer Vision and Pattern Recognition, pp. 19830–19839 (2023)

11. Lu, M.Y., Williamson, D.F., Chen, T.Y., Chen, R.J., Barbieri, M., Mahmood, F.: Data-efficient and weakly supervised computational pathology on whole-slide images. Nat. Biomed. Eng. **5**(6), 555–570 (2021)
12. Shao, Z., , et al.: Transmil: Transformer based correlated multiple instance learning for whole slide image classification. Adv. Neural Inf. Process. Syst. **34**, 2136–2147 (2021)
13. Shi, B., Liu, X., Zhang, F.: MLCN: metric learning constrained network for whole slide image classification with bilinear gated attention mechanism. In: International Workshop on Computational Mathematics Modeling in Cancer Analysis, pp. 35–46 (2022)
14. Shi, X., et al.: Loss-based attention for interpreting image-level prediction of convolutional neural networks. IEEE Trans. Image Process. **30**, 1662–1675 (2020)
15. Tellez, D., Litjens, G., van der Laak, J., Ciompi, F.: Neural image compression for gigapixel histopathology image analysis. IEEE Trans. Pattern Anal. Mach. Intell. **43**(2), 567–578 (2019)
16. Wang, H., et al.: Iteratively coupled multiple instance learning from instance to bag classifier for whole slide image classification. arXiv preprint arXiv:2303.15749 (2023)
17. Xiang, H., Shen, J., Yan, Q., Xu, M., Shi, X., Zhu, X.: Multi-scale representation attention based deep multiple instance learning for gigapixel whole slide image analysis. Med. Image Anal. **89**, 102890 (2023)
18. Xiong, C., Chen, H., Sung, J.J., King, I.: Diagnose like a pathologist: transformer-enabled hierarchical attention-guided multiple instance learning for whole slide image classification. arXiv preprint arXiv:2301.08125 (2023)
19. Zhang, H., et al.: DTFD-MIL: double-tier feature distillation multiple instance learning for histopathology whole slide image classification. In: IEEE Conference on Computer Vision and Pattern Recognition, pp. 18802–18812 (2022)
20. Zhu, Z., Yu, L., Wu, W., Yu, R., Zhang, D., Wang, L.: MURCL: multi-instance reinforcement contrastive learning for whole slide image classification. IEEE Trans. Med. Imag. **42**(5), 1337–1348 (2022)

DHSampling: Diversity-Based Hyperedge Sampling in GNN Learning with Application to Medical Imaging Classification

Jiameng Liu[1], Furkan Pala[2], Islem Rekik[2(✉)], and Dinggang Shen[1,3,4(✉)]

[1] School of Biomedical Engineering and State Key Laboratory of Advanced Medical Materials and Devices, ShanghaiTech University, Shanghai 201210, China
[2] BASIRA Lab, Imperial-X (I-X) and Department of Computing, Imperial College London, London W12 7TA, UK
Islem.Rekik@gmail.com
[3] Shanghai United Imaging Intelligence Co., Ltd., Shanghai 200230, China
[4] Shanghai Clinical Research and Trial Center, Shanghai 201210, China
Dinggang.Shen@gmail.com

Abstract. Graph Neural Networks (GNNs) have become increasingly essential in modeling complex clinical data, thereby facilitating heterogeneous data-based disease diagnosis. However, the application of GNNs to large-scale clinical data faces challenges due to their exponentially increasing computational costs and memory requirements, which restrict their effectiveness in medical image classification. Dividing large-scale graphs into subgraphs through partition methods emerges as a significant strategy for reducing computational resource consumption in graph learning. Nonetheless, this subgraph partition method requires traversing all subgraphs during training, significantly prolonging model convergence. To address these issues, in this study, we proposed a topology and embedding diversity-based sampling strategy, along with a hyperedge-based graph partition framework (*DHSampling*) to enhance the classification performance of subgraph-based GNNs. First, unlike traditional edges connecting only two nodes for each edge, we randomly assign nodes to hyperedges for forming a hypergraph, which connects multiple nodes simultaneously, allowing for the representation of complex relationships involving multiple entities. Then, we sample a subset of hyperedges with the highest diversity in both topology and embeddings to train the GNNs, providing accelerated training while maintaining minimal performance drops when sampling a subset for training. To the best of our knowledge, we are the first to utilize hyperedges in conjunction with diversity-based sampling to address the challenges faced by GNNs when applied to large-scale clinical data. Extensive experiments on two large-scale medical image classification benchmark datasets demonstrate that our *DHSampling* strategy can *not only* markedly reduce the model training time, *but also* achieve excellent classification performance compared to existing representative methods without increasing computational resource occupancy excessively. Our *DHSampling* code is available at https://github.com/basiralab/DHSampling.

Keywords: Topology and Embedding Diversity · Diversity-based Sampling · Efficient GNN Learning

1 Introduction

Graph Neural Networks (GNNs) stand as the most widely used techniques for handling unstructured data and have demonstrated success in various applications, such as drug discovery and recommendation systems [4,9]. In light of the efficacy of GNNs in handling heterogeneous data, their application extends to clinical domains [2], particularly in diagnosing brain diseases through brain functional and structural connectivity graphs [16], as well as in pathology image classification [19]. However, the utilization of GNNs on large-scale graphs is constrained due to their substantial memory demands for storing node and edge features at every layer, coupled with resource-intensive recursive computation for aggregating neighboring nodes, leading to a phenomenon called neighborhood expansion problem [3]. Consequently, applying GNNs to large-scale medical data is particularly difficult, limiting their scalability for such datasets.

Considerable efforts have been devoted in the literature to achieve efficient memory and computation for GNNs when learning on large-scale graphs [1,18]. These efforts generally fall into two main categories: (1) data-level and (2) model-level optimizations. Data-level optimizations primarily involve sampling techniques [17], while model-level optimizations encompass various strategies such as model quantization [8] and simplification of the forward pass [22]. *DHSampling* belongs to the category of data-level optimization. In this context, we will discuss conventional sampling methods, which can be categorized into three aspects: (1) node-wise, (2) layer-wise, and (3) subgraph-wise.

Node-wise Sampling targets the selection of a fixed number of neighbors during the neighborhood aggregation phase for each node in the original training graph [5,10]. This strategy aims to reduce memory overload by reducing the number of nodes that need to be loaded into memory. However, the recursive approach of node-wise sampling introduces redundancy in embedding calculations. When multiple nodes share the same sampled neighbor, the embedding calculation for this neighbor is duplicated. This redundancy becomes increasingly pronounced with the growth in the number of layers. Although node-wise sampling helps mitigate the neighborhood expansion problem, it does not offer a complete solution.

Layer-Wise Sampling, introduced by Chen *et al.* [6] in their study on Fast-GCN, presents an innovative approach to address the neighborhood expansion problem and expedite learning on graphs. Unlike node-wise sampling, which selects neighboring nodes directly from the training graphs, layer-wise sampling opts for selecting nodes from the layers of GCN after constructing the computation graph, which allows FastGCN to scale linearly with the number of layers.

However, the independent sampling in each layer may compromise the model's ability to capture between-layer relations, potentially limiting its expressiveness.

Subgraph-Wise Sampling strategy aims to reduce the size of the original graph by employing well-designed partition methods such as METIS [12], Graclus [15] to enhance communication efficiency and minimize memory usage during training. For example, Chiang et al. [7], introduced Cluster-GCN, which partitions the graph into distinct clusters using METIS [12]. Subsequently, these subgraphs are employed to train a GNN model, thereby reducing memory usage during training and yielding superior performance compared to training on the entire graph. However, it's noteworthy that the nodes within each subgraph in Cluster-GCN are independent and non-overlapping, potentially introducing significant bias into the graph learning process. Zeng et al. [24] proposed node, edge, and random walk samplers to generate subgraphs in each batch with overlapped nodes among different subgraphs to reduce the bias in Cluster-GCN sampling. While these methods have greatly enhanced efficient GNN learning, the graph partition strategy necessitates the traversal of all subgraphs during the training process, thereby significantly increasing model training time.

All these sampling strategies circumvent the large-scale GNN training by overlooking the nature of graphs since they introduce a bias by creating independent samples. However, the nature of the graph undermines the relational dependency between nodes. To address this, we aim to preserve the relational structure and topology of the graph as much as possible while introducing efficient, scalable GNN training. This motivation led us to utilize the hypergraph structure with hyperedges connecting multiple nodes rather than just a pair of nodes, allowing for the preservation of complex relational patterns in the graph. By sampling these hyperedges, we boost the expressiveness of the model, resulting in significant gains in terms of computation budget and minimal performance drop.

Building on this insight, in this study, we propose a Diversity-based Hyperedge Sampling (*DHSampling*), which integrates the hyperedge-based graph partition strategy with the sampling strategy for efficient GNNs learning. In the sampling framework, hyperedges are ranked based on their diversity, and we prioritize sampling the most diverse ones. This ensures that we capture a wide range of relational patterns with only a small number of samples, optimizing both efficiency and effectiveness in GNN training. Our framework consists of two distinct stages including, (I) the hyperedge construction stage, and (II) diversity-based hyperedge sampling training. In Stage I, we distribute nodes into various centers, allowing for the existence of overlapping nodes, thereby forming hyperedges of each center. This approach mitigates data bias often encountered in conventional graph partitioning methods that yield non-overlapping subgraphs. In Stage II, we implement a diversity-based strategy to sample a subset of hyperedges characterized by maximal topology and embedding diversity for training GNNs. This strategy expedites model training while upholding performance standards by selecting hyperedges with the highest diversity, ensuring the inclusion of comprehensive information.

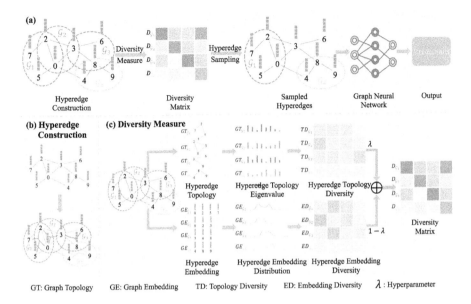

Fig. 1. Illustration of our proposed *DHSampling*. (a) Overview of the diversity-based hyperedge sampling framework; (b) Hyperedge construction; (c) Detailed illustration of hyperedge topology and embedding diversity.

2 Method

The overview of our proposed *DHSampling* framework is shown in Fig. 1 (a), comprising two stages: 1) hyperedge construction (Fig. 1 (b)), and 2) hyperedge sampling based on the topology and embedding diversity (Fig. 1 (c)). The key idea of *DHSampling* is that we only take a subset of hyperedges to achieve comparable even better performance than previous methods trained on the entire graph with shorter training time. The following paragraph describes the hyperedge construction process (Sec. 2.1), and the process of how to sample a subset of hyperedges based on the topology and embedding diversity of hyperedges (Sec. 2.2).

2.1 Hyperedge Construction

Given a graph $\mathcal{G} = (\mathcal{V}, \mathcal{E})$, which consists of N nodes $\mathcal{V}$ with features $\mathbf{X} \in \mathbb{R}^{N \times F}$, and edges $\mathcal{E}$ with corresponding adjacent matrix $\mathbf{A} \in \mathbb{R}^{N \times N}$, its corresponding target label is represented as $\mathbf{Y} \in \mathbb{R}^{N \times F}$ with F classes. For the nodes $\mathcal{V} = [v_1, v_2, ..., v_N]$ in the graph $\mathcal{G}$, we first randomly assign two labels between 1 and C, where C denotes the number of hyperedges, and then group the nodes sharing the same label into the same hyperedge, constructing the new hypergraph:

$$\hat{\mathcal{G}} = [\{\mathcal{V}_1, \mathcal{E}_1\}, \{\mathcal{V}_2, \mathcal{E}_2\}, ..., \{\mathcal{V}_C, \mathcal{E}_C\}] \tag{1}$$

with corresponding adjacent matrix $\hat{\mathbf{A}} = [\{\mathbf{A}_1, \mathbf{A}_2, ..., \mathbf{A}_C\}]$. Here, each $\mathcal{E}_i$ only contains the links within nodes in $\mathcal{V}_i$. Notably, since nodes are overlapping between different hyperedges, the sum of node numbers of each hyperedge is larger than the total node number N. Then, we use each hyperedge as a mini-batch to train the GNNs.

Algorithm 1. Topology and Embedding Diversity-based Hyperedge Sampling

---
Input: Graph $\mathcal{G}(\mathcal{V}, \mathcal{E})$; feature: $\mathbf{X}$; labels: $\mathbf{Y}$; # of hyperedges: C; # of sampled hyperedges: c;
Output: GNN model $\mathcal{M}$ with well-trained weight;
1: Convert entire graph $\mathcal{G}$ into hypergraph $\hat{\mathcal{G}}$ with C hyperedges;
2: $\mathcal{G}_i(\mathcal{V}_i, \mathcal{E}_i) \leftarrow$ Randomly sampling single hyperedge from hypergraph $\hat{\mathcal{G}}$;
3: $D_{\mathcal{G}_i} \leftarrow$ Calculate topology and embedding diversities according to Eq. 2 and Eq. 3;
4: $\mathcal{G}_S \leftarrow$ Sample $c - 1$ hyperedges $\mathcal{G}_s$ with largest diversity to $\mathcal{G}_i$ based on $D_{\mathcal{G}_i}$ to form sampled hyperedges;
5: **for** each epoch **do**
6:    $g \leftarrow$ compute gradient $\nabla \mathcal{L}(\mathcal{M}(\mathcal{G}_S), \mathbf{Y}_{\mathcal{G}_S})$ based on selected hyperedges;
7:    Conduct Adam update using gradient g
8: **end for**

2.2 Diversity-Based Sampling

Unlike conventional graph partitioning methods, our *DHSampling* framework trains GNNs exclusively on subsets characterized by c (where $c < C$) hyperedges, thereby substantially reducing time usage for the model training. Specifically, we first calculate the topology (Sec. 2.2) and embedding (Sec. 2.2) diversity across different hyperedges. This allows us to ensure that the sampled hyperedges contain as much information as possible, reducing the information loss caused by sampling. Our *DHSampling* algorithm is presented in Algorithm 1.

Topology Diversity Measure. We take the distribution of eigenvalues of each hyperedge as the graph topology representation. Specifically, given a hyperedge $\mathcal{G}_i$, we first obtain the Laplacian matrix $L_{\mathcal{G}_i} = DM_{\mathcal{G}_i} - A_{\mathcal{G}_i}$, where $DM_{\mathcal{G}_i}$ and $A_{\mathcal{G}_i}$ represent the degree matrix and the adjacent matrix of $\mathcal{G}_i$, respectively. Then, the eigenvalue distribution $GT_{\mathcal{G}_i}$ of $\mathcal{G}_i$ is calculated from the $DM_{\mathcal{G}_i}$.

Given two arbitrary hyperedge $\mathcal{G}_i$ and $\mathcal{G}_j$ (where $i \neq j$), we measure the Wasserstein distance [20] between eigenvalue distributions of hyperedges as topology diversity $TD_{(\mathcal{G}_i, \mathcal{G}_j)}$, which can be formulated as:

$$TD_{(\mathcal{G}_i, \mathcal{G}_j)} = \text{Wasserstein\_Distance}(GT_{\mathcal{G}_i}, GT_{\mathcal{G}_j}), \text{where } i \neq j \quad (2)$$

The higher the value of $TD_{(\mathcal{G}_i, \mathcal{G}_j)}$, the greater the topology diversity.

Table 1. Details of datasets of OrganCMNIST and OrganSMNIST

Dataset	Nodes	Edges	Features	Task	Class	Train/Val/Test
OrganCMNIST	23147	7080511	784	Multi-class	11	70%/10%/20%
OrganSMNIST	24819	9219888	784	Multi-class	11	70%/10%/20%

Embedding Diversity Measure. To evaluate the diversity of graph feature embedding, we utilize the maximum mean discrepancy (MMD) to measure the distance between graph feature embedding distributions. Given the feature embedding $X_{\mathcal{G}_i}$ and $X_{\mathcal{G}_j}$ of hyperedge $\mathcal{G}_i$ and $\mathcal{G}_j$ (where $i \neq j$), the embedding diversity $ED_{(\mathcal{G}_i, \mathcal{G}_j)}$ can be formula as:

$$ED_{(\mathcal{G}_i, \mathcal{G}_j)} = \text{MMD}(X_{\mathcal{G}_i}, X_{\mathcal{G}_j}), \text{where } i \neq j \tag{3}$$

The higher the value of $ED_{(\mathcal{G}_i, \mathcal{G}_j)}$, the greater the embedding diversity. We take hyperparameter λ to combine the topology and embedding diversity to represent the diversity $D_{(\mathcal{G}_i, \mathcal{G}_j)}$ with $D_{(\mathcal{G}_i, \mathcal{G}_j)} = \lambda T D_{(\mathcal{G}_i, \mathcal{G}_j)} + (1 - \lambda) ED_{(\mathcal{G}_i, \mathcal{G}_j)}$, where $\lambda \in (0, 1)$.

In our sampling process, we begin by randomly choosing a hyperedge $\mathcal{G}_i$ from the hypergraph $\hat{\mathcal{G}}_j$. Next, we calculate the diversity between $\mathcal{G}_i$ and each of the other $C - 1$ hyperedges, denoted as $\mathcal{G}_j$ where $j \in \{1, 2, ..., C\}$ and $i \neq j$. Subsequently, we select the top $c - 1$ hyperedges by sorting them in decreasing order according to the calculated diversity. It is noteworthy that our *DHSampling* framework can incorporate with any state-of-the-art (SOTA) GNN model to efficient model training while maintaining model performance.

3 Experiments

3.1 Dataset and Evaluation Metrics

In this work, we train and evaluate our proposed framework on two publicly available biomedical image classification benchmark datasets from MedMNIST [23]: OragnCMNIST and OrganSMNIST. These datasets were collected from abdominal CT scans of liver tumors at coronal and sagittal views, with dimensions of 28×28. We first normalize the intensities of each image to a standard distribution using z-score normalization before sending it to the GNN model. Then, the datasets are represented as a weighted graph, with each image considered as a node and pairwise correlations between images as edges. Edges are constructed based on cosine similarity between nodes, with only the top 20% edges being kept to reduce complexity and mitigate the influence of extreme values on similarity. The details of these datasets are presented in Table 1. The dataset is randomly divided into three subsets: 1) 70% for training, 2) 10% for validation, and 3) 20% for testing. The classification performance is comprehensively evaluated using three metrics: Precision, Recall, and F1-score [11].

Table 2. Quantitative comparisons with the SOTA method conducted on two datasets: OrganCMNIST, OrganSMNIST.

Dataset	Method	Precision (%) ↑	Recall (%) ↑	F1-Score (%) ↑	Time ↓	Mem ↓
OrganCMNIST	Cluster-GCN	79.11 ± 1.25	78.13 ± 0.58	78.34 ± 0.71	113 s	274 MB
	DHSampling	**80.52 ± 0.89**	**78.91 ± 0.91**	**79.37 ± 0.58**	93 s	458 MB
OrganSMNIST	Cluster-GCN	46.68 ± 1.09	39.93 ± 0.95	41.11 ± 0.80	114 s	292 MB
	DHSampling	**49.42 ± 1.00**	**40.42 ± 1.79**	**41.21 ± 0.68**	104 s	532 MB

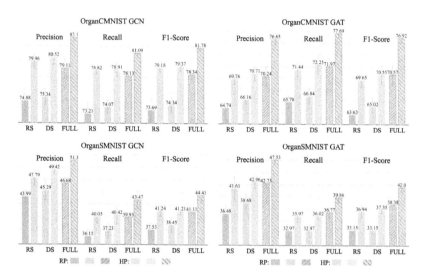

Fig. 2. Experiments results of incorporating our proposed *DHSampling* with several GNNs (GCN & GAT); RP: Random Partition; HP: Hyperedge Partition; RS: Random Sampling; DS: Diversity-based Sampling.

3.2 Implementation Details

The entire framework is implemented using PyTorch 1.7.1 and trained on a workstation equipped with a single NVIDIA RTX 20280Ti GPU. We adopt GCN and GAT models as GNNs to evaluate the effectiveness of our proposed *DHSampling*. We utilize the Adam optimizer [13] with a momentum of 0.99 and weight decay of 0.01 to optimize network parameters under the supervision of CrossEntropy. The initial learning rate is set to 0.001. We configured the number of hyperedges C to 20, with 30% of these hyperedges sampled for training our framework. Framework performance is evaluated by calculating the F1 score on the validation dataset at the end of each epoch, and we select the epoch with the highest F1 score as the best model parameter. To facilitate further research the source code our *DHSampling* is publicly available at https://github.com/basiralab/DHSampling.

3.3 Results and Discussion

To demonstrate the efficacy of our *DHSampling*, we conduct a comprehensive experiment by incorporating our diversity-based sampling strategy with two advanced GNN models, including 1) GCN [14], the most widely used graph convolution model; and 2) GAT [21], the first application of attention to GNNs. For each of these baselines, we conduct ablation studies by 1) extending the random graph partition (RP) to the hyperedge partition (HP), and 2) random sampling (RS) to the diversity-based sampling (DS). As listed in Fig 2, we summarize quantitative evaluation for all the methods for medical image classification on OrganCMNIST and OrganSMNIST datasets.

Effectiveness of Hyperedge Partition. In Fig. 2, it is evident that the hyperedge-based partitioning results in superior classification performance compared to random partitioning. This superiority stems from the capacity of hyperedges to facilitate the overlapping of nodes among different partitions, thereby offering greater flexibility to the model. By contrast, random partitioning generates subgraphs with non-overlapping nodes, introducing bias into the training of GNNs. We can conclude that the hyperedges play a crucial role in preserving the relationships between nodes, thereby enhancing the model's ability to capture complex interdependencies.

Effectiveness of Diversity-Based Sampling. While training on the entire set of partitioned subgraphs leads to optimal performance, it comes with a significant time overhead. A random sampling of these subgraphs offers a trade-off between computation time and performance but results in a drop in performance with shorter training times. By employing diversity-based sampling, we can bridge this performance gap while still training the GNNs within a shorter timeframe. Furthermore, as seen in Fig. 2, hyperedge partitioning with diversity-based sampling emerges as the optimal approach, even outperforming scenarios where the same GNN is trained on the entirety of partitioned subgraphs. Furthermore, in Table. 2, we summarize the quantitative comparison results of our *DHSampling* and Cluster-GCN for two dataset. We can find that, compared to the Cluster-GCN, our *DHSampling* obtains the highest performance across all evaluation metrics, as well as the lowest time consuming, which demonstrates the efficacy of our *DHSampling* in GNNs learning with shorter training time.

Generalizability Across Different GNNs. Through empirical evaluation on both GCN and GAT architectures (Fig. 2), we highlight its effectiveness, thereby establishing its generalizability and versatility across various graph neural network frameworks.

4 Conclusion and Discussion

In this study, we present a simple yet effective framework, named *DHSampling*, to markedly reduce the model training time while achieving excellent classification performance compared to the existing representative methods without increasing computational resource occupancy excessively. Experimental results show that this method can be incorporated into any advanced GNNs, *i.e.*, GCN, and GAT, for efficient GNN training. This enables us to extend our framework for training ultra-large-scale graph data in the clinic, facilitating the use of clinically heterogeneous multimodal data in disease diagnosis.

Acknowledgments. This work was supported in part by National Natural Science Foundation of China (No. 62131015, 62250710165, 62203355, and U23A20295), the STI 2030-Major Projects (No. 2022ZD0209000), Shanghai Municipal Central Guided Local Science and Technology Development Fund (No. YDZX20233100001001), and The Key R&D Program of Guangdong Province, China (No. 2023B0303040001 and 2021B0101420006).

Disclosure of Interests. The authors have no competing interests to declare that are relevant to the content of this article.

References

1. Adnel, C., Rekik, I.: Affordable graph neural network framework using topological graph contraction. In: Xue, z, et al. (eds.) Medical Image Learning with Limited and Noisy Data: Second International Workshop, MILLanD 2023, Held in Conjunction with MICCAI 2023, Vancouver, BC, Canada, October 8, 2023, Proceedings, pp. 35–46. Springer Nature Switzerland, Cham (2023). https://doi.org/10.1007/978-3-031-44917-8_4
2. Bessadok, A., Mahjoub, M.A., Rekik, I.: Graph neural networks in network neuroscience. IEEE Trans. Pattern Anal. Mach. Intell. **45**(5), 5833–5848 (2022)
3. Bojchevski, A., et al.: Is pagerank all you need for scalable graph neural networks. In: ACM KDD, MLG Workshop (2019)
4. Bongini, P., Bianchini, M., Scarselli, F.: Molecular generative graph neural networks for drug discovery. Neurocomputing **450**, 242–252 (2021)
5. Chen, J., Zhu, J., Song, L.: Stochastic training of graph convolutional networks with variance reduction. arXiv preprint arXiv:1710.10568 (2017)
6. Chen, J., Ma, T., Xiao, C.: FastGCN: fast learning with graph convolutional networks via importance sampling. arXiv preprint arXiv:1801.10247 (2018)
7. Chiang, W.L., Liu, X., Si, S., Li, Y., Bengio, S., Hsieh, C.J.: Cluster-GCN: an efficient algorithm for training deep and large graph convolutional networks. In: Proceedings of the 25th ACM SIGKDD international conference on knowledge discovery and data mining, pp. 257–266 (2019)
8. Ding, M., et al.: VQ-GNN: a universal framework to scale up graph neural networks using vector quantization. In: Ranzato, M., Beygelzimer, A., Dauphin, Y., Liang, P., Vaughan, J.W. (eds.) Adv. Neural Inf. Process. Syst. **34**, 6733–6746. Curran Associates, Inc. (2021). https://proceedings.neurips.cc/paper_files/paper/2021/file/3569df159ec477451530c4455b2a9e86-Paper.pdf
9. Fan, W., Ma, Y., Li, Q., He, Y., Zhao, E., Tang, J., Yin, D.: Graph neural networks for social recommendation. In: The World Wide Web Conference, pp. 417–426 (2019)
10. Hamilton, W., Ying, Z., Leskovec, J.: Inductive representation learning on large graphs. Adv. Neural Inf. Process. Syst. **30** (2017)

11. Hossin, M., Sulaiman, M.N.: a review on evaluation metrics for data classification evaluations. Int. J. Data Min. Knowl. Manage. Process **5**(2), 1 (2015)
12. Karypis, G., Kumar, V.: METIS: a software package for partitioning unstructured graphs, partitioning meshes, and computing fill-reducing orderings of sparse matrices (1997)
13. Kingma, D.P., Ba, J.: Adam: a method for stochastic optimization. arXiv preprint arXiv:1412.6980 (2014)
14. Kipf, T.N., Welling, M.: Semi-supervised classification with graph convolutional networks. arXiv preprint arXiv:1609.02907 (2016)
15. Kulis, B., Guan, Y.: Graclus–efficient graph clustering software for normalized cut and ratio association on undirected graphs, 2008 (2010)
16. Liu, M., Zhang, H., Shi, F., Shen, D.: Hierarchical graph convolutional network built by multiscale atlases for brain disorder diagnosis using functional connectivity. IEEE Trans. Neural Netw. Learn. Syst. (2023)
17. Liu, X., Yan, M., Deng, L., Li, G., Ye, X., Fan, D.: Sampling methods for efficient training of graph convolutional networks: a survey. IEEE/CAA J. Automatica Sinica **9**(2), 205–234 (2022). https://doi.org/10.1109/JAS.2021.1004311
18. Liu, X., et al.: Survey on graph neural network acceleration: an algorithmic perspective. arXiv preprint arXiv:2202.04822 (2022)
19. Pati, P., et al.: Hierarchical graph representations in digital pathology. Med. Image Anal. **75**, 102264 (2022)
20. Vallender, S.: Calculation of the Wasserstein distance between probability distributions on the line. Theory Probab. Appl. **18**(4), 784–786 (1974)
21. Veličković, P., Cucurull, G., Casanova, A., Romero, A., Lio, P., Bengio, Y.: Graph attention networks. arXiv preprint arXiv:1710.10903 (2017)
22. Wu, F., Souza, A., Zhang, T., Fifty, C., Yu, T., Weinberger, K.: Simplifying graph convolutional networks. In: International Conference on Machine Learning, pp. 6861–6871. PMLR (2019)
23. Yang, J., et al.: medmnist v2-a large-scale lightweight benchmark for 2D and 3D biomedical image classification. Sci. Data **10**(1), 41 (2023)
24. Zeng, H., Zhou, H., Srivastava, A., Kannan, R., Prasanna, V.: Graphsaint: graph sampling based inductive learning method. arXiv preprint arXiv:1907.04931 (2019)

Author Index

A

Abolmaesumi, Purang I-361
Aggarwal, Hemant Kumar I-187
Aguiar-Fernández, Pablo I-331
Al-Battal, Abdullah F. I-320
Al-Belmpeisi, Rami II-222
Ali, Sharib II-43
Alomar, Antonia II-148
An, Cheolhong I-320
Andersen, Thomas Lund II-222
Antalek, Matthew I-1
Anwar, Syed Muhammad I-73
Anwar, Syed Muhammed II-148

B

Bagci, Ulas I-1, I-208
Baumgartner, Michael II-22
Berens, Philipp II-53
Beutler, Anna I-94
Bi, Lei I-238
Biswas, Koushik I-1
Borhani, Amir I-1
Boudissa, Selma I-228
Broos, Kenneth I-94
Bu, Zhenyu I-42

C

Cai, Xiran I-340
Cao, Xiaohuan II-191
Cao, Yan II-1
Cao, Zehong I-166
Carass, Aaron I-248
Cencini, Matteo II-128
Cetin, Ahmet Enis I-208
Chaitanya, Krishna II-201
Charvát, František I-124
Chatterjee, Sudhanya I-187
Chen, Haoyuan I-22
Chen, Long I-290
Chen, Rusi II-232
Chen, Tianqi I-259

Chen, Yanlin II-232
Chen, Yiming I-300
Chen, Ziyang I-269
Cheng, Jian I-114
Cheng, Pujin I-350
Chong, Yosep I-143
Chouinard, Paul I-300
Cobo, Miriam II-12
Connors, Theressa II-74
Cui, Zhiming I-218
Cula, Oana Gabriela II-201

D

D'Souza, Niharika S. I-300
Dahl, Anders Bjorholm II-222
Damasceno, Pablo F. II-201
Damasceno, Pablo I-94
Dasgupta, Prokar I-42
Deden-Binder, Lucas J. II-74
Deman, Michael I-94
Devadharshiniinst, S. I-372
Dewey, Blake E. I-248
Dey, Neel I-300
Deyer, Louisa II-95
Deyer, Timothy II-95
Dong, Fajin II-1
Dou, Haoran II-1
Du, Siyi I-382
Du, Tongchun I-156
Duan, Yaofei II-232
Dubey, Shikha I-143
Duhme, Christof II-108
Duncan, James S. I-259
Durak, Gorkem I-1

E

Eghbali, Reza I-310
Elhabian, Shireen Y. I-143, II-117
Emerson, Monica Jane II-222

F

Fadnavis, Shreyas II-201
Fan, Wenxin I-114
Farré-Melero, Arnau I-331
Fay, Louisa II-53
Fayad, Zahi A. II-95
Fernandez, Pablo Blasco II-74
Fischer, Maximilian II-22
Fooladgar, Fahimeh I-361
Fountoulakis, Nicholas I-94
Fu, Huazhu II-85
Fu, Weiheng I-392
Fu, Yihang I-269

G

Gao, Zijun I-94
Gatidis, Sergios II-53
Ghanem, Louis R. II-201
Gilany, Mahdi I-361
Gluud, Lise Lotte II-222
Goh, Rick Siow Mong II-85
Golbabaee, Mohammad II-128
Goldberger, Jacob II-159
Golland, Polina I-300
Gopinath, Karthik II-74
Gösche, Erik I-310
Gou, Zhongshan I-52
Govind, Darshana I-94
Granados, Alejandro I-42
Gundogdu, Batuhan I-208

H

Han, Tong I-52
Hari, K.V.S. I-187
Harmanani, Mohamed I-361
Hassani, Atefe II-32
He, Kelei I-197
He, Yichu I-166
Henninger, Patrick I-300
Herisse, Rogeny II-74
Hoerr, Verena II-108
Holzschuh, Julius C. II-22
Hou, Jun I-259
Hsia, Elizabeth I-94
Hu, Qiuting I-350
Huang, Yuhao I-52, II-1, II-232
Huo, Jiayu I-42
Huver, Sean II-95
Hyman, Bradley II-74

I

Iglesias, Juan Eugenio II-74
Ipšić, Ivo II-63
Isensee, Fabian II-22
Iyer, Krithika II-117

J

Jamzad, Amoon I-361
Janiczek, Robert I-94
Jeon, Byunghwan I-63
Jha, Debesh I-1, I-208
Jiang, Caiwen I-83
Jiang, Pengbo I-340
Jiang, Wentao I-156
Jiang, Xiaoyi II-108
Jiang, Zhifan II-148
Jiang, Zhiguo I-133
Jie, Biao I-156
Jiřík, Radovan I-228
Jung, Sunghee I-63

K

Kainz, Bernhard II-180
Kang, Solha II-169
Kanli, Georgia I-228
Karani, Neerav I-300
Karri, Meghana I-1
Kashyap, Satyananda I-300
Keunen, Olivier I-228
Kim, Jinman I-238
Knobloch, Catherine II-22
Knoll, Florian I-310
Knudsen, Beatrice I-143
Koch, Lisa M. II-53
Kozanno, Liana II-74
Kozel, Jiří I-124
Küstner, Thomas II-53
Kybic, Jan I-124

L

Ladner, Daniela I-1
Lakshmanan, Manojkumar I-372
Lam, Van Khanh I-73
Langs, Georg II-138
Larsen, Peter Hjørringgaard II-222
Ledesma-Carbayo, María J. II-148
Li, Cheng I-114, I-280
Li, Kang I-392
Li, Meiyu I-22
Li, Ruochen I-31

Li, Wen   I-156
Li, Yonghao   I-22
Li, Zhenhui   I-22
Liang, Hongqin   I-197
Liang, Yong   I-114
Lin, Li   I-350
Linguraru, Marius George   I-73, II-148
Lippe, Chris   II-108
Liu, Chen   I-197
Liu, Chi   I-259
Liu, Jiameng   I-166, I-402, II-191
Liu, Lian   I-52
Liu, Lianli   I-177
Liu, Mianxin   I-83
Liu, Yang   I-42
Liu, Yong   II-85
Liu, Zelong   II-95
Liu, Zhe   I-52
Lloret Iglesias, Lara   II-12

M

Maier-Hein, Klaus H.   II-22
Maik, Vivek   I-372
Mandepally, Akshith   I-300
Manojlović, Teo   II-63
Mansi, Tommaso   II-201
Mayo, Perla   II-128
Medetalibeyoglu, Alpay   I-1
Mei, Lanzhuju   II-191
Mei, Xueyan   II-95
Mei, Yunhao   II-95
Menéndez Fernández-Miranda, Pablo   II-12
Menze, Bjoern H.   II-128
Menzel, Marion I.   II-128
Merhof, Dorit   I-290
Meyer, Craig S.   I-94
Michalčová, Patricie   I-124
Mitic, Branko   II-138
Mobadersany, Pooya   II-201
Mousavi, Parvin   I-361
Müller, Johanna P.   II-180
Müller, Sarah   II-53
Murphy, Adam B.   I-208
Murphy, Philip S.   I-94

N

Nabilou, Puria   II-222
Neeraja, R.   I-372
Nguyen, Truong Q.   I-320

Ni, Dong   I-52, II-1, II-232
Ni, Juncheng   I-31
Niñerola-Baizán, Aida   I-331
Noonan, Lenore   I-94

O

Oakley, Derek   II-74
Ochoa-Ruiz, Gilberto   II-43
Osman, Yousuf Babiker M.   I-280
Ou, Zaixin   I-83
Ourselin, Sebastien   I-42
Ozbulak, Utku   II-169

P

Pakizer, David   I-124
Pal, Ridam   I-1
Pala, Furkan   I-402
Pan, Hongyi   I-208
Pan, Jiazhen   I-31
Parida, Abhijeet   I-73, II-148
Parmar, Chaitanya   I-94, II-201
Patel, Nikhil   II-95
Patel, Shaswat   I-1
Patil, Uday   I-187
Peng, Jingjing   I-42
Penso, Coby   II-159
Pérez del Barrio, Amaia   II-12
Perlo, Daniele   I-228
Pirkl, Carolin M.   II-128
Prince, Jerry L.   I-248
Prosch, Helmut   II-138
Purayath, Aparna   I-372

Q

Qian, Xiaohua   I-12
Qian, Xuejun   I-22
Qin, Chen   I-382
Qiu, Liang   I-177
Qu, Jiaqi   I-12
Qu, Ruowen   I-104

R

Rauschecker, Andreas M.   I-310
Rekik, Islem   I-402, II-32
Remedios, Samuel W.   I-248
Reza, Amit   I-1
Rizq, Raed   I-300
Rokuss, Maximilian R.   II-22
Rooney, Terence   I-94
Roshanitabrizi, Pooneh   II-148

Ross, Ashley I-208
Roth, Holger R. II-148
Rueckert, Daniel I-31

S

Sanz Bellón, Pablo II-12
Scherer, Emily I-94, II-201
Seeböck, Philipp II-138
Shah, Mubarak I-73
Shanbhag, Dattesh I-187
Shao, Wei II-212
Shaukat, Furqan I-73
Shen, Dinggang I-22, I-83, I-166, I-340, I-402, II-191
Shi, Dongzi I-104
Shi, Feng I-166
Shi, Jun I-133
Shi, Xiaoshuang I-392
Shi, Yangyang II-212
Silva, Wilson II-12
Simon, Walsh I-83
Sivaprakasam, Mohanasankar I-372
Školoudík, David I-124
Sørensen, Kristine Aavild II-222
Soultanidis, George II-95
Sparks, Rachel I-42
Štajduhar, Ivan II-63
Standish, Kristopher I-94, II-201
Straub, Jennifer II-138
Sun, Dongdong I-133
Sundgaard, Josefine Vilsbøll II-222
Surace, Lindsey II-201
Syeda-Mahmood, Tanveer F. I-300

T

Tan, Boyuan I-238
Tan, Tao II-232
Tang, Van Ha I-320
Tang, Xiaoying I-350
Tapp, Austin II-148
Teevno, Mansoor Ali II-43
Tieu, Andrew II-95
To, Minh Nguyen Nhat I-361
Tosetti, Michela II-128
Truong, Steven Q. H. I-320
Turkbey, Baris I-208

U

Ulrich, Constantin II-22

V

Vankerschaver, Joris II-169
Velichko, Yury I-1, I-208
Venkateswaran, N. I-372

W

Wald, Tassilo II-22
Wan, Peng II-212
Wang, Fang I-52
Wang, Haoshen I-218
Wang, Jian I-52
Wang, Kaini I-42
Wang, Runqi I-166
Wang, Shanshan I-114, I-280
Wang, Wei I-133
Wang, Xiaodong I-218
Wang, Xingang I-197
Wang, Ying II-95
Wang, Yinsong I-382
Wang, Yuping I-133
Wei, Xunbin I-12
Williams-Ramirez, John II-74
Wilson, Paul I-361
Wodlinger, Brian I-361
Wu, Haibo I-133
Wu, Jie I-392
Wu, Ruoyou I-114
Wu, Wenxuan I-104
Wu, Zejun I-248
Wu, Zhaoxiang I-156

X

Xia, Yong I-269
Xiang, Lei I-197
Xie, Huidong I-259
Xing, Lei I-177
Xing, Xiaodan I-83
Xing, Xiaofen I-104
Xing, Xiaohan I-177
Xiong, Jiuli II-191
Xiong, Tong I-104
Xu, Meilan I-392
Xu, Qi I-22
Xu, Xiangmin I-104
Xu, Xinxing II-85
Xu, Yanyu II-85
Xu, Ziyue II-148
Xue, Yuxin I-238
Xue, Zhong I-340, II-191

Author Index

Y

Yamamoto, Shinobu I-94
Yang, Guang I-83
Yang, Jing I-114
Yang, Xin I-52, II-1, II-232
Yang, Yang I-156
Yang, Yichen I-340
Ye, Yiwen I-269
Yip, Stephen S. F. I-94
Yoon, Jongum I-63
Young, Sean I. II-74
Yu, Junxuan II-232
Yu, Lequan I-177
Yu, Shoujun I-280

Z

Zachary, Marcos I-300
Zemlianskaia, Natalia I-94, II-201
Zemlyanker, Dina II-74
Zeng, Zhiming I-197
Zenk, Maximilian II-22
Zhang, Daoqiang II-212
Zhang, Jiadong I-22
Zhang, Xin I-104
Zhang, Zheng I-197
Zheng, Hairong I-280
Zheng, Shaoming I-382
Zheng, Yushan I-133
Zhou, Alexander II-95
Zhou, Bo I-259
Zhou, Guangquan I-42
Zhou, Han II-1, II-232
Zhou, Yinchi I-259
Zhou, Yongsong II-232
Zhou, Zechen I-197
Zhu, Lei II-85
Zhu, Lingting I-177
Zhu, Qi II-212
Zhu, Xiaofeng I-392
Zhu, Xiliang I-52
Zhu, Xin I-208
Zhu, Youxiang I-31
Zhu, Zhiqing I-197
Zou, Juan I-114
Zuo, Yingli II-212

Printed in the USA
CPSIA information can be obtained
at www.ICGtesting.com
CBHW050526281024
16500CB00004B/130